# Polyhedral Structures, Symmetry, and Applications

## Special Issue Editor
Egon Schulte

MDPI • Basel • Beijing • Wuhan • Barcelona • Belgrade

**MDPI**

*Special Issue Editor*
Egon Schulte
Northeastern University
USA

*Editorial Office*
MDPI AG
St. Alban-Anlage 66
Basel, Switzerland

This edition is a reprint of the two Special Issues published online in the open access journal *Symmetry* (ISSN 2073-8994 and ISSN 2073-8994) from 2012–2017 (available at:
http://www.mdpi.com/journal/symmetry/special_issues/polyhedral_structures
and http://www.mdpi.com/journal/symmetry/special_issues/polyhedra).

For citation purposes, cite each article independently as indicated on the article page online and as indicated below:

Author 1; Author 2. Article title. *Journal Name* **Year**, *Article number*, page range.

**First Edition 2017**

**ISBN 978-3-03842-689-9 (Pbk)**
**ISBN 978-3-03842-688-2 (PDF)**

# Table of Contents

## Section A: Polyhedra, Tilings, and Crystallography

# Section B: Abstract Polyhedra, Maps on Surfaces, and Graphs

# Section C: Polyhedral Structures, Arts, and Architecture

# Section D: Combinatorial Geometry

# About the Special Issue Editor

**Egon Schulte**, Professor, received his Ph.D. in 1980 and his Habilitation in 1985, both in Mathematics, from the University of Dortmund, Germany. He served as Wissenschaflicher Assistent at the University of Dortmund in 1980–1983 and 1984–1985; Visiting Assistant Professor at the University of Washington in Seattle in 1983-1984; Privat Dozent at the University of Dortmund in 1985–1987; and Visiting Assistant Professor at Massachusetts Institute of Technology in 1987–1989. Professor Schulte joined the Mathematics faculty at Northeastern University in Boston as an Associate Professor in 1989, advancing to Professor in 1992. He has been at Northeastern University ever since. Professor Schulte's research interests include discrete and combinatorial geometry, combinatorics, and group theory. He is coauthor of the research monograph "Abstract Regular Polytopes" and has published well over 100 research articles.

# Preface to "Polyhedral Structures, Symmetry, and Applications"

The study of polyhedra and symmetry has a long and fascinating history tracing back to the early days of geometry. With the passage of time, various notions of polyhedra have attracted attention and have brought to light new exciting classes of symmetric structures, including the well-known Platonic and Archimedean solids, the Kepler-Poinsot star polyhedra, the Petrie-Coxeter sponge polyhedra, and the Grünbaum-Dress regular polyhedra, as well as the more recently discovered chiral skeletal polyhedra and regular polygonal complexes.

Over time, we can observe a shift from the classical approach of viewing a polyhedron as a solid, to topological and algebraic approaches focussing on the underlying maps on surfaces, to combinatorial approaches highlighting the underlying incidence structures, and recently, to discrete geometric approaches featuring a polyhedron as a skeletal figure in space.

The flexibility inherent in the concept proves an important point—polyhedra and symmetry have shown an enormous potential for revival! One explanation for this phenomenon is the appearance of symmetric polyhedra in many contexts that a priori seem to have little apparent relation to symmetry, such as the occurrence of many figures in nature as crystals. In addition, their internal beauty appeals to the artistic senses and sparks the desire for a rigorous mathematical analysis and understanding of the figures themselves, as well as of their relationships with other sciences.

This special issue book features a collection of nine-teen articles about discrete geometric and combinatorial polyhedral structures, with symmetry as the unifying theme. These articles have appeared in two related special issues of Symmetry, on "Polyhedra" in 2012/2013 and on "Polyhedral Structures" in 2016/2017. The collection presents an attractive mix of topics and covers both theory and applications.

**Egon Schulte**
*Special Issue Editor*

# Section A:
# Polyhedra, Tilings, and Crystallography

*symmetry*

MDPI

*Article*

# Hexagonal Inflation Tilings and Planar Monotiles

**Michael Baake** [1], **Franz Gähler** [1] **and Uwe Grimm** [2,*]

[1]   Fakultät für Mathematik, Universität Bielefeld, Postfach 100131, Bielefeld 33501, Germany;
      mbaake@math.uni-bielefeld.de (M.B.); gaehler@math.uni-bielefeld.de (F.G.)

[2]   Department of Mathematics and Statistics, The Open University, Walton Hall, Milton Keynes MK7 6AA, UK

*   Author to whom correspondence should be addressed; u.g.grimm@open.ac.uk.

Received: 2 September 2012; in revised form: 8 October 2012; Accepted: 14 October 2012;
Published: 22 October 2012

**Abstract:** Aperiodic tilings with a small number of prototiles are of particular interest, both theoretically and for applications in crystallography. In this direction, many people have tried to construct aperiodic tilings that are built from a single prototile with nearest neighbour matching rules, which is then called a monotile. One strand of the search for a planar monotile has focused on hexagonal analogues of Wang tiles. This led to two inflation tilings with interesting structural details. Both possess aperiodic local rules that define hulls with a model set structure. We review them in comparison, and clarify their relation with the classic half-hex tiling. In particular, we formulate various known results in a more comparative way, and augment them with some new results on the geometry and the topology of the underlying tiling spaces.

**Keywords:** Euclidean monotiles; aperiodicity; local rules; inflation

---

## 1. Introduction

A well-known inflation rule with integer inflation factor is the half-hex inflation from [1], Exercise 10.1.3 and Figure 10.1.7, which we show in Figure 1. As such, it is a lattice substitution (or inflation) in the sense of [2,3]. Moreover, it is a face to face *stone inflation* (in the sense of Danzer), which means that each inflated tile is precisely dissected into copies of the prototile so that the final tiling is face to face. This rule defines an aperiodic tiling of the plane, but it does not originate from an aperiodic prototile set (for the terminology, we refer to [4] and references therein). In principle, the procedure of [5] can be applied to add local information to the prototile and to the inflation rule (via suitable markers and colours), until one arrives at a version with an aperiodic prototile set. However, to our knowledge, this has never been carried out, as it (most likely) would result in a rather large prototile set.

**Figure 1.** Half-hex inflation rule.

Interestingly, two different inflation rules for hexagonally shaped prototiles have independently been constructed, namely one by Roger Penrose [6] and one by Joan Taylor [7] (see also [8]), each defining a tiling hull that can also be characterised by aperiodic local rules. Viewed as dynamical systems under the translation action of $\mathbb{R}^2$, they both possess the continuous half-hex hull (and also the arrowed half-hex, to be introduced later) as a topological factor, though with subtle differences. They may be considered as covers of the half-hex that comprise just enough local information to admit an aperiodic prototile set. In fact, both examples are again lattice inflations. They were found in the attempt to construct an aperiodic planar monotile, which loosely speaking is a single prototile together with some local rules that tiles the plane, but only non-periodically. Let us mention that Taylor's original inflation tiling can be embedded into a slightly larger tiling space that still possesses aperiodic local rules [8], but is no longer minimal (see below for more on minimality).

All three, the half-hex, the Penrose and the Taylor tiling, are structures that can be described as model sets; see [9–11] for background on model sets. For the half-hex tiling, this was first shown in [12]. Observing that two half-hexes always join to form a regular hexagon, with edge length 1 say, one obtains a hexagonal packing where three types of hexagons are distinguished by a single diagonal line. We represent each hexagon by a single point located at its centre, of type $\ell \in \{0,1,2\}$, where $\ell$ corresponds to a diagonal that is rotated by $\ell\pi/3$ against the horizontal. This gives a partition

$$H_0 \cup H_1 \cup H_2 \,=\, \Gamma \,=\, \sqrt{3} \left\langle \xi, \xi^3 \right\rangle_{\mathbb{Z}}$$

with $\xi = e^{\pi i/6}$, where $\Gamma$ is a triangular lattice of density $\frac{2}{9}\sqrt{3}$. When starting from a fixed point tiling of the half-hex inflation rule, the fixed point equations for the three point sets lead to the solution

$$H_\ell \,=\, A_\ell \cup \bigcup_{n \geq 0} 2^n (2\Gamma + \sqrt{3}\,\xi^{3+2\ell}) \tag{1}$$

where the $A_\ell$ are empty sets except for one, which is the singleton set $\{0\}$. Where the latter occurs depends on the seed of the selected fixed point. This means that there are precisely three possibilities, corresponding to the three possible choices for the central hexagon; see [9] for the detailed derivation.

This description establishes the model set structure, with the 2-adic completion of $\Gamma$ as internal space. Note that the point 0 is the unique limit point of any of the three unions in Equation (1) in the 2-adic topology. The structure of the union over expanded and shifted copies of $\Gamma$ is also called a *Toeplitz structure*. It is an example of a limit-periodic system [13] with pure point diffraction (and, equivalently, with pure point dynamical spectrum [14–16]). The diffraction measure of a weighted Dirac comb on the half-hex tiling can be calculated from Equation (1) via the Poisson summation formula, by an application of the methods explained in [9,17]. This description of the half-hex tiling will be the key observation to also identify the two covers (by Penrose and by Taylor) as model sets.

Below, we discuss the two inflation tilings due to Penrose and Taylor in some detail. To be more precise with the latter case, we only consider the minimal part of the tiling space considered by Socolar and Taylor in [8]. This minimal part is the tiling LI (local indistinguishability) class defined by Taylor's original stone inflation rule [7]. To distinguish the two hulls, we use the term *Taylor tiling* for the minimal inflation hull and refer to the elements of the larger tiling space as the *Socolar–Taylor tilings*.

We assume the reader to be familiar with the concept of *mutual local derivability* (MLD), which was introduced in [18]; see also [9,19]. When the derivation rules commute with all symmetries of the tilings under consideration (respectively their hulls), they are called symmetry preserving. The corresponding equivalence classes are called SMLD classes [19].

The *hull* of a (planar) tiling $\mathcal{T}$ with *finite local complexity* (FLC) in $\mathbb{R}^2$ is defined as the orbit closure in the local topology, $\mathbb{X}(\mathcal{T}) = \{t + \mathcal{T} \mid t \in \mathbb{R}^2\}$. Here, two tilings are $\varepsilon$-close when they agree on the ball $B_{1/\varepsilon}(0)$, possibly after (globally) translating one of them by an element from $B_\varepsilon(0)$. Due to the FLC property, the hull is compact [16], with continuous action of the group $\mathbb{R}^2$ via translation. Consequently, the pair $(\mathbb{X}(\mathcal{T}), \mathbb{R}^2)$ is a topological dynamical system. It is called *minimal* when the translation orbit of every element of the hull is dense in it. Our examples below will be minimal hulls of FLC tilings, or of equivalent representatives of the corresponding MLD class.

A hull $\mathbb{X}$ is called *aperiodic* when no element of it possesses non-trivial periods. In other words, $\mathbb{X}$ is aperiodic when, for every $X \in \mathbb{X}$, the equation $t + X = X$ only holds for $t = 0$. A hull is said to have *local rules* when it is specified by a finite list of legal local configurations, for instance in the form of a finite atlas of patches. If a set of local rules specifies an aperiodic hull, the rules themselves are called *aperiodic*. When a set of rules specifies a hull that is minimal, they are called *perfect*. Of special interest now are local rules that are aperiodic and perfect, such as the well-known arrow matching conditions of the classic rhombic Penrose tiling; compare [1].

A single prototile, assumed compact and simply connected, is called a *monotile* (in the strict sense) when a set of aperiodic perfect local rules exists that can be realised by nearest neighbour matchings only. A subtle question in this context is whether one allows reflected copies of the prototile or not. Quite often, geometric matching conditions are replaced by suitable decorations of the prototile together with rules how these decorations have to form local patterns in the tiling process. The latter need not be restricted to conditions for nearest neighbour tiles, in which case one speaks of a *functional monotile* to indicate the slightly more general setting. Our planar examples below are of the latter type, or even a further extension of it.

This article, which is a brief review together with some new results on the two tiling spaces, grew out of a meeting on discrete geometry that was held at the Fields Institute in autumn 2011. As such, it is primarily written for a readership with background in discrete geometry, polytopes and tilings. We also try to provide the concepts and methods for readers with a different background, though this is often only possible by suitable pointers to the existing literature. In particular, where correct and complete proofs are available, we either refer to the original source or sketch how the arguments have to be applied to suit our formulation. As is often the case in discrete geometry, following a proof might need some pencil and paper activity on the side of the reader; compare the introduction and the type of presentation in [1]. This is particularly true of arguments around local derivation rules, inflation properties and aperiodic prototile sets.

When we prepared this manuscript, we rewrote known results on both tiling spaces in a way that emphasises their similarities, and mildly extended them, for instance by the percolation property of two derived parity patterns. Moreover, we calculated several topological invariants of the tiling spaces under consideration, which (as far as we are aware) were not known before. In order not to create an imbalance, we only describe how to do that in principle (again with proper references) and then state the results. We also include the dynamical zeta functions for the inflation action on the hulls, which are the generating functions for the corresponding fixed point counts. They turn out to be particularly useful for deriving the structure of the hull. Since several examples of a similar nature have recently been investigated in full detail, compare [20–22], we felt that this short account is adequate (in particular, as the explicit results are the outcome of a computer algebra program).

The paper is organised as follows. In Section 2, we begin with a discussion of the $(1 + \varepsilon + \varepsilon^2)$-tiling due to Penrose, together with various other elements of the MLD class defined by it. Section 3 contains the corresponding material on the Taylor (and the Socolar–Taylor) tilings and their "derivatives", which we discuss in slightly more detail, including the percolation result on the parity patterns of both tilings. The topological invariants and various other quantities for a comparison of the tilings are presented in Section 4, which is followed by some concluding remarks and open problems.

## 2. Penrose's Aperiodic Hexagon Tiling and Related Patterns

The $(1 + \varepsilon + \varepsilon^2)$-tiling due to Roger Penrose is built from three prototiles, up to Euclidean motions (including reflections). The tiles and the inflation rule (with linear inflation multiplier 2) are shown in Figure 2. The name $(1 + \varepsilon + \varepsilon^2)$-tiling refers to the three prototiles as the 1-tile (hexagon), the $\varepsilon$-tile (edge tile) and the $\varepsilon^2$-tile (corner tile). The $\varepsilon$-tile has a definite length, but can be made arbitrarily thin, while the $\varepsilon^2$-tile can be made arbitrarily small. This inflation rule is not primitive, but still defines a unique tiling LI class in the plane via a fixed point tiling with a hexagon at its centre. A patch of such a tiling is shown in Figure 3. There are 12 fixed point tilings of this type, each defining the same LI class. These fixed points form a single orbit under the $D_6$ symmetry of the LI class.

The following result was shown in [6] by the composition-decomposition method; see [23] for a detailed description of this method.

**Proposition 1** *The inflation rule of Figure 2 defines a unique tiling LI class with perfect aperiodic local rules. The latter are formulated via an aperiodic prototile set, which consists of the three tiles from Figure 2 (together with rotated and reflected copies). The rules are then realised as purely geometric matching conditions of the tiles.*

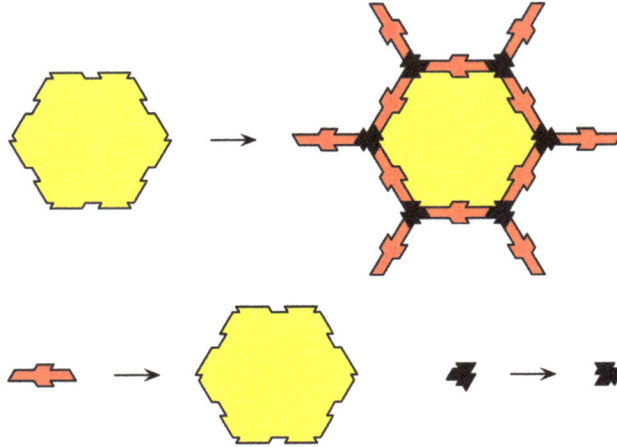

**Figure 2.** Inflation rule for Penrose's $(1 + \varepsilon + \varepsilon^2)$-tiling.

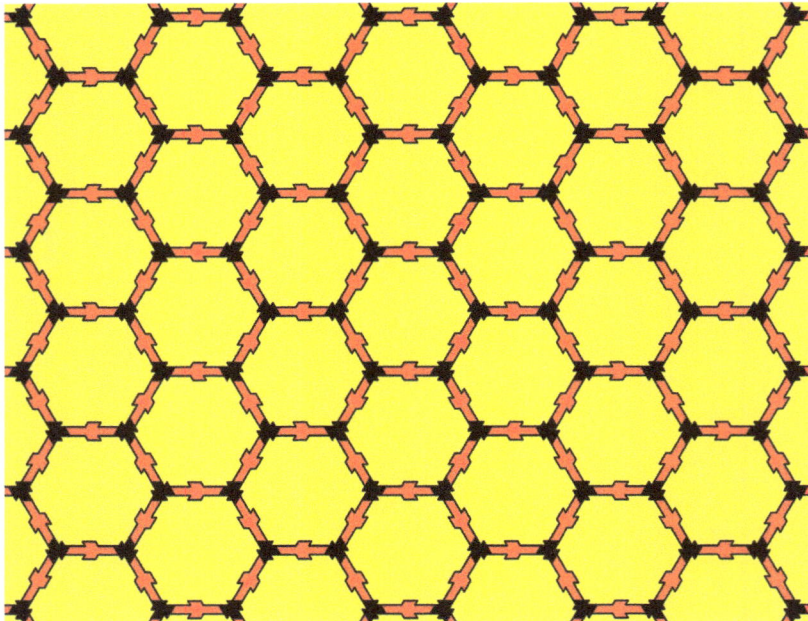

**Figure 3.** Patch of Penrose's $(1 + \varepsilon + \varepsilon^2)$-tiling.

In the original publication [6], it was argued that this system comes "close" to an aperiodic monotile in the sense that it is essentially a marked hexagon tiling with matching conditions that are realised by "key tiles", which can be made thin and small. A transformation to an equivalent version was only sketched briefly at the end of the article, and subsequently substantiated in [24] in the form of a puzzle and its solution. The key idea is to change from the $(1 + \varepsilon + \varepsilon^2)$-tiling to the double hexagon tiling of Figure 4, which is possible by the local derivation rule sketched in Figure 5, when read from left to right.

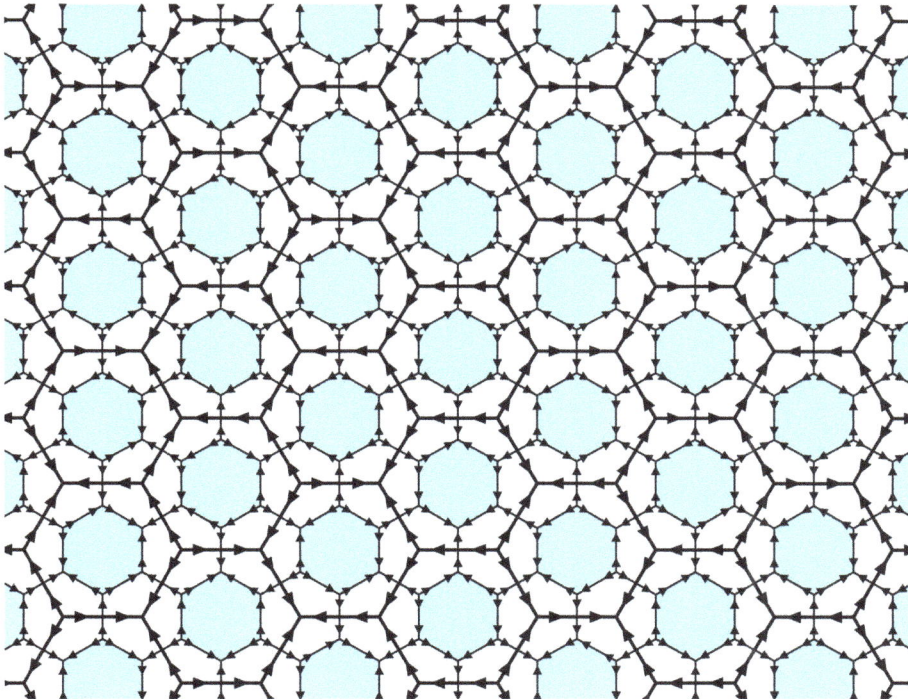

**Figure 4.** A patch of the double hexagon tiling, exactly corresponding to that of Figure 3.

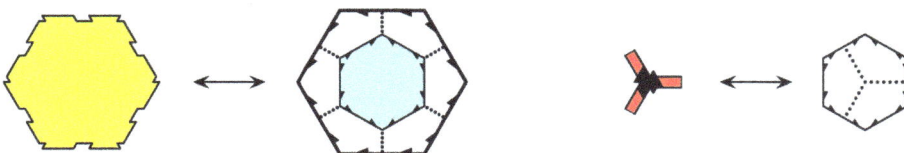

**Figure 5.** Rules for the mutual local derivation between $(1 + \varepsilon + \varepsilon^2)$-tilings and double hexagon tilings.

The actual rule consists of two steps. In the first, each hexagonal 1-tile is replaced by a double hexagon as shown in the left panel of Figure 5. This produces a complete tiling of the larger hexagons (with matching arrows) with fully oriented inscribed hexagons, while all remaining small hexagons, namely those around the vertices of the larger hexagons, are still incomplete (thin dotted lines). Their missing orientations derive consistently from the $\varepsilon^2$-tiles via the rule shown in the right panel.

Note that all hexagons (which means on both scales) have the same type of arrow pattern. In particular, each hexagon has precisely one pair of parallel edges that are oriented in the same direction, while all other arrows point towards the two remaining (antipodal) vertices. Note that this is the same edge orientation pattern as seen at the boundary of the hexagonal 1-tile in Figures 2 and 3.

The derivation rule specified in Figure 5 is clearly local in both directions, and commutes with the translation action as well as with all symmetry operations of the group $D_6$ (the symmetry group of the regular hexagon). The following result is thus obvious.

**Proposition 2** *The LI classes of the $(1 + \varepsilon + \varepsilon^2)$-tiling of Figure 3 and the Penrose double hexagon pattern of Figure 4 are SMLD.*

A closer inspection of the prototile set of Figure 2 reveals that the hexagonal 1-tile occurs in two chiralities with six orientations each. Keeping track of the chiralities only (by two colours, white and grey say) and disregarding all other structural elements provides a local derivation from the $(1 + \varepsilon + \varepsilon^2)$-tiling to an ensemble of 2-colourings of the hexagonal packing. An example of the latter is illustrated in Figure 6. We call the elements of this new ensemble the *parity patterns* of the $(1 + \varepsilon + \varepsilon^2)$-tilings. This parity pattern was introduced in [6], but (as far as we know) has not been further investigated. Our motive to do so will become clear from the comparison with the llama tilings in the next section. By construction, the parity patterns form a single LI class. A little surprising is the following property.

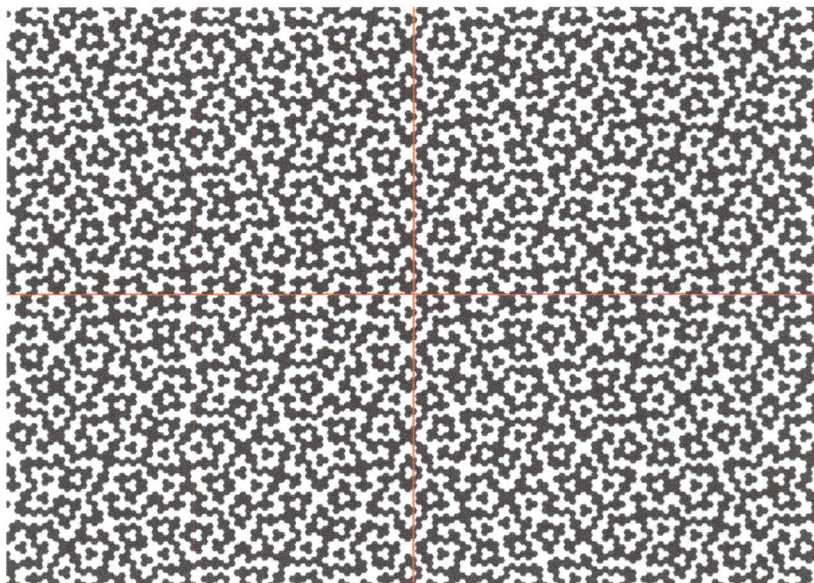

**Figure 6.** Parity pattern of a $(1 + \varepsilon + \varepsilon^2)$-tiling, as derived from a fixed point tiling of the inflation rule of Figure 2. This particular pattern (also the infinite one) possesses an almost colour reflection symmetry with respect to the indicated lines; see text for details.

**Theorem 1** *The LI class of the $(1 + \varepsilon + \varepsilon^2)$-tiling of Figure 3 and that of the corresponding parity patterns of Figure 6 are SMLD.*

Sketch of Proof. The determination of the parity pattern that belongs to a $(1 + \varepsilon + \varepsilon^2)$-tiling is clearly local and preserves all symmetries, so that this direction is clear.

Conversely, starting from a parity pattern, the corresponding $(1 + \varepsilon + \varepsilon^2)$-tiling is locally reconstructed via hexagonal coronae of order 3. A simple computer search shows that such coronae uniquely specify the decorated hexagon that corresponds to its centre, once again in a symmetry-preserving way.

Let us note that, as a result of the relation between two fixed point tilings with mirror image central hexagons, the particular parity patch of Figure 6 shows an almost reflection colour symmetry for the reflections in the two lines indicated. More precisely, under reflection and colour inversion, the patch is mapped onto itself, except for some hexagons along the reflection line.

**Remark 1** *The left panel of Figure 5 shows the building block of the double hexagon tiling without arrows on the dashed lines. As mentioned above, they are added by the local rule how to complete the vertex configurations. In this sense, one has a single prototile template together with a set of local rules that specify how to put them together. Any double hexagon tiling of the plane that everywhere satisfies the rules is an element of the double hexagon LI class, and in this sense one has a functional monotile template. We suggest calling this a weak functional monotile, as it stretches the monotile concept to some extent.*

Let us return to the structure of the $(1 + \varepsilon + \varepsilon^2)$-tilings. As mentioned before, each hexagonal 1-tile possesses a unique pair of parallel edges that are oriented (by the arrows) in the same direction. If we divide it now into two half-hexes along the diagonal that is parallel to this edge pair, we can locally derive a half-hex tiling from any $(1 + \varepsilon + \varepsilon^2)$-tiling. An inspection of Figure 2 confirms that this derivation rule indeed induces the half-hex inflation of Figure 1. This is a local derivation of sliding block map type on the underlying hexagonal packing, hence continuous in the local topology.

**Proposition 3** *The LI class of the half-hex tiling as defined by the stone inflation rule of Figure 1 defines a minimal topological dynamical system under the translation action of $\mathbb{R}^2$ that is a topological factor of the LI class of the $(1 + \varepsilon + \varepsilon^2)$-tilings defined by the inflation rule of Figure 2. The corresponding factor map is one-to-one almost everywhere, but the two tiling spaces define distinct MLD classes.*

Sketch of Proof. The first claim follows from our above description of the local derivation rule via the additional diagonal line in the hexagonal 1-tiles. The two LI classes cannot be MLD (and hence also not SMLD) because the existence of aperiodic local rules is an invariant property of an MLD class, hence shared by all LI classes that are MLD. It is well-known [1,12] that the half-hex hull has no such set of rules, because any finite atlas of patches can still be part of a periodic arrangement.

Since the $(1 + \varepsilon + \varepsilon^2)$-tilings have perfect aperiodic local rules by Proposition 1, the last claim is clear. The statement on the multiplicity of the mapping is a consequence of the model set structure, which we prove below in Theorem 5.

The (regular) model set structure of the half-hex tiling, as spelled out in Equation (1) for the fixed points under the inflation rule, implies that there is a "torus parametrisation" map onto a compact Abelian group [16,25]. Here, it is a factor map onto the two-dimensional dyadic solenoid $\mathbb{S}_2^2$, which is almost everywhere one-to-one by Theorem 5 of [25]. Since the half-hex LI class is the image of the $(1 + \varepsilon + \varepsilon^2)$ LI class under a factor map that is itself one-to-one almost everywhere (which follows by the same argument that we use below to prove Theorem 5), we know (via concatenation) that there exists an almost everywhere one-to-one factor map from the $(1 + \varepsilon + \varepsilon^2)$ LI class onto $\mathbb{S}_2^2$. Then, Theorem 6 from [25] implies the following result.

**Corollary 1** *The $(1 + \varepsilon + \varepsilon^2)$ LI class has a model set structure, with the same cut and project scheme as derived for the half-hex tilings.*

In summary, the $(1 + \varepsilon + \varepsilon^2)$ LI class has all magical properties: It can be defined by an inflation rule, by a set of perfect aperiodic local rules, and as a regular model set. Moreover, it comes close to solving the quest for a (functional) monotile.

Let us turn our attention to a later (though completely independent) attempt of a similar kind that improves the monotile state-of-affairs.

### 3. Taylor's Inflation Tiling

Consider the primitive stone inflation rule of Figure 7. It is formulated with 14 prototiles (up to similarity) of half-hex shape that are distinguished by colour and a decoration (with points and lines). Each half-hex occurs in two chiralities and six orientations, so that the total number of prototiles (up to translation) is 168. In the original paper [7], the 7 colours are labelled $A, B, C, \ldots, G$. The C-type tiles (light blue in our version) are special in the sense that they are in the centre of any fixed point tiling under the inflation rule. They are also more frequent than the other types. As mentioned above, we call the elements of the tiling space (or hull) defined by this inflation rule the *Taylor tilings*. The following result is immediate from Figure 7, in comparison with Figure 1, by simply removing decorations and colour.

**Lemma 1** *The half-hex LI class is a topological factor of the LI class of the Taylor tilings, where the latter is again minimal. In particular, the Taylor LI class is aperiodic.*

As in the case of the Penrose tiling above, one can locally derive a hexagonal parity pattern from every Taylor tiling. To this end, one considers the natural half-hex pairs, disregards their colours, and applies a grey/white coding of the two chiralities. The resulting two-coloured hexagonal packings are called *llama tilings*, see Figure 8 for an illustration. The name refers to the shape of the smallest island (of either colour), and was coined by Taylor. As in the previous section, the parity pattern still contains the full local information. Also, the llama tiling has the same type of almost colour reflection symmetry that we encountered in the previous section for the Penrose parity pattern.

**Theorem 2** *The LI class of Taylor's inflation tiling and that of the llama tiling of Figure 8 are SMLD. In particular, the llama tiling is aperiodic.*

Sketch of Proof. As in the case of Theorem 1, the derivation of the llama tiling, which is the parity pattern of the Taylor tiling, is obviously local and symmetry preserving.

For the converse direction, there are three proofs known. The first is based on an idea by Joan Taylor, and is spelled out in detail in [26]. A related argument uses the local information contained in the llamas together with the correspondence of coloured and marked hexagons with local parity patterns, as indicated in Figure 9; see [9] for details. Finally, as for Theorem 1, one can reconstruct the complete decoration of any hexagon from the order-3 coronae of the llama tiling, which preserves the symmetry.

The final claim follows from the aperiodicity of the Taylor tiling by standard arguments.

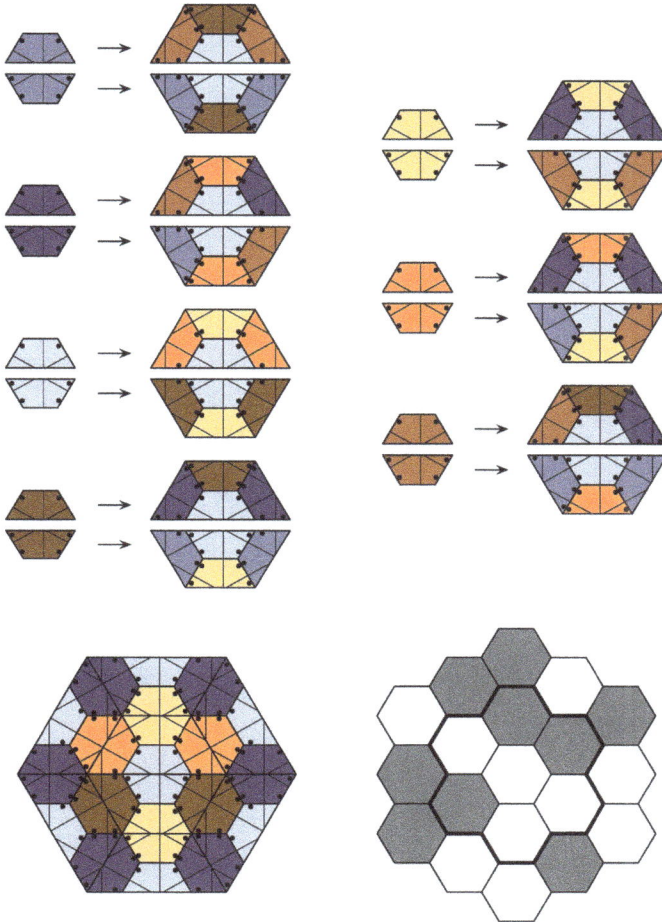

**Figure 7.** The primitive inflation rule of Taylor's half-hex inflation (top). The central patch of a fixed point tiling is shown on the lower left panel, with its parity pattern to the right.

Before we continue with our general discussion, let us mention an interesting property of the llama tilings, which also holds for the Penrose parity pattern.

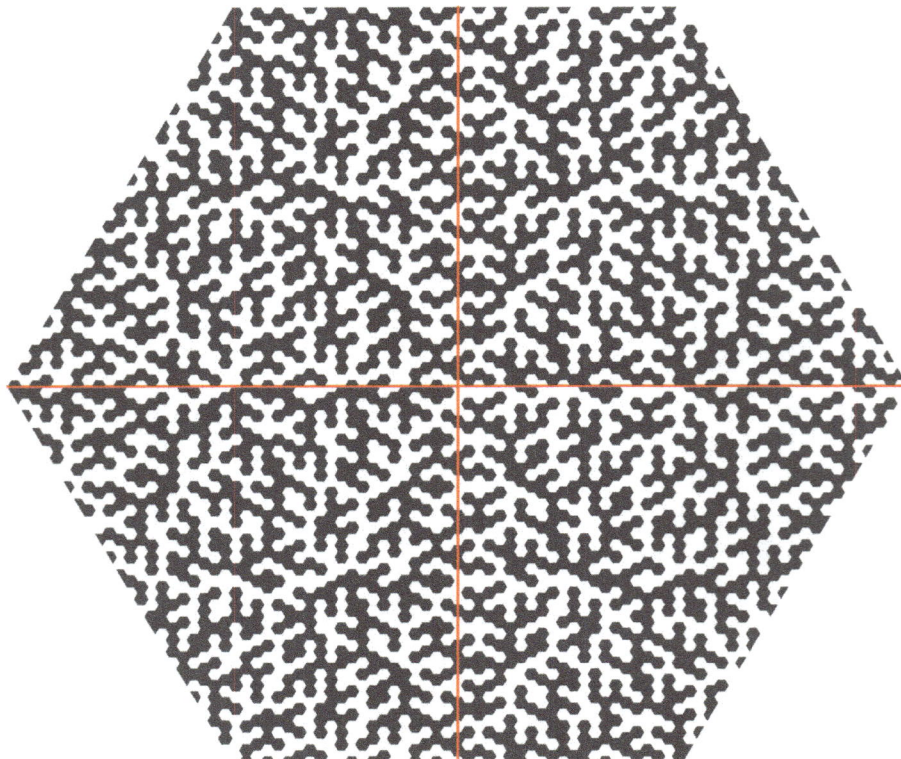

**Figure 8.** Patch of a llama tiling, as derived from a fixed point tiling of the inflation rule of Figure 7. This particular pattern (and its infinite extension) possesses an almost colour reflection symmetry with respect to the indicated lines; see text for details.

**Theorem 3** *The llama tiling of Figure 8 possesses connected components of either colour of unbounded size. The same conclusion also holds for the Penrose parity pattern of Figure 6. Moreover, this property extends to every element of the respective LI classes.*

Proof. The patch of the Taylor tiling shown in Figure 8 is derived from one inflation fixed point with a hexagon (originally of type *C*) of positive chirality as a seed, which we denote as pattern P1. Another fixed point pattern, P2, can be obtained from the corresponding hexagon of opposite chirality, which has grey and white colours interchanged relative to P1. Nevertheless, P1 and P2 are LI, so that arbitrarily large patches of either pattern occur in the other.

Now, assume that P1 does not contain connected patches of white hexagons of unbounded size, where two hexagons are called connected when they share an edge. If so, there must be a maximal white "island" in P1, of diameter $r$ say, which must then be surrounded by a connected "belt" of grey hexagons (which is possibly part of an even larger connected patch). Then, the same belt exists in P2, this time as a white belt around a grey island, and this belt has diameter $> r$ by construction. Since P1 and P2 are LI, this patch from P2 must also occur somewhere in P1, in contradiction to the assumption, and our claim follows.

The argument for the Penrose parity pattern is completely analogous to that for the llama tilings, while the final claim is obvious.

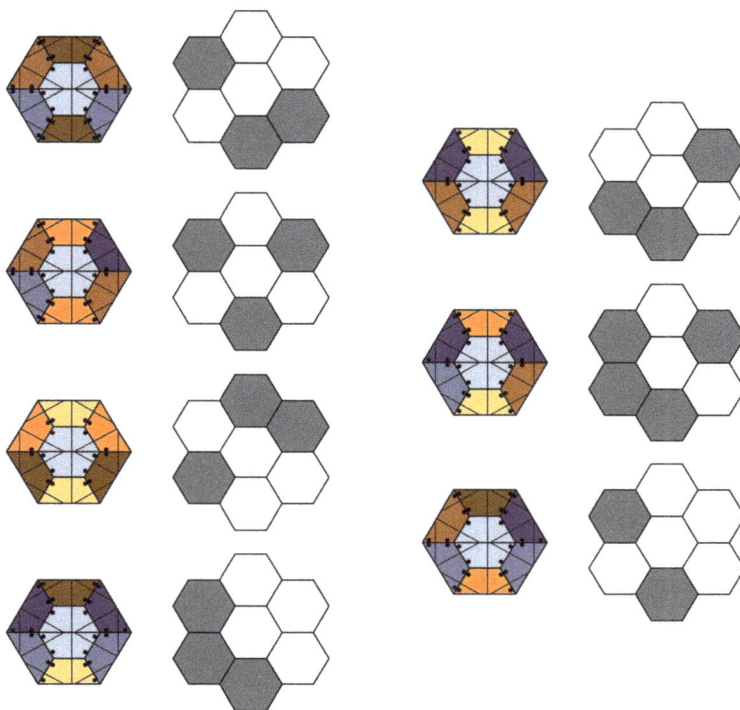

**Figure 9.** Correspondence between fully decorated (half) hexagons, with central hexagon of type *C*, and their parity patterns.

The llama tiling as well as Penrose's parity pattern are thus interesting examples of deterministic aperiodic structures with *percolation*. In particular, an element with an infinite connected cluster must exist in either LI class. Since each class is a compact space, this claim follows from a compactness argument, because any sequence of tilings with connected clusters of increasing diameters around the origin must contain a subsequence that converges in the local topology to a tiling with an infinite cluster. Note, however, that the above argument does not imply the existence of a sequence of islands of growing size, which has been conjectured for the llama LI class [8] and seems equally likely in the other LI class as well, both on the basis of inflation series of suitable patches.

Between any Taylor tiling and the corresponding llama tiling (which are SMLD) is another version that still shows all line and point markings (and hence the chirality of the hexagons), but not the seven colours. Clearly, also this version, which we call the *decorated llama tiling*, is in the same SMLD class. One can now formulate three local rules for the corresponding prototile set, which consists of 12 tiles (up to translations).

R1.   The hexagons must match at common edges in the sense that the decoration lines do not jump on crossing the common edge;

R2.   The point markers must satisfy the edge transfer rule sketched in the middle panel of Figure 10, as indicated by the arrow, for any pair of hexagons separated by a single edge. The two points at

13

the corners adjacent to that edge have to be in the same position, and this rule applies irrespective of the chirality types of the tiles;

R3.   No vertex configuration is allowed to have adjacent points in a threefold symmetric arrangement, such as the one shown in the right panel of Figure 10, or its rotated and reflected versions.

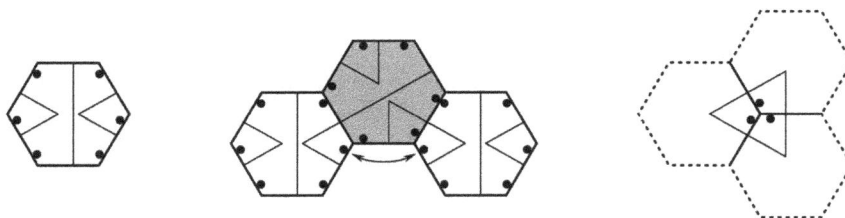

**Figure 10.** Taylor's functional monotile (left), a sketch of the edge transfer rule R2 (central) and the forbidden threefold vertex seed (right); see text for details.

In [7,8], the following result is shown by the composition-decomposition method, with immediate consequences for other members of the corresponding MLD class.

**Theorem 4** *The rules R1–R3 constitute perfect aperiodic local rules for the LI class of the decorated llama tiling. Consequently, a corresponding set of perfect aperiodic local rules also exists for the LI class of the Taylor tilings and for the llama LI class.*

The decorated llama tiling was selected in the MLD class for the following reason.

**Corollary 2** *The hexagon of the left panel of Figure 10, together with its reflected copy, provides a functional monotile under the rules R1–R3. It defines the LI class of the decorated llama tiling.*

Let us mention that one can relax the three rules by omitting R3. As is proved in [8], this still provides a set of aperiodic local rules, this time for the larger space of the Socolar–Taylor tilings. This space is not minimal, and additionally contains two patterns with global threefold rotation symmetry (and their translates).

## 4. Topological Invariants and the Structure of the Hulls

In this section, we derive and compare further details of the hulls of the various hexagonal tilings, where we go beyond the material that has appeared in the literature so far. For this purpose, it proves useful to generate the tilings by lattice inflations in which the tiles are represented by the points of a triangular lattice, with the tile type attached as a label to each point. Geometrically, such lattice inflation tilings consist of labelled hexagon tiles only. Key tiles like the ones in the $(1 + \varepsilon + \varepsilon^2)$-tiling have to be absorbed into the type of the tiles (via suitable decorations).

When passing from half hexagons to full hexagons, one easily obtains the overlapping inflation (also called pseudo inflation) shown in Figure 11 (left) for the half-hex tiling. The outer ring of hexagons is shared between the inflations of neighbouring hexagons. A pseudo inflation has to be *consistent* in the sense that it agrees on the overlap regions. From this pseudo inflation, a standard inflation can be obtained by replacing each hexagon by the four shaded hexagons only, independently of the orientation of the original hexagon. Conversely, each hexagon in the inflated tiling is assigned to a unique supertile hexagon. The inflation rule obtained in this way is not rotation covariant, but this is only a minor disadvantage. Also, the fixed point tiling obtained under the iterated inflations no longer covers the entire plane, but only a 120-degree wedge, which is not a problem either. Since the inflation rule is derived from a pseudo inflation, it determines the tiling also in the other sectors, and there is a unique continuation to the rest of the plane. Technically speaking, the inflation rule *forces the border* [27], a property that simplifies the computation of topological invariants considerably.

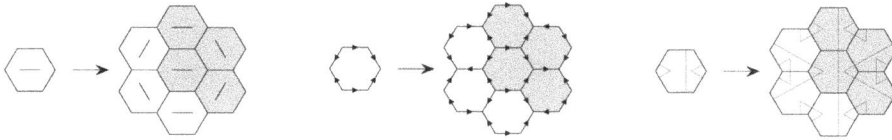

**Figure 11.** Hexagon (pseudo) inflation for the half-hex (left) and for two (locally equivalent) variants of the arrowed half-hex (centre and right). If only the shaded hexagons are retained, a standard inflation for a wedge with opening angle $\frac{2\pi}{3}$ is obtained; see text for details.

A natural generalisation of the half-hex tiling is the *arrowed half-hex tiling*, obtained by either of the two variants of (pseudo) inflations shown in Figure 11 (centre and right). Clearly, the two decorations are locally equivalent and lead to tilings that are MLD. Just as for the half-hex, a pseudo inflation is obtained first, from which a standard inflation is then derived by taking only the shaded tiles. The arrow pattern on the hexagons shown in Figure 11 (centre) is the same as that of the hexagon tile of the $(1 + \varepsilon + \varepsilon^2)$-tiling. In fact, it is easy to see that the arrowed half-hex is locally derivable from the latter, and thus is a factor. Similarly, the hexagon decoration of Figure 11 (right) is part of the decoration of the Taylor hexagons, so that the arrowed half-hex is also a factor of the Taylor tiling.

**Corollary 3** *The hull of the half-hex tiling, viewed as a dynamical system under the action of* $\mathbb{R}^2$, *is a topological factor of both the* $(1 + \varepsilon + \varepsilon^2)$-*tiling hull and the Taylor tiling hull. The same property holds for the hull of the arrowed half-hex tiling.*

A larger patch of the arrowed half-hex tiling is shown in Figure 12 (in the Taylor decoration variant). The fixed point tilings generated from the six different orientations of the arrowed hexagon differ only along three rows of hexagons crossing in the centre of the figure. We call these the *singular rows*. The six fixed point tilings project to the same point on the solenoid $\mathbb{S}_2^2$. Moreover, pushing the fixed point centre along one of the singular rows towards infinity, pairs of tilings that are mirror images of each other are obtained, except on the one singular row that remains. Each such pair of tilings also projects to the same points of the solenoid $\mathbb{S}_2^2$. In fact, each singular row gives rise to a 1d sub-solenoid $\mathbb{S}_2^1$, onto which the projection is 2-to-1 (except for the crossing point of these sub-solenoids). Three such sub-solenoids $\mathbb{S}_2^1$, onto which the projection is 2-to-1, clearly exist in the $(1 + \varepsilon + \varepsilon^2)$-tiling and in the Taylor tiling as well. However, for the (naked) half-hex tiling, there are *no* singular rows. In this case, there are just three fixed point tilings, differing in the orientation of the central hexagon only, which project to the same point on the solenoid $\mathbb{S}_2^2$. We refer to [9] for an explicit derivation of the model set coordinates of the arrowed half-hex tiling.

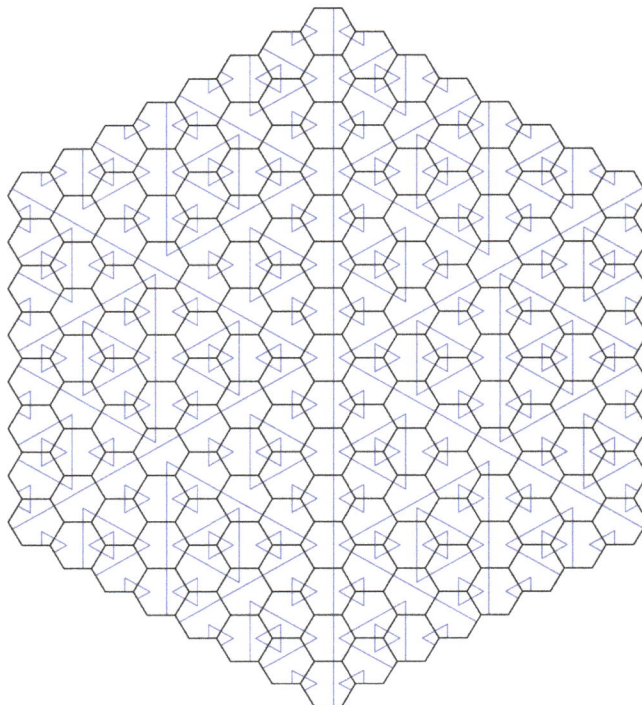

**Figure 12.** Patch of the arrowed half-hex tiling. The blue decoration lines form an infinite hierarchy of triangles of all sizes. The blue triangles of a given size and orientation form a triangular lattice.

The hexagon of the arrowed half-hex tiling has still one remaining line of mirror symmetry. Both the $(1 + \varepsilon + \varepsilon^2)$-tiling and the Taylor tiling break this remaining symmetry, alongside with a splitting into 7 subtypes of hexagons, which differ in their behaviour under the inflation. For the Taylor tiling, a pseudo inflation for fully asymmetric hexagons follows easily from Figure 7, which can then be converted into a standard lattice inflation as before.

For the $(1 + \varepsilon + \varepsilon^2)$-tiling, we replace the $\varepsilon$- and $\varepsilon^2$-tiles by the equivalent line decoration for the hexagons shown in Figure 13, which was introduced already in [6]. In addition to the blue lines of the arrowed half-hex, which encode the arrows of the $\varepsilon$-tiles along the hexagon edges, there is now also a second set of red lines, which encode the information contained in the $\varepsilon^2$-tiles and in the asymmetry across the $\varepsilon$-tiles. The latter is represented by the piece of red line parallel to the hexagon edge. The matching condition of the $\varepsilon^2$-tiles is replaced by the requirement that red lines must continue across tile edges. In the interior, each hexagon carries an X-shaped pair of red line angles, and a red line belt, which breaks the remaining mirror symmetry of the blue line decoration. The inflation rule for the 7 types of decorated hexagons is shown in Figure 14.

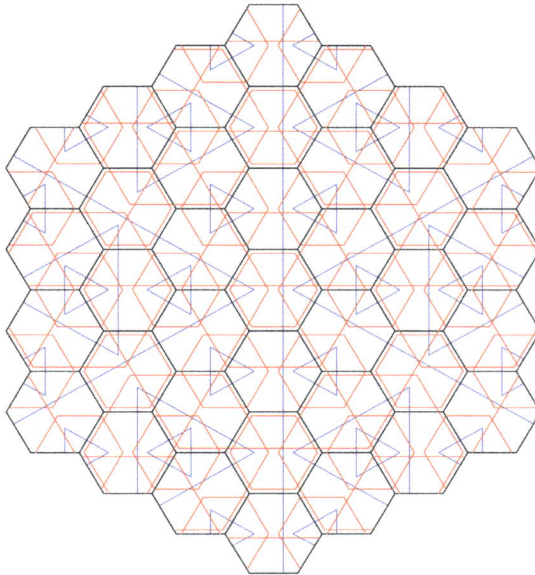

**Figure 13.** Patch of the $(1 + \varepsilon + \varepsilon^2)$-tiling in the variant with line decorations. In addition to the hierarchy of blue triangles from the half-hex tiling, there is now also a hierarchy of red hexagons. Hexagons of each size form a lattice periodic array, with lattices of different densities.

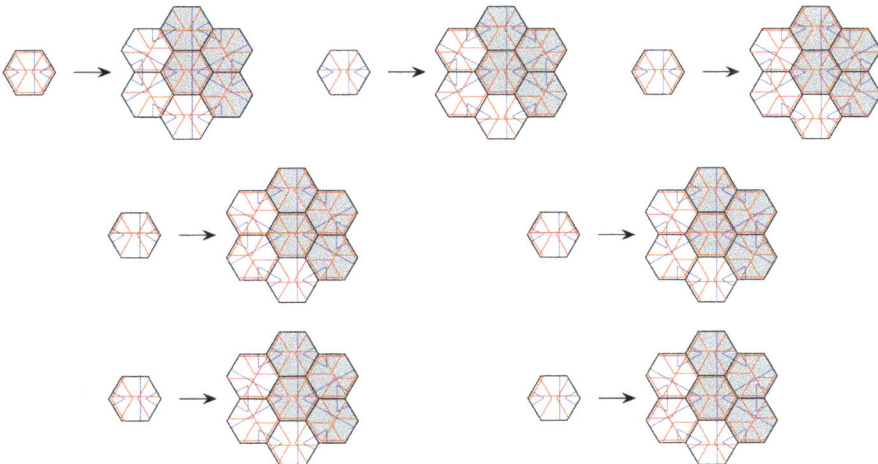

**Figure 14.** Inflation rules for $(1 + \varepsilon + \varepsilon^2)$-tiling in the variant with line decorations. If a hexagon is replaced by 7 hexagons, a pseudo inflation is obtained. Taking only the 4 shaded hexagons results in a standard inflation rule (for one sector).

From Figure 13, it is clear that the red line decoration breaks the horizontal mirror symmetry of the blue arrowed half-hex decoration. Therefore, there will be three additional 1d sub-solenoids $\mathbb{S}^1_2$ in three further directions, to which the projection is 2-to-1 rather than 1-to-1. These 1d solenoids, as well as those of the arrowed half-hex, all intersect in the single point to which the now 12 fixed point tilings project. Moreover, the three "red" sub-solenoids intersect also in two further points at corners of the hexagons, which are fixed points under the square of the inflation. On these points, the projection to the solenoid is 6-to-1.

The set of points on the solenoid $\mathbb{S}_2^2$ to which the projection of the Taylor tiling is not 1-to-1 is completely analogous. The corresponding ambiguities of the decoration, starting with a fixed point tiling with full $D_6$ or $D_3$ symmetry, has been derived in [8]. For an independent approach, see [26]. A summary of the projection situation is sketched and explained in Figure 15.

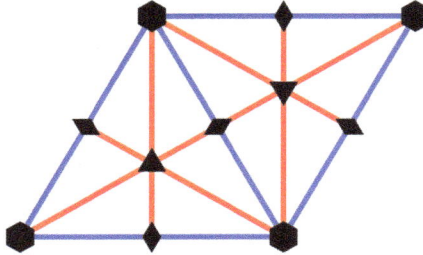

**Figure 15.** Schematic representation of the hulls of the different hexagon tilings via a (closed) toral slice of the underlying solenoid. The projection fails to be 1-to-1 on the lines as follows. For the *half-hex*, the projection is 3-to-1 precisely at the points marked by a hexagon, and 1-to-1 elsewhere. For the *arrowed half-hex*, the 1d sub-solenoids where the projection is 2-to-1 are shown as blue lines. At the hexagon points, three blue sub-solenoids intersect, and the projection is 6-to-1. For the *Taylor* and the *Penrose* hexagon tilings, there are three additional sub-solenoids (red lines) along which the projection is 2-to-1. These intersect at two inequivalent points (marked by triangles) where the projection is 6-to-1 (at such points, three infinite order supertiles meet). The projection is 12-to-1 at the hexagon points (which are centres of infinite-order supertiles). Note that, at the points marked by diamonds, not only two but in fact all six sub-solenoids intersect. This is due to the dyadic structure of the hull. As these points have half-integer coordinates with respect to the hexagonal lattice, they are equivalent (in the solenoid) to the points marked by hexagons.

The first benefit of having represented all four tilings as lattice inflations is the following.

**Theorem 5** *The half-hex, arrowed half-hex, $(1 + \varepsilon + \varepsilon^2)$ and Taylor tilings are all model sets, and as such have pure point dynamical and diffraction spectrum.*

Sketch of Proof. As the half-hex and arrowed half-hex tilings are both factors of the other two tilings, we only need to consider the $(1 + \varepsilon + \varepsilon^2)$- and Taylor tilings. Both have one "preferred" tile type, which occurs twice as often as the other tile types. For the Taylor tiling, it is type C (light blue in Figure 7), and for the $(1 + \varepsilon + \varepsilon^2)$-tiling it is the first type in Figure 14. In both cases, the centres of the preferred tiles form a triangular sublattice of index 4. Also the supertiles (of any fixed order) of the preferred tile types form a periodic array (disregarding orientation). Moreover, the preferred tiles are the seeds of the 12 fixed point tilings, which differ from each other only along the 6 mirror lines of the $D_6$ symmetry group. A sufficiently high order supertile therefore contains tiles which are the same in all preferred supertiles, independently of their orientation. Therefore, any Taylor or $(1 + \varepsilon + \varepsilon^2)$-tiling contains a lattice-periodic subset of tiles. The results of Theorem 3 from [3] then imply that they must be model sets and hence have pure point dynamical and diffraction spectrum by [15].

Another benefit of dealing with a lattice inflation is that the Anderson–Putnam approach [27] to computing the Čech cohomology of the hull becomes relatively easy to implement, since tiles are represented as labelled lattice points whose environments are easy to determine. Moreover, as the lattice inflation rules are derived from overlapping pseudo inflation rules, they have the property of forcing the border, which allows one to avoid the complication of using collared tiles; see [27] for details.

In this approach, from the local environments of the tiling a finite approximant cell complex is constructed, whose points represent cylinder sets of tilings. The full tiling space is then obtained as the inverse limit space of the inflation acting on the approximant cell complex. Correspondingly, the Čech cohomology of the hull is the direct limit of the inflation action on the cohomology of the approximant complex. For further details, we refer to [27,28]; compare also [20] for some explicitly worked-out examples. For our four tiling spaces, the following results are obtained.

**Theorem 6** *The four hexagon tiling spaces have the following Čech cohomologies:*

$$
\begin{array}{lllll}
Half-hex & H^2 = \mathbb{Z}[\tfrac{1}{4}] \oplus \mathbb{Z}^2, & H^1 = \mathbb{Z}[\tfrac{1}{2}]^2, & H^0 = \mathbb{Z}; \\
Arrowed\,half-hex & H^2 = \mathbb{Z}[\tfrac{1}{4}] \oplus \mathbb{Z}[\tfrac{1}{2}]^3, & H^1 = \mathbb{Z}[\tfrac{1}{2}]^2 \oplus \mathbb{Z}, & H^0 = \mathbb{Z}; \\
1+\varepsilon+\varepsilon^2: & H^2 = \mathbb{Z}[\tfrac{1}{4}] \oplus \mathbb{Z}[\tfrac{1}{2}]^6 \oplus \mathbb{Z}^5 \oplus \mathbb{Z}_3, & H^1 = \mathbb{Z}[\tfrac{1}{2}]^2 \oplus \mathbb{Z}^2, & H^0 = \mathbb{Z}; \\
Taylor & H^2 = \mathbb{Z}[\tfrac{1}{4}] \oplus \mathbb{Z}[\tfrac{1}{2}]^6 \oplus \mathbb{Z}^7, & H^1 = \mathbb{Z}[\tfrac{1}{2}]^2 \oplus \mathbb{Z}^4, & H^0 = \mathbb{Z}.
\end{array}
$$

*In particular, the $H^2$-groups are distinct.*

Since the $(1+\varepsilon+\varepsilon^2)$- and the Taylor tiling have different cohomology, it immediately follows that they cannot be in the same MLD class.

**Corollary 4** *The $(1+\varepsilon+\varepsilon^2)$ LI class and the Taylor LI class define distinct MLD classes.*

As a side result of the cohomology calculation, the Artin–Mazur dynamical zeta function of the inflation action on the hull can also be obtained. It is defined as

$$
\zeta(z) = \exp\left( \sum_{m=1}^{\infty} \frac{a_m}{m} z^m \right)
$$

where $a_m$ is the number of points in the hull invariant under an $m$-fold inflation. We note that, if the hull consists of two (or more) components for which the periodic points can be counted separately, $a_m = a'_m + a''_m$, the corresponding partial zeta functions have to be multiplied: $\zeta(z) = \zeta'(z)\cdot\zeta''(z)$. This will turn out to be useful shortly. For the action of the mapping $x \mapsto 2x$ on the dyadic solenoids $\mathbb{S}_2^1$ and $\mathbb{S}_2^2$, the zeta function coincides with that of the corresponding toral endomorphism, which are

$$
\frac{1-z}{1-2z} \quad\text{and}\quad \frac{(1-2z)^2}{(1-z)(1-4z)}
$$

by an application of the results from [29]. The corresponding fixed point counts read $a_m^{(1)} = 2^m - 1$ and $a_m^{(2)} = (2^m - 1)^2$, for $m \in \mathbb{N}$.

Anderson and Putnam [27] have shown that the dynamical zeta function can be computed from the action of the inflation on the cochain groups of the approximant complex. Here, we rather express it in terms of the action of the inflation on the rational cohomology groups of the hull. If $A^{(m)}$ is the matrix of the inflation action on the $m$-th rational cohomology group, the dynamical zeta function is given by

$$
\zeta(z) = \frac{\prod_{k\,\text{odd}} \det\left(1 - zA^{(d-k)}\right)}{\prod_{k\,\text{even}} \det\left(1 - zA^{(d-k)}\right)} = \frac{\prod_{k\,\text{odd}} \prod_i \left(1 - z\lambda_i^{(d-k)}\right)}{\prod_{k\,\text{even}} \prod_i \left(1 - z\lambda_i^{(d-k)}\right)}
$$

where the latter equality holds when the matrices $A^{(m)}$ are diagonalisable with eigenvalues $\lambda_i^{(m)}$. The additional terms in the expressions from [27] cancel between numerator and denominator.

19

From the cohomology and the eigenvalues $\lambda_i^{(m)}$, the dynamical zeta function is easily obtained. For the half-hex, we get

$$\zeta(z) = \frac{(1-2z)^2}{(1-z)(1-4z)(1-z)^2} = \frac{(1-2z)^2}{(1-z)(1-4z)} \cdot \frac{1}{(1-z)^2} \tag{2}$$

which we have written as the product of the zeta function of the 2d solenoid $\mathbb{S}_2^2$ and the generating function of two additional fixed points. This is in line with our observation that the projection to $\mathbb{S}_2^2$ is 1-to-1 except at one point, where it is 3-to-1, wherefore there are two extra fixed points beyond the one already present in $\mathbb{S}_2^2$.

For the arrowed half-hex, we find

$$\zeta(z) = \frac{(1-2z)^2(1-z)}{(1-z)(1-4z)(1-2z)^3} = \frac{(1-2z)^2}{(1-z)(1-4z)} \cdot \left(\frac{1-z}{1-2z}\right)^3 \cdot \frac{1}{(1-z)^2} \tag{3}$$

which is the product of the zeta functions of a 2d solenoid $\mathbb{S}_2^2$, three 1d solenoids $\mathbb{S}_2^1$, and two additional fixed points. As discussed above, we have three 1d solenoids $\mathbb{S}_2^1$ where the projection to $\mathbb{S}_2^2$ is 2-to-1, hence the three extra copies of $\mathbb{S}_2^1$. The projection from the fixed points of the arrowed half-hex tiling is 6-to-1 (they form a $D_3$-orbit), so that there must be two extra fixed points in the zeta function, in addition to those contained in the 4 solenoids.

Finally, both for the $(1+\varepsilon+\varepsilon^2)$- and the Taylor tilings, we obtain

$$\begin{aligned}\zeta(z) &= \frac{(1-2z)^2(1-z)^{2(+2)}}{(1-z)(1-4z)(1-2z)^6(1-z)^{3(+2)}(1+z)^2} \\ &= \frac{(1-2z)^2}{(1-z)(1-4z)} \cdot \left(\frac{1-z}{1-2z}\right)^6 \cdot \frac{1}{(1-z)^5} \cdot \frac{1}{(1-z^2)^2}\end{aligned} \tag{4}$$

which is the product of the zeta functions of a 2d solenoid $\mathbb{S}_2^2$, six 1d solenoids $\mathbb{S}_2^1$, five additional fixed points, and two extra 2-cycles. Even though the two tilings have different cohomology, the additional terms in the zeta function of the Taylor tiling (indicated by the extra exponents in parentheses) cancel each other. As discussed above, both hulls contain six 1d solenoids onto which the projection is 2-to-1. The 12 fixed points, forming a $D_6$-orbit, all project to the same point, where the six 1d sub-solenoids intersect. In addition to the 7 fixed points in the altogether 7 solenoids, there must hence be 5 further fixed points, which indeed show up in the zeta function. Finally, the three "red" 1d solenoids intersect also at two types of corners of the hexagon tiles. These points form 2-cycles under the inflation, and the projection to them is 6-to-1 (two $D_3$-orbits). Hence, two extra 2-cycles are present in the zeta function. The zeta functions derived above confirm that our analysis of the set where the projection to $\mathbb{S}_2^2$ fails to be 1-to-1 must have been complete.

**Corollary 5** *The Artin–Mazur zeta functions for the inflation action on the half-hex and on the arrowed half-hex hull are given by Equations (2) and (3), while the $(1+\varepsilon+\varepsilon^2)$-hull and the Taylor tiling hull have the same zeta function, as given by Equation (4).*

*The corresponding fixed point counts, for $m \geq 1$, are given by $a_m^{(\mathrm{hh})} = (2^m-1)^2+2$, by $a_m^{(\mathrm{ahh})} = (2^m-1)^2+3(2^m-1)+2$, and by $a_m^{(\mathrm{P/T})} = (2^m-1)^2+6(2^m-1)+5+2(1+(-1)^m)$.*

It is a rather amazing fact that the Penrose and the Taylor tiling, despite defining distinct MLD classes, share the same dynamical zeta function for the respective inflation action, and have a projection to the 2d solenoid $\mathbb{S}_2^2$ with exactly the same multiplicities.

## 5. Outlook and Open Problems

The two tiling spaces due to Penrose and due to Taylor, which are both model sets (see also [26]), show amazing similarities, though they are certainly not MLD. Whether there is a local derivation in

*Symmetry* **2012**, *4*, 581–602

one direction is still not fully clear, but unlikely. It is an open problem where and what exactly is the difference between the two tilings.

In the two parity patterns, beyond the percolation structure, one can see islands of growing size in both tilings. Based on the inflation structure, it is thus natural to conjecture that both classes of parity patterns contain islands of unbounded size, though a proof does not seem obvious. Also, the emergence via inflation series seems slightly different in the two tilings.

The quest for a true monotile in the plane is not settled yet, because the rules cannot be realised by nearest neighbour conditions, and hence not by simple markings alone (unless one admits a prototile version with disconnected parts). There are other attempts to find an example, for instance via polyominoes and related objects; compare [30] and references therein.

An entirely different situation is met in 3-space, where the famous SCD prototile [31,32] establishes a mechanism that is truly three-dimensional. Indeed, the non-periodicity here is a result of a screw axis with an incommensurate rotation, wherefore the repetitive cases are aperiodic, but not strongly aperiodic; see [4] for a discussion. Also, unlike the situation above, the local rules for the SCD tile have to explicitly exclude the use of a reflected version, which is perhaps not fully satisfactory either.

In summary, some progress was made in the quest for an aperiodic monotile in recent years, but the search is certainly not over yet!

**Acknowledgments:** We are grateful to Roger Penrose and Joan Taylor for important comments and suggestions, to Robert Moody and Egon Schulte for helpful discussions, and to two anonymous reviewers for various constructive comments. This work was supported by the German Research Council (DFG), within the CRC 701. UG is grateful to the Fields Institute for financial support.

# References

1. Grünbaum, B.; Shephard, G.C. *Tilings and Patterns*; Freeman: New York, NY, USA, 1987.
2. Frettlöh, D.; Sing, B. Computing modular coincidences for substitution tilings and point sets. *Discrete Comput. Geom.* **2007**, *37*, 381–407. [CrossRef]
3. Lee, J.-Y.; Moody, R.V. Lattice substitution systems and model sets. *Discrete Comput. Geom.* **2001**, *25*, 173–201. [CrossRef]
4. Baake, M.; Grimm, U. On the notions of symmetry and aperiodicity for Delone sets. *Symmetry* **2012**, *4*, 566–580. [CrossRef]
5. Goodman-Strauss, C. Matching rules and substitution tilings. *Ann. Math.* **1998**, *147*, 181–223. [CrossRef]
6. Penrose, R. Remarks on tiling: Details of a $(1 + \varepsilon + \varepsilon^2)$-aperiodic set. In *the Mathematics of Long-Range Aperiodic Order*; Moody, R.V., Ed.; NATO ASI series 489; Kluwer: Dordrecht, the Netherlands, 1997; pp. 467–497.
7. Taylor, J.M. Aperiodicity of a functional monotile. Available online: http://www.math.uni-bielefeld.de/sfb701/preprints/view/420 (accessed on 16 October 2012).
8. Socolar, J.E.S.; Taylor, J.M. An aperiodic hexagonal tile. *J. Comb. Theory A* **2011**, *118*, 2207–2231. [CrossRef]
9. Baake, M.; Grimm, U. *Theory of Aperiodic Order: A Mathematical Invitation*; Cambridge University Press: Cambridge, UK, in preparation.
10. Moody, R.V. Meyer sets and their duals. In *the Mathematics of Long-Range Aperiodic Order*; Moody, R.V., Ed.; NATO ASI series 489; Kluwer: Dordrecht, the Netherlands, 1997; pp. 403–441.
11. Moody, R.V. Model sets: A survey. In *from Quasicrystals to More Complex Systems*; Axel, F., Dénoyer, F., Gazeau, J.P., Eds.; Springer: Berlin, Germany, 2000; pp. 145–166.
12. Frettlöh, D. Nichtperiodische Pflasterungen mit ganzzahligem Inflationsfaktor. Ph.D. thesis, University Dortmund, Dortmund, Germany, 2002.
13. Gähler, F.; Klitzing, R. The diffraction pattern of self-similar tilings. In *the Mathematics of Long-Range Aperiodic Order*; Moody, R.V., Ed.; NATO ASI series 489; Kluwer: Dordrecht, the Netherlands, 1997; pp. 141–174.
14. Baake, M.; Lenz, D. Dynamical systems on translation bounded measures: Pure point dynamical and diffraction spectra. *Ergod. Theory Dyn. Syst.* **2004**, *24*, 1867–1893. [CrossRef]
15. Lee, J.-Y.; Moody, R.V.; Solomyak, B. Pure point dynamical and diffraction spectra. *Ann. Henri Poincaré* **2002**, *3*, 1003–1018. [CrossRef]

*Symmetry* **2012**, *4*, 581–602

16. Schlottmann, M. Generalised model sets and dynamical systems. In *Directions in Mathematical Quasicrystals*; Baake, M., Moody, R.V., Eds.; AMS: Providence, RI, USA, 2000; pp. 143–159.

17. Baake, M.; Moody, R.V. Weighted Dirac combs with pure point diffraction. *J. Reine Angew. Math. (Crelle)* **2004**, *573*, 61–94. [CrossRef]

18. Baake, M.; Schlottmann, M.; Jarvis, P.D. Quasiperiodic patterns with tenfold symmetry and equivalence with respect to local derivability. *J. Phys. A* **1991**, *24*, 4637–4654. [CrossRef]

19. Baake, M. A guide to mathematical quasicrystals. In *Quasicrystals—An Introduction to Structure, Physical Properties and Applications*; Suck, J.-B., Schreiber, M., Häussler, P., Eds.; Springer: Berlin, Germany, 2002; pp. 17–48.

20. Baake, M.; Gähler, F.; Grimm, U. Spectral and topological properties of a family of generalised Thue-Morse sequences. *J. Math. Phys.* **2012**, *53*, 032701:1–24.

21. Baake, M.; Gähler, F.; Grimm, U. Examples of substitution systems and their factors. Unpublished work. 2012.

22. Gähler, F. Substitution rules and topological properties of the Robinson tilings. Unpublished work. 2012.

23. Gähler, F. Matching rules for quasicrystals: The composition-decomposition method. *J. Non-Cryst. Solids* **1993**, *153–154*, 160–164. [CrossRef]

24. Penrose, R. Supplement to remarks on tiling: Details of a $(1 + \varepsilon + \varepsilon^2)$-aperiodic set. In *Roger Penrose Collected Works*; Oxford University Press: Oxford, UK, 2011; Volume 6.

25. Baake, M.; Lenz, D.; Moody, R.V. Characterisation of model sets by dynamical systems. *Ergod. Theory Dyn. Syst.* **2007**, *27*, 341–382. [CrossRef]

26. Lee, J.-Y.; Moody, R.V. Taylor-Socolar hexagonal tilings as model sets. *Symmetry* **2012**. submitted for publication. [CrossRef]

27. Anderson, J.E.; Putnam, I.F. Topological invariants for substitution tilings and their associated C*-algebras. *Ergod. Theory Dyn. Syst.* **1998**, *18*, 509–537. [CrossRef]

28. Sadun, L. *Topology of Tiling Spaces*; AMS: Providence, RI, USA, 2008.

29. Baake, M.; Lau, E.; Paskunas, V. A note on the dynamical zeta function of general toral endomorphisms. *Monatsh. Math.* **2009**, *161*, 33–42. [CrossRef]

30. Rhoads, G.C. Planar tilings by polyominoes, polyhexes, and polyiamonds. *J. Comput. Appl. Math.* **2005**, *174*, 329–353.

31. Baake, M.; Frettlöh, D. SCD patterns have singular diffraction. *J. Math. Phys.* **2005**, *46*, 033510:1–10. [CrossRef]

32. Danzer, L. A family of 3D-spacefillers not permitting any periodic or quasiperiodic tiling. In *Aperiodic '94*; Chapuis, G., Paciorek, W., Eds.; World Scientific: Singapore, 1995; pp. 11–17.

*symmetry*

MDPI

*Article*

# Regular and Irregular Chiral Polyhedra from Coxeter Diagrams via Quaternions

**Nazife Ozdes Koca \* and Mehmet Koca †**

Department of Physics, College of Science, Sultan Qaboos University, P.O. Box 36, Al-Khoud 123, Muscat, Oman
* Correspondence: nazife@squ.edu.om; Tel.: +968-2414-1445
† Retired professor: mehmetkocaphysics@gmail.com

Received: 6 July 2017; Accepted: 2 August 2017; Published: 7 August 2017

**Abstract:** Vertices and symmetries of regular and irregular chiral polyhedra are represented by quaternions with the use of Coxeter graphs. A new technique is introduced to construct the chiral Archimedean solids, the snub cube and snub dodecahedron together with their dual Catalan solids, pentagonal icositetrahedron and pentagonal hexecontahedron. Starting with the proper subgroups of the Coxeter groups $W(A_1 \oplus A_1 \oplus A_1)$, $W(A_3)$, $W(B_3)$ and $W(H_3)$, we derive the orbits representing the respective solids, the regular and irregular forms of a tetrahedron, icosahedron, snub cube, and snub dodecahedron. Since the families of tetrahedra, icosahedra and their dual solids can be transformed to their mirror images by the proper rotational octahedral group, they are not considered as chiral solids. Regular structures are obtained from irregular solids depending on the choice of two parameters. We point out that the regular and irregular solids whose vertices are at the edge mid-points of the irregular icosahedron, irregular snub cube and irregular snub dodecahedron can be constructed.

**Keywords:** Coxeter diagrams; irregular chiral polyhedra; quaternions; snub cube; snub dodecahedron

---

## 1. Introduction

In fundamental physics, chirality plays a very important role. A Weyl spinor describing a massless Dirac particle is either in a left-handed state or in a right-handed state. Such states cannot be transformed to each other by the proper Lorentz transformations. Chirality is a well-defined quantum number for massless particles. Coxeter groups and their orbits [1] derived from the Coxeter diagrams describe the molecular structures [2], viral symmetries [3,4], crystallographic and quasi crystallographic materials [5–7]. Chirality is a very interesting topic in molecular chemistry. Certain molecular structures are either left-oriented or right-oriented. In three-dimensional Euclidean space, chirality can be defined as follows: if a solid cannot be transformed to its mirror image by proper isometries (proper rotations, translations and their compositions), it is called a chiral object. For this reason, the chiral objects lack the plane and/or central inversion symmetries. In two earlier publications [8,9], we studied the symmetries of the Platonic–Archimedean solids and their dual solids, the Catalan solids, and constructed their vertices. Two Archimedean solids, the snub cube and snub dodecahedron as well as their duals are chiral polyhedral, whose symmetries are the proper rotational subgroups of the octahedral group and the icosahedral group, respectively. Non-regular, non-chiral polyhedra have been discussed earlier [10]. Chiral polytopes in general have been studied in the context of abstract combinatorial form [11–14]. The chiral Archimedean solids, snub cube, snub dodecahedron and their duals have been constructed by employing several other techniques [15,16], but it seems that the method in what follows has not been studied earlier in this context.

We follow a systematic method for the construction of the chiral polyhedra. Let $G$ be a rank-3 Coxeter graph where $W(G)^+$ represents the proper rotation subgroup of the Coxeter group $W(G)$.

For the snub cube and snub dodecahedron, the Coxeter graphs are the $B_3$ and $H_3$, respectively. To describe the general technique, we first begin with simpler Coxeter diagrams $A_1 \oplus A_1 \oplus A_1$ and $A_3$, although they describe the achiral polyhedra such as the families of regular and irregular tetrahedron and icosahedron, respectively. We explicitly show that achiral polyhedra possess larger proper rotational symmetries transforming them to their mirror images. We organize the paper as follows. In Section 2 we introduce quaternions and construct the Coxeter groups in terms of quaternions [17]. We extend the group $W(A_1 \oplus A_1 \oplus A_1)$ to the octahedral group by the symmetry group $Sym(3)$ of the Coxeter diagram $A_1 \oplus A_1 \oplus A_1$. In Section 3 we obtain the proper rotation subgroup of the Coxeter group $W(A_1 \oplus A_1 \oplus A_1)$ and determine the vertices of an irregular tetrahedron. In Section 4 we discuss a similar problem for the Coxeter-Dynkin diagram $A_3$ leading to an icosahedron and again prove that it can be transformed by the group $W(B_3)^+$ to its mirror image, which implies that neither the tetrahedron nor icosahedron are chiral solids. We focus on the irregular icosahedra constructed either by the proper tetrahedral group or its extension pyritohedral group and construct the related dual solids tetartoid and pyritohedron. We also construct the irregular polyhedra taking the mid-points of edges of the irregular icosahedron as vertices. Section 5 deals with the construction of irregular and regular snub cube and their dual solids from the proper rotational octahedral symmetry $W(B_3)^+$ using the same technique employed in the preceding sections. The chiral polyhedron taking the mid-points as vertices of the irregular snub cube is also discussed. In Section 6 we repeat a similar technique for the constructions of irregular snub dodecahedra and their dual solids using the proper icosahedral group $W(H_3)^+$, which is isomorphic to the group of even permutations of five letters $Alt(5)$. The chiral polyhedra whose vertices are the mid-points of the edges of the irregular snub dodecahedron are constructed. Irregular polyhedra transform to regular polyhedra when the parameter describing irregularity turns out to be the solution of certain cubic equations. Section 7 involves the discussion of the technique for the construction of irregular chiral polyhedra.

## 2. Quaternionic Constructions of the Coxeter Groups

Let $q = q_0 + q_i e_i$, $(i = 1, 2, 3)$ be a real unit quaternion with its conjugate defined by $\bar{q} = q_0 - q_i e_i$, and the norm $q\bar{q} = \bar{q}q = 1$. The quaternionic imaginary units satisfy the relations:

$$e_i e_j = -\delta_{ij} + \varepsilon_{ijk} e_k, \ (i, j, k = 1, 2, 3) \tag{1}$$

where $\delta_{ij}$ and $\varepsilon_{ijk}$ are the Kronecker and Levi-Civita symbols, and summation over the repeated indices is understood. The unit quaternions form a group isomorphic to the special unitary group $SU(2)$. Quaternions generate the four-dimensional Euclidean space with the scalar product

$$(p, q) := \frac{1}{2}(\bar{p}q + \bar{q}p) = \frac{1}{2}(p\bar{q} + q\bar{p}) \tag{2}$$

The Coxeter diagram $A_1 \oplus A_1 \oplus A_1$ can be represented by its quaternionic roots as shown in Figure 1 where $\sqrt{2}$ is just the norm.

**Figure 1.** The Coxeter diagram $A_1 \oplus A_1 \oplus A_1$ with quaternionic simple roots.

The Cartan matrix and its inverse are given as follows:

$$C = \begin{bmatrix} 2 & 0 & 0 \\ 0 & 2 & 0 \\ 0 & 0 & 2 \end{bmatrix}, C^{-1} = \frac{1}{2}\begin{bmatrix} 1 & 0 & 0 \\ 0 & 1 & 0 \\ 0 & 0 & 1 \end{bmatrix} \tag{3}$$

The simple roots $\alpha_i$ and the weight vectors $\omega_i$ for a simply laced root system satisfy the scalar product [18,19]

$$(\alpha_i, \alpha_j) = C_{ij}, \ (\omega_i, \omega_j) = \left(C^{-1}\right)_{ij}, \ (\alpha_i, \omega_j) = \delta_{ij}, (i, j = 1, 2, 3). \tag{4}$$

Note that one can express the roots in terms of the weights or vice versa:

$$\alpha_i = C_{ij}\omega_j, \ \omega_i = \left(C^{-1}\right)_{ij}\alpha_j \tag{5}$$

If $\alpha_i$ is an arbitrary quaternionic simple root and $r_i$ is the reflection generator with respect to the plane orthogonal to the simple root $\alpha_i$, then the reflection of an arbitrary quaternion $\Lambda$, an element in the weight space $\omega_1, \omega_2, \omega_3$ can be represented as [20]:

$$r_i\Lambda = -\frac{\alpha_i}{\sqrt{2}} \overline{\Lambda} \frac{\alpha_i}{\sqrt{2}} := \left[\frac{\alpha_i}{\sqrt{2}}, -\frac{\alpha_i}{\sqrt{2}}\right]^* \Lambda. \tag{6}$$

It is then straight forward to show that $r_i\omega_j = \omega_j - \delta_{ij}\alpha_j$. We will use the notations $[p,q]^*\Lambda := p\overline{\Lambda}q$ and $[p,q]\Lambda := p\Lambda q$ for the rotoreflection (rotation and reflection combined) and the proper rotation, respectively, where $p$ and $q$ are arbitrary unit quaternions. If $\Lambda$ is pure imaginary quaternion, then $[p,q]^*\Lambda := p\overline{\Lambda}q = [p,-q]\Lambda$.

The Coxeter group $W(A_1 \oplus A_1 \oplus A_1) = \ < r_1, r_2, r_3 >$ can be generated by three commutative group elements:

$$r_1 = [e_1, -e_1]^*, r_2 = [e_2, -e_2]^*, r_3 = [e_3, -e_3]^*. \tag{7}$$

They generate the elementary abelian group $W (A_1 \oplus A_1 \oplus A_1) \approx C_2 \times C_2 \times C_2 := 2^3$ of order 8. The scaled root system $(\pm e_1, \pm e_2, \pm e_3)$ represents the vertices of an octahedron and has more symmetries than the Coxeter group $W (A_1 \oplus A_1 \oplus A_1)$. The automorphism group of the root system can be obtained by extending the Coxeter group of order 8 by the Dynkin diagram symmetry $Sym(3)$ of the Coxeter diagram in Figure 1. This is obvious from the diagram where the generators of the symmetric group $Sym(3)$ of order 6 can be chosen as:

$$s = \left[\frac{1}{2}(1 + e_1 + e_2 + e_3), \frac{1}{2}(1 - e_1 - e_2 - e_3)\right], d = [\frac{1}{\sqrt{2}}(e_1 - e_2), -\frac{1}{\sqrt{2}}(e_1 - e_2)]^*,$$

$$s^3 = d^2 = 1, \ dsd = s^{-1}. \tag{8}$$

It is clear that the generators permute the quaternionic imaginary units as:

$$s: e_1 \rightarrow e_2 \rightarrow e_3 \rightarrow e_1, \ d: e_1 \leftrightarrow e_2, e_3 \rightarrow e_3,$$

and satisfy the relations $sr_is^{-1} = r_{i+1}$, $dr_1d = r_2$, $dr_3d = r_3$ where the indices are considered modulo 3.

It is clear that the Coxeter group $< r_1, r_2, r_3 >$ is invariant under the permutation group $Sym(3)$ by conjugation so that the automorphism group of the root system $(\pm e_1, \pm e_2, \pm e_3)$ is an extension of the group $< r_1, r_2, r_3 >$ by the group $Sym(3)$ which is the octahedral group $2^3 : Sym(3)$ of order 48 (here: stands for the semi-direct product). It is clear from this notation that the Coxeter group $2^3$ is an invariant subgroup of the octahedral group. We use a compact notation for the octahedral group in terms of quaternions as the union of subsets:

$$O_h = [T, \pm\overline{T}] \cup [T', \pm\overline{T'}] \tag{9}$$

where the sets of quaternions $T$ and $T'$ are given by:

$$T = \{\pm 1, \pm e_1, \pm e_2, \pm e_3, \tfrac{1}{2}(\pm 1 \pm e_1 \pm e_2 \pm e_3)\},$$

$$T' = \{\tfrac{1}{\sqrt{2}}(\pm 1 \pm e_1), \tfrac{1}{\sqrt{2}}(\pm e_2 \pm e^3), \tfrac{1}{\sqrt{2}}(\pm 1 \pm e_2), \tfrac{1}{\sqrt{2}}(\pm e_3 \pm e_1), \tfrac{1}{\sqrt{2}}(\pm 1 \pm e_3), \tfrac{1}{\sqrt{2}}(\pm e_1 \pm e_2)\}.$$

(10)

Here $T$ and $T \cup T'$ represent the binary tetrahedral group and the binary octahedral group, respectively. We have used a short-hand notation for the designation of the groups, e.g., $[T, \pm \overline{T}]$ means the set of all $[t, \pm \overline{t}] \in [T, \pm \overline{T}]$ with $t \in T$. The maximal subgroups of the octahedral group can be written as [21]:

$$\text{Chiral octahedral group}: \ O = [T, \overline{T}] \cup [T', \overline{T'}] \approx W(B_3)^+,$$

$$\text{Tetrahedral group}: \ T_d = [T, \overline{T}] \cup [T', -\overline{T'}],$$

(11)

$$\text{Pyritohedral group}: \ T_h = [T, \pm \overline{T}].$$

The octahedral group, as we will see in what follows, can also be obtained as the $Aut(A_3) \approx W(B_3)$.

The proper rotation subgroup of the Coxeter group $W(A_1 \oplus A_1 \oplus A_1)$ is the Klein four-group $C_2 \times C_2$ represented by the elements:

$$I = [1, 1], r_1 r_2 = [e_3, -e_3], \ r_2 r_3 = [e_1, -e_1], \ r_3 r_1 = [e_2, -e_2]$$

(12)

We also note in passing that the Klein four-group in (12) is invariant by conjugation under the group generated by $s$, with $s^3 = 1$. The Klein four-group together with the group element $s$ generate the tetrahedral rotation group, which is represented by $[T, \overline{T}]$ in our notation.

Next, we use the tetrahedral group $T_d \approx W(A_3)$. Its Coxeter-Dynkin diagram $A_3$ with its quaternionic roots is shown in Figure 2.

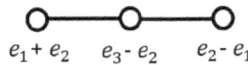

$$e_1 + e_2 \qquad e_3 - e_2 \qquad e_2 - e_1$$

**Figure 2.** The Coxeter diagram $A_3$ with quaternionic simple roots.

The Cartan matrix of the Coxeter diagram $A_3$ and its inverse matrix are given by the respective matrices:

$$C = \begin{bmatrix} 2 & -1 & 0 \\ -1 & 2 & -1 \\ 0 & -1 & 2 \end{bmatrix}, C^{-1} = \frac{1}{4} \begin{bmatrix} 3 & 2 & 1 \\ 2 & 4 & 2 \\ 1 & 2 & 3 \end{bmatrix}$$

(13)

The generators of the Coxeter group $W(A_3)$ are given by:

$$r_1 = [\frac{1}{\sqrt{2}}(e_1 + e_2), -\frac{1}{\sqrt{2}}(e_1 + e_2)]^*$$

$$r_2 = [\frac{1}{\sqrt{2}}(e_3 - e_2), -\frac{1}{\sqrt{2}}(e_3 - e_2)]^*$$

(14)

$$r_3 = [\frac{1}{\sqrt{2}}(e_2 - e_1), -\frac{1}{\sqrt{2}}(e_2 - e_1)]^*$$

$$\omega_1 = \frac{1}{2}(e_1 + e_2 + e_3), \ \omega_2 = e_3, \ \omega_3 = \frac{1}{2}(-e_1 + e_2 + e_3).$$

The generators in (14) generate the Coxeter group [22–24] which is isomorphic to the tetrahedral group $T_d$ of order 24. The automorphism group $Aut(A_3) \approx W(A_3) : C_2 \approx O_h$ where $C_2$ is generated by the Dynkin diagram symmetry $\gamma = [e_1, -e_1]^*$, which exchanges the first and the third simple roots and leaves the second root intact in Figure 2.

The Coxeter diagram $B_3$ leading to the octahedral group $W(B_3) \approx Sym(4) \times C_2 \approx O_h$ is shown in Figure 3.

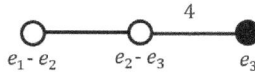

**Figure 3.** The Coxeter diagram $B_3$ with quaternionic simple roots.

The Cartan matrix of the Coxeter diagram $B_3$ and its inverse matrix are given by:

$$C = \begin{bmatrix} 2 & -1 & 0 \\ -1 & 2 & -2 \\ 0 & -1 & 2 \end{bmatrix}, C^{-1} = \frac{1}{2} \begin{bmatrix} 2 & 2 & 2 \\ 2 & 4 & 4 \\ 1 & 2 & 3 \end{bmatrix} \tag{15}$$

The generators below generate the octahedral group given in (9) and the weight vectors are given by:

$$r_1 = [\tfrac{1}{\sqrt{2}}(e_1 - e_2), -\tfrac{1}{\sqrt{2}}(e_1 - e_2)]^*$$
$$r_2 = [\tfrac{1}{\sqrt{2}}(e_2 - e_3), -\tfrac{1}{\sqrt{2}}(e_2 - e_3)]^* \tag{16}$$
$$r_3 = [e_3, -e_3]^*$$

$$\omega_1 = e_1, \omega_2 = e_1 + e_2, \omega_3 = \frac{1}{2}(e_1 + e_2 + e_3)$$

The group generated by the rotations $r_1 r_2$ and $r_2 r_3$ is isomorphic to the rotational octahedral group $< r_1 r_2, r_2 r_3 > \approx \left\{ [T, \overline{T}] \cup [T', \overline{T'}] \right\}$.

The Coxeter diagram $H_3$ leading to the icosahedral group is shown in Figure 4 with the quaternionic simple roots:

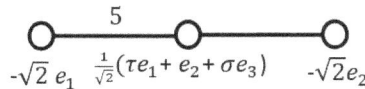

**Figure 4.** The Coxeter diagram of $H_3$ with quaternionic simple roots.

Here $\tau = \frac{1+\sqrt{5}}{2}$ is the golden ratio and $= \frac{1-\sqrt{5}}{2}$. The Cartan matrix of the diagram $H_3$, its inverse and the weight vectors are given as follows:

$$C = \begin{bmatrix} 2 & -\tau & 0 \\ -\tau & 2 & -1 \\ 0 & -1 & 2 \end{bmatrix}, C^{-1} = \frac{1}{2} \begin{bmatrix} 3\tau^2 & 2\tau^3 & \tau^3 \\ 2\tau^3 & 4\tau^2 & 2\tau^2 \\ \tau^3 & 2\tau^2 & \tau+2 \end{bmatrix} \tag{17}$$

$$\omega_1 = \frac{\tau}{\sqrt{2}}(\sigma e_1 - \tau e_3), \omega_2 = -\sqrt{2}\tau e_3, \omega_3 = \frac{\tau}{\sqrt{2}}(\sigma e_2 - e_3)$$

The quaternionic generators of the icosahedral group $I_h \approx Alt(5) \times C_2 \approx W(H_3)$ are given by:

$$r_1 = [e_1, -e_1]^*,$$
$$r_2 = \left[ \tfrac{1}{2}(\tau e_1 + e_2 + \sigma e_3), -\tfrac{1}{2}(\tau e_1 + e_2 + \sigma e_3) \right]^*, \tag{18}$$
$$r_3 = [e_2, -e_2]^*,$$

or shortly, $W(H_3) = [I, \pm \bar{I}]$ where $I$ is the set of 120 quaternionic elements of the binary icosahedral group generated by the quaternions $e_1$ and $\frac{1}{2}(\tau e_1 + e_2 + \sigma e_3)$ [20]. The icosahedral rotation group is represented by the proper rotation subgroup $W(H_3)^+ = [I, \bar{I}] \approx Alt(5)$. All finite subgroups of the groups $O(3)$ and $O(4)$ in terms of quaternions can be found in the references [17,25].

A general vector in the dual space is represented by the vector $\Lambda = a_1 \omega_1 + a_2 \omega_2 + a_3 \omega_3$. We will use the notation $\Lambda = a_1 \omega_1 + a_2 \omega_2 + a_3 \omega_3 : (a_1 a_2 a_3)$, which are called the Dynkin indices in the Lie algebraic representation theory [26]. We use the notation $W(G)\Lambda := (a_1 a_2 a_3)_G$ for the orbit of the Coxeter group $W(G)$ generated from the vector $\Lambda$ where the letter $G$ represents the Coxeter diagram. A few examples could be useful to illuminate the situation by recalling the identifications [8]:

$$(100)_{A_3}(\text{tetrahedron}), (111)_{A_3}(\text{truncated octahedron}), (100)_{B_3}(\text{octahedron}),$$

$$(001)_{B_3}(\text{cube}), (100)_{H_3}(\text{dodecahedron}), (001)_{H_3}(\text{icosahedron}), \text{and} (010)_{H_3},$$

$$(\text{icosidodecahedron}).$$

## 3. The Orbit $C_2 \times C_2 (a_1 a_2 a_3)$ as an Irregular Tetrahedron

The proper rotation subgroup $C_2 \times C_2$ of the Coxeter group $W(A_1 \oplus A_1 \oplus A_1)$ transforms a generic vector $\Lambda$ as follows:

$$\begin{aligned}
\Lambda &= \tfrac{1}{2}(a_1 e_1 + a_2 e_2 + a_3 e_3) \\
r_1 r_2 \Lambda &= \tfrac{1}{2}(-a_1 e_1 - a_2 e_2 + a_3 e_3), \\
r_2 r_3 \Lambda &= \tfrac{1}{2}(a_1 e_1 - a_2 e_2 - a_3 e_3), \\
r_3 r_1 \Lambda &= \tfrac{1}{2}(-a_1 e_1 + a_2 e_2 - a_3 e_3).
\end{aligned} \tag{19}$$

These four vectors define the vertices of an irregular tetrahedron with four identical scalene triangles with edge lengths $\sqrt{a_1{}^2 + a_2{}^2}$, $\sqrt{a_2{}^2 + a_3{}^2}$, and $\sqrt{a_3{}^2 + a_1{}^2}$. An irregular tetrahedron with $a_1 = 1, a_2 = 2, a_3 = 3$ is depicted in Figure 5.

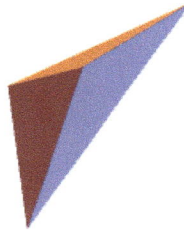

**Figure 5.** An irregular tetrahedron with scalene triangular faces.

Mid-points of the edges of an irregular tetrahedron are given by the vectors $\pm a_1 e_1, \pm a_2 e_2, \pm a_3 e_3$ forming an irregular octahedron with eight identical scalene triangles of sides $\sqrt{a_1{}^2 + a_2{}^2}$, $\sqrt{a_2{}^2 + a_3{}^2}$, $\sqrt{a_3{}^2 + a_1{}^2}$ and the corresponding interior angles, say, $\alpha$, $\beta$, $\gamma$. Four triangles with identical face-angles $\alpha$ surround the vertex $a_1 e_1$. The remaining 4 triangles similarly meet at the opposite vertex $-a_1 e_1$. This is true for every face-angle around every vertex.

The mirror image of an irregular tetrahedron is obtained by applying any one of the reflection generators on these vectors, which lead to the vectors:

$$r_1\Lambda = \tfrac{1}{2}(-a_1e_1 + a_2e_2 + a_3e_3),$$

$$r_2\Lambda = \tfrac{1}{2}(a_1e_1 - a_2e_2 + a_3e_3),$$

$$r_3\Lambda = \tfrac{1}{2}(a_1e_1 + a_2e_2 - a_3e_3),\tag{20}$$

$$r_1r_2r_3\Lambda = -\tfrac{1}{2}(a_1e_1 + a_2e_2 + a_3e_3).$$

The vectors in (20) can also be obtained from those in (19) by quaternion conjugation.

The eight vectors in (19) and (20) form a rectangular prism with edge lengths $a_1, a_2$ and $a_3$ possessing the symmetry $(C_2 \times C_2) : C_2 \approx D_{2h}$ of order 8. Assume now that we apply the one of the Dynkin diagram-symmetry operators $d$ on vector $\Lambda$ and assume that:

$$\Lambda = d\Lambda = \frac{1}{2}(ae_1 + ae_2 + a_3e_3).\tag{21}$$

If we apply the rotation generators as in (19), we obtain an irregular tetrahedron with identical four isosceles triangles. The mirror copy of these vectors can be obtained similar to the procedure in (20) and when all are combined, we obtain a square prism with edge lengths $a$ and $a_3$.

A more symmetric case is obtained by assuming:

$$\Lambda = s\Lambda = \frac{1}{2}a(e_1 + e_2 + e_3).\tag{22}$$

In this case, the symmetry is the chiral tetrahedral group, and the vertices in (19) will represent a regular tetrahedron of edge length $\sqrt{2}a$ which is also invariant under the larger tetrahedral symmetry $T_d = [T, \overline{T}] \oplus [T', -\overline{T'}]$, as expected. A mirror image of the regular tetrahedron is obtained either by reflections as described by (20) or by a rotation of $180°$ around the $e_1 - e_2$ axis, which can be obtained by the group element:

$$\left[ \frac{1}{\sqrt{2}}(e_1 - e_2), -\frac{1}{\sqrt{2}}(e_1 - e_2) \right] \in [T', \overline{T'}].\tag{23}$$

A tetrahedron is not a chiral solid since it can be converted to its mirror image by a rotation such as the one in (23). A regular tetrahedron with its mirror image constitutes a cube which has the full octahedral symmetry of order 48.

## 4. The Regular and Irregular Icosahedron Derived from the Orbit $W(A_3)^+(a_1a_2a_3)$

The tetrahedral rotational subgroup $W(A_3)^+$ of the Coxeter group $W(A_3)$ is the tetrahedral group of order 12 isomorphic to even permutations of four letters, $Alt(4)$, which can be generated by the generators $a = r_1r_2$ and $b = r_2r_3$ satisfying the generation relations $a^3 = b^3 = (ab)^2 = 1$. Let $\Lambda = (a_1a_2a_3)$ be a general vector in the weight space of $A_3$. The following sets of vertices form two equilateral triangles.

$$(\Lambda, r_1r_2\Lambda, r_2r_1\Lambda) \text{ and } (\Lambda, r_2r_3\Lambda, r_3r_2\Lambda)\tag{24}$$

with respective edge lengths $\sqrt{2(a_1^2 + a_1a_2 + a_2^2)}$ and $\sqrt{2(a_2^2 + a_2a_3 + a_3^2)}$. Three more triangles can be obtained by joining the vertex $r_1r_3\Lambda = r_3r_1\Lambda$ to the vertices $\Lambda$, $r_1r_2\Lambda$ and $r_3r_2\Lambda$ and $r_2r_1\Lambda$ to $r_2r_3\Lambda$. The new edges are of the following lengths.

$$|r_1r_2\Lambda - r_1r_3\Lambda| = \sqrt{2(a_2^2 + a_2a_3 + a_3^2)},$$

$$|r_3r_2\Lambda - r_1r_3\Lambda| = \sqrt{2(a_1^2 + a_1a_2 + a_2^2)},\tag{25}$$

$$|\Lambda - r_1r_3\Lambda| = |r_2r_1\Lambda - r_2r_3\Lambda| = \sqrt{2(a_1^2 + a_3^2)}$$

The vertices joined to vector $\Lambda$ are illustrated in Figure 6.

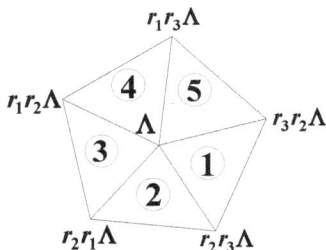

**Figure 6.** The vertices connected to the general vertex $\Lambda$ (note that all vertices are not in the same plane).

Factoring by an overall factor $a_2^2 \neq 0$ and defining the parameters $x = \frac{a_1}{a_2}$ and $y = \frac{a_3}{a_2}$ we obtain three classes of triangles. After dropping the overall factor $a_2$ we obtain,

$$\text{equilateral triangle with edge length}: \ A = \sqrt{2(1 + x + x^2)},$$

$$\text{equilateral triangle with edge length}: \ B = \sqrt{2(1 + y + y^2)},$$

$$3 \text{ scalene triangles with edge lengths}: \ A, B, C = \sqrt{2(x^2 + y^2)}$$

as shown in Figure 6. The sum of the face-angles at the vertex $\Lambda$ (at every vertex indeed) is $\frac{2\pi}{3} + (\theta + \vartheta + \varphi) = \frac{5\pi}{3}$ where $\theta$, $\vartheta$, and $\varphi$ are the interior angles of the scalene triangles. The angular deficiency $\delta = 2\pi - \frac{5\pi}{3} = \frac{\pi}{3}$ is the same as in the regular icosahedron. Using Descartes' formula, the number of vertices can also be obtained as in [27].

$$N_0 = \frac{4\pi}{\delta} = 12. \tag{26}$$

The vector $\Lambda$ can be written in terms of quaternions as $\Lambda = \alpha e_1 + \beta e_2 + \gamma e_3$ where the new parameters are defined by

$$\alpha = \frac{x - y}{2}, \ \beta = \frac{x + y}{2}, \ \gamma = \beta + 1. \tag{27}$$

The orbit of the chiral tetrahedral group $W(A_3)/C_2 \approx [T, \overline{T}]$ generated from the vector $\Lambda$ can be written as:

$$[T, \overline{T}]\Lambda = \{\pm \alpha e_1 \pm \beta e_2 \pm \gamma e_3, \ \pm \beta e_1 \pm \gamma e_2 \pm \alpha e_3, \ \pm \gamma e_1 \pm \alpha e_2 \pm \beta e_3\}$$
$$\text{(even number of } (-) \text{ signs)}. \tag{28}$$

The quaternions in (28) with an odd number of $(-)$ signs constitute the mirror image. A rotation element, e.g., the one in (23), transforms the quaternions in (28) to their mirror images. This proves that the set in (28) does not represent a chiral polyhedron. Before we discuss the irregular icosahedral structures, we point out that for $x > 0$, $y = 0$ or $y > 0$, $x = 0$, the set of 12 vectors describes a truncated tetrahedron. For $x = y = 0$, it describes an octahedron. More interesting cases arise as we will discuss below.

(1) $A = B = C$, $1 + x + x^2 = 1 + y + y^2 = x^2 + y^2$

Here, all the edges are equal leading to the solution $x = y = x^2 - 1 \Rightarrow x^2 - x - 1 = 0$, $\Rightarrow$ $x = \tau$, or $x = \sigma$. Substituting the first solution for $= \tau$ which results in $\alpha = 0$, $\beta = \tau$, $\gamma = \tau^2$, the set of vectors in (28) can be written as:

$$[T, \overline{T}]\Lambda = \tau\{\pm e_1 \pm \tau e_2, \ \pm e_2 \pm \tau e_3, \pm e_3 \pm \tau e_1\}. \tag{29}$$

These vertices, which are also invariant under the pyritohedral symmetry, represent a regular icosahedron. Note that the vector $\Lambda = \tau(e_2 + \tau e_3)$ is invariant under the 5-fold rotation by $s = [\frac{1}{2}(\sigma + e_2 + \tau e_3), \frac{1}{2}(\sigma - e_2 - \tau e_3)]$ while the other vectors in (29) are transformed to each other. This proves that the pyritohedral group $[T, \pm \overline{T}]$ can be extended to the icosahedral group $[I, \pm \overline{I}]$ by the generators so that the set of vertices in (29) is invariant under the icosahedral group of order 120. We emphasize that although (29) has a larger symmetry of the icosahedral group, it is obtained from its chiral tetrahedral subgroup. Its mirror image can be obtained by a rotation of $180°$ around the vector $e_1 - e_2$ implying that icosahedron is not a chiral solid.

There is another trivial solution for $A = B = C$ where $x = -1$, $y = 0$. The number of vertices in (28) reduces to 4 representing a regular tetrahedron.

(2)   $A = B \neq C, y = \frac{1}{2}\left(-1 \pm \sqrt{1 + 4(x^2 + x)}\right)$

For various values of $x$ and the corresponding $y$, one obtains an irregular icosahedron with $(4+4)$ equilateral triangles and 12 isosceles triangles. An interesting case would be $x = y = -\tau$ with 12 vertices of $\{\pm \tau e_1 \pm \sigma e_2, \pm \tau e_2 \pm \sigma e_3, \pm \tau e_3 \pm \sigma e_1\}$ which represent an irregular icosahedron with $(4+4)$ equilateral triangles with edges of length 2 and 12 isosceles triangles of sides 2, $2\tau$, $2\tau$ (Robinson triangles). Any irregular icosahedron with $x = y \neq 0$ has a pyritohedral symmetry of order 24 as pointed out in (11). Another solution of the quadratic equation $x = -\tau$, $y = -\sigma$ leads to an irregular icosahedron of $(4+4)$ equilateral triangles with edge lengths 2 and 12 isosceles triangles of edge lengths 2, 2, $\sqrt{6}$.

(3)   $A = C \neq B$

This is another case with 12 isosceles triangles but here the 4 sets of equilateral triangles are not equal to the other set of 4 equilateral triangles. We have $x = y^2 - 1$, $x \neq \tau$ or $\sigma$ and $x \neq 0$, $y \neq 0$. For the values $y = -\tau$, $x = \tau$, the irregular icosahedron consists of 4 equilateral triangles with edge length 2, 4 equilateral triangles with edge length $2\tau$ and 12 Robinson triangles with edge lengths 2, $2\tau$, $2\tau$ as discussed in (2).

(4)   $A \neq C = B$

Here, $y = x^2 - 1$, $x \neq \tau$ or $\sigma$. For $x = -\tau$, $y = \tau$, the faces of the irregular icosahedron are identical to the one in case (3).

(5)   $A \neq C \neq B$

Now we have $x \neq y^2 - 1$ and $y \neq x^2 - 1$. Taking $x = 1, y = 2$ the irregular icosahedron will consist of 4 equilateral triangles with edge of length $\sqrt{6}$, 4 equilateral triangles of edge length $\sqrt{14}$, and 12 scalene triangles with edges $\sqrt{6}$, $\sqrt{10}$ and $\sqrt{14}$.

## 4.1. Dual of an Irregular Icosahedron

Now we discuss the construction of a dual of an irregular icosahedron. A dual of an irregular polyhedron can be obtained by determining the vectors orthogonal to its faces. Referring to Figure 6, vectors orthogonal to the equilateral faces #1 and #3 can be taken as $b_1 := w_1$ and $b_3 := w_3$ as they are invariant under the rotations $r_2 r_3$ and $r_1 r_2$, respectively. The vectors orthogonal to the faces #2, #4 and #5 can be determined as:

$$b_2 = n_1 e_1 + n_2 e_2 + n_3 e_3,$$
$$b_4 = r_1 r_2 b_2 = -n_3 e_1 - n_1 e_2 + n_2,$$
$$b_5 = r_3 r_2 b_2 = n_3 e_1 + n_1 e_2 + n_2 e_3,$$

(30)

where $n_1 = y - x$, $n_2 = x + y + 2xy$, $n_3 = x + y$.

These vectors should be rescaled in order to determine the plane orthogonal to the vector $\Lambda$. Let us redefine the vectors $d_1 := \lambda w_1$, $d_2 = b_2$, $d_3 = \varrho b_3$, $d_4 = b_4$, $d_5 = b_5$. The scale factors can be determined as:

$$\lambda = \frac{2(x+y+2)(x+y+2xy)}{3x+y+2}$$
$$\varrho = \frac{2(x+y+2)(x+y+2xy)}{(x+3y+2)}. \tag{31}$$

The dual solid is an irregular dodecahedron with the sets of vertices:

$$[T, \overline{T}]d_1 = \tfrac{\lambda}{2}(\pm e_1 \pm e_2 \pm e_3), \text{ (even number of } (-) \text{ signs)}$$
$$[T, \overline{T}]d_3 = \tfrac{\varrho}{2}(\pm e_1 \pm e_2 \pm e_3), \text{ (odd number of } (-) \text{ signs)}$$
$$[T, \overline{T}]d_2 = \{\pm n_1 e_1 \pm n_2 e_2 \pm n_3 e_3, \pm n_2 e_1 \pm n_3 e_2 \pm n_1 e_3, \pm n_3 e_1 \pm n_1 e_2 \pm n_2 e_3\}$$
$$\text{(even number of } (-) \text{ signs)}. \tag{32}$$

We will not discuss in detail how an irregular dodecahedron varies with five different cases discussed above. A few examples would be sufficient in the order of increasing symmetry toward regularity. The case (5) above corresponds to the invariance under the chiral tetrahedral group. Substituting $x = 1$, $y = 2$ in (32), we obtain an irregular dodecahedron called a tetartoid corresponding to the mineral cobaltite. The vertices $d_1, d_2, d_3, d_4$ and $d_5$ form an irregular pentagon with three different edge lengths. The tetartoid with its dual irregular icosahedron are depicted in Figure 7.

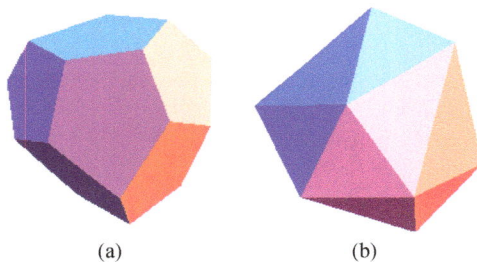

(a)                    (b)

**Figure 7.** (**a**) A face transitive irregular dodecahedron (tetartoid) under rotational tetrahedral symmetry; (**b**) the dual irregular icosahedron.

Since under the pyritohedral symmetry $x = y$, the vertices in (32) take a simpler form where:

$$\lambda = \varrho = \frac{4x(x+1)^2}{(2x+1)}, \; n_1 = 0, \; n_2 = 2x(x+1), \; n_3 = 2x., \tag{33}$$

The edge lengths of the irregular pentagon take the values:

$$|d_1 - d_2| = |d_2 - d_3| = |d_3 - d_4| = |d_1 - d_5|$$
$$= \left|\frac{2x}{(2x+1)}\right| \sqrt{(3x^4 + 6x^3 + 7x^2 + 4x + 1)} \tag{34}$$
$$|d_4 - d_5| = 4|x|.$$

The irregular dodecahedron with the vertices of (32) with $\lambda$ and $\varrho$ from (33) is called the pyritohedron. They can also be rearranged and the set can be given in its standard form:

$$(\pm e_1 \pm e_2 \pm e_3),$$
$$\{\pm(1-h^2)e_1 \pm (1+h)e_2\},$$
$$\{\pm(1-h^2)e_2 \pm (1+h)e_3\},$$
$$\{\pm(1-h^2)e_3 \pm (1+h)e_1\},$$

$$(35)$$

where $h = \dfrac{x}{x+1}$.

A special case $x = y = 5$ discussed in (2) corresponding to $h = \dfrac{5}{6}$ in (35) is plotted in Figure 8.

**Figure 8.** The pyritohedron.

In the limit of either $x = \tau$ or $x = \sigma$, the dodecahedron is regular and the pentagon turns out to be regular with all edges equal either $4\tau$ or $-4\sigma$. After rescaling by $4\tau^2$, the set of vertices of the dodecahedron with $x = \tau$ are:

$$\frac{1}{2}(\pm e_1 \pm e_2 \pm e_3), \frac{1}{2}\{\pm \tau e_1 \pm \sigma e_2, \pm \tau e_2 \pm \sigma e_3, \pm \tau e_3 \pm \sigma e_1\}. \tag{36}$$

Each set above is invariant under the pyritohedral group $[T, \pm \overline{T}]$. The 20 vertices of (36) form a dodecahedron with regular faces. As such, they possess a larger icosahedral symmetry $[I, \pm \overline{I}]$.

The first 8 vertices of (36) represent a cube and the second set of 12 vertices, as we recall from previous discussions, represents an irregular icosahedron with $(4+4)$ equilateral triangles and 12 Robinson triangles. The dual of the irregular icosahedron in second set of 12 vertices in (36) is another irregular dodecahedron whose vertices are the union of a cube and a regular icosahedron albeit with different magnitudes of vectors.

## 4.2. Regular and Irregular Icosidodecahedron

It is well known that mid-points of the edges of a regular icosahedron or dodecahedron form the Archimedean solid icosidodecahedron with 30 vertices 32 faces (12 pentagons + 20 triangles) and 60 edges which can be obtained from the Coxeter graph of $H_3$ as an orbit $(010)_{H_3}$ [28]. An irregular icosidodecahedron consists of irregular pentagonal faces and scalene triangles in the most general case and will be derived from the chiral tetrahedral group and will be extended by pyritohedral group representing a larger symmetry. One can define five vectors as follows representing the vertices of the irregular pentagon which is orthogonal to the vertex $\Lambda$:

$$c_1 = \frac{\Lambda + r_1 r_2 \Lambda}{2}, \quad c_2 = \frac{\Lambda + r_2 r_1 \Lambda}{2}, \quad c_3 = \frac{\eta(\Lambda + r_2 r_3 \Lambda)}{2}, \quad c_4 = \frac{\eta(\Lambda + r_3 r_2 \Lambda)}{2},$$

$$c_4 = \frac{\eta(\Lambda + r_3 r_2 \Lambda)}{2}, \quad c_5 = \frac{\kappa(\Lambda + r_1 r_3 \Lambda)}{2},$$

$$\eta = \frac{x^2 + 3y^2 + 2xy + 2x + 4y + 2}{y^2 + 3x^2 + 2xy + 4x + 2y + 2},$$

$$\kappa = \frac{x^2 + 3y^2 + 2xy + 2x + 4y + 2}{(x+y+2)^2}.$$

$$(37)$$

In terms of $x$ and $y$ the vertices read:

$$c_1 = \frac{1}{2}[-(y+1)e_1 + ye_2 + (x+y+1)e_3],$$

$$c_2 = \frac{1}{2}[-ye_1 + (x+y+1)e_2 + (y+1)e_3],$$

$$c_3 = \frac{\eta}{2}[xe_1 + (x+y+1)e_2 + (x+1)e_3], \qquad (38)$$

$$c_4 = \frac{\eta}{2}[(x+1)e_1 + xe_2 + (x+y+1)e_3],$$

$$c_5 = \kappa(\beta+1)e_3.$$

Here, $c_1$ with $c_2$, $c_3$ with $c_4$ and $c_5$ define three orbits under the chiral tetrahedral symmetry of sizes 12, 12 and 6, respectively. Imposing the pyritohedral group invariance ($x = y$) and moreover, letting $x = \tau$ and dividing each vector by a scale factor $2\tau^2$, one obtains the usual quaternionic vertices of the icosidodecahedron [28]

$$\left\{ \tfrac{1}{2}(\pm e_1 \pm \sigma e_2 \pm \tau e_3), \tfrac{1}{2}(\pm \sigma e_1 \pm \tau e_2 \pm e_3), \tfrac{1}{2}(\pm \tau e_1 \pm e_2 \pm \sigma e_3) \right\}, \qquad (39)$$

$$\pm e_1, \pm e_2, \pm e_3$$

They constitute a subset of the quaternionic binary icosahedral group $I$. An icosidodecahedron consists of regular pentagons and equilateral triangles as shown in Figure 9a. A general irregular icosidodecahedron consists of 12 pentagons of edges $a$, $b$, $c$ as shown in Figure 9b. In addition, it has 4 equilateral triangles with edges $a$, 4 equilateral triangles with edges $b$, 12 scalene triangles with edges $a, b, c$ where $a, b, c$ can be expressed in terms of $x$ and $y$.

(a)          (b)

**Figure 9.** (a) Regular icosadodecahedron($x = y = \tau$); (b) irregular icosidodecahedron ($x = 1, y = 2$).

## 5. The Regular and Irregular Snub Cubes Derived from $W(B_3)^+(a_1a_2a_3)$

The snub cube is a chiral Archimedean solid with 24 vertices, 60 edges and 38 faces (8 squares, $6 + 24$ equilateral triangles). Its vertices and its dual can be determined by employing the same method described in Sections 3 and 4. The proper rotational subgroup of the Coxeter group $W(B_3)$ is the rotational octahedral group $Sym(4) \approx \left\{ [T, \overline{T}] \cup [T', \overline{T'}] \right\}$, of order 24, which permutes the diagonals of a cube [21]. The group is generated by two rotation generators $a = r_3r_2$ and $b = r_2r_1$ satisfying the generation relations $a^4 = b^3 = (ab)^2 = 1$. When $\Lambda$ is taken as a general vector, the following sets of vertices form an equilateral triangle and a square, respectively,

$$(\Lambda, \, r_2r_1\Lambda, \, (r_2r_1)^2\Lambda), \; (\Lambda, \, r_3r_2\Lambda, \, (r_3r_2)^2\Lambda, \, r_2r_3\Lambda), \qquad (40)$$

with respective edge lengths $\sqrt{2(a_1^2 + a_1a_2 + a_2^2)}$ and $\sqrt{2\left(a_2^2 + a_2a_3 + \frac{1}{2}a_3^2\right)}$. With the vertex $r_1r_3\Lambda = r_3r_1\Lambda$, we obtain a figure consisting of 7 vertices as shown in Figure 10.

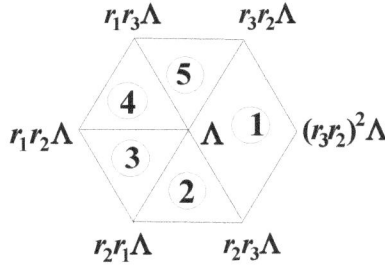

**Figure 10.** The vertices connected to the vertex $\Lambda$.

There are 3 different edge lengths among 11 edges which are given as:

$$|r_1r_2\Lambda - r_1r_3\Lambda| = \sqrt{2(a_2^2 + a_2a_3 + \tfrac{1}{2}a_3^2)},$$

$$|r_3r_2\Lambda - r_1r_3\Lambda| = \sqrt{2(a_1^2 + a_1a_2 + a_2^2)}, \tag{41}$$

$$|\Lambda - r_1r_3\Lambda| = |r_2r_1\Lambda - r_2r_3\Lambda| = \sqrt{2(a_1^2 + \tfrac{1}{2}a_3^2)}$$

Factoring by $a_2 \neq 0$ redefining $x = \frac{a_1}{a_2}$, $y = \frac{a_3}{a_2}$ and $\alpha = x + \frac{y}{2} + 1$, $\beta = \frac{y}{2} + 1$, $\gamma = \frac{y}{2}$ and dropping $a_2$, one obtains the following classes of faces of the irregular snub cube:

6 equilateral triangles with edge length: $A = \sqrt{2(1 + x + x^2)}$,
8 squares with edge length: $B = \sqrt{2(1 + y + y^2)}$,

24 scalene triangles with edge lengths: $A, B, C = \sqrt{2\left(x^2 + \tfrac{1}{2}y^2\right)}$.

Since the angular deficiency is $\delta = \frac{\pi}{6}$, the number of vertices of an irregular snub cube is 24.
The vertex $\Lambda = a_1\omega_1 + a_2\omega_2 + a_3\omega_3$ can be written in terms of quaternions as $\Lambda = \alpha e_1 + \beta e_2 + \gamma e_3$. The orbit generated by the chiral octahedral group reads:

$$W(B_3)^+\Lambda = \{\pm\alpha e_1 \pm \beta e_2 \pm \gamma e_3, \ \pm\beta e_1 \pm \gamma e_2 \pm \alpha e_3, \ \pm\gamma e_1 \pm \alpha e_2 \pm \beta e_3\},$$

$$W(B_3)^+\Lambda\prime = \{\pm\beta e_1 \pm \alpha e_2 \pm \gamma e_3, \ \pm\alpha e_1 \pm \gamma e_2 \pm \beta e_3, \ \pm\gamma e_1 \pm \beta e_2 \pm \alpha e_3\}, \tag{42}$$

where $\Lambda\prime = r_1\Lambda$ is the mirror image of $\Lambda$. The vertices of (42) represent (1) an octahedron for $a_1 = 1, a_2 = a_3 = 0$; (2) truncated octahedron for $x = 1$, $y = 0$; (3) cube for $a_1 = a_2 = 0, a_3 = 1$ and (4) cuboctahedron for $a_1 = a_3 = 0, a_2 = 1$.

The irregular chiral convex solid will have 24 vertices 38 faces (8 square, 6 equilateral triangles and 24 scalene triangles) and 60 edges (24 of length $A$, 24 of length $B$, 12 of length $C$). In what follows we classify them according to their edge lengths of triangles and squares.

(1) The snub cube: $A = B = C$, $1 + x + x^2 = 1 + y + y^2 = x^2 + \tfrac{1}{2}y^2$

When all edges are equal, one eliminates the variable satisfying $y = x^2 - 1$ and obtains the cubic equation $x^3 - x^2 - x - 1 = 0$ which can also be written as $x + x^{-3} = 2$ The solution is the tribonacci constant $x = \lim_{n\to\infty} \frac{F_{n+1}}{F_n} \approx 1.8393$ where $F_n$ is the $n$-th term in the Tribonacci series $0, 0, 1, 1, 2, 4, 7, 13, 24, 44, \ldots$.

After dropping an overall factor $\frac{1}{2}(x^2+1)$, vertices of the snub cube and its mirror image are given by:

$$W(B_3)^+\Lambda = \{\pm xe_1 \pm e_2 \pm x^{-1}e_3, \pm e_1 \pm x^{-1}e_2 \pm xe_3, \pm x^{-1}e_1 \pm xe_2 \pm e_3\}$$

$$W(B_3)^+\Lambda\prime = \{\pm e_1 \pm xe_2 \pm x^{-1}e_3, \ \pm xe_1 \pm x^{-1}e_2 \pm e_3, \ \pm x^{-1}e_1 \pm e_2 \pm xe_3\}. \tag{43}$$

For $x = 1.8393$, the snub cube is shown in Figure 11.

**Figure 11.** The snub cube.

(2)    $A = B \neq C, y = -1 \pm \sqrt{(1 + 2(x^2 + x))}$

For $x = 1$, $y = -2\tau$, the irregular snub cube consists of 6 equilateral triangles and 8 squares of edges $\sqrt{6}$ respectively and 24 isosceles triangles of edges $\sqrt{6}$, $\sqrt{6}$, $\sqrt{2(2\tau + 3)}$ as shown in Figure 12.

**Figure 12.** The irregular snub cube with squares, equilateral triangles, isosceles triangles and squares (for $x = 1$, $y = -2\tau$).

(3)    $A = C \neq B, x = \frac{1}{2}y^2 - 1$ and $x^3 - x^2 - x - 1 \neq 0$

For $y = 4$ and $x = 7$ corresponding to $\alpha = 10$, $\beta = 3$, $\gamma = 2$, the irregular snub cube consists of squares of edges $\sqrt{13}$, equilateral triangles of edges $\sqrt{57}$ and isosceles triangles with edges $\sqrt{57}$, $\sqrt{57}$ and $\sqrt{13}$; all edges scaled by $\sqrt{2}$.

(4)    $A \neq B = C, y = x^2 - 1$ and $x^3 - x^2 - x - 1 \neq 00$

For $x = 3$, $y = 8$ the irregular snub cube after rescaling by $\sqrt{2}$ has equilateral triangles of sides $\sqrt{13}$, squares of sides $\sqrt{41}$, and isosceles triangles of edges $\sqrt{41}$, $\sqrt{41}$, $\sqrt{13}$ as faces.

(5)    $A \neq C \neq B$

The variables should satisfy the inequalities $y \neq x^2 - 1$ and $y^2 \neq 2(x + 1)$.
For $x = 2$ and $y = 4$, the irregular snub cube consists of scaled equilateral triangles of sides $\sqrt{7}$, squares of sides $\sqrt{13}$ and the scalene triangles of sides $\sqrt{7}$, $\sqrt{13}$ and $\sqrt{10}$ as shown in Figure 13.

**Figure 13.** The irregular snub cube with equilateral triangles, scalene triangles and squares (for $x = 2$ and $y = 4$).

*5.1. Dual of the Irregular Snub Cube*

The vectors orthogonal to the faces in Figure 10 can be determined as:

$$d_1 = \mu e_1,$$
$$d_2 = v(n_1 e_1 + n_2 e_2 + n_3 e_3),$$
$$d_3 = \tfrac{1}{2}(e_1 + e_2 + e_3), \tag{44}$$
$$d_4 = v(n_3 e_1 + n_1 e_2 + n_2 e_3),$$
$$d_5 = v(n_1 e_1 + n_3 e_2 - n_2 e_3).$$

Here the parameters are given by:

$$n_1 = (x+1)y + x, \; n_2 = x, \; n_3 = x(y+1),$$
$$\mu = \tfrac{2x+3y+4}{4x+2y+4}, \; v = \tfrac{1}{2}\tfrac{2x+3y+4}{(xy+x+y)(2x+y+2)+x(y+2)+yx(y+1)} \tag{45}$$

Vertices of the dual of the irregular snub cube can be written as three sets of orbits under the chiral octahedral group $\left\{[T,\,\overline{T}] \oplus [T',\overline{T'}]\right\}$,

$$\mu(\pm e_1, \pm e_2, \pm e_3),$$
$$\tfrac{1}{2}(\pm e_1 \pm e_2 \pm e_3) \tag{46}$$
$$v\{(\pm n_1 e_1 \pm n_2 e_2 \pm n_3 e_3), \; (\pm n_2 e_1 \pm n_3 e_2 \pm n_1 e_3), \; (\pm n_3 e_1 \pm n_1 e_2 \pm n_2 e_3)\}.$$

The mirror image can be obtained from (46) by exchanging $e_1 \leftrightarrow e_2$.

The dual of the irregular snub cube consists of 24 irregular pentagons with three different edge lengths in general. However, for the special case $y = x^2 - 1$ and $x^3 - x^2 - x - 1 = 0$ which corresponds to the regular snub cube, the parameters are given by $\mu = \tfrac{x}{2}, v = \tfrac{x-3}{2}, n_1 = x(2x+1), n_2 = x, n_3 = x^3$. The lengths of the edges of the pentagon satisfy the relations

$$|d_1 - d_2| = |d_1 - d_5| = \sqrt{\tfrac{x^2-1}{4x}}, \tag{47}$$
$$|d_2 - d_3| = |d_3 - d_4| = |d_4 - d_5| = \sqrt{2-x},$$

where $x \approx 1.8393$. The dual of the snub cube is shown in Figure 14.

**Figure 14.** Dual of the snub cube (pentagonal icositetrahedron) with its pentagonal face. If we substitute $x = 1, y = 0$ in (46) we obtain the Catalan solid tetrakis hexahedron (dual of the truncated octahedron).

We shall not discuss all duals of the irregular snub cubes. They can be obtained by substituting $x$ and $y$ in (45) and (46) corresponding to each case above. We will illustrate only the case for $x = 2$ and $y = 4$ where the pentagon has three different edge lengths. This is the dual of the irregular snub cube corresponding to the case of (5). It is depicted in Figure 15 with its pentagonal faces consisting of three different edge lengths satisfying the relations $|d_1 - d_2| = |d_1 - d_5|$, $|d_2 - d_3| = |d_3 - d_4|$, $|d_4 - d_5|$.

**Figure 15.** Dual of the irregular snub cube of Figure 13.

### 5.2. Chiral Polyhedra with Vertices at the Edge Mid-Points of the Irregular Snub Cube

With a similar discussion to the case of irregular icosidodecahedron in Section 4, we may construct a chiral polyhedron assuming the mid-points of edges of an irregular snub cube as vertices. The vectors whose tips constitute the plane orthogonal to the vector $\Lambda$ in Figure 10 which are constructed similar to (37) and (38) and are given by:

$$c_1 = [(x+y+1)e_1 + (x+y+2)e_2 + (y+1)e_3],$$
$$c_2 = [(x+y+2)e_1 + (y+1)e_2 + (x+y+1)e_3]$$
$$c_3 = \mu\prime[(2x+y+2)e_1 + e_2 + (y+1)e_3],$$
$$c_4 = \mu\prime[(2x+y+2)e_1 + (y+1)e_2 - e_3], \tag{48}$$
$$c_5 = \nu\prime(x+y+2)(e_1 + e_2),$$
$$\mu\prime = \frac{(\alpha+\beta)\alpha + (\beta+\gamma)\beta + (\gamma+\alpha)\gamma}{2\alpha^2 + \beta^2 + \gamma^2}, \quad \nu\prime = \frac{(\alpha+\beta)\alpha + (\beta+\gamma)\beta + (\gamma+\alpha)\gamma}{(\alpha+\beta)^2}.$$

The vertices $c_1$ and $c_2$ are in the same orbit of size 24 under the chiral octahedral group $\{[T, \overline{T}] \cup [T\prime, \overline{T\prime}]\}$, and the orbit involving $c_3$ and $c_4$ is of size 24. The orbit of $c_1$ consists of cyclic permutations of the coefficients of the unit quaternions with all possible sign changes. The same argument is valid for the orbit of $c_3$. The orbit of $c_5$ comprises of 12 quaternions obtained as the cyclic permutations of the pair of quaternions with all sign changes. Actually, the orbit of $c_5$ by itself represents the vertices of a cuboctahedron. For all allowed $x$ and $y$, the chiral polyhedra with 60 vertices can be displayed. However, we will only display the polyhedron corresponding to the special case where the 60 vertices obtained from (48) represent the mid-points of edges of the regular case, namely the snub cube. This is obtained, as we discussed before, substituting $y = x^2 - 1$ and using the real solution of the cubic equation $x^3 - x^2 - x - 1 = 0$. Then one obtains $\mu\prime = \nu\prime = 1$ and the square lengths of sides of the irregular pentagon obtained from (48) are given by:

$$|c_1 - c_2| = |c_2 - c_3| = |c_4 - c_5| = |c_5 - c_1| = \frac{|c_3 - c_4|}{\sqrt{2}} = \sqrt{2x^3} \approx 3.528 \tag{49}$$

The vector orthogonal to this pentagon is represented by the vector $\Lambda$ and around this pentagon there are 4 equilateral triangles of sides $\sqrt{2x^3}$ and 1 square of sides $2\sqrt{x^3}$. With this pentagonal face, the chiral polyhedron is shown in Figure 16. It has 60 vertices, 62 faces (6 squares, $(8 + 24)$ equilateral triangles and 24 irregular pentagons) and 120 edges (96 of sides 3.528 and 24 of length 4.989).

**Figure 16.** Chiral polyhedron consisting of equilateral triangles, squares and irregular pentagons.

## 6. The Regular and Irregular Snub Dodecahedron Derived from $W(H_3)^+(a_1 a_2 a_3)$

The snub dodecahedron is a chiral Archimedean solid with 60 vertices, 92 faces (20 pentagons +80 triangles) and 150 edges. We will discuss how to obtain the snub dodecahedron from an irregular snub dodecahedron. The vertices of an irregular snub dodecahedron, its dual and the chiral polyhedron obtained from the mid-points of edges will be constructed by employing the same technique described in Section 5. The proper rotational subgroup of the icosahedral Coxeter group is also called chiral icosahedral group $W(H_3)^+ = [I, \bar{I}] \approx Alt(5)$ which is a simple group of order 60. They can be generated by the generators $a = r_1 r_2$, $b = r_2 r_3$ which satisfy the generation relations $a^5 = b^3 = (ab)^2 = 1$. Let $\Lambda = a_1 \omega_1 + a_2 \omega_2 + a_3 \omega_3$ be a general vector where $\omega_i$, $(i = 1, 2, 3)$ can be obtained from (17). The following sets of vertices form a regular pentagon and an equilateral triangle, respectively:

$$(\Lambda, r_1 r_2 \Lambda, (r_1 r_2)^2 \Lambda, (r_1 r_2)^3 \Lambda, (r_1 r_2)^4 \Lambda), (\Lambda, r_2 r_3 \Lambda, (r_2 r_3)^2 \Lambda) \tag{50}$$

with respective edge lengths $\sqrt{2(a_1^2 + \tau a_1 a_2 + a_2^2)}$ and $\sqrt{2(a_2^2 + a_2 a_3 + a_3^2)}$. With the addition of the vertex $r_1 r_3 \Lambda = r_3 r_1 \Lambda$ as shown in Figure 17, there are three edge lengths including $\sqrt{2(a_1^2 + a_3^2)}$. The discussions of Sections 4 and 5 will be repeated to obtain the regular and irregular snub dodecahedra. The dual and the chiral polyhedron based on the mid-points of edges follow the same technique.

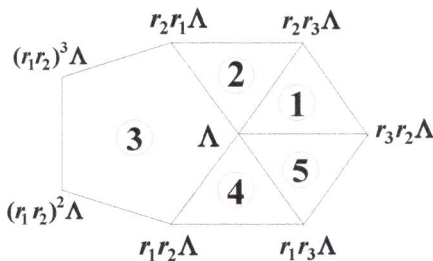

**Figure 17.** The vertices connected to the vertex $\Lambda$.

Numbering of the faces has been done according to the $H_3$ diagram of Figure 4. Factoring by $a_2 \neq 0$, redefining $x = \frac{a_3}{a_2}$, $y = \frac{a_1}{a_2}$ and dropping $\frac{a_2}{\sqrt{2}}$ one obtains an irregular snub dodecahedron with 12 equilateral triangles with edge length: $A = \sqrt{2(1 + x + x^2)}$, 20 pentagons with edge length: $B = \sqrt{2(1 + \tau y + y^2)}$, 60 scalene triangles with edge lengths: $A, B, C = \sqrt{2(x^2 + y^2)}$. The triangles numbered as 2, 4 and 5 are identical scalene triangles, but number 1 is an equilateral triangle. The angular deficiency is $\delta = 2\pi - \left( \frac{\pi}{3} + \frac{3\pi}{5} + (\theta + \vartheta + \varphi) \right) = \frac{\pi}{15}$ where $\theta$, $\vartheta$ and $\varphi$ are the interior angles of the scalene triangles. Descartes' formula verifies the number of vertices $N_0 = 60$.

The orbit $[I, \bar{I}]\Lambda$ generated from the vector $\Lambda = -ye_1 - xe_2 - \tau(\tau y + x + 2)e_3$ involves 60 vectors representing the vertices of an irregular snub dodecahedron. The mirror image $[I, \bar{I}]\Lambda\prime$ can be obtained from the vector $\Lambda\prime = r_1\Lambda = ye_1 - xe_2 - \tau(\tau y + x + 2)e_3$. A few remarks are in order before we discuss the usual classification. For some special values of the parameters $x$ and $y$, we obtain some of the Archimedean solids. For example,

(i)  $x = 1$ and $y = 0$ : Truncated icosahedron,
(ii)  $x = 0$ and $y = 1$ : Truncated dodecahedron,
(iii)  $x = y = 0$ : Icosidodecahedron.

Let us follow the sequence of Section 5 to discuss the regular and irregular snub dodecahedra.

(1)  The snub dodecahedron: $A = B = C$, $1 + x + x^2 = 1 + \tau y + y^2 = x^2 + y^2$

Since all edges are equal, it leads to the relations $y = \sigma(1 - x^2)$ and $y^2 = x + 1$ resulting in the cubic equation $x^3 - x^2 - x - \tau = 0$. The real solution is approximately $x = 1.943$. The snub dodecahedron is shown in Figure 18.

**Figure 18.** The snub dodecahedron.

(2)  $A = B \neq C$, $y = \frac{1}{2}\left(-\tau \pm \sqrt{(\tau^2 + 4(x^2 + x))}\right)$

For $x = -1$, $y = -\tau$, the irregular snub dodecahedron consists of 12 equilateral triangles and 20 pentagons of sides $\sqrt{2}$ and 60 isosceles triangles of sides $\sqrt{2}$, $\sqrt{2}$, $\sqrt{2(\tau + 2)}$

(3)  $A = C \neq B$, $y^2 = x + 1$ and $x^3 - x^2 - x - \tau \neq 0$

For $y = 2$ and $x = 3$, the corresponding irregular snub dodecahedron consists of pentagons of edge length $\sqrt{2(2\tau + 5)}$, equilateral triangles of edge length $\sqrt{26}$ and isosceles triangles with edge lengths $\sqrt{26}$, $\sqrt{26}$ and $\sqrt{2(2\tau + 5)}$

(4)  $A \neq B = C$, $x^2 = \tau y + 1$ and $x^3 - x^2 - x - \tau \neq 0$

For $y = 1$, $x = \tau$, the irregular snub dodecahedron has pentagons of sides $\sqrt{2(\tau + 2)}$, equilateral triangles of sides $\sqrt{2\tau}$, and isosceles triangle of sides $\sqrt{2(\tau + 2)}$, $\sqrt{2(\tau + 2)}$ and $\sqrt{2\tau}$.

(5)  $A \neq C \neq B$

For $x = 2$ and $y = 3$, the irregular snub dodecahedron consists of equilateral triangles of sides $\sqrt{14}$, pentagons of sides $\sqrt{2(3\tau + 10)}$ and the scalene triangles of sides $\sqrt{14}$, $\sqrt{2(3\tau + 10)}$ and $\sqrt{26}$ as shown in Figure 19.

**Figure 19.** Irregular snub dodecahedron with $x = 2$, $y = 3$.

*6.1. Dual of the Irregular Snub Dodecahedron*

We obtain 5 vectors as follows to determine the plane orthogonal to the vertex $\Lambda$ :

$$d_1 = \sqrt{2}\omega_1 = \tau(\sigma e_1 - \tau e_3),$$

$$d_2 = v[(y - \sigma)x e_1 - \sigma y(x + 1)e_2 + (\tau xy + \tau x + y)e_3]$$

$$d_3 = \sqrt{2}\mu\omega_3 = \mu(-e_2 - \tau e_3),$$

$$d_4 = v[\sigma x e_1 - \sigma y e_2 + (2xy + \tau x + y)e_3]$$

$$d_5 = v[-\sigma x e_1 + \sigma y e_2 + (2xy + \tau x + y)e_3],$$

$$\mu = \frac{(3y + \tau x + 2\tau)}{\tau y + (\sigma + 2)x + 2}, \quad v = \frac{(3y + \tau x + 2\tau)}{(\sigma x - y + 2\sigma)(2xy + \tau x + y)}. \tag{51}$$

The orbits generated from these five vectors represent the vertices of the dual solid of the irregular snub dodecahedron. Faces of the dual consist of irregular pentagons, one of which is represented by the vertices of (51). Note that the orbits $[I, \bar{I}\,]d_1$ and $[I, \bar{I}\,]d_3$ represent the orbits of sizes 20 and 12, respectively. The other three vertices are in the same orbit; therefore, one single notation $[I, \bar{I}\,]d_2$ for this orbit of size 60 suffices. Therefore, the dual consists of 92 vertices, 60 irregular pentagons and 150 edges. The regular case is obtained by substituting $y = \sigma(1 - x^2)$ and $x = 1.943$ in (51). The dual of the snub dodecahedron, the pentagonal hexecontahedron, is shown Figure 20.

**Figure 20.** The pentagonal hexecontahedron (dual of snub dodecahedron).

We also illustrate the dual of the irregular snub dodecahedron obtained from 5) by substituting the values $x = 2$ and $y = 3$ in (51). The 92 vertices with this substitution describe the chiral solid depicted in Figure 21.

**Figure 21.** The dual of irregular snub dodecahedron with $x = 2$ and $y = 3$.

*6.2. Chiral Polyhedra with Vertices at the Edge Mid-Points of the Irregular Snub Dodecahedron*

Similar to the previous discussions, we can construct a chiral polyhedron possessing the mid-points of the edges of the irregular snub cube as vertices. The vectors whose tips constitute the plane orthogonal to the vector $\Lambda$ in Figure 17 are given by:

$$c_1 = \tfrac{1}{2}[-(2y+\tau)e_1 + e_2 - \tau(2\tau y + 2x + \tau + 2)e_3],$$

$$c_2 = \tfrac{1}{2}\left[-(2y+\tau x+\tau)e_1 - (x+1)e_1 - \tau(2\tau y + 2x + \tau + 2)e_3\right],$$

$$c_3 = \tfrac{\alpha}{2}\left[-\tau(\tau y+1)e_1 - (\tau y + 2x + 1)e_2 - \tau(\tau^2 y + 2x + \tau + 2)e_3\right],$$

$$c_4 = \tfrac{\alpha}{2}[\tau e_1 - (2x+1)e_2 - \tau(2\tau y + 2x + \tau + 2)e_3],$$

$$c_5 = -\beta\tau(\tau y + x + 2)e_3,$$

$$\alpha = \frac{3y^2 + 2\tau xy + x^2 + 4\tau y + x(\tau+2) + \tau + 2}{\tau^2 y^2 + 2\tau xy + x^2(\sigma+2) + (3\tau+1)y + 4x + \tau + 2},$$

$$\beta = \frac{3y^2 + 2\tau xy + x^2 + 4\tau y + x(\tau+2) + \tau + 2}{(\tau y + x + 2)^2}$$

(52)

In the limit $y = \sigma(1-x^2)$, $x^3 - x^2 - x - \tau = 0$ and $\alpha = \beta = 1$ ($x = 1.943$); for a snub dodecahedron, the vertices in (52) are simplified, and the edge lengths of the pentagon satisfy the relations:

$$|c_1 - c_2| = |c_2 - c_3| = |c_4 - c_5| = |c_5 - c_1| = \frac{|c_3 - c_4|}{\tau} = \sqrt{x^2 + x + 1}$$

(53)

This chiral polyhedron has 150 vertices and 152 faces (12 regular pentagons + 60 irregular pentagons and $(20+60)$ equilateral triangles all surrounding 60 irregular pentagons), a typical one of which is represented with the relations of vertices in (53). The chiral polyhedron with 300 edges is illustrated in Figure 22.

**Figure 22.** Chiral polyhedron whose vertices are the mid-points of the edges of the snub dodecahedron.

## 7. Concluding Remarks

In this paper we presented a systematic construction of regular and irregular chiral polyhedra, the snub cube, the snub dodecahedron and their duals using proper rotational subgroups of the octahedral group and the icosahedral group. Chiral polyhedra whose vertices are the mid-points of the chiral polyhedra were also constructed. We used the Coxeter diagrams $B_3$ and $H_3$ to obtain the quaternionic description of the relevant chiral groups. Employing the same technique for the diagrams $A_1 \oplus A_1 \oplus A_1$ and $A_3$, the irregular tetrahedra and icosahedra have been included although they are not chiral as they can be transformed to their mirror images by the proper rotational subgroup of the octahedral group.

The dual solids of the irregular icosahedron, the tetartoid and the pyritohedron are also constructed representing the chiral symmetries of chiral tetrahedral group and the pyritohedral group, respectively.

This method can be extended to higher dimensional Coxeter groups to determine the irregular-regular chiral polytopes. For example, the snub 24-cell, a chiral polytope in the 4D Euclidean space can be determined using the $D_4$ Coxeter diagram [29], and its irregular form can be obtained using the same technique used for the irregular icosahedron.

**Author Contributions:** The authors N.O.K. and M.K. equally contributed to the paper in conceiving, programming and writing.

**Conflicts of Interest:** The authors declare no conflict of interest.

## References

1. Coxeter, H.S.M.; Moser, W.O.J. *Generators and Relations for Discrete Groups*; Springer: Berlin, Germany, 1965.
2. Cotton, F.A.; Wilkinson, G.; Murillo, C.A.; Bochmann, M. *Advanced Inorganic Chemistry*, 6th ed.; Wiley-Interscience: New York, NY, USA, 1999.
3. Caspar, D.L.D.; Klug, A. Cold spring harbor symp. *Quant. Biol.* **1962**, *27*, 1. [CrossRef]
4. Twarock, R. Mathematical virology: A novel approach to the structure and assembly of viruses. *Philos. Trans. R. Soc.* **2006**, *364*, 3357–3373. [CrossRef] [PubMed]
5. Jaric, M.V. (Ed.) *Introduction to the Mathematics of Quasicrystals*; Academic Press: New York, NY, USA, 1989.
6. Senechal, M. *Quasicrystals and Geometry*; Cambridge University Press: Cambridge, UK, 1995.
7. Suck, J.B.; Schreiber, M.; Haussler, P. (Eds.) *Quasicrystals (An Introduction to Structure, Physical Properties, and Applications)*; Springer: Berlin/Heidelberg, Germany, 2002.
8. Koca, M.; Koc, R.; Al-Ajmi, M. Polyhedra obtained from Coxeter groups and quaternions. *J. Math. Phys.* **2007**, *48*, 113514. [CrossRef]
9. Koca, M.; Koca, N.O.; Koc, R. Catalan solids derived from 3D-root systems. *J. Math. Phys.* **2010**, *51*, 043501. [CrossRef]
10. Koca, M.; Al-Ajmi, M.; Al-Shidhani, S. Quasi regular polyhedra and their duals with Coxeter symmetries represented by quaternions II. *Afr. Rev. Phys.* **2011**, *1006*, 53.
11. McMullen, P.; Schulte, E. *Abstract Regular Polytopes*; Cambridge University Press: Cambridge, UK, 2002.
12. Schulte, E.; Weiss, A.I. Chiral polytopes. In *Applied Geometry and Discrete Mathematics (The Victor Klee Festschrift)*; DIMACS Series in Discrete Mathematics and Theoretical Computer Science; Gritzmann, P., Sturmfels, B., Eds.; American Mathematical Society: Providence, RI, USA; The Association for Computing Machinery: New York, NY, USA, 1991; Volume 4, pp. 493–516.
13. Schulte, E.; Weiss, A.I. Chirality and projective linear groups. *Discret. Math.* **1994**, *131*, 221–261. [CrossRef]
14. Schulte, E.; Weiss, A.I. Free extensions of chiral polytopes. *Can. J. Math.* **1995**, *47*, 641–651. [CrossRef]
15. Huybers, P.; Coxeter, H.S.M. A new approach to the chiral Archimedean solids. *Math. Rep. Acad. Sci. Can.* **1979**, *1*, 259–274.
16. Weissbach, B.; Martini, H. *Beitrage zur Algebra und Geometrie (Contributions to Algebra and Geometry)*; Springer: Berlin/Heidelberg, Germany, 2002; Volume 43, p. 121.
17. Conway, J.H.; Smith, D.A. *On Quaternion's and Octonions: Their Geometry, Arithmetics, and Symmetry*; Peters, A.K., Ltd.: Natick, MA, USA, 2003.

18. Carter, R.W. *Simple Groups of Lie Type*; John Wiley & Sons Ltd.: Hoboken NJ, USA, 1972.
19. Humphreys, J.E. *Reflection Groups and Coxeter Groups*; Cambridge University Press: Cambridge, UK, 1990.
20. Koca, M.; Koc, R.; Barwani, M.A. Non-crystallographic Coxeter Group $H_4$ in $E_8$. *J. Phys. A Math. Gen.* **2001**, *11201*, A34.
21. Koca, N.O.; Al-Mukhaini, A.; Koca, M.; Al-Qanobi, A. Symmetry of the pyritohedron and lattices. *SQU J. Sci.* **2016**, *21*, 140–150. [CrossRef]
22. Koca, M.; Koc, R.; Al-Barwani, M. Quaternionic roots of SO(8), SO(9), $F_4$ and the related Weyl groups. *J. Math. Phys.* **2003**, *44*, 3123. [CrossRef]
23. Koca, M.; Koc, R.; Al-Barwani, M. Quaternionic root systems and subgroups of the *Aut* ($F_4$). *J. Math. Phys.* **2006**, *47*, 043507. [CrossRef]
24. Koca, M.; Koc, R.; Al-Barwani, M.; Al-Farsi, S. Maximal subgroups of the Coxeter group $W(H_4)$ and quaternions. *Linear Algebra Appl.* **2006**, *412*, 441. [CrossRef]
25. Du Val, P. *Homographies, Quaternions, and Rotations*; Oxford University Press: Oxford, UK, 1964.
26. Slansky, R. Group theory for unified model building. *Phys. Rep.* **1981**, *79*, 1–128. [CrossRef]
27. Coxeter, H.S.M. *Regular Polytopes*, 3rd ed.; Dover Publications: New York, NY, USA, 1973.
28. Koca, M.; Koc, R.; Al-Ajmi, M. Group theoretical analysis of 4D polytopes 600-cell and 120-cell with quaternions. *J. Phys. A Math. Theor.* **2007**, *40*, 7633. [CrossRef]
29. Koca, M.; Koca, N.O.; Al-Barwani, M. Snub 24-Cell derived from the Coxeter-Weyl group $W(D_4)$. *Int. J. Geom. Methods Mod. Phys.* **2012**, *9*, 15. [CrossRef]

![symmetry logo] *symmetry*

MDPI

Article

# Taylor–Socolar Hexagonal Tilings as Model Sets

**Jeong-Yup Lee [1],\* and Robert V. Moody [2]**

[1]   Department of Mathematics Education, Kwandong University, Gangneung, Gyeonggi-do 210-701, Korea
[2]   Department of Mathematics and Statistics, University of Victoria, Victoria, British Columbia V8W 3P4,
      Canada; rmoody@uvic.ca
\*   Author to whom correspondence should be addressed; jylee@kwandong.ac.kr; Tel.: +82-33-649-7776;
      Fax: +82-33-642-7716.

Received: 7 August 2012; in revised form: 6 December 2012; Accepted: 7 December 2012;
Published: 28 December 2012

**Abstract:** The Taylor–Socolar tilings are regular hexagonal tilings of the plane but are distinguished in being comprised of hexagons of two colors in an aperiodic way. We place the Taylor–Socolar tilings into an algebraic setting, which allows one to see them directly as model sets and to understand the corresponding tiling hull along with its generic and singular parts. Although the tilings were originally obtained by matching rules and by substitution, our approach sets the tilings into the framework of a cut and project scheme and studies how the tilings relate to the corresponding internal space. The centers of the entire set of tiles of one tiling form a lattice $Q$ in the plane. If $X_Q$ denotes the set of all Taylor–Socolar tilings with centers on $Q$, then $X_Q$ forms a natural hull under the standard local topology of hulls and is a dynamical system for the action of $Q$. The $Q$-adic completion $\overline{Q}$ of $Q$ is a natural factor of $X_Q$ and the natural mapping $X_Q \longrightarrow \overline{Q}$ is bijective except at a dense set of points of measure 0 in $\overline{Q}$. We show that $X_Q$ consists of three LI classes under translation. Two of these LI classes are very small, namely countable $Q$-orbits in $X_Q$. The other is a minimal dynamical system, which maps surjectively to $\overline{Q}$ and which is variously 2 : 1, 6 : 1, and 12 : 1 at the singular points. We further develop the formula of what determines the parity of the tiles of a tiling in terms of the coordinates of its tile centers. Finally we show that the hull of the parity tilings can be identified with the hull $X_Q$; more precisely the two hulls are mutually locally derivable.

**Keywords:** monotile tiling; Taylor–Socolar tiling; model sets; pure point spectrum; parity tiling

---

## 1. Introduction

This paper concerns the aperiodic hexagonal mono-tilings created by Joan Taylor. We learned about these tilings from the unpublished (but available online) paper of Joan Taylor [1], the extended paper of Socolar and Taylor [2], and a talk given by Uwe Grimm at the KIAS conference on aperiodic order in September, 2010 [3]. These tilings are in essence regular hexagonal tilings of the plane, but there are two forms of marking on the hexagonal tile (or if one prefers, the two sides of the tile are marked differently). We refer to this difference as **parity** (and eventually distinguish the two sides as being sides 0 and 1), and in terms of parity the tilings are aperiodic. In fact the parity patterns of tiles created in this way are fascinating in their apparent complexity, see Figures 1 and 2.

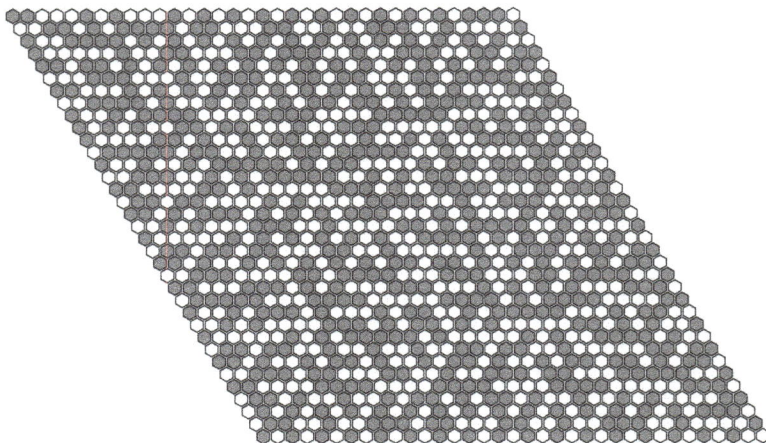

**Figure 1.** A section of a Taylor–Socolar tiling showing the complex patterning arising from the two sides of the hexagonal tile, here indicated in white and gray. Notice that there are islands (Taylor and Socolar call the llamas) both of white and gray tiles.

The two Taylor–Socolar tiles are shown in Figure 3, the main features being the black lines, one of which is a stripe across the tile, and the three colored diameters, one of which is split in color. (Note that the two tiles here are not mirror images of each other, unless one switches color during the reflection. In [2] there is an alternative description of the tiles in which the diagonals have flags at their ends, and in this formulation the two tiles are mirror images of each other.) The difference in the two tiles is only in which side of the color-split diameter the stripe crosses. In the figure the tiles are colored white and gray to distinguish them, but it is the crossing-color of the black stripe that is the important distinguishing feature.

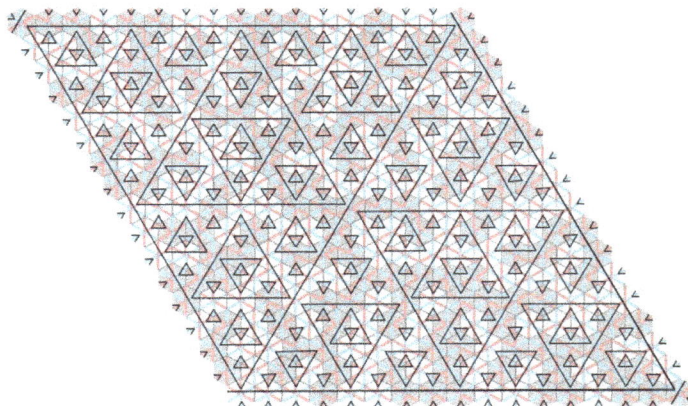

**Figure 2.** The figure shows a pattern of triangles emerging from the construction indicated in Section 2, manifesting the rule **R1**. The underlying hexagonal tiling is indicated in light and dark shades, which indicate the parity of the hexagons. The underlying diagonal shading on the hexagons manifests the rules **R2**.

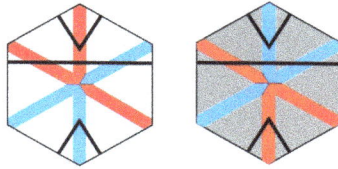

**Figure 3.** The two basic hexagonal tiles. One is a white tile and the other a light gray. These are colored with red and blue diameters. The rotational position of the tiles is immaterial. Note how the tiles are identical as far as the red diagonal and blue diagonal are concerned. The distinction is in which color of the red-blue diagonal cuts the black stripe.

Taylor–Socolar tilings can be defined by following simple matching rules **R1,R2** and can also be constructed by substitution (the scaling factor being 2). In this paper it is the matching rules that are of importance.

**R1** the black lines must join continuously when tiles abut;
**R2** the ends of the diameters of two hexagonal tiles that are separated by an edge of another tile must be of opposite colors, Figure 4.

The paper [2] emphasizes the tilings from the point of view of matching rules, whereas [1] emphasizes substitution (and the half-hex approach). There is a slight mismatch between the two approaches, see [3], which we will discuss later.

If one looks at part of a tiling with the full markings of the tiles made visible, then one is immediately struck by how the black line markings of the tiles assemble to form nested equilateral triangles, see Figure 2. Although these triangles are slightly shrunken (which ultimately is important), we see that basically the vertices of the triangles are tied to the centers of the hexagons, and the triangle side-lengths are $1, 2, 4, 8, \cdots$ in suitable units. This triangle pattern is highly reminiscent of the square patterns that underlie the famous Robinson tilings [4,5], which also appear in sizes that scale up by factors of 2. These tilings are limit-periodic tilings and can be described by model sets whose internal spaces are 2-adic spaces. The Taylor–Socolar tilings are also limit-periodic and it seems natural to associate some sort of 2-adic spaces with them and to give a model-set interpretation of the picture.

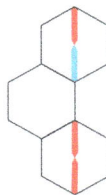

**Figure 4.** Rule **R2**: Two hexagon tiles separated by the edge of another hexagon tile. Note that the diameter colors of the two hexagons are opposite at the two ends of the separating edge. It makes no difference whether or not the diameters color-split—the diameters must have different colors where they abut the separating edge.

One purpose of this paper is to do this, and it has the natural consequence that the tilings are pure point diffractive. It is convenient to base the entire study on a fixed standard hexagonal tiling of the coordinate plane $\mathbb{R}^2$. The centers of the hexagonal tiles can then be interpreted as a lattice in the plane (with one center at $(0,0)$). The internal space of the cut and project scheme that we shall construct is based on a 2-adic completion $\overline{Q}$ of the group $Q$ consisting of all translation vectors between the centers of the hexagons. We shall show that there is a precise one-to-one correspondence between triangulations and elements of $\overline{Q}$. But the triangulation is not the whole story.

The set of all Taylor–Socolar tilings associated with a fixed standard hexagonal tiling of the plane form a tiling hull $X_Q$. This hull is a dynamical system (with group $Q$) and carries the standard topology of tiling hulls. Each tiling has an associated triangulation, but the mapping $\xi : X_Q \longrightarrow \overline{Q}$ so formed, while generically $1 - 1$, is not globally $1 - 1$. What lies behind this is the question of backing up from the triangulations to the actual tilings themselves. The question is how are the tile markings deduced from the triangulations so as to satisfy the rules **R1,R2**? There are two aspects to this. The triangulations themselves are based on hexagon centers, whereas in an actual tiling the triangles are shrunken away from vertices. This shrinking moves the triangle edges and is responsible for the off-centeredness of the black stripe on each hexagon tile. How is this shrinking (or edge shifting, as we call it) carried out? The second feature is the coloring of the diagonals of the hexagons. What freedom for coloring exists, given that the coloring rule **R2** must hold?

In this paper we explain this and give a complete description of the hull and the mapping $\xi$, which coordinatises members of $X_Q$ through $\overline{Q}$, Theorem 6.9. There are numerous places at which $\xi$ is singular (not bijective); in fact the set of singular points in $\overline{Q}$ is dense. Two special classes of singular points are those corresponding to the central hexagon triangulations (**CHT**) (see Figure 5) and the infinite concurrent $w$-line tilings (**iCw-L**) (see Figure 6). In both cases there is 3-fold rotational symmetry of the *triangulation* and in both cases the mapping $\xi$ is many-to-one. These two types of tilings play a significant role in [2].

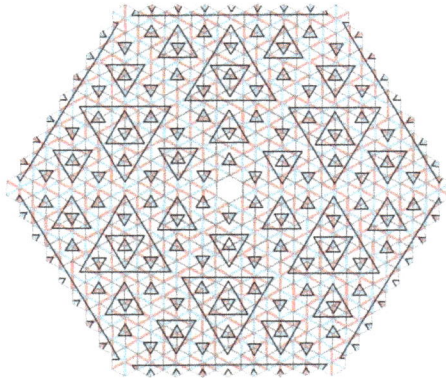

**Figure 5.** The (central part of a) central hexagon (**CHT**) tiling. Full (edge-shifted) triangles of levels $0, 1, 2$ are shown. At the outside edges one can see the beginnings of triangles of level 3. The rays from the central hexagon in the six $a$-directions will have infinite $a$-lines in them. However the edge shifting rules cannot be applied to them because they are of infinite level—they are not composed of edges of finite triangles. In the end a full tiling is obtained by placing a fully decorated tile into the empty central hexagon. There are 12 ways to do this, and each way then determines the rest of the tiling completely. These tilings violate both forms of generic condition.

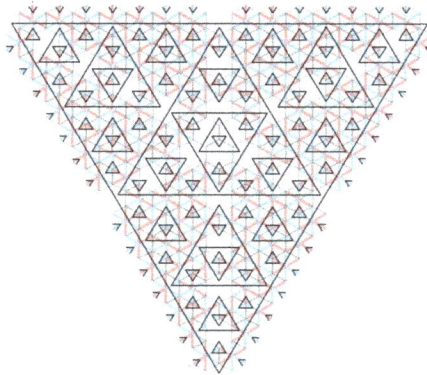

**Figure 6.** The **iCw-L** tilings. The triangulation is generic-a but not generic-w. Of course the partial tiling shown is perfectly consistent with generic tilings—in fact all Taylor–Socolar tilings contain this type of patch of tiles. However, if the pattern established in the picture is maintained at all scales, then indeed the result is a not a generic tiling since it fails generic-$w$.

The hull has a minimal invariant component of full measure and this is a single LI class. There are two additional orbits, whose origins are the **iCw-L** triangulations, and although they perfectly obey the matching rules they are not in the same LI class as all the other tilings. On the other hand the **CHT** tilings (those lying over the **CHT** triangulations) are in the main LI class and, because of the particular simplicity of the unique one whose center is $(0,0)$, the question of describing the parity (which tiles are facing up and which are facing down) becomes particularly easy. Here we reproduce the parity formula for this **CHT** tiling as given in [2] (with some minor modifications in notation). We use this to give parity formulas for all the tilings of $X_Q$.

A couple of comments about earlier work on aperiodic hexagonal tilings are appropriate here. D. Frettlöh [6] discusses the half-hex tilings (created out of a simple substitution rule) and proves that natural point sets associated with these can be expressed as model sets. Half-hexes do not play an explicit role in this paper, though the hull of the half-hex tilings is a natural factor of $X_Q$ lying between $X_Q$ and $\overline{Q}$ [6–8]. They were important to Taylor's descriptions of her tilings and are implicitly embedded in them.

In [9], Penrose gives a fine introduction to aperiodic tilings and then goes on to create a class of aperiodic hexagonal tilings, which he calls $1 + \epsilon + \epsilon^2$-tilings in which there are three types of tiles that assemble by matching rules. The main tiles are hexagonal, with keyed edges. The other two are a linear-like tile with an arbitrarily small width ($\epsilon$ tiles) that fit along the hexagon edges, and some very tiny tiles ($\epsilon^2$) that fit at the corners of the tiles. Further musings on Penrose's approach to his hexagonal tilings can be found in [10].

The Penrose hexagonal tilings are closely related to the Taylor–Socolar tilings, though a comparison of the corresponding parity tilings, our Figures 1 and 25 of [9], makes it clear that they cannot be the same. In fact, extensive computational work of F. Gähler shows that the cohomology of their hulls are different, so the two tilings are quite distinct from one another. Nonetheless both tiling hulls have the half-hex hull and $\overline{Q}$ as factors, and amazingly both have the same dynamical zeta functions. A report on this work appears in this same volume [11] of *Symmetry*, where the Penrose tiling and the Taylor tilings are carefully compared. The approach there is based on the construction of the tilings through inflations, and the complications of the singular points of the hull arise from observing the special symmetries possessed by certain of the tilings. In our approach, which is more algebraic, we begin with $\overline{Q}$ and the singularities arise as obstructions to the process of reconstructing a tiling from its corresponding pattern of triangles.

*Symmetry* **2013**, *5*, 45–85

There is an algorithmic computation for determining that certain classes of substitution tilings have pure-point spectrum. This has been used to confirm in yet another way that the Taylor–Socolar substitution tilings have pure point spectrum or, equivalently, are regular model sets [12].

## 2. The Triangulation

In principle the tilings that we are interested in are not connected to the points of lattices and their cosets in $\mathbb{R}^2$, but are only point sets that arise in Euclidean space $\mathbb{E}$ as the vertices and centers of tilings. However, our objective here is to realize tiling vertices in an algebraic context and for that we need to fix an origin and a coordinate system so as to reduce the language to that of $\mathbb{R}^2$. Let $Q$ be the triangular lattice in $\mathbb{R}^2$ defined by

$$Q := \mathbb{Z}a_1 + \mathbb{Z}a_2$$

where $a_1 = (1,0)$ and $a_2 = \left(-\frac{1}{2}, \frac{\sqrt{3}}{2}\right)$. Then $P := \mathbb{Z}w_1 + \mathbb{Z}w_2$ where $w_1 = \frac{2}{3}a_1 + \frac{1}{3}a_2$ and $w_2 = \frac{1}{3}a_1 + \frac{2}{3}a_2$ is a lattice containing $Q$ as sublattice of index 3, see Figure 7. For future reference we note that $|a_1|=|a_2|=|a_1+a_2|=1$ and $|w_1|=|w_2|=|w_2-w_1|=1/\sqrt{3}$.

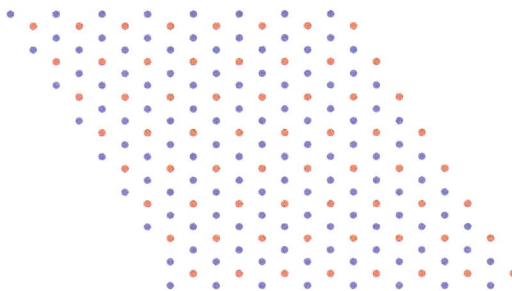

**Figure 7.** The figure shows the standard triangular lattice $Q$ (the red points) and the larger lattice $P$ (red and blue points) in which $Q$ lies with index 3. The points of $Q$ may be viewed as the vertices of a triangularization of the plane by equilateral triangles of side length 1. The blue points are the centres of these triangles. The color here has nothing to do with the coloring of the diagonals of the tiles—it only distinguishes the two cosets.

Joining the points of $Q$ that lie at distance 1 from one another creates a triangular tiling. Inside each of the unit triangles so formed there lies a point of $P$, and indeed $P$ consists of three $Q$ cosets: $Q$ itself, the centroids of the "up" triangles (those with a vertex above a horizontal edge), and the "down" triangles (those with a vertex below a horizontal edge), see Figure 8. What we aim to do is to create a hexagonal tiling of $\mathbb{R}^2$. When this tiling is complete, the points of $Q$ will be the centers of the hexagonal tiles and the points of $P$ immediately surrounding the points of $Q$ will make up the vertices of the tiles [13].

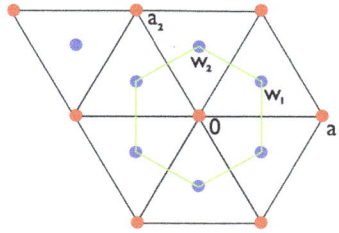

**Figure 8.** The generators $a_1, a_2$ of $Q$ and the generators $w_1, w_2$ of $P$, showing how the cosets of $Q$ in $P$ split into the points of $Q$ and the centroids (centers) of the up and down triangles. Around the point $0 \in Q$ we see the hexagonal tile centered on 0 with vertices in $P \backslash Q$.

Each of the hexagonal tiles will be marked by colored diagonals and a black stripe, see Figure 3. These markings divide the tiles into two basic types, and it is describing the pattern made from these two types in model-set theoretical terms that is a primary objective of this paper (see Figure 2). The other objective is to describe the dynamical hull that encompasses all the tilings that belong to the Taylor–Socolar tiling family.

We let the coset of up (respectively down) points be denoted by $S_1^{\uparrow} = w_1 + Q$ and $S_1^{\downarrow} = w_2 + Q$ respectively:

$$P = Q \cup S_1^{\uparrow} \cup S_1^{\downarrow}$$

**Remark 2.1** *There are three cosets of $Q$ in $P$. In our construction of the triangle patterns we have taken the point of view that $Q$ itself will be used for triangle vertices and the other two cosets for triangle centroids. However, we could use any of the three cosets as the triangle vertices and arrive at a similar situation. This amounts to a translation of the plane by $w_1$ or $w_2$. We come back to this point in Section 9.*

We now wish to re-triangularize the plane still using points of $Q$ as vertices, but this time making triangles of side length equal to 2 using as vertices a coset of $2Q$ in $Q$. There are four cosets of $2Q$ in $Q$ and they lead to four different ways to make the triangularization. Figure 9 shows the four types of triangles of side length 2. The lattices generated by the points of any one of these triangles is a coset of $2Q$ and together they make up all four cosets of $2Q$ in $Q$.

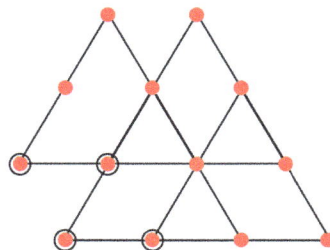

**Figure 9.** Four superimposed triangles, each indicated by its circled bottom lefthand vertex. The vertices of each triangle generate a different coset of $Q$ modulo $2Q$.

Choose one of these cosets, call it $q_1 + 2Q$, where $q_1 \in Q$, and thereby triangulate the plane with triangles of side length 2. The centroids of the new triangles are a subset of the original set of centroids and, in fact, together with the vertices $q_1 + 2Q$ they form the coset $q_1 + 2P$. This is explained in the Figure 10, which also explains the important fact that the new centroids, namely those of the new edge-length-2 triangles of $q_1 + 2Q$, make up two cosets of $2Q$ in $q_1 + 2P$ depending on the orientation of the new triangles, and these orientations are *opposite* to those that these points originally had. Thus we obtain $S_2^\uparrow = q_1 + 2w_1 + 2Q$ (which is in $w_2 + Q$ !), $S_2^\downarrow = q_1 + 2w_2 + 2Q$ (which is in $w_1 + Q$), and the coset decomposition

$$q_1 + 2P = (q_1 + 2Q) \cup S_2^\uparrow \cup S_2^\downarrow$$

with $S_2^\uparrow \subset S_1^\downarrow$ and $S_2^\downarrow \subset S_1^\uparrow$.

We now repeat this whole process. There are four cosets of $4Q$ in $q_1 + 2Q$ and we select one of them, say $q_1 + q_2 + 4Q$, with $q_2 \in 2Q$, and this gives us a new triangulation with triangles of side length 4. Their centroids in $q_1 + q_2 + 4P$ form $4Q$-cosets $S_3^\uparrow \subset S_2^\downarrow$ and $S_3^\downarrow \subset S_2^\uparrow$, and we have the decomposition

$$q_1 + q_2 + 4P = (q_1 + q_2 + 4Q) \cup S_3^\uparrow \cup S_3^\downarrow$$

Continuing this way we obtain $q_1, q_2, q_3, \ldots$ with $q_k \in 2^{k-1}Q$, and sets $S_k^\uparrow, S_k^\downarrow$ with $S_{k+1}^\uparrow \subset S_k^\downarrow$ and $S_{k+1}^\downarrow \subset S_k^\uparrow$ for all $k = 1, 2, \ldots$, and the partition

$$q_1 + \cdots + q_k + 2^k P = \left( q_1 + \cdots + q_k + 2^k Q \right) \cup S_{k+1}^\uparrow \cup S_{k+1}^\downarrow \qquad (1)$$

We have

$$S_{k+1}^\uparrow \;=\; q_1 + \cdots + q_k + 2^k (w_1 + Q) \qquad (2)$$

$$S_{k+1}^\downarrow \;=\; q_1 + \cdots + q_k + 2^k (w_2 + Q)$$

Explicit formulas for $2^k w_1$ and $2^k w_2$ are given in Lemma 3.2.

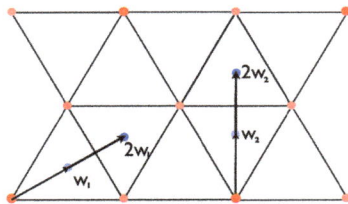

**Figure 10.** The figure shows how the centroids (indicated with solid blue dots) of the new side-length-2 triangles (indicated with solid red dots) are obtained as vectors from $q_1 + 2P$. The two $2Q$-cosets of $q_1 + 2P$ which are not $q_1 + 2Q$ itself indicate the centroids of the new up and down triangles. Notice that the orientations of the new triangles, and hence the orientations associated with the new centroids, are opposite to the orientations associated with these points when they were viewed as centroids of the original triangulation. This explains why $S_2^\uparrow \subset S_1^\downarrow$ and $S_2^\downarrow \subset S_1^\uparrow$.

We now carry out the entire construction based on an arbitrary infinite sequence

$$(q_1, q_2, \ldots, q_k, \ldots)$$

where $q_k \in 2^{k-1}Q$ for all $k$. This results in a pattern of overlapping triangulations based on triangles of edge lengths $1, 2, 4, 8, \ldots$ (these are referred to as being triangles of levels $0, 1, 2, 3, \ldots$). In Section 4 we shall make our tiling out of this pattern. But certain features of the entire pattern are clear:

- all points involved as vertices of triangles are in $Q$;
- all triangle centroids are in $P \backslash Q$;
- there is no translational symmetry.

The last of these is due to the fact that there are triangles of all scales, and no translation can respect all of these scales simultaneously.

A point $x \in P\backslash Q$ is said to **have an orientation** (up or down) if there is a positive integer $k$ such that for all $k' > k$, $x \notin S_{k'}^{\uparrow} \cup S_{k'}^{\downarrow}$. Every element of $P\backslash Q$ is in $S_k^{\uparrow}$ or $S_k^{\downarrow}$ for $k = 1$, and some for other values of $k$ as well. For the elements $x$ which have an orientation there is a largest such $k$ for which this is true. We call this $k$ the **level of its orientation**. If there is no such $k$ we shall say that $x$ is **not oriented**. We shall see below (Proposition 3.3) what it means for a point not to have an orientation [14].

## 3. The Q-Adic Completion

In this section we create and study a completion of $P$ under the $Q$-adic topology. The $Q$-adic topology is the uniform topology based on the metric on $P$ defined by $d(x,y) = 2^{-k}$ if $x - y \in 2^k Q\backslash 2^{k+1}Q$ and $d(x,y) := 2$ when $x,y$ are in different cosets of $Q$. This metric is $Q$-translation invariant. $\overline{P}$ is the completion of $P$ in this topology and $\overline{Q}$ is the closure of $Q$ in $\overline{P}$, which is also the completion of $Q$ in the $Q$-adic topology.

$\overline{P}$ may be viewed as the set of sequences

$$(b_1, b_2, \ldots)$$

where $b_k \in P$ for all $k$ and $b_{k+1} \equiv b_k \bmod 2^k Q$.

$\overline{P}$ is a group under component-wise addition and $\overline{Q}$ is the subgroup of all such sequences with all components in $Q$. There is the obvious coset decomposition

$$\overline{P} = \overline{Q} \cup (w_1 + \overline{Q}) \cup (w_2 + \overline{Q})$$

so $\overline{Q}$ has index 3 in $\overline{P}$. We note that $\overline{Q}$ and $\overline{P}$ are compact topological groups.

We have $i : P \longrightarrow \overline{P}$ via

$$b \mapsto (b, b, b, \ldots)$$

We often identify $P$ as a subgroup of $\overline{P}$ via the embedding $i$.

Note that the construction of expanding triangles of Section 2 depends on the choice of the element $(q_1, q_2, \ldots)$, where $q_k \in 2^{k-1}Q$. Then we can obtain the compatible sequence

$$\mathbf{q} = (q_1, q_1 + q_2, \ldots, q_1 + q_2 + \cdots + q_k, \ldots) \in \overline{Q}$$

and thus we can identify each possible construction with an element of $\overline{Q}$. Let $\mathcal{T}(\mathbf{q})$ denote the pattern of triangles arising from $\mathbf{q} \in \overline{Q}$.

Let $\mu$ denote the unique Haar measure on $\overline{P}$ for which $\mu(\overline{P}) = 1$. The key feature of $\mu$ is that $\mu(p + 2^k\overline{Q}) = 2^{-k}/3$ for all $p \in \overline{P}$. We note that $P \subset \overline{P}$ is countable and has measure 0, and that $\mu(\overline{Q}) = \frac{1}{3}$ and $\mu(S_k^{\uparrow}) = \mu(S_k^{\downarrow}) = 2^{-k+1}/3$.

**Remark 3.1** *We should note a subtle point here. In $\overline{Q}$ one can divide by 3. In fact, for all $\mathbf{x} \in \overline{Q}$,* $-\lim_{k\to\infty}(\mathbf{x} + 4\mathbf{x} + 4^2\mathbf{x} + \cdots + 4^k\mathbf{x})$ *exists since $4^k\mathbf{x} \in 2^{2k}\overline{Q}$, and*

$$-3 \lim_{k\to\infty}\left(\mathbf{x} + 4\mathbf{x} + 4^2\mathbf{x} + \cdots + 4^k\mathbf{x}\right) = \lim_{k\to\infty}(1-4)\left(\mathbf{x} + 4\mathbf{x} + 4^2\mathbf{x} + \cdots + 4^k\mathbf{x}\right) = \lim_{k\to\infty}\left(1 - 4^{k+1}\right)\mathbf{x} = \mathbf{x}.$$

*Thus we can find an element $\mathbf{w_1}$ of $\overline{Q}$ corresponding to $w_1 = \frac{2}{3}a_1 + \frac{1}{3}a_2$ and similarly $\mathbf{w_2} \in \overline{Q}$ corresponding to $w_2$. However, our view is that $P = Q \cup (w_1 + Q) \cup (w_2 + Q)$ and $\overline{P}$ is the $Q$-adic completion of this, with each of the three cosets leading to a different coset of $\overline{Q}$ in $\overline{P}$. Thus $w_1 - \mathbf{w_1} \neq 0$ but $3(w_1 - \mathbf{w_1}) = 0$ and we conclude that $\overline{P}$ has 3-torsion.*

Two examples of this are important in what follows. Define $s_1^{(-1)} := 0$ and $s_1^{(k)} := a_1 + 4a_1 + 4^2 a_1 + \cdots + 4^k a_1)$ for $k = 0, 1, \ldots$, and similarly $s_2^{(k)}$ based on $a_2$. Their limits are denoted by $\mathbf{s_1}, \mathbf{s_2}$ respectively. They lie in $\overline{Q}$.

**Lemma 3.2** *For all $k = 0, 1, \ldots$,*

$$
\begin{aligned}
2^{2k} w_1 &= w_1 + s_2^{(k-1)} + 2 s_1^{(k-1)} \\
2^{2k+1} w_1 &= w_2 + s_1^{(k)} + 2 s_2^{(k-1)} .
\end{aligned}
$$

*Similarly for $2^m w_2$, interchanging the indices $1, 2$.*

*In particular $\lim_{k \to \infty} 2^{2k} w_1 = w_1 + \mathbf{s_2} + 2\mathbf{s_1}$ and $\lim_{k \to \infty} 2^{2k+1} w_1 = w_2 + \mathbf{s_1} + 2\mathbf{s_2}$. Furthermore, $3(w_1 + \mathbf{s_2} + 2\mathbf{s_1}) = 0 = 3(w_2 + \mathbf{s_1} + 2\mathbf{s_2})$.*

**Proof:** From the definitions, $2w_1 = w_2 + a_1$ and $2w_2 = w_1 + a_2$. This gives the case $k = 0$ of the Lemma. Now proceeding by induction,

$$
2^{2k} w_1 = 2\left( w_2 + s_1^{(k-1)} + 2 s_2^{(k-2)} \right) = w_1 + a_2 + 2 s_1^{(k-1)} + 4 s_2^{(k-2)} = w_1 + s_2^{(k-1)} + 2 s_1^{(k-1)} ,
$$

as required. Similarly

$$
2^{2k+1} w_1 = 2\left( w_1 + s_2^{(k-1)} + 2 s_1^{(k-1)} \right) = w_2 + a_1 + 2 s_2^{(k-1)} + 4 s_1^{(k-1)} = w_2 + s_1^{(k)} + 2 s_2^{(k-1)} .
$$

Taking the limits and using the formula for multiplication by 3 at the beginning of Remark 3.1, we find that $-3(\mathbf{s_2} + 2\mathbf{s_1}) = a_2 + 2a_1 = 3w_1$ and similarly with the indices $1, 2$ interchanged.

Consider what happens if there is a point $x \in P \backslash Q$ that does not have orientation. This means that there is an infinite sequence $k_1 < k_2 < \cdots$ with $x \in S_{k_j}^{\uparrow} \cup S_{k_j}^{\downarrow}$. Then from (2),

$$
x \in \left( \left( q_1 + \cdots + q_{k_j - 1} + 2^{k_j - 1}(w_1 + Q) \right) \right) \cup \left( \left( q_1 + \cdots + q_{k_j - 1} + 2^{k_j - 1}(w_2 + Q) \right) \right) \text{ for each } k_j. \text{ This}
$$

means $x = \mathbf{q} + w_1 + \mathbf{s_2} + 2\mathbf{s_1}$ or $x = \mathbf{q} + w_2 + \mathbf{s_1} + 2\mathbf{s_2}$.

**Proposition 3.3** $\mathcal{T}(\mathbf{q})$ *has at most one point without orientation. A point without orientation can occur if and only if $\mathbf{q} \in -\mathbf{s_2} - 2\mathbf{s_1} + Q$ or $\mathbf{q} \in -\mathbf{s_1} - 2\mathbf{s_2} + Q$. These two families are countable and disjoint.*

**Proof:** If $x \in P \backslash Q$ does not have an orientation, then either $x = \mathbf{q} + w_1 + \mathbf{s_2} + 2\mathbf{s_1}$ and $-w_1 + x \in Q$, which gives one of the cases; or $x = \mathbf{q} + w_2 + \mathbf{s_1} + 2\mathbf{s_2}$, which gives the other. Conversely, in either case we have points without orientation. Since in one case $x \in w_1 + Q$ and in the other case $x \in w_2 + Q$, we see that the two families are disjoint.

**Remark 3.4** *We do not need to go into the exact description of the orientations of triangles, but confine ourselves to a few remarks here. For any fixed $\mathbf{q}$, define the sequence of sets $W_k^{\uparrow}$ and $W_k^{\downarrow}$, $k = 1, 2, \ldots$, inductively by $W_1^{\uparrow} = S_1^{\uparrow}$ and*

$$
W_{k+1}^{\uparrow} = \left( W_k^{\uparrow} \backslash S_{k+1}^{\downarrow} \right) \cup S_{k+1}^{\uparrow} ,
$$

*and similarly for $W_k^{\downarrow}$. In other words we put together into $W_k^{\uparrow}$ all the points which are oriented upwards at step $k$, and likewise all that are oriented downwards at step $k$.*

*Since $S_{k+1}^{\downarrow}$ and $S_{k+1}^{\uparrow}$ have measure $2^{-k}/3$, we see that the sets $W_k^{\uparrow}$ change by less and less as $k$ increases. Furthermore it is clear that $\mu\left( W_k^{\uparrow} \right) = 1/3$ for all $k$.*

**Proposition 3.5** *For all k the sets* $\overline{W_k^\uparrow}$ *and* $\overline{W_k^\downarrow}$ *are clopen and disjoint. They each have measure* $1/3$.

For each $\mathbf{q} \in \overline{Q}$ we define $W^\uparrow(\mathbf{q}) := \overline{\{x : x \text{ which have up orientation}\}}$, and similarly for $W^\downarrow(\mathbf{q})$.

**Proposition 3.6** $\overline{P} = \overline{Q} \cup W^\uparrow(\mathbf{q}) \cup W^\downarrow(\mathbf{q})$ *where* $\overline{Q}$ *is disjoint from* $W^\uparrow(\mathbf{q}) \cup W^\downarrow(\mathbf{q})$, *and*

$$W^\uparrow(\mathbf{q}) \cap W^\downarrow(\mathbf{q}) = \{\mathbf{q} + w_1 + \mathbf{s}_2 + 2\mathbf{s}_1, \mathbf{q} + w_2 + \mathbf{s}_1 + 2\mathbf{s}_2\}.$$

$W^\uparrow(\mathbf{q})$ *is the union of an open set and* $\{\mathbf{q} + w_1 + \mathbf{s}_2 + 2\mathbf{s}_1, \mathbf{q} + w_2 + \mathbf{s}_1 + 2\mathbf{s}_2.\}$ *The same goes for* $W^\downarrow(\mathbf{q})$. *In particular* $W^\uparrow(\mathbf{q})$ *and* $W^\downarrow(\mathbf{q})$ *are the closures of their interiors. Both* $W^\uparrow(\mathbf{q})$ *and* $W^\downarrow(\mathbf{q})$ *are sets of measure* $1/3$.

## 4. The Tiles

Let us assume that we have carried out a triangulation $\mathcal{T}(\mathbf{q})$ as described in Section 2. We now have an overlaid pattern of equilateral triangles of side lengths $1, 2, 4, \cdots$. Each of these triangles has vertices in $Q$ and its centroid in $P \backslash Q$. The points of the two cosets of $P$ different from $Q$ (shown as blue points in Figure 7) form the vertexes of a tiling of hexagons made from the triangulation, see Figure 11. This tiling, with the tiles suitably marked, is the tiling that we wish to understand. Our objective is to give each hexagon of the tiling markings in the form of a black stripe and three colored diagonals as shown in Figure 3.

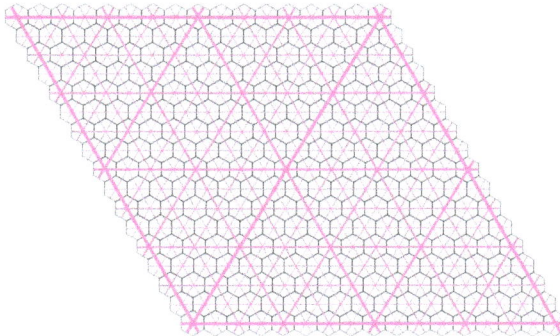

**Figure 11.** A partial triangulation of the plane overlaid on the basic lattice of hexagons which will make up the tiles. The levels of the triangles are indicated by increasing thickness. One can clearly see triangles of levels $0, 1, 2, 3, 4$ and one can also see how triangle edges of level $k$ ultimately become edges passing through the interior of triangles of level $k + 1$. This will be used to make the shifting of edges later on.

Apart from the lines of the triangulation (which give rise to short diagonals of the hexagons of the tiling) we also have the lines on which the long diagonals of the hexagons lie and which carry the color. To distinguish these sets of lines we call the triangulation lines $a$-**lines** (since they are in the directions $a_1, a_2, a_1 + a_2$) and the other set of lines $w$-**lines** (since they are in the directions $w_1, w_2, w_2 - w_1$). We also call the $w$-lines **coloring lines**, since they are the ones carrying the colors red and blue. The $w$-lines pass through the centroids of the triangles of the triangulation. We say that a $w$-line has **level** $k$ if there are centroids of level $k$ triangles on it, but none of any higher level. We shall discuss the possibility of $w$-lines that do not have a level in this sense below. Note that every point of $P \backslash Q$ is the centroid of some triangle, some of several, or even many!

There are two steps required to produce the markings on the tiles. One is to shift triangle edges off center so as to produce the appropriate stripes on the tiles. We refer to this step as *edge shifting*. The second is to appropriately color the main diagonals of each tile. This we refer to as *coloring*. The two steps can be made in either order. However, each of the two steps requires certain generic aspects of the triangulation to be respected in order to be carried out to completion. We first discuss the nature of these generic conditions and then finish this section by showing how edge shifting is carried out.

We need to understand the structure of the various lines (formed from the edges of the various sized triangles) that pass through each hexagon. Let us say that an *element* of $Q$ is of **level** $k$ if it is a vertex of a triangle of edge length $2^k$ but is not a vertex of any longer edge length. Similarly an *edge* of a triangle is of **level** $k$ if it is of length $2^k$, and an $a$-line (made up of edges) is of **level** $k$ if the longest edges making it up are of length $2^k$. All lines of all levels are made from the original set of lines arising from the original triangulation by triangles of edge length 1, so a line of level $k$ has edges of lengths $1, 2, \cdots, 2^k$ on it.

The word "level" occurs in a variety of senses in the paper. These are summarized in Table 1. There are two types of generic assumptions that we need to consider.

**Table 1.** Uses of the word "level" $k$ and section number where it is defined. If there is no such $k$ the level is infinite.

| | |
|---|---|
| of a triangle | Section 2 $k$ if the side length is $2^k$, where a side length $1 = 2^0$ is the length of $a_1$ and $a_2$ |
| of orientation of $x \in P$ | Section 2 $k$ at which $x$ stops switching between $S_k^\uparrow$ and $S_k^\downarrow$ |
| of a $w$-line | Section 4 max. $k$ of centroids of level $k$ triangles on it |
| of a point of $Q$ | Section 4 max. $k$ for which it is a vertex of a triangle of level $k$ |
| of a triangle edge | Section 4 $k$ for which it is an edge of a level $k$ triangle |
| of an $a$-line | Section 4 max. $k$ for $k$-edges on this line |

**Definition 4.1** *A triangulation (or the value of* **q** *associated with it) in which every w-line has a finite level is called* **generic-w**. *This means that for every w-line there is a finite bound on the levels of the centroids (points of $P \backslash Q$) that lie on that line. In this case for any ball of any radius anywhere in the plane, there is a level beyond which no w-lines of higher level cut through that ball. See Figure 6 for an example that shows failure of the generic-w condition.*

*A triangulation (or the value of* **q** *associated with it) is said to be* **generic-a** *if every a-line has a finite level. This means for every a-line there is a finite bound on the levels of edges that lie in that line. In this case for any ball of any radius anywhere in the plane, there is a level beyond which no lines of the triangulation of higher level cut through that ball. See Figures 6 and 12.*

*A tiling is said to be* **generic** *if it is both generic-w and generic-a. All other tilings (or elements* **q** $\in Q$*) are called* **singular**. *One case of the failure of generic-w is discussed in Proposition 3.3 above. The only way for one of our generic conditions to fail is that there are a-lines or w-lines of infinite level. This situation is discussed in Section 6.*

Every element of $Q$ has a hexagon around it and three lines passing through it in the directions $\pm a_1, \pm a_2, \pm(a_1 + a_2)$. These lines pass through pairs of opposite edges of the hexagon at right-angles to those edges. We shall call these lines **short diameters**. These short diameters arise out of the edges of the triangles of the triangulations that we have created. Each triangle edge is part of a line which is a union of edges, all of the same level. As we have pointed out, the line (and its edges) have level $k$ if they occur at level $k$ (and no higher). The original triangulation has level 0. One should note that a line may occur as part of the edges of many levels of triangles, but under the assumption of generic-a there will be a highest level of triangles utilizing a given line, and it is this highest level that gives the line its level and determines the corresponding edges.

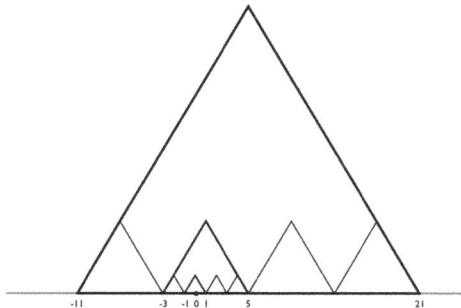

**Figure 12.** This shows a sketch of how a tiling with an infinite *a*-line (the horizontal line) can be constructed so that it is generic-*w*. Here $\mathbf{q} = za_1$, where $z$ is the 2-adic integer $(1, 1, 5, 5, 21, 21, 85, 85, \ldots)$ (the mod $2, 4, 8, 16, 32, 64, 128, 256, \ldots$ values). Some triangles of edge lengths $2, 8, 32$ are shown. There are triangles of arbitrary large side lengths on the horizontal line, but the triangulation does not admit a second infinite *a*-line and cannot admit an infinite *w*-line since $\mathbf{q}$ is of the wrong form.

In looking at the construction of level 1 triangles out of the original triangulation of level 0 triangles, we note immediately that every point of $Q$ has at least one line of level 1 through it (though by the time the triangularization is complete this line may have risen to higher level), see Figure 9. The vertices of the level 1 triangles have three lines of level 1 through them, and the rest (the mid-points of the sides of the level 1 triangles) have just one of level 1 and the other two of level 0. Thus at this stage of the construction each hexagon has either one short diameter from a level 1 line or it has 3 short diameters all of level 1.

This is the point to remember: At each stage of determining the higher level triangles, we find that the hexagon around each element of $Q$ is of one of two kinds: It either has three short diameters of which two have equal level and the third a higher level, or three short diameters all of the same level $k$. The latter only occurs when the element of $Q$ is a vertex of a triangle of level $k$. Since we are in the generic-*a* case, there is no element of $Q$ that is a vertex of triangles of unbounded scales, and the second condition cannot hold indefinitely. Once an element of $Q$ is not a vertex at some level then it never becomes a vertex at any other higher level (all vertices of triangles at each level are formed from vertices of triangles at the previous level).

We conclude ultimately that in the generic-*a* cases every hexagon has three short diameters of which two are of one level and one of a higher level. See Figure 11.

**Lemma 4.2** *For* $\mathbf{q}$ *satisfying generic-a each hexagonal tile of* $\mathcal{T}(\mathbf{q})$ *has three short diameters of which exactly one has the largest level and the other two equal but lesser levels.*

We now describe edge shifting. Fix any $\epsilon$ with $0 < \epsilon \leq 1/4$. This $\epsilon$ is going to be the distance by which lines are shifted. It is fixed throughout, but it exact value plays no role in the discussion. Take a tiling based on $\mathbf{q}$.

Now consider any edge that has level $k < \infty$ but does not occur as part of an edge of higher level. This edge occurs as an edge *inside* some triangle $T$ of level $k + 1$, and this allows us to distinguish two sides of that edge. The side of the edge on which the centroid of $T$ lies is called the **inner** side of the edge, and the other side its **outer** side. This edge (but not the entire line) is shifted inwards (*i.e.*, towards the centroid of $T$) by the distance $\epsilon$. Note that the shifting distance $\epsilon$ is independent of $k$. This shifted edge then becomes the *black* stripe on the hexagonal tiles through which this edge cuts, see

Figure 13. Figure 14 shows how edge shifting works. At the end of shifting, each hexagon has on it a pattern made by the shifted triangle edges that looks like the one shown in Figure 13.

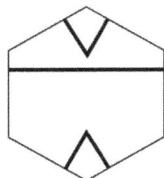

**Figure 13.** The basic hexagon with its markings arising from shrunken triangles.

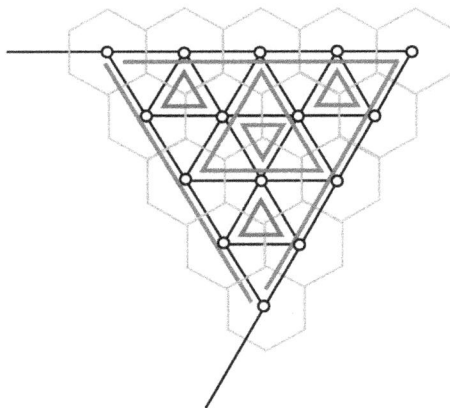

**Figure 14.** The figure shows how edge shifting is done. Part of a triangulation is shown in thin black lines. The shifted edges are shown in thicker gray lines. The extended black lines indicate that the largest (level 2) triangle sits in the top right corner of a level 3 triangle, which is not shown in full. Note how the edges of the level 2 triangle shift.

In the case that **q** satisfies generic-a, the edges of every line of the triangulation are of bounded length. Thus every edge undergoes a shift by the prescription above. Thus,

**Proposition 4.3** *If $\mathcal{T}(\mathbf{q})$ satisfies the condition generic-a, then there is a uniquely determined edge shifting on it.*

## 5. Color

So far we have constructed a triangulation from our choice of **q**, and have shown how edges can be shifted to produce the corresponding hexagonal tiling with the tiles suitably marked by black stripes. We wish now to show how the (long) diagonals of the hexagons are to be colored. This amounts to producing a color (red, blue, or red-blue) for each of the long diagonals of each hexagon of the tiling. The only requirement is that the overall coloring obey the rule **R2** that is used to make Taylor–Socolar tilings.

As we mentioned above, coloring is made independently of shifting in the sense that the two processes can be done in either order. In fact, in this argument we shall suppose that the stripes have not been shifted, so they still run through the centroids of the tiles.

We shall show that for $\mathbf{q} \in \overline{\mathbf{Q}}$ in the generic case, there is exactly one allowable coloring.

Assume that we have a generic tiling (this means both *a and w* generic). Now consider any hexagon of the tiling. We note from Lemma 4.2 that it has three short diameters, one of which is uniquely of highest level, and it is this last short diameter that determines (after shifting) the black stripe for this hexagon. We will refer to this short diameter as the stripe, even though in this discussion it has not been shifted. The other two colored (long) diameters are a red one that lies at $\pi/6$ clockwise of the stripe and a blue one that lies $\pi/6$ counterclockwise of the stripe. The red-blue diameter cuts the stripe at right-angles, but which way around it is (red-blue or blue-red) is not determined yet.

Consider Figure 15 in which we see two complete level 1 triangles overlaid on the basic level 0 triangles. Tiles of the hexagonal tiling are shown on points of $Q$ with the hexagons at the vertices of the level 1 triangles shown in green. These latter are points of $q_1 + 2Q$. At each point of $Q$ there are three edge lines running through it. But notice that at the midpoints of the sides of the level 1 triangles (white hexagons), the edge belonging to the level 1 triangle has higher level than the other two. This is the edge that will become the stripe for the hexagon at that point. This stripe *forces* the red and blue diameters for this hexagon.

The idea behind coloring is based on extrapolating this argument to $w$-lines passing through midpoints of higher level triangles. Consider Figure 16. The point $u$ is the midpoint of an edge of a triangle $T'$ of level 3. Drawing the $w$-line $L$ towards the centroid $d$ of the top left corner triangle $T$ of level 2 we see first of all that the edge of the level 3 triangle through $u$ is the highest level edge through $u$ and hence the coloring along the $w$-line $L$ starts off red, as shown. Now the rule **R2** forces the next part of the coloring to be blue and we come to the hexagon center $e$. This has three edges through it, but the one that our $w$-line crosses at right-angles has the highest level, and so will produce the stripe for the corresponding hexagon. The color must switch at the stripe, and so we see the next red segment as we come to $d$.

And so it goes, until we reach the point $v$. Here $L$ meets the midpoint of the edge of another level 3 riangle. This edge produces the stripe for the hexagon at $v$, but it is not at right-angles to $L$, so there is no color change on $L$ at $v$. Since $v$ is the midpoint of this level 3 triangle, the same argument that we used at $u$ shows that the coloring should start off blue, as indeed we have seen it does. At this point one can see by the glide reflection symmetry along $L$ that the entire line $L$ will ultimately be colored so as to fully respect the rule **R2**. For a full example where one can see the translational symmetry take over, the reader can fill in the coloring on the gray line through $y$.

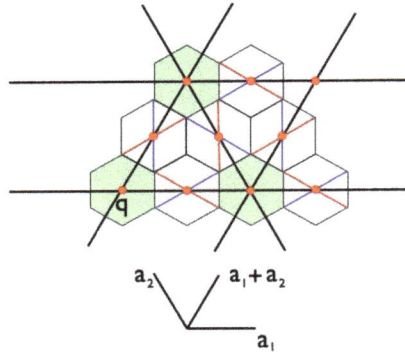

**Figure 15.** The figure shows some hexagonal tiles, each centered on a point of $Q$. The point $q$ is assumed to be in the coset $q_1 + 2Q$ and the gray hexagons are those in the picture whose centers are in this coset. The white hexagons are centered at points from all three of the remaining cosets of $Q$ relative to $2Q$. These are the midpoints of the edges of the level 1 triangles. Notice in each, the red and blue diagonals clockwise and counterclockwise of the direction of the black stripes. At the bottom we see the three vectors $a_1, a_2, a_1 + a_2$. The centers of the white hexagons are, reading left to right and bottom to top, $q_1 + a_1, q_1 + 3a_1; q_1 + a_1 + a_2, q_1 + 2a_1 + a_2, q_1 + 3a_1 + a_2; q_1 + 3a_1 + 2a_2$. The picture manifests the rule **R2** and shows that elements of the same coset carry the same orientation of diameters. Note that from the rotational symmetry of the process and the fact that the hexagons centered on $q_1 + a_1$ and $q_1 + 3a_1$ have identically aligned diagonals, we can infer that this property is retained across each of the cosets $q_1 + a_1 + 2Q, q_1 + a_2 + 2Q, q_1 + a_1 + a_2 + 2Q$.

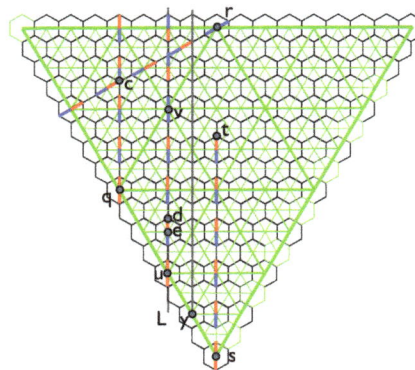

**Figure 16.** Coloring of lines. Colors are forced on $w$-lines as they pass through the midpoint of a triangle directed towards the centroid of one of its corner triangles.

One can see a similar $w$-line coloring of the $w$-line passing through $q$ and $c$. This time the point $q$ is the midpoint of an edge of a level 4 triangle and $c$ is the centroid of one of the level 3 corner triangles of this level 4 triangle. The pair $r, c$ produces another example, with this time the first color out of $r$ being blue.

Finally, we show part of a potential line coloring starting at $s$ towards $t$. We say "potential" because from the figure we do not know how the level 5 triangles lie. If $s$ is a midpoint of an edge of a level 5 triangle, then the indicated $w$-line is colored as shown. If $s$ is not a midpoint then this $w$-line is not yet colorable.

We can thus continue in this way indefinitely. The important question is, does every tile get fully colored in the process? Using condition generic-w, the answer is yes. To see this note that each element of $p$ of $P\backslash Q$ has three coloring lines through it. It will suffice to prove that the process described above will color these three coloring lines.

Now assuming the condition generic-w we know that $p$ has an orientation. This means that it is the centroid of some triangle $T$ of level $k$ in the triangulation, and it is not the centroid of any higher level triangle. The triangle $T$ then sits as one of the corner triangles in a triangle $T'$ of level $k+1$. Up to orientation, the situation is that shown in Figure 17. The colors of the two hexagons shown are then determined because the edges of $T'$ produces stripes on them. Thus the two corresponding coloring lines that pass through $p$ are indeed colored. Thus the colorings of these two coloring lines through $p$, the ones that pass through the mid-points of two sides of $T'$, are forced.

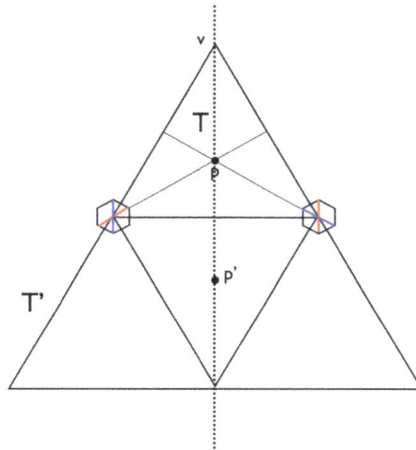

**Figure 17.** The figure shows how the centroid of the triangle $T$, which is in the top corner of the main triangle, is on two coloring lines. The third coloring line through $p$ is the dotted line through $v$. This passes through the centroid $p'$ of $T'$.

What about the third line $l$ through $p$ (shown as the dotted line in Figure 17)? We wish to see this as a $w$-line through a midpoint of an edge, just as we saw the other two lines. We look at the centroid $p'$ of $T'$, since the coloring line $l$, which is through $v$ and $p$, is the same as the line through $v$ and $p'$ and the centroid $p'$ is of higher level than $p$ and also has an orientation. We can repeat the process we just went through with $p$ with $p'$ instead, to get a new triangle $T''$ of which $p'$ is the centroid, and a triangle $T'''$ in which $T''$ sits as one of its corners ($p'$ is the centroid of $T'$ but it may be the centroid of higher level triangles as well).

If this still fails to pick up the line $l$ then it must be that $l$ still passes through a vertex of $T'''$ (as opposed to through the midpoint of one of its edges) and the line $l$ passes through the centroid $p'''$ of $T'''$. However, $p'''$ is of higher level still than that of $p'$. The upshot of this is that if we never reach a forced coloring of $l$ (so that it remains forever uncolored in our coloring process) then we have on the line $l$ centroids of triangles of unbounded levels. This violates condition generic-w. Thus in the generic situation the coloring does reach every coloring line and the coloring is complete in the limit.

This completes the argument that there is one and only one coloring for each generic triangularization.

If one is presented with a triangularization and wishes to put in the colors, then one sees that the coloring becomes known in stages, looking at the triangles (equivalently cosets) of ever increasing levels. Figure 18 shows the amount of color information that can be gleaned at level $k = 2$.

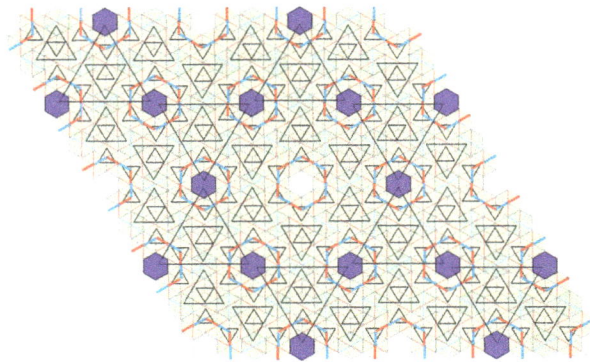

**Figure 18.** This figure shows how the coloring appears if one determines the coloring by the information in increasing coset levels. This figure corresponds to the process at $k = 2$. The triangle vertices and their corresponding hexagons are indicated at levels $0, 1, 2$ and the corresponding partial coloring is noted.

**Proposition 5.1** *Any generic tiling is uniquely colorable.*

We note that in the generic situation, the shifting and coloring are determined locally. That is, if one wishes to create the marked tiles for a finite patch of a generic tiling, one need only examine the tiling in a finite neighbourhood of that patch. That is because the shifting and coloring depend only on knowing levels of lines, and what the levels of various points on them are. Because of the generic conditions, these levels are all bounded in any finite patch and one needs only to look a finite distance out from the patch in order to pick up all the appropriate centroids and triangle edges to decide on the coloring and shifting within the patch. Of course the radii of the patches are not uniformly bounded across the entire tiling.

Here we offer a different proof of a result that appears in [2]:

**Proposition 5.2** *In any generic tiling and at any point p that is a hexagon vertex, the colors of the three concurrent diagonals of the three hexagons that surround p are not all the same (where they meet at p).*

**Proof:** The point $p$ is the centroid of come corner triangle of one of the triangles of the triangulation. Figure 17 shows how the coloring is forced along two of the medians of the corner triangle and that they force opposite colorings at $p$. See also Figure 19.

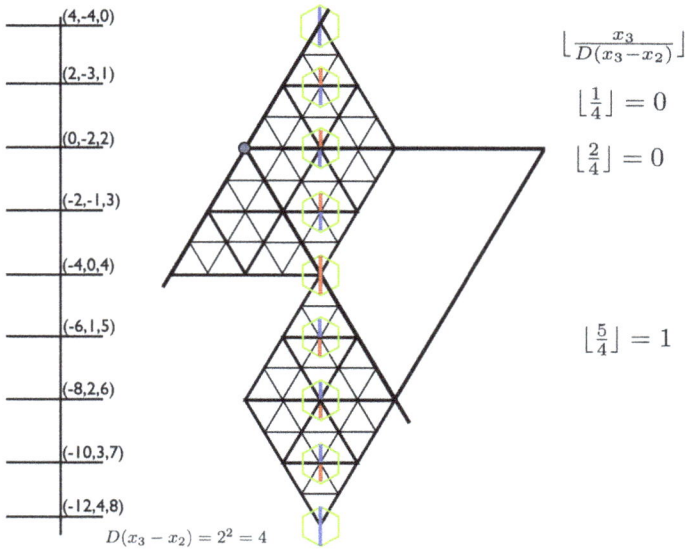

Coordinate labels on the axis (top to bottom): $(4,-4,0)$, $(2,-3,1)$, $(0,-2,2)$, $(-2,-1,3)$, $(-4,0,4)$, $(-6,1,5)$, $(-8,2,6)$, $(-10,3,7)$, $(-12,4,8)$

$$D(x_3 - x_2) = 2^2 = 4$$

$$\left\lfloor \frac{x_3}{D(x_3 - x_2)} \right\rfloor$$

$$\left\lfloor \tfrac{1}{4} \right\rfloor = 0$$

$$\left\lfloor \tfrac{2}{4} \right\rfloor = 0$$

$$\left\lfloor \tfrac{5}{4} \right\rfloor = 1$$

**Figure 19.** In the figure the small circle indicates $(0,0)$. A vertical color line is shown, which meets the $a_3$ axis in the point $p = (4, -4, 0)$. The points on this axis are all of the form $(u, -u, 0)$, $u \in \mathbb{Z}$. In the **CHT** tiling set-up, $D(u)$ indicates that the point is a vertex of a level $\log_2 D(u)$ triangle, in this case $D(4) = 4$, so we are on a level 2 triangle ($2^2 = 4$). We already saw that the color starts with a full blue diagonal at $(4, -4, 0)$. Moving down the line to the next point decreases $x_1$ by 2 and increases $x_2, x_3$ by 1. We note that $x_3 - x_2$ remains constant, and $D(x_3 - x_2) = 4$. At the second step we cross, at right-angles into another triangle of level 2, and the color proceeds without interruption. At the 4th step we are at a vertex that is the midpoint of the side of a triangle of level 3 and looking up our vertical line we can see that it is passing into a corner triangle of level 4—namely where we just came from, and we see a forced full red diagonal. In the first three steps the diameters are all red-blue (top to bottom), whereas after the red-red diagonal the next three steps are blue-red diameters, so there is a switch that affects parity. We can ignore the points with full diameters (they get sorted out in a different w-direction). We note that $\left\lfloor \frac{x_3}{D(x_3 - x_2)} \right\rfloor$ maintains the value 0 on steps 1, 2, 3 and maintains the value 1 on steps 5, 6, 7, showing that the formula notices the change of diameters correctly. If we continue $\left\lfloor \frac{x_3}{D(x_3 - x_2)} \right\rfloor = 2$ on the next three step sequence, but modulo 2 this is the same as 0.

## 5.1. Completeness

We have now shown how to work from a triangularization to a tiling satisfying the matching rules **R1,R2**. Does this procedure produce all possible tilings satisfying these rules? The answer is yes, and this is already implicit in [2]. We refer the reader to the paper for details, but the point is that in creating a tiling following the rules a triangle pattern emerges from the stripes of the hexagons. This triangularization can be viewed as the edge-shifting of a triangulation $\mathcal{T}$ conforming to our edge shifting rule. Thus we know that working with all triangulations, as we do, we are bound to be able to produce the same shrunken triangle pattern as appears in $T$.

In the generic cases, the coloring that we impose on this triangulation is precisely that forced by **R2**. When we discuss the non-generic cases in Section 6.2, we shall see that for non-generic triangulations there are actually choices for the colorings of some lines, but these choices exhaust the possibilities allowed by the rule **R2**. Thus the tiling $T$ must be among those that we construct from $\mathcal{T}$ and so we see that our procedure does create all possible tilings conforming to the matching rules.

When we determine the structure of the hull in Section 6.4 we shall also see that it is comprised of a minimal hull and two highly exceptional countable families of tilings. The former contains triangulations of all types and, as the terminology indicates, the orbit closure of any one of its tilings contains all the others in the minimal hull, so in a sense if you have one then you have them all. (The two exceptional families of tilings appear in the rule based development of the tilings, but do not appear in the inflation rule description.)

## 6. The Hull

### 6.1. Introducing the Hull

Let $X_Q$ denote the set of all Taylor–Socolar hexagonal tilings whose hexagons are centered on the points of $Q$ and whose vertices are the points of $P \backslash Q$. The group $Q$ (with the discrete topology) acts on $X_Q$ by translations. We let $X$ be the set of all translations by $\mathbb{R}^2$ of the elements of $X_Q$. We call $X_Q$ and $X$ the **hulls** of the Taylor–Socolar tiling system. We give $X_Q$ and $X$ the usual local topologies—two tilings are close if they agree on a large ball around the origin allowing small shifts. In the case of $X_Q$ one can do away with "the small shifts" part. See [15] for the topology.

In fact it is easy to see that, although we have produced it out of $X_Q$, $X$ is just the standard hull that one would expect from the set of all Taylor–Socolar tilings when they have not been anchored onto the points of $Q$. Thus $X$ is compact, and since $X_Q$ is a closed subset of it, it too is compact.

The translation actions of $Q$ on $X_Q$ and $\mathbb{R}^2$ on $X$ are continuous. We note that the hulls $X_Q$ and $X$ are invariant under six-fold rotation and under complete interchange of the two tile types. Our task is to provide some understanding of $X_Q$ and $X$. Here we shall stick primarily to $X_Q$ since the corresponding results for $X$ are easily inferred. See also [11] the structure of the hull $X$ is determined (in a totally different way). The 2-adic group $Q$ is then replaced by the two dimensional dyadic solenoid.

We let $X_Q^{gen}$ denote the set of all the generic tilings in $X_Q$.

Each element $\Lambda \in X_Q$ produces a triangularization of the plane, and the triangularizations are parameterized precisely by elements in $\overline{Q}$. In particular there is an element $\mathbf{q}(\Lambda) \in \overline{Q}$ corresponding to $\Lambda$, and we have a surjective mapping

$$\zeta : X_Q \longrightarrow \overline{Q} \tag{3}$$

$$\Lambda \mapsto \mathbf{q}(\Lambda)$$

**Proposition 6.1** *The mapping $\zeta$ is continuous (with respect to the local topology on $X_Q$ and the $Q$-adic topology on $\overline{Q}$). Furthermore $\zeta$ is $1-1$ on $X_Q^{gen}$.*

**Proof:** Any $\mathbf{q} \in \overline{Q}$ is determined by its congruence classes modulo $2\overline{Q}, 4\overline{Q}, \cdots$, which are represented equally well by the congruence classes of $Q$ modulo $2Q, 4Q, \cdots$. These congruence classes are the sets of vertices of the triangles of increasing sizes, starting with those of level 1. Now any patch of tiles containing a ball $B_R$, $R > 2$, will determine part of the triangulation with triangles of all levels $1, 2, \cdots n$ for some $n = n(R)$, and we have $n(R) \to \infty$ as $R \to \infty$. The larger the patch the more congruence classes we know, and this is the continuity statement.

In the case of a generic tiling $\Lambda$, $\mathbf{q}(\Lambda)$ already determines the entire markings of the tiles and hence determines $\Lambda$. Thus $\zeta$ is $1-1$ on $X_Q^{gen}$.

Below we shall see that with respect to the Haar measure on $\overline{Q}$ the set of singular (*i.e.*, non-generic) $\mathbf{q}$ is of measure 0. A consequence of this is [16], Theorem 6:

**Corollary 6.2** *$X_Q$ is uniquely ergodic and the elements of $X_Q^{gen}$ are regular model sets.*

We shall make the model sets rather explicit in Section 7.

*6.2. Exceptional Cases*

We now consider what happens in the case of non-generic tilings. To be non-generic a tiling must violate either generic-a or generic-w. We consider these two situations in turn.

6.2.1. Violation of Generic-a

In the case of violation of generic-a, there is an *a*-line of infinite level (that is, it does not have a level as we have defined it). Let $\overline{\mathbb{Z}_2}$ be the $Q$-adic completion of $\mathbb{Z}$.

**Proposition 6.3** *A tiling $\Lambda$, where $\xi(\Lambda) = \mathbf{q}$, has an a-line of infinite level if and only if $\mathbf{q} \in x + \overline{\mathbb{Z}_2}a$ for some $a \in \{a_1, a_2, a_1 + a_2\}$ and some $x \in Q$. Furthermore when this happens the points of $Q$ lying on the infinite-level-line are those of the set $x + \mathbb{Z}a$.*

**Proof:** All lines of the triangulation are in the directions $\pm a_1, \pm a_2, \pm(a_1 + a_2)$ and all lines of the triangulation contain edges of all levels from 0 up to the level of the line itself. Thus the points of $Q$ on any line $l$ of the triangulation are always a set of the form $x + \mathbb{Z}a$ where $a \in \{a_1, a_2, a_1 + a_2\}$ and $x \in Q \cap l$.

Suppose that we have a line $l$ of infinite level and its intersection with $Q$ is contained in $x + \mathbb{Z}a$. The line $l$ has elements $y_1, y_2, \ldots$ where $y_k$ is a vertex of a triangle of level $k$. This means that $y_1 \in q_1 + 2Q, y_2 \in q_1 + q_2 + 4Q, \ldots$. We conclude that $\{y_k\} \to \mathbf{q}$. Furthermore $y_{k+1} - y_k \in 2^k Q \cap \mathbb{Z}a = 2^k \mathbb{Z}a$. This is true for all $k \geq 0$ if we define $y_0 = x$. Writing $y_{k+1} - y_k = 2^k u_k a$ with $u_k \in \mathbb{Z}$ and $\mathbf{u} = \left(0, u_1, \ldots, \sum_{j=1}^{k} u_j 2^j, \ldots\right)$, we have

$$y_{k+1} = x + \left(\sum_{j=1}^{k} u_j 2^j\right) a$$

where $\mathbf{u} \in \overline{\mathbb{Z}_2}$. Thus $\mathbf{q} = x + \mathbf{u}a$. This proves the only if part of the proposition.

Going in the reverse direction, if $\mathbf{q} = x + \mathbf{u}a$ then this is a prescription for a line of points in $Q$ that have vertices of all levels. Then the line is of infinite level.

**Proposition 6.4** *If a tiling $\Lambda \in X_Q$ has an infinite a-line then it is in $X_Q \backslash X_Q^{gen}$ and has either precisely one infinite a-line or three infinite a-lines that are concurrent. The latter case occurs if and only if $\mathbf{q} \in Q$, where $\mathbf{q} = \xi(\Lambda)$.*

**Proof:** Let $\Lambda \in X_Q$ have an infinite a-line $l$. We already know that $\Lambda \in X_Q \backslash X_Q^{gen}$ and $\mathbf{q} = x + \mathbf{u}a$ for some $x \in Q, \mathbf{u} \in \overline{\mathbb{Z}_2}$, and some $a \in \{a_1, a_2, a_1 + a_2\}$ from Proposition 6.3. If it has a second (different) infinite line $l'$ then similarly $\mathbf{q} = y + \mathbf{v}b$ where $y \in Q$, $\mathbf{v} \in \overline{\mathbb{Z}_2}$, and $b \in \{a_1, a_2, a_1 + a_2\}$. Certainly $a \neq b$; otherwise the two lines are parallel and this leads to overlapping triangles of arbitrarily large size, which cannot happen. But we have $y - x \in Q \cap \left(\overline{\mathbb{Z}_2}a + \overline{\mathbb{Z}_2}b\right) = \mathbb{Z}a + \mathbb{Z}b$. Since $a, b$ are linearly independent over $\overline{\mathbb{Z}_2}$ and $y - x = \mathbf{u}a - \mathbf{v}b$ we see that $\mathbf{u}$ and $\mathbf{v}$ are actually in $\mathbb{Z}$. Then $\mathbf{q} \in Q$. We will indicate this by writing $q$ for $\mathbf{q}$.

In this case, since $q \equiv q_1 + \cdots + q_k \bmod 2^k Q$ we find that $q$ is a vertex of a level $k$ triangle, for all $k$. Since this is true for all $k$, $q$ is a point through which infinite level lines in all three directions $\{a_1, a_2, a_1 + a_2\}$ pass. Thus the existence of two infinite lines implies the existence of three concurrent lines.

In the other direction, if $\mathbf{q} \in Q$, then as we have just seen there will be three concurrent infinite lines passing through it.

The case of a tiling with three concurrent infinite *a*-lines in Proposition 6.4 is called a central hexagon tiling (**CHT** tiling) in [2] (see Figure 5). We also refer to them as **iCa-L** tilings. Edge shifting is not defined along these three lines, and we shall see that we have the freedom to shift them arbitrarily to produce legal tilings. The tilings in which there is one infinite *a*-line are designated as **ia-L** tilings.

### 6.2.2. Violation of Generic-w

The case of violation of generic-w is somewhat similar, though it takes more care. One aspect of this is to avoid problems of 3-torsion in $\overline{P}$, which we shall do by staying inside $\overline{Q}$ where this problem does not occur. Thus in the discussion below the quantity $3w$, where $w \in \{w_1, w_2, w_2 - w_1\}$, is of course in $Q$, but when we see it with coefficients from $\overline{\mathbb{Z}}_2$ we shall understand it as being in $\overline{Q}$ (as opposed to being in $\overline{P}$). Another problem is that the violation of generic-w is not totally disjoint from the violation of generic-a, as we shall see.

**Proposition 6.5** $\Lambda(\mathbf{q})$ *has a w-line of infinite level if and only if* $\mathbf{q} \in x + \overline{\mathbb{Z}}_2 \, 3w$ *for some* $w \in \{w_1, w_2, w_2 - w_1\}$ *and some* $x \in Q$. *Furthermore when this happens the points of $Q$ lying on the infinite-level-line are those of the set* $x + \mathbb{Z} \, 3w$.

**Proof:** All *w*-lines deriving from the triangulation are necessarily in the directions $w \in \{\pm w_1, \pm w_2, \pm(w_2 - w_1)\}$. Of course $3w \in Q$, and $mw \in Q$ iff $3 \mid m$. It really makes no difference which of the six choices $w$ is, but for convenience in presentation we shall take herewith $w = w_2$ so that $3w = a_1 + 2a_2$. This is in the vertical direction in the plane.

All *w*-lines contain centroids of levels up to the level of the line itself. Furthermore if a *w*-line contains a centroid of level $k$ then it also contains one of the vertices of the corresponding triangle and so also at least one point of $Q$ of level $k$. It follows that for any *w*-line $l$ in the direction $w$ there is an $x \in Q$ so that the set of points of $Q$ on $l$ is the set $x + \mathbb{Z} \, 3w = x + \mathbb{Z}(a_1 + 2a_2)$.

Suppose that we have a w-line $l$ of infinite level and its intersection with $Q$ is $x + \mathbb{Z}(a_1 + 2a_2)$. Then the line $l$ has elements $y_1, y_2, \ldots$ where $y_k$ is a vertex of a triangle of level $k$. This means that $y_1 \in q_1 + 2Q, y_2 \in q_1 + q_2 + 4Q, \ldots$. We conclude that $\{y_k\} \to \mathbf{q}$. Furthermore $y_{k+1} - y_k \in 2^k Q \cap \mathbb{Z}(a_1 + 2a_2)$. This is true for all $k \geq 0$ if we define $y_0 = x$. Writing $y_{k+1} - y_k = 2^k u_k (a_1 + 2a_2)$ with $u_k \in \mathbb{Z}$ and $\mathbf{u} = \left( 0, u_1, \ldots, \sum_{j=1}^{k} u_j, \ldots \right) \in \overline{\mathbb{Z}}_2$, we have

$$ y_{k+1} = x + \left( \sum_{j=0}^{k} u_j 2^j \right) (a_1 + 2a_2) \to x + \mathbf{u} \, (a_1 + 2a_2) $$

Thus $\mathbf{q} = x + \mathbf{u} \, (a_1 + 2a_2)$. This proves the only if part of the proposition.

Going in the reverse direction, if $\mathbf{q} = x + \mathbf{u} \, (a_1 + 2a_2)$ then this is a prescription for a line of points in $Q$ that have vertices of all levels. The corresponding *w*-line has centroids of unbounded levels, so the line is a *w*-line of infinite level.

**Proposition 6.6** *If a tiling* $\Lambda \in X_Q$ *has an infinite w-line then* $\Lambda \in X_Q \backslash X_Q^{gen}$ *and it has either precisely one infinite w-line or three infinite w-lines that are concurrent. The latter case occurs if and only if the point of concurrency is either a point of infinite level (discussed in Proposition 6.4) or a non-orientable point (discussed in Proposition 3.3).*

**Proof:** Let $\Lambda \in X_Q$ have an infinite *w*-line $l$. We again take this to be in the direction of $w_2$. Then $\mathbf{q} := \xi(\Lambda) = x + \mathbf{u} \, (a_1 + 2a_2)$ for some $x \in Q, \mathbf{u} \in \overline{\mathbb{Z}}_2$.

Suppose that it has a second (different) infinite line $l'$. Then similarly $\mathbf{q} = y + \mathbf{v} \, 3w'$ where $y \in Q$, $\mathbf{v} \in \overline{\mathbb{Z}}_2$, and $3w' \in \{2a_1 + a_2, a_1 + 2a_2, a_2 - a_1\}$. As above, we note that $w \neq w'$ because if $w = w'$, the two lines are parallel and each of the two lines contains vertices of arbitrarily large levels. But the parallel lines through vertices and centroids of level $k$ are spaced at a distance of $2^{k-1}$ apart. Thus no two distinct parallel $w$-lines can both carry centroids of arbitrary level.

Again, for concreteness we shall take a specific choice for $w'$, namely $w' = w_1 = 2a_1 + a_2$. Other choices lead to similar results.

There are two scenarios. Either the two lines $l, l'$ meet at a point of $Q$ or not. Suppose that they meet in a point of $Q$. Then we can choose $x = y$ and obtain

$$x + \mathbf{u}\,(a_1 + 2a_2) = \mathbf{q} = x + \mathbf{v}\,(2a_1 + a_2)$$

Since $a_1$ and $a_2$ are independent in $\overline{Q}$ over $\overline{\mathbb{Z}_2}$, we obtain $\mathbf{u} = 2\mathbf{v}$ and $2\mathbf{u} = \mathbf{v}$. The only solution to this in $\overline{\mathbb{Z}_2}$ is $\mathbf{u} = \mathbf{v} = 0$. Thus $\mathbf{q} = x \in Q$. This puts us in the situation of Proposition 6.4, the point of intersection of the two lines is actually a vertex of infinite level, and this is a **CHT** tiling.

The alternative is that $l, l'$ meet at a point $p$ of $P \backslash Q$. In this case we go back to the discussion of coloring given in Section 5. The point $p$ is a centroid and it either has infinite level, in which case it has no orientation and we go to Proposition 3.3, or it has a finite level in which two of the three $w$-lines through it have forced color and finite level, which is a contradiction. This proves the result.

As shown in Table 2, infinite level $a$-lines occur if and only if $\mathbf{q} \in x + \overline{\mathbb{Z}_2}a$ and infinite level $w$-lines occur if and only if $\mathbf{q} \in x + \overline{\mathbb{Z}_2}w$, with $a, w$ being in the basic $a$ and $w$ directions respectively. Three concurrent $a$-lines occur if and only if $\mathbf{q} \in Q$, whereupon the condition for three concurrent $w$-lines also is true. These are the **CHT** tilings.

**Table 2.** Summary of infinite level $a$-lines and $w$-lines.

| Type | Single | Three concurrent |
|---|---|---|
| infinite $a$-line | $\mathbf{q} \in Q + \overline{\mathbb{Z}_2}a$ | $\mathbf{q} \in Q$ **CHT** |
| infinite $w$-line | $\mathbf{q} \in Q + \overline{\mathbb{Z}_2}w$ | $\mathbf{q} \in Q$ **CHT** $\mathbf{q} \in -\mathbf{s}_2 - 2\mathbf{s}_1 + Q$ or $\mathbf{q} \in -\mathbf{s}_1 - 2\mathbf{s}_2 + Q$ |

Since the singular elements of $\overline{Q}$ lie on a countable union of lines, it is clear that their total measure is 0.

**Lemma 6.7** *The set of singular* $\mathbf{q} \in Q$ *has Haar measure 0.*

**Lemma 6.8** *If a triangularization* $\mathcal{T}(\mathbf{q})$ *has both an infinite level $a$-line and an infinite level $w$-line, then their point of intersection is a point of concurrence of three infinite level $w$-lines and three infinite level $a$-lines, and the tiling is a* **CHT** *tiling.*

**Proof:** By Propositions 6.3 and 6.5,

$$\mathbf{q} \in x_1 + \overline{\mathbb{Z}_2}a \quad \text{and} \quad \mathbf{q} \in x_2 + \overline{\mathbb{Z}_2}3w$$

for some $x_1, x_2 \in Q$ and $a \in \{a_1, a_2, a_1 + a_2\}$, $w \in \{w_1, w_2, w_2 - w_1\}$. Putting these together,

$$\mathbf{q} = x_1 + \mathbf{z}_1 a = x_2 + \mathbf{z}_2 3w$$

for some $\mathbf{z}_1, \mathbf{z}_2 \in \overline{\mathbb{Z}_2}$. However $a$ and $3w$ are independent elements of $Q$ (over $\mathbb{Z}$), and hence are also independent over $\overline{\mathbb{Z}_2}$. Since $x_2 - x_1 \in Q$, this forces $\mathbf{z}_1, \mathbf{z}_2 \in \mathbb{Z}$. Thus $\mathbf{q} \in Q$, which is the condition for simultaneous concurrency of three $a$-lines and three $w$-lines (Proposition 6.4).

## 6.3. Coloring for the iCw-L Tilings

According to Proposition 3.3 we have a point of no orientation precisely when $\mathbf{q} \in -\mathbf{s}_2 - 2\mathbf{s}_1 + Q$ or $\mathbf{q} \in -\mathbf{s}_1 - 2\mathbf{s}_2 + Q$. In these cases, by Proposition 6.6, we have three concurrent $w$-lines and their intersection is a point of no orientation. This point of intersection is $x = \mathbf{q} + w_1 + \mathbf{s}_2 + 2\mathbf{s}_1$ or $x = \mathbf{q} + w_2 + \mathbf{s}_1 + 2\mathbf{s}_2$. The former can be anywhere in $w_1 + Q$ and the latter anywhere in $w_2 + Q$. The triangulation can be described as a set of nested triangles of levels $0, 1, 2, 3, \ldots$ (and all the lesser level triangles that occur within them) all of which have the centroid $x$. The level $k = 0$ triangle is an up triangle in the $w_1$ case and a down triangle in the $w_2$ case. The infinite $l$-lines are in the directions $w_1, w_2, w_2 - w_1$ through $x$ and these three lines have no forced colorings.

We call these tilings the **iCw-L** tilings (infinite concurrent $w$-line tilings). We also refer to the underlying triangulations with the same terminology. See Figure 6.

The symmetry belongs to the triangulation, not necessarily to the tilings themselves. The colorings of the three exceptional lines of an **iCw-L** tiling can be made in an arbitrary way without violating the tiling conditions **R1,R2** [2]. Of the 8 possible colorings the two truly symmetric ones (the ones that give an overall 3-fold rotational symmetry—including color symmetry—to the actual tiling) are exceptional in the sense that no other tilings in the Taylor–Socolar system have a point $p \in P \backslash Q$ (i.e., a tile vertex) with the property that the three hexagon diagonals emanating from it are all of the same color (see Proposition 5.2). These exceptional symmetric **iCw-L** tilings are called **SiCw-L** tilings. In [2] these tilings are described as having a "defect" at this point, and indeed they are not LI to any other tilings except other **SiCw-L** tilings.

Thus there are 2 exceptional colorings for any **iCw-L** triangulation. In the other 6 colorings there are at each hexagon vertex two diameters of the same color and one of the opposite color, and we shall soon prove that they all occur in $X_Q$.

Tilings for which there is just one infinite $w$-line in the triangulation are called **iw-L** tilings.

## 6.4. The Structure of the Hull

In this subsection we describe the hull $X_Q$ in more detail. We note that the only symmetries of $X_Q$ which we discuss are translational symmetries (not rotational). These translational symmetries are the elements of $Q$. Of course none of the elements of $X_Q$ has any non-trivial translational symmetry; it is only the hull itself that has them. When we discuss LI classes below we mean local indistinguishability classes under translational symmetry.

**Theorem 6.9** $X_Q$ *consists of three LI classes,* $X_Q^b$, $X_Q^r$, *and* $X_Q^t$. *Of these* $X_Q^b$ *is the countable set of* **SiCw-L** *tilings with three blue-red (blue first) diameters emanating from some hexagon vertex* $q$, *which form a single* $Q$-*orbit in* $X_Q$, *and* $X_Q^r$ *is the companion orbit with red-blue diameters. Both of these orbits are dense in* $X_Q$.

$X_Q^t$ *is the orbit closure of* $X_Q^{gen}$ *and contains all other tilings, including all the* **iCw-L** *tilings that are not color symmetric. Restricted to the minimal hull* $X_Q^t$, *the mapping* $\xi$ *defined in* (3) *is:*

(i)   *1 : 1 on* $X_Q^{gen}$;
(ii)  *6 : 1 at* **iCw-L** *points except* **SiCw-L** *points;*
(iii) *12 : 1 at* **CHT** *points;*
(iv)  *2 : 1 at all other non-generic points.*

**Remark 6.10** *The images of* $\xi$ *of the set of singular points (non-generic points) is dense in* $\overline{Q}$. *For instance, the triangulations with three concurrent a-lines are parameterized by* $Q$, *which is a dense subset of* $\overline{Q}$, *and these tilings produce the* **CHT** *tilings (or* **iCa-L** *tilings) described above. Both* $X_Q^{gen}$ *and* $X_Q^t$ *are of full measure in* $X_Q$.

**Proof:** First, we consider generic tilings. Let $\Lambda$, where $\xi(\Lambda) = \mathbf{q}$, be any generic tiling and let $B_R$ be the ball of radius $R$ centered on 0. Let $\mathcal{T}(\Lambda)$ be the triangulation determined by $\Lambda$ (with edges not displaced) and let $\mathcal{T}_R(\Lambda)$ be the part of the triangulation that is determined by $B_R$.

Because we are in a generic situation, to know how to shift an edge of level $k$ we need only that edge to appear as an inner edge of a triangle of level $k + 1$. To determine the coloring of a $w$-line we need to know its level (which is finite). So to know all this information for $\mathcal{T}_R$ we need only choose $r$ large enough so that $B_r$ contains all the appropriate triangles.

Now if generic $\Lambda'$, where $\xi(\Lambda') = \mathbf{q}'$, produces the same pattern of triangles in $B_r$, then it is indistinguishable from $\Lambda$ in $B_R$. In particular if $\mathbf{q}'$ satisfies $\mathbf{q}' - \mathbf{q} \in 2^k Q$ for large enough $k$ then $\Lambda$ and $\Lambda'$ must agree (as tilings) on $B_R$. This proves that convergence of $\mathbf{q}'$ to $\mathbf{q}$ produces corresponding convergence in $X_Q$.

With this it is easy to see that any two generic elements of $X$ are LI. Let $\xi(\Lambda) = \mathbf{q}$ and $\xi(\Lambda') = \mathbf{q}'$ be generic. Let $\mathbf{q}$ correspond to $q_1, q_2, \ldots$ and $\mathbf{q}'$ correspond to $q_1', q_2', \ldots$. Then we can construct the tiling sequence $q_1' - q_1 + \Lambda, q_1' + q_2' - (q_1 + q_2) + \Lambda, \ldots$ and it converges to $\Lambda'$.

This same argument can be used to show that the orbit closure of any tiling contains all of $X_Q^{gen}$. Let $\Lambda$ be any tiling with $\mathbf{q} = \xi(\Lambda)$ and $\Lambda'$ be a generic tiling with $\mathbf{q}' = \xi(\Lambda')$. Then one simply forms a sequence of translates of $\Lambda$ that change $\mathbf{q}$ into $\mathbf{q}'$. The convergence of the triangulation on increasing sized patches forces convergence of the color and we see $\Lambda'$ in the orbit closure of $\Lambda$.

Second, we consider the **iCw-L** cases, where $x := \mathbf{q} + w_2 + \mathbf{s_1} + 2\mathbf{s_2} \in w_2 + Q$ or $x := \mathbf{q} + w_1 + \mathbf{s_2} + 2\mathbf{s_1} \in w_1 + Q$. In these cases $x$ is a non-orientable point and there exists a nested sequence of triangles of all levels centered on $x$. This sequence begins either with an up triangle of level 0 or a down triangle of level 0. In either case everything about the triangulation is known and the entire tiling is determined except for the coloring of the three $w$-lines through $x$. In fact all of the 8 potential colorings of these three lines are realizable as tilings, as we shall soon see.

Of these **iCw-L** triangulations we have the **SiCw-L** tilings in which the colors of the diagonals of the three hexagons of which $x$ is a vertex start off the same—all red or all blue. This arrangement at a hexagon vertex never arises in a generic tilings, and it is for this reason that these tilings produce different LI classes than the one that the generic tilings lie in: one "red" LI class and one "blue" LI class. As pointed out in [2] these **SiCw-L** tilings have the amazing property that they are completely determined once the three hexagons around $x$ have been decided (It is also pointed out in [3] that the **SiCw-L** tilings do not arise in the substitution tiling process originally put forward in Taylor's paper. However, they do arise as legal tilings from the matching rule perspective, though they could be trivially removed by adding in a third rule to forbid them. A similar situation has been shown to occur with the Robinson tilings for which there is a matching rule and also a substitution scheme that result in a hull and its minimal component [5]. As pointed out in [2], this is different from tilings like the Penrose rhombic tiling where the matching rules determine the minimal hull). The form of the points $x$ with no orientation shows that there are just two $Q$ orbits of them, one for each of the two non-trivial cosets of $Q$ in $P$.

What about the other 6 color arrangements around such a point $x$? Here we can argue that they all exist in the following way. Since in any triangulation there are tile centroids of any desired level $k$, we can start with any generic $\Lambda$ and form a sequence of translates of it that have centroids of ever increasing level at 0. The sequence has at least one limit point and this is an **iCw-L** tiling. Since each element of the sequence has a unique coloring and coloring in generic tiles is locally determined by local conditions, there must be a subsequence of these tilings that converges to one of some particular coloring. This must produce a coloring of diameters with two diameters of one color and one of the other color since we are using only generic tilings in the sequence. Now the rotational three-fold symmetry and the color symmetry of $X_Q$ shows that all 6 possibilities for the coloring will exist. This also shows that all these tilings are in the orbit closure of $X_Q^{gen}$.

Third, the **CHT/iCa-L** triangulations have the form $\mathcal{T}(q)$ where $q \in Q$. They have three concurrent a-lines and three concurrent w-lines at $q$ and leave the central tile completely undetermined. This tile can be placed in any way we wish, and this fixes the entire tiling. There are a total of 12 ways to place this missing tile (6 for each parity), whence $\xi$ is $12 : 1$ over $q$.

Finally, apart from the **iCw-L** and **CHT/iCa-L** tilings, the remaining singular values of **q** correspond to the **ia-L** and **iw-L** triangulations where there is either a single infinite level $a$-line or a single infinite level $w$-line, Section 6.3. Fortunately these two things cannot happen at the same time, see Lemma 6.8. That means that there is only one line open to question and there is only one line on which either the shift or color is not determined. In the **ia-L** case there is an $a$-line for which edge shifting is un-defined and we wish to show that all the two potentially available shifts lead to valid tilings. Likewise in the **iw-L** case there is a $w$-line to which no color can be assigned, and we wish to prove that both coloring options are viable.

Suppose one starts with a **CHT** tiling $\Lambda$ centered at 0. If one forms a sequence $\{q_1 + \cdots + q_k + \Lambda\}$ of **CHT** tilings and if $\{q_1 + \cdots + q_k\}$ converges to a point of on the line $\overline{\mathbb{Z}_2}a_1$ that is not in $Q$ then the point of **CHT** concurrence has vanished and one is left only with the $x$-axis as an single infinite level $a$-line, and it will have the shifting induced by the original shifting along the $x$-axis in $\Lambda$ (which can be either of the two potential possibilities). Of course one can do this in any of the $a$ directions. A similar type of procedure works to produce all of the **iw-L** tilings. This concludes the proof of the theorem.

## 7. Tilings as Model Sets

In this section we consider Taylor–Socolar tilings, and in particular the parity sets of such tilings, from the point of view of model sets. There are a number of advantages to establishing that point sets are model sets since there is a very extensive theory for them, including fundamental theorems regarding their intricate relationship to their autocorrelation measures and their pure point diffractiveness [16–18]. In fact there are various ways in which one can establish that the vertices or the tile centres of a Taylor–Socolar tiling always form a model set. We have already pointed out different ways in [12] and Corollary 6.2. There are also other ways pointed out in [11]. The first is through the almost everywhere one-to-one mapping from the minimal dynamical hull of the Taylor–Socolar tilings to the minimal dynamical hull of half-hex tilings. It is known that the half-hex tilings give model sets [6], and it follows that the Taylor–Socolar tilings do too. The second is through checking a modular coincidence on a Taylor–Socolar tiling which is a fixed point of a substitution. One can observe that the Taylor–Socolar tiling contains a periodic lattice of hexagons of type $C$ and $\overline{C}$ (given in [2]) in different orientations. It can be checked that for some order $n$ of supertiling, all supertiles of $C$ and $\overline{C}$-type hexagons contain a same type of hexagon at exactly the same relative position of these supertiles. This indicates the modular coincidence introduced in [19,20]. In our present paper the set-up that we have created makes it easy to see the model set construction rather explicitly, and this is the purpose of this section.

The basic pre-requisite for the cut and project formalism is a cut and project scheme. Most often, especially in mathematical physics, the cut and project schemes have real spaces (*i.e.*, spaces of the form $\mathbb{R}^n$) as embedding spaces and internal spaces. But the theory of model sets is really part of the theory of locally compact Abelian groups [21]. In the case of limit-periodic sets, some sort of "adic" space is the natural ingredient for the internal space. In our case the internal space is $\overline{P}$, see [19].

### 7.1. The Cut and Project Scheme

Form the direct product of $\mathbb{R}^2$ and $\overline{P}$. The subset $\mathcal{P} = \{(x, i(x)) \in \mathbb{R}^2 \times \overline{P} : x \in P\}$ is a lattice in $\mathbb{R}^2 \times \overline{P}$ (that is, $\mathcal{P}$ is discrete and $((\mathbb{R}^2 \times \overline{P})/\mathcal{P}$ is compact) with the properties that the projection mappings

$$\mathbb{R}^2 \xleftarrow{\pi_1} \mathbb{R}^2 \times \overline{P} \xrightarrow{\pi_2} \overline{P}$$
$$\cup$$
$$P \xleftrightarrow{\cong} \mathcal{P} \tag{4}$$

satisfy $\pi_1|_{\mathcal{P}}$ is injective and $\pi_2(\mathcal{P})$ is dense in $\overline{P}$. The setup of (4) is called a **cut and project scheme**. Then $P = \pi_1(\mathcal{P}) \subset \mathbb{R}^2$ and the "star mapping" $(\cdot)^\star : P \longrightarrow \overline{P}$ defined by $\pi_2 \circ (\pi_1|_{\mathcal{P}})^{-1}$ is none other than the embedding $i$ defined above.

Let $W \subset \overline{P}$ which satisfies $W^\circ \subset W \subset \overline{W^\circ} = \overline{W}$ with $\overline{W}$ compact, we define

$$\curlywedge(W) := \{x \in P : x^\star \in W\}$$

This is the **model set** defined by the **window** $W$. Most often we wish to have the additional condition that the boundary $\partial W := \overline{W} \backslash W^\circ$ of $W$ has Haar measure $0$ in $\overline{P}$. In this case we call $\curlywedge(W)$ a **regular model set**.

As an illustration of how the cut and project scheme is used to define the model sets, we give here a model set interpretation for the sets $W^\uparrow(\{\mathbf{q}\})$ and $W^\downarrow(\{\mathbf{q}\})$ of Proposition 3.6:

$$\curlywedge^\uparrow(\{\mathbf{q}\}) := \{x \in P : x^\star \in W^\uparrow(\{\mathbf{q}\})\} \tag{5}$$

$$\curlywedge^\downarrow(\{\mathbf{q}\}) := \{x \in P : x^\star \in W^\downarrow(\{\mathbf{q}\})\}$$

The windows $W^\uparrow(\mathbf{q})$ and $W^\downarrow(\mathbf{q})$ are compact and the closures of their interiors, so these two sets are pure point diffractive model sets, and clearly they are basically the points of $P\backslash Q$ which have orientation up and down respectively. In the case of values of $\mathbf{q}$ treated in Proposition 3.3 there will be one point without orientation. It is on the common boundary of $W^\uparrow(\mathbf{q})$ and $W^\downarrow(\mathbf{q})$.

However, our intention here is not to interpret features of the triangulation in terms of model sets (which is more or less obvious) but to understand parity, which is a more subtle feature depending on edge-shifting and color, in terms of model sets. In this paper we will need only to deal with model sets lying in $Q$, and for this it is useful to restrict the cut and project scheme above to the lattice $\mathcal{Q} = \{(x, i(x)) \in \mathbb{R}^2 \times \overline{Q} : x \in Q\}$ in $\mathbb{R}^2 \times \overline{Q}$:

$$
\begin{array}{ccccc}
\mathbb{R}^2 & \xleftarrow{\pi_1} & \mathbb{R}^2 \times \overline{Q} & \xrightarrow{\pi_2} & \overline{Q} \\
 & & \cup & & \\
Q & \xleftrightarrow{\;\simeq\;} & \mathcal{Q} & &
\end{array}
\tag{6}
$$

Of course we shall not be looking for just one window and one model set, but rather two windows, one for each of the two choices of parity.

### 7.2. Parity in Terms of Model Sets: The Generic Case

Each tiling in $X_Q$ is composed of hexagons centered at points of $Q$ that are of one of the two types shown in Figure 3. We call them white or gray according to the coloring shown in the figure. At the beginning we shall work only with the generic cases, since for them the tiling is completely represented by its value in $\overline{Q}$.

Let $\Lambda$ be a generic tiling for which $\xi(\Lambda) = \mathbf{q} \in \overline{Q}$. We define $Q(\mathbf{q})^{wh}$ (resp. $Q(\mathbf{q})^{gr}$) to be the set of points of $Q$ whose corresponding tiles in $\Lambda$ are white (resp. gray), so we have a partition

$$Q = Q(\mathbf{q})^{wh} \cup Q(\mathbf{q})^{gr}$$

We shall show that each of $Q(\mathbf{q})^{wh}$ and $Q(\mathbf{q})^{gr}$ is a union of a countable number of $2^k Q$-cosets (for various $k$) of $Q$. If this is so then since the closure of a coset $x + 2^k Q$ is $x + 2^k \overline{Q}$ which is clopen in $\overline{Q}$, we see that $\overline{Q(\mathbf{q})^{wh}}$ contains the open set $U^{wh}$ consisting of the union of all the clopen sets coming from the closures of the cosets of $Q(\mathbf{q})^{wh}$, and $\overline{Q(\mathbf{q})^{wh}}$ is the closure of $U^{wh}$. Similarly $\overline{Q(\mathbf{q})^{gr}}$ contains an open set $U^{gr}$. We note that $U^{wh}$ and $U^{gr}$ are disjoint since they are the unions of disjoint cosets, and their union contains all of $Q$.

We also point out that $U^{wh}$ is the interior of $\overline{Q(\mathbf{q})}^{wh}$ since any open set in $\overline{Q}$ is a union of clopen sets of the form $x + 2^k\overline{Q}$ with $x \in Q$ (they are a basis for the topology of $\overline{Q}$) and each of these is either in $U^{wh}$ or $U^{gr}$. But no point of $U^{gr}$ is a limit point of $U^{wh}$ and so $U^{gr} \cap \overline{Q(\mathbf{q})}^{wh} = \varnothing$. Similarly $U^{gr}$ is the interior of $\overline{Q(\mathbf{q})}^{gr}$.

Evidently $\overline{Q(\mathbf{q})}^{wh} \cap \overline{Q(\mathbf{q})}^{gr}$ is a closed set with no interior, since $U^{wh}$ and $U^{gr}$ are disjoint. Thus $\overline{Q(\mathbf{q})}^{wh} \cap \overline{Q(\mathbf{q})}^{gr}$ lies in the boundaries of each set and contains no points of $Q$. Each of the sets $\overline{Q(\mathbf{q})}^{wh}$ and $\overline{Q(\mathbf{q})}^{gr}$ is compact and each is the closure of its interior. The boundaries of the two sets are both of measure 0 since $U^{wh}$ and $U^{gr}$ can account for the full measure of $\overline{Q}$. Finally, $\overline{Q} = \lambda\left(\overline{Q(\mathbf{q})}^{wh}\right) \cup \lambda\left(\overline{Q(\mathbf{q})}^{gr}\right)$.

**Theorem 7.1** *Let $\Lambda$ be a generic tiling for which $\xi(\Lambda) = \mathbf{q} \in \overline{Q}$, where $\mathbf{q} = (q_1, q_1 + q_2, \ldots, q_1 + \cdots + q_k, \ldots) \in \overline{Q}$. We have the model-set decomposition for white and gray points of the hexagon centers of $\Lambda$:*

$$Q(\mathbf{q})^{wh} = \lambda\left(\overline{Q(\mathbf{q})}^{wh}\right) \qquad (7)$$

$$Q(\mathbf{q})^{gr} = \lambda\left(\overline{Q(\mathbf{q})}^{gr}\right)$$

$$\overline{Q} = \lambda\left(\overline{Q(\mathbf{q})}^{wh}\right) \cup \lambda\left(\overline{Q(\mathbf{q})}^{gr}\right)$$

*where $\overline{Q(\mathbf{q})}^{wh}$ is the closure of the union of the clopen sets in $\overline{Q}$. Thus these sets are regular model sets.*

**Proof:** We have to show that $Q(\mathbf{q})^{wh}$ and $Q(\mathbf{q})^{gr}$ are each unions of a countable number of $2^k Q$-cosets (for various $k$) of $Q$. There are two components that enter into the white/gray coloring: the diameter coloring of the tiles and the edge shifting. The generic condition guarantees that both coloring and shifting are completely unambiguous. Our argument deals with coloring first, and shifting second. Finally both parts are brought together.

Let $\Omega := \{\pm w_1, \pm w_2, \pm(w_2 - w_1)\}$ and $\Omega^+ := \{w_1, w_2, w_2 - w_1\}$. For $w \in \Omega$ and $k = 1, 2, \ldots$, define

$$U_k := \bigcup_{w \in \Omega} q_1 + \cdots q_k + 2^k Q + \mathbb{Z}3w$$

Recall that for $w \in \Omega$, $3w \in Q\backslash 2Q$. The points of $q_1 + \cdots + q_k + 2^k Q$ are the vertices of the level $k$ triangles and the sets $U_k$ are composed of the points of $Q$ on the $w$-lines that pass through such vertices. We have $Q = U_1 \supset U_2 \supset U_3 \supset \cdots$.

A point of $Q$ may be a vertex of many levels of triangles, but we wish to look at the highest level vertex that lies on a given $w$-line. Thus we define $V_k := U_k \backslash U_{k+1}, k = 1, 2, \ldots$. The sets $V_k$ are mutually disjoint. See Figure 20.

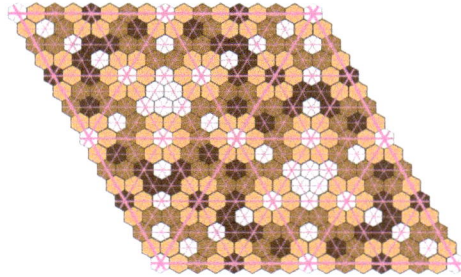

**Figure 20.** The tiles associated with points of $V_1, V_2, V_3$ are indicated by increasingly dark shades.

The sets $U_k$ are made up of various elements $q_1 + \cdots q_k + 2^k u + 3nw$ where $u \in Q$ and $n \in \mathbb{Z}$. However we can restrict $n$ in the range $0 \leq n \leq 2^{k-1}$ since $3\,2^k w \in 2^k Q$ and

$$q_1 + \cdots q_k + 2^k u + 3\,2^{k-1} w = q_1 + \cdots q_k + 2^k u + 3\,2^k w + 3\,2^{k-1}(-w)$$

which changes $w$ to $-w$ at the expense of a translation in $2^k Q$. To make things unique we shall assume that $w \in \Omega^+$ in the extreme cases when $n = 0$ or $n = 2^{k-1}$.

We claim that under the condition generic-w we have $Q = \bigcup_{k=1}^{\infty} V_k$. The only way that $x \in Q$ can fail to be in some $V_k$ is that $x \in U_k$ for all $k$. Then $x$ is on w-lines through vertices of arbitrarily high level triangles. At least one $w \in \Omega$ occurs infinitely often in this. Fix such a $w$. The vertex of a level $k$ triangle is always the vertex of 6 such triangles around that vertex. So whenever the w-line passes through a vertex of a level $k$ triangle it also passes into the interior of one of the level $k$ triangles of which this is a vertex and then through the centroid of this level $k$ triangle. Thus the w-line $x + \mathbb{Z}w$ has centroids of arbitrary level on it, violating generic-w.

Let $x = q_1 + \cdots q_k + 2^k u + 3nw = x_0 + 3nw \in V_k$ for some $k$, where $n$ satisfies our conventions noted above on the values it may take. Then by the definition of $V_k$, $x_0$ is not the vertex of any edge of a triangle $T$ of side length $2^{k+1}$ and so $x_0$ is the mid-point of an edge $f$ of such a $T$. In particular $x_0 \notin V_k$. Figures 21 and 22 indicate, up to orientation, what all this looks like. The edge $f$ is on the highest level line through $x_0$ and so determines the stripe of the hexagon at $x_0$ and, more importantly, the coloring of the w-line that we are studying. The coloring at $x_0$ in the direction $w$ starts red in the case of Figure 22a and blue in the case of Figure 22b. The color then alternates along the line in the manner illustrated in Figure 16.

The color pattern determined here repeats modulo $2^{k+1}Q$ (not $2^k Q$), so $V_k$ splits into subsets, each of which is a union of cosets of $2^{k+1}Q$ in $Q$,

$$V_k = V_k^r \cup V_k^b = \bigcup_{w \in \Omega} \bigcup_{n=0}^{2^k - 1} V_k^r(w, n) \cup V_k^b(w, n)$$

corresponding to the red-blue configurations and corresponding also to which $w \in \Omega$ is involved. Here $V_k^r(w, n)$ is the set of points $q_1 + \cdots + q_k + 2^k u + 3nw \in V_k$ which *start* in the direction $w$ with the color red, where $0 \leq n < 2^{k-1}$ with the boundary conditions on $n$ established as above. The situation with $V_k^b(w, n)$ is the same except red is replaced by blue. Notice that each $V_k^r(w, n)$ (or $V_k^b(w, n)$) is a union of $2^{k+1}Q$-cosets for various values of $k$.

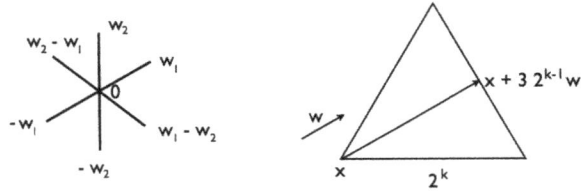

**Figure 21.** $\Omega$ and a line through a vertex $x$ in the direction $w$ meeting the opposite edge at $3\,2^{k-1}w$.

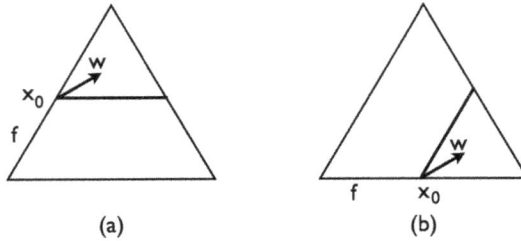

(a)                                    (b)

**Figure 22.** Showing $x_0$ as a midpoint of an edge $f$ of a triangle of level $2^{k+1}$ and the direction of $w$ from it.

Now we need to look at the other aspect to determining the white and gray tiles, namely edge shifting. For this we assume the condition generic-a. Every $x \in Q$ lies on an edge of level $k$ for some $k$. Recall that this means that $x$ is on the edge of a triangle of level $k$ but not one of level $k + 1$. The condition generic-a says that such an edge must exist for $x$. Such an edge must then appear as the inner edge $e$ of a corner triangle of a triangle $T$ of level $k + 1$. The edge $e$ then shifts towards the centroid of $T$, say in the direction $w \in \Omega$, carrying corresponding diameters in along with it.

Let

$$L_l(w) := \{x \in Q \ : \ x \text{ lies on an edge of level } l \text{ which shifts in the direction } w\}\,.$$

Then $Q = \bigcup_{l=1}^{\infty} \bigcup_{w \in \Omega} L_l(w)$. As one sees from Figure 23, $L_l(w)$ is a union of $2^{l+1}Q$-cosets (but not $2^l Q$-cosets).

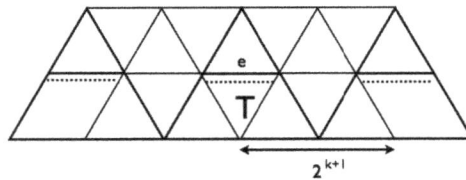

**Figure 23.** Showing a level $k$ edge inside a triangle $T$ of level $k + 1$ and its corresponding shift (dotted line). Note how edge shifting for edges of level $k$ repeats modulo $2^{k+1}$.

We now put these two types of information together. We display the results in the form of two tables: any $x \in Q$ satisfies

$$x \in V_k^c(w, n) \cap L_l(w') \tag{8}$$

for some $c \in \{r, b\}$, $k, l \geq 1$, $w \in \Omega$, $w' \in \Omega$.

Notice that $V_k^c(w,n) \cap L_l(w')$ is a union of $2^m Q$-cosets for $m = \max\{k+1, l+1\}$. So we finally obtain that each of $Q^{wh}$ and $Q^{gr}$ is the union of such cosets. This is what we wanted to show and concludes the proof of the theorem.

*7.3. Parity in Terms of Model Sets: The Non-Generic Case*

For non-generic sets there are two situations to consider. First of all, let us consider tilings of the minimal hull. Any such tiling $\Lambda'$ can be viewed as the limit of translates of a generic tiling, $\Lambda$. Let $\xi(\Lambda) = \mathbf{q}$ and let $W^{wh}$ and $W^{gr}$ denote the two closed windows that define the parity point sets of the tile centers of $\Lambda$. Translation $t + \Lambda$ of $\Lambda$ by $t \in Q$, amounts to translation by $t^*$ of $W^{wh}$ and $W^{gr}$. This is in fact just translation by $t$ but with $t$ seen as an element of $\overline{Q}$. Translation does not affect the type of the tiling (**iw-L**, **iCa-L**, *etc.*).

Convergence of a sequence of translates $t_1 + \Lambda, t_1 + t_2 + \Lambda, \ldots$ to $\Lambda'$ in the hull topology implies $Q$-adic convergence of $t_1 + t_2 + \cdots$, say to $\mathbf{t} \in \overline{Q}$. The translated sets then also converge to $\mathbf{t} + W^c$, $c = \{wh, gr\}$. However, if $\mathbf{t} \notin Q$ then we will not necessarily have $\Lambda' = \lambda(\mathbf{t} + W^c)$.

Here is what happens. If $u \in \Lambda'$ then for large enough $n$, $u \in t_1 + t_2 + \cdots + t_n + \Lambda$ and $u^* \in t_1 + \cdots + t_n + W^c$. Thus $u^* \in \mathbf{t} + W^c$ and we have that $\Lambda' \subset \lambda(\mathbf{t} + W^c)$. On the other hand we have $\Lambda' \supset \lambda(\mathbf{t} + (W^c)^\circ)$. For suppose that $x \in \lambda(\mathbf{t} + (W^c)^\circ))$. Then $x^* \in (\mathbf{t} + (W^c)^\circ) \cap Q^*$ and using the convergence $t_1 + t_2 + \cdots \to \mathbf{t}$ we see that for large $n$, $x^* \in (t_1 + t_2 + \cdots + t_n + (W^c)^\circ)) \cap Q^*$. Thus for large $n$, $x \in \lambda(t_1 + t_2 + \cdots + t_n + W^c) = t_1 + t_2 + \cdots + t_n + \Lambda$, and so $x \in \Lambda'$. We conclude that $\Lambda' = \lambda(Z)$ for some window $Z$ satisfying $\mathbf{t} + (W^c)^\circ \subset Z \subset \mathbf{t} + W^c$. This shows that $\Lambda'$ is a model set since $Z$ lies between its interior and the (compact) closure of its interior. Also $\partial Z \subset \partial(\mathbf{t} + W^c)$ has Haar measure 0 as it was explained in the beginning of Section 7.2.

The remaining cases are the **SiCw-L** tilings. Let $\Lambda$ be such a tiling, which we may assume to be associated with $\xi(\Lambda) = 0 \in Q$. Comparing the **SiCw-L** tiling $\Lambda$ with an **iCw-L** tiling $\Gamma$ for which $\xi(\Gamma) = 0$, we notice that the only difference between $\Lambda$ and $\Gamma$ is on the lines through 0 in the $w$-directions where $w \in \Omega$. The total index is introduced in [19]. Notice that it is enough to compute that the total index of the set of all points off these $w$-lines is 1 (see cite[LM1]). Because the set of points off the lines of $w$-directions is the disjoint union of cosets $V_k$ (we have seen this earlier), we only need to show that the total index of $\cup_1^\infty V_k$ is 1, *i.e.*,

$$\sum_{k=1}^{\infty} c(V_k) = 1$$

Following the construction of $V_k$, $k \geq 1$, already discussed above, we compute the coset index of $V_k$. Within each $V_k$ we need to divide the point set $V_k$ into two sets. One is the point set whose points are completely within the $(k+1)$-th level triangles and the other is the point set whose points are lying on the lines of the $(k+1)$-th level triangles. We note that

$$c(V_1) = \frac{6}{4 \cdot 2^2}$$

$$c(V_2) = \frac{(2^1 - 1) \cdot 2 \cdot 3 \cdot 2}{4 \cdot 2^2 \cdot 2^2} + \frac{6}{4 \cdot 2^2 \cdot 2^2}$$

$$c(V_3) = \frac{(2^2 - 1) \cdot 2 \cdot 3 \cdot 2}{4 \cdot 2^2 \cdot 2^2 \cdot 2^2} + \frac{6}{4 \cdot 2^2 \cdot 2^2 \cdot 2^2}$$

$$\vdots$$

$$c(V_k) = \frac{(2^{k-1} - 1) \cdot 2 \cdot 3 \cdot 2}{4 \cdot (2^2)^k} + \frac{6}{4 \cdot (2^2)^k}$$

So

$$\sum_{k=1}^{\infty} c(V_k) = \frac{6}{4}\left(\frac{1}{2^2} + \frac{1}{2^2 \cdot 2^2} + \cdots + \frac{1}{(2^2)^k} \cdots\right) + 3\left(\frac{2}{(2^2)^2} + \frac{2^2}{(2^2)^3} + \cdots + \frac{2^{k-1}}{(2^2)^k} + \cdots\right)$$
$$-3\left(\frac{1}{(2^2)^2} + \frac{1}{(2^2)^3} + \cdots + \frac{1}{(2^2)^k} + \cdots\right)$$
$$= \frac{6}{4}\left(\frac{1/4}{1-1/4}\right) + 3\left(\frac{1/8}{1-1/2}\right) - 3\left(\frac{1/16}{1-1/4}\right)$$
$$= 1$$

## 8. A Formula for the Parity

In this section we develop formulae for tilings that determine the parity of each tile of a tiling from the coordinates of the center of that tile. We begin with the formula for parity derived in [2] for the **CHT** tilings centered at $(0,0)$. These correspond to the triangulation for $\mathbf{q} = 0$. The parity formula for a tile is based on the coordinates of the center of the hexagonal tile. Due to the non-uniqueness of the **CHT** tilings along the 6-rays at angles $2\pi k/6$ emanating from the origin, the basic formula is valid only off these rays. Later we show how to adapt this formula to arbitrary $\mathbf{q}$.

The parity of a tile depends on the relationship of its main stripe to the diameter at right-angles to this stripe. In terms of the triangulation, the parity of a tile depends on two things: The way the triangle edge on which the stripe is located is shifted and the order of the two colors of the color line as it passes through the tile: red-blue or blue-red. Changing the shift or the color order changes the parity, changing them both retains the parity. Thus the parity can be expressed as the modulo 2 sum of two binary, *i.e.*, $\{0,1\}$, variables representing the shift and the color order. Which parity belongs to which type of tile is an arbitrary decision. In our case we shall make it so that the white tile has parity 1 and the gray tile has parity 0.

We introduce here a special coordinate system for the plane (which we really only use for elements of Q). We take three axes through $(0,0)$, in the directions of $a_1, a_2, -(a_1 + a_2)$ with these three vectors as unit vectors along each. For convenience we define $a_3 := -(a_1 + a_2)$. Each $x \in Q$ is given the coordinates $(x_1, x_2, x_3)$ where $x_1$ is the $a_2$ coordinate where the line parallel to $a_1$ through $x$ meets the $a_2$-axis. Similarly $x_2$ is the $a_3$ coordinate where the line parallel to $a_2$ through $x$ meets the $a_3$-axis, and $x_3$ is the $a_1$ coordinate where the line parallel to $a_3$ through $x$ meets the $a_1$-axis. This is shown in Figure 24. Notice that $x_1 + x_2 + x_3 = 0$. We call these coordinates the **triple coordinates**.

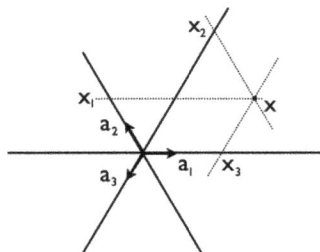

**Figure 24.** The three coordinate system of labelling points in the plane.

The redundant three label coordinate system that we use has the advantage that one can just cycle around the coordinates to deal with each of the three $w$-directions. Counterclockwise rotation through $2\pi/3$ amounts to replacing $(x_1, x_2, x_3)$ by $(x_3, x_1, x_2)$.

We note that $x \in Q$ if and only if $x_1, x_2, x_3 \in \mathbb{Z}$. Let $v : \mathbb{Z} \longrightarrow \mathbb{Z}$ be the 2-adic valuation defined by $v(z) = k$ if $2^k || z$, *i.e.*, if $2^k$ divides $z$ but $2^{k+1}$ does not divide $z$. We define $v(0) = \infty$. Finally we define $D(z) = 2^{v(z)}$. Note that $D(-z) = D(z)$. When levels appear, they are related to $\log_2(D(z))$.

Now for $x \in Q$ we note that, except for $x = 0$, exactly two of $D(x_1), D(x_2), D(x_3)$ are equal and the remaining one is larger. This is a consequence of $x_1 + x_2 + x_3 = 0$.

*8.1. CHT Formula*

In this section we derive the formula for parity for the **CHT** tiling [2]. The **CHT** tiling has the advantage that all the shifting due to the choice of the triangulation is taken out of the way, and this makes it easier to see what is going on. Our notation and use of coordinates is different from that in [2], but the argument is essentially the same. Figure 25 shows how the **CHT** triangulation looks around its center $(0,0)$. The formula for parity is made of two parts each of which corresponds to one the two features which combine to make parity: edge shifting and the color.

First consider the shifting part of the formula. We consider a horizontal line of the **CHT** triangulation, different from the $a_1$-axis. This line meets the $a_3$-axis at a point $\left(n2^k, -n2^k, 0\right)$ for some non-negative integer $k$ and some odd integer $n$. This point is the apex of a level $k$ triangle and is the midpoint of an edge from a triangle of level $k + 1$ (though the $a_3$-axis itself is of infinite level here). As such we see that the horizontal edge to the right from $\left(n2^k, -n2^k, 0\right)$ is shifted downwards. As the edge passes into the next level $k + 1$ triangle we see that the shift is upwards. This down-up pattern extends indefinitely both to the right and to the left. In Figure 26 $n = 1$ and $k$ is unspecified, but the underlying idea does not depend on the value of $n$. We now note that the points along the horizontal edge rightwards from $\left(2^k, -2^k, 0\right)$ are $\left(2^k, -2^k - 1, 1\right), \left(2^k, -2^k - 2, 2\right), \dots,$ or $x = \left(2^k, -2^k - x_3, x_3\right)$ in general. Now we note that

$$\left\lfloor \frac{x_3}{D(x_3 + x_2)} \right\rfloor = \begin{cases} 0 & \text{if } x \text{ corresponds to a shift down edge} \\ 1 & \text{if } x \text{ corresponds to a shift up edge} \end{cases} \tag{9}$$

Thus this is the formula for edge downwards (0) and edge upwards (1). This formula is not valid if $x_3 \equiv 0 \bmod 2^k$. What distinguishes these bad values is that for these, and these only, $D(x_2) \neq D(x_3)$. We see that the fact that we are dealing with a horizontal line (in the direction of the $a_1$ axis) is related to the fact that $D(x_1)$ is the largest of $\{D(x_1), D(x_2), D(x_3)\}$, and whenever that condition fails the above formula fails. But then of course we should use the appropriate formula with the indices cycled.

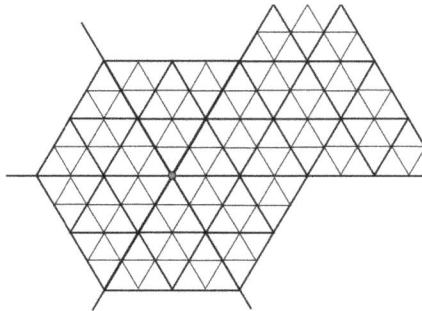

**Figure 25.** Part of the triangulation corresponding to **q** centered at 0 (indicated by the dot). Triangles of scales $1, 2, 4$ are shown, as well as part of a triangle of scale 8.

Next we explain the color component of the formula. The underlying idea is much the same, but, as to be expected, the details are a little more complicated. The color lines are the $w$-lines and are oriented in one of the three $w$ directions. We treat here the case of color lines that are in the vertical direction. The formula utilizes the same three coordinate formulation above. For other $w$-directions one cycles the three components around appropriately.

Consider the sector of the **CHT** tiling as indicated in Figure 27. The figure indicates how the color must be on the $a_3$-axis as we proceed in the vertical direction.

Most of the explanation for the color part of the formula appears in the caption to Figure 19. Although that picture seems tied to the point of intersection of the vertical line and the $a_3$-axis having the special form $(4, -4, 0)$, we note that the same applies whenever $D(x_3 - x_2) = 4$. It is $D(x_3 - x_2)$ that determines the level of the triangle that we are looking at and thus how the stepping sequence will modify the hexagon diameters. In the case where it is 4 here there are 3-step sequences of one diagonal type followed by 3-step sequences of the other type. If $D(x_3 - x_2) = 2^k$ these become $2^k - 1$ step sequences, and still $\left\lfloor \frac{x_3}{D(x_3-x_2)} \right\rfloor$ changes by 1 each time we move from one $2^k - 1$ step sequence to the next. All sequences start from the full blue diameter with $x_3 = 0$ and $x_3$ increases by 1 at each step.

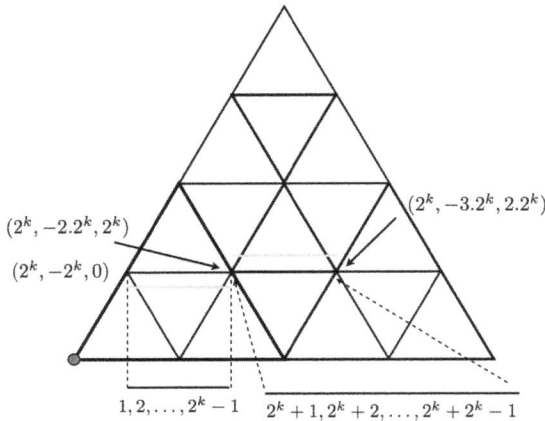

**Figure 26.** The horizontal line through the point $\left(2^k, -2^k, 0\right)$ is seen as passing through midpoints of consecutive $2^{k+1}$ triangles. The corresponding edges along this line shift downwards and upwards alternately. The points $\left(2^k, -2^k - m, m\right)$ with $m = 1, 2, \ldots, 2^k - 1$ are on a shift-down edge. The next set for $m = 2^k + 1, \ldots, 2^k + 2^k - 1$ are on a shift-up edge. The next set, $m = 2.2^k + 1, \ldots 2.2^k + 2^k - 1$ are on a shift-down edge, *etc.* The formula $\left\lfloor \frac{x_3}{D(x_3+x_2)} \right\rfloor = \left\lfloor m / \left(2^k\right) \right\rfloor$ accounts for this precisely, varying between 0 and 1 according to down and up.

In putting the two formulas together, we note first of all that although the formulas have been derived along specific $a$ and $w$ axes, the formulas remain unchanged if the same configurations are rotated through an angle of $\pm 2\pi/3$. Likewise the coloring and shifting rules depend on the geometry and not the orientation modulo $\pm 2\pi/3$. The final formula is then effectively just the sum of the two formulas that we have derived, and it is only a question of determining which color of tile belongs to parity 0.

**Theorem 8.1** [2] *In the* **CHT** *tilings centered at* $(0,0)$, *the parity of a hexagonal tile centered on* $x = (x_1, x_2, x_3)$ *is*

$$\mathbb{P}(x) = \left\lfloor \frac{x_{j+2}}{D(x_{j+2} - x_{j+1})} \right\rfloor + \left\lfloor \frac{x_{j+2}}{D(x_{j+2} + x_{j+1})} \right\rfloor \bmod 2$$

*provided that* $D(x_j)$ *is the maximum of* $D(x_j), D(x_{j+1}), D(x_{j+2})$ *and* $x_{j+1} \pm x_{j+2} \neq 0$ *(subscripts j are taken modulo 3).*

**Proof:** Referring to Figure 19, we check the parity of the tile at $(2, -3, 1)$. In this case $j$ of the theorem is 1 and the displayed formula gives the value 0. On the other hand, the edge shift is down at $(2, -3, 1)$ and the shifted edge meets the blue part of the hexagonal diameter, whence the tile is gray. This establishes the parity formula everywhere.

**Remark 8.2** *Recall that in the paper we have the convention that white corresponds to 1 and gray corresponds to 0.*

**Remark 8.3** *Notice that in the* **CHT** *triangulation centered at* $(0, 0)$ *the hexagon diameters along the three axes defined by* $a_1, a_2, a_3$ *have no shift forced upon them and can be shifted independently either way to get legal tilings. These are the hexagons centered on the points excluded by the condition* $x_{j+1} + x_{j+2} \neq 0$. *Similarly the three w-lines through the origin have no coloring pattern forced upon them and can independently take either. The centers of the hexagons that lie on these lines are excluded by the condition* $x_{j+1} - x_{j+2} \neq 0$. *In the* **CHT** *tiling, the central tile can be taken to be either of the two hexagons and in any of its six orientations. Having chosen one of these 12 options for the central tile the rest of the missing information for tiles is automatically completed. The parity function* $\mathbb{P}$ *can be then extended to a function* $\mathbb{P}^e$ *so as to take the appropriate parity values on the 6 lines that we have just described.*

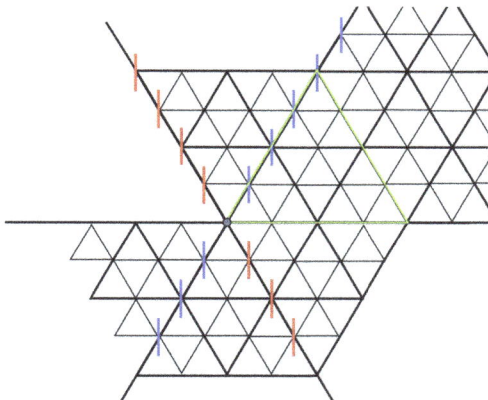

**Figure 27.** Vertical color lines start with a full blue diameter or full red diameter as shown. The green triangle indicates why the second blue line segment up from the origin is in fact blue. The vertical line is centered at the mid-point of an edge of a level 4 triangle and passes into one of the level 2 corner triangles of this level 4 triangle. The discussion on color shows that it must be blue. All the other blue line segments are explained in the same way. We cannot assign color at the origin itself since the origin is not the mid-point of any edge.

## 8.2. Parity for Other Tilings

We can create a formula for arbitrary triangulations $\mathcal{T}(\mathbf{q})$ by the following argument. First of all consider what happens if we shift the center of our triple coordinate system to some new point $c = (c_1, c_2, c_3)_0 \in Q$ (we are in the triple coordinate system centered at the origin and have indicated this with the subscript 0). Then relative to the new center $c$, still using axes in the directions $a_1, a_2, a_3$, the triple coordinates of $x = (x_1, x_2, x_3)_0$ are $(y_1, y_2, y_3)_c = (x_1 - c_1, x_2 - c_2, x_3 - c_3)$. Thus, if consider

$\mathbf{q} = c$ (or more properly $\mathbf{q} = (c, c, c, \ldots)$) so we are looking at the **CHT** triangulation now centered at $c$, then the formulae above become

$$\mathbb{P}_c(x) = \left\lfloor \frac{x_j - c_j}{D(x_{j+2} - c_{j+2} - x_{j+1} + c_{j+1})} \right\rfloor + \left\lfloor \frac{x_j - c_j}{D(x_{j+2} - c_{j+1} + x_{j+1} - c_{j+2})} \right\rfloor \text{ mod } 2$$

Consider now an element $\mathbf{q} = \lim_{k \to \infty} q_1 + q_2 + \cdots + q_k$ and the corresponding sequence of triangulations $\{\mathcal{T}(q_1 + q_2 + \cdots + q_k)\}$. These converge to $\mathcal{T}(\mathbf{q})$, and with them also we get convergence of edge shifting and color.

It is also true that for any fixed $x = (x_1, x_2, x_3)_0$ in the plane the $D$-values of the three triple components of $y = (y_1, y_2, y_3) := x - (q_1 + q_2 + \cdots + q_k)$, as well their various pairwise sums and differences, do not change once $k$ is high enough, since if the 2-content of a number $n$ is $2^m$ then so also is the 2-content of $n + 2^p$ for any $p > m$. Thus $\mathbb{P}(x - (q_1 + q_2 + \cdots + q_k))$ is constant once $k$ is large enough, and we can denote this constant value by $\mathbb{P}_{\mathbf{q}}(x) := \mathbb{P}(x - \mathbf{q})$. This defines the parity function $\mathbb{P}_{\mathbf{q}}$ for $\mathcal{T}(\mathbf{q})$. Although $\{\mathcal{T}(q_1 + q_2 + \cdots + q_k)\}$ is a **CHT** triangulation, its limit $\mathcal{T}(\mathbf{q})$ need not be. In fact we know that the translation orbit of any of the **CHT** tilings centered at $(0, 0)$ is dense in the minimal hull, and so we can compute a parity function $\mathbb{P}_{\mathbf{q}}$ of any tiling of the minimal hull in this way. In the case of generic $\mathbf{q}$ this results in a complete description of the parity of the tiling. In the case that there is convergence of either $a$-lines or $w$-lines (so one is not in a generic case) one can still start with one of the extension functions $\mathbb{P}^e$ and arrive at a complete parity description $\mathbb{P}_{\mathbf{q}}^e$ of any of the possible tilings associated to $\mathcal{T}(\mathbf{q})$.

**Corollary 8.4** *The parity function for a generic tiling is* $\mathbb{P}_{\mathbf{q}}$.

## 9. The Hull of Parity Tilings

A Taylor–Socolar tiling is a hexagonal tiling with two tiles (if we allow rotations). With the appropriate markings (not the ones we use in this paper), the two tiles can be considered as reflections of each other. If we just consider the tiling as a tiling by two types of hexagons, white and gray, then we get the striking parity tilings, for example, of Figure 1. We may consider the hull $Y_Q$ created by these parity tilings. Evidently $Y_Q$ is a factor of $X_Q$. In this section we show that in fact the factor mapping is one-to-one—in other words, when we discard all the information of the marked tiles except the colors white and gray—no information is lost, we can recover the fully marked tiles if we know the full parity tiling. The argument uses a tool that is central to the original work of Taylor, but has only played an implicit role in our argument: The Taylor–Socolar tilings have an underlying scaling inflation by a scale factor of 2. One form of this scaling symmetry is especially obvious from the point of view of the $Q$-adic triangularization.

### 9.1. Scaling

Suppose that we have a Taylor–Socolar tiling with $\mathcal{T}(\mathbf{q})$, where $\mathbf{q} = (q_1, q_1 + q_2, \ldots, q_1 + \cdots + q_k, \ldots)$. Then $q_1 + 2Q$ is the set of triangle vertices of all triangles of level at least 1. To make things quite specific, which we need to do to go on, we choose $q_1 \in \{0, a_1, a_2, a_1 + a_2\}$. In the same way we shall assume $q_2 \in 2\{0, a_1, a_2, a_1 + a_2\}$, and so on. We can view $q_1 + 2Q$ as being a new lattice (even though it may not be centered at 0) and then we note that $\mathbf{q}' := (0, q_2, q_2 + q_3, \ldots, q_2 + \cdots + q_k, \ldots)$ is another triangularization (now of this larger scale lattice, and taken relative to an origin located at $q_1$) that determines a Taylor–Socolar tiling with hexagonal tiles of twice the size. Each of these new double-sized hexagons is centered on a hexagon of the original tiling which itself is centered at a vertex of a level 1 triangle.

The new triangularization has its own edge shifting, and since all that has happened is that all the lines of the triangularization that do not pass through a vertex of a level 1 triangle have gone, we can see that the shifting rules mean that the remaining lines still shift exactly as they did before. Also the whole process of coloring the new double-size hexagons goes just as it did before. Figure 28 shows why the coloring of the new double-size hexagons is the same as the coloring of the original size hexagons on which they are centered.

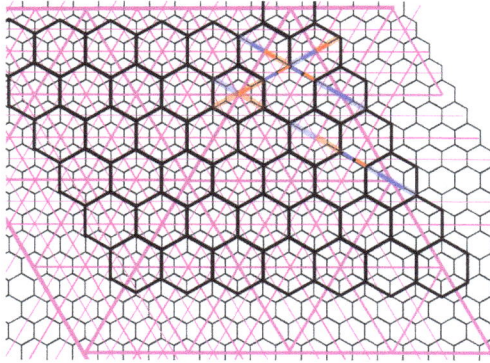

**Figure 28.** This shows how the scaled up hexagon inherits the colors of the smaller hexagons on which they are centered. The coloring is carried out by the same procedure that was used to color the original sized hexagons, though now the diameter sizes have doubled and only the lines of level at least 1 are used.

What happens if the Taylor–Socolar tiling $\mathcal{T}(\mathbf{q})$ is non-generic? A tiling is non-generic if and only if it has an $a$-line or a $w$-line of infinite level. From the definitions, one sees that just rescaling so that the level 0 triangles vanish and all other triangles are now lowered in level by 1 still leaves infinite level lines intact and so we are still in a non-generic case. We can also see this in detail. Non-generic tilings happen if and only if $\mathbf{q} \in x + \overline{\mathbb{Z}_2}\, a$ or $\mathbf{q} \in x + \overline{\mathbb{Z}_2}\, 3w$ for some $x \in Q$. Consider $\mathbf{q}' = (0, q_2, q_2 + q_3, \ldots, q_2 + \cdots + q_k, \ldots) \in 2\overline{Q}$. For definiteness, take the second case. Then we can write

$$\mathbf{q}' \in -q_1 + x + \alpha 3w + \overline{\mathbb{Z}_2} 3(2w)$$

where $\alpha = 0$ if $-q_1 + x \in 2Q$ and $\alpha = 1$ if $-q_1 + x \notin 2Q$. This gives $\mathbf{q}' \in x' + \overline{\mathbb{Z}_2} 3(2w)$ for some $x' \in 2Q$, and we are in the non-generic case again. The same thing happens in the other case.

This scaling self-similarity of the Taylor–Socolar tilings is an important part of Taylor's original construction of the tilings. Here we see self-similarity arise from the simple procedure of "left shifting" the $Q$-adic number $\mathbf{q}$. Geometrically, shifting means subtracting $q_1$ and dividing by 2. We have restricted $q_1, q_2, \ldots$ to be specific coset representatives so that this process of subtraction/division is uniquely prescribed.

We can also work this process in the other direction. Given $\mathbf{q}$ there are four ways in which to choose a coset from $\left(\frac{1}{2}Q\right)/Q$ and for each of these choices we get a new element of the $Q$-adic completion of $\frac{1}{2}Q$. Thus reducing the scale by a factor of 2 we obtain new triangularizations and hexagonal tilings [22]. The original tiling reappears via the scaling up by a factor of 2, as we have just seen. Also note that the new tilings are non-generic if and only if the original tiling was since infinite $a$ or $w$ lines remain as such in the new tilings. Thus we see that scaling has no effect on the generic or non-generic nature of the tilings in question. It is also clear that **CHT** tilings, **iCw-L** tilings, and even **SiCw-L** tilings all transform into tilings of the same type.

### 9.2. From Parity to Taylor–Socolar Tilings

Creating a full Taylor–Socolar tiling from a parity tiling is carried out in two steps: first work out the triangularization, then add the color. For generic tilings the second step is superfluous. Even for non-generic tilings, the fact that the tiling is non-generic is an observable property of the triangularization and then knowing the parity makes it straightforward to recover the colouring along the ambiguous line or lines. The color is not a significant issue and we discuss only the recovery of the triangularization here.

So let us assume that we have a parity tiling $\mathcal{T}_P$ that arises from some tiling $\mathcal{T}(\mathbf{q})$ which at this point we do not know. To obtain the triangularization, *i.e.*, $\mathbf{q}$, we need to work out the translated lattices $q_1 + 2Q, q_2 + 4Q, \ldots$. These are the sets of vertices of the triangles of levels $\geq 1, \geq 2$, *etc.* Suppose that we have a method that can recognize the translated lattice $q_1 + 2Q$ (which is a subset of the hexagon centers of the parity tiling). Then, in effect, we know $q_1$ (at least modulo $2Q$, which is all we need to know about it). We can now imagine changing our view point to double the scale by redrawing the parity tiling with double sized hexagonal tiles at the points of $q_1 + 2Q$ and while retaining the color. We know that this new tiling will be the parity tiling of the scaled up tiling $q_1 + \mathcal{T}(\mathbf{q}')$ where $\mathbf{q}' = (0, q_2, q_2 + q_3, \ldots, q_2 + \cdots + q_k, \ldots)$.

At this point we can proceed by induction to determine $q_2 + 4Q$, rescale again and get $q_3 + 8Q$, and so on. So what is needed is only to determine the vertices of the level $\geq 1$ triangles, or equivalently determine which hexagons of the parity tiling lie on such vertices.

Let us call a patch of 7 tiles which consists of a tile in the center and its 6 surrounding tiles a *basic patch of 7 tiles*. The key observation is that in a Taylor–Socolar-tiling it never happens that a basic patch of 7 tiles whose center tile is a corner tile of a triangle of level $\geq 1$ has 5 surrounding tiles all of the same color. Thus in the parity tiling the hexagons centered on the (as yet unknown) vertices of the level $\geq 1$ triangles, cannot appear as shown in Figure 29 or in any of their rotated forms. On the other hand, as we shall see, for all the other cosets of $Q$ modulo $2Q$ there are points around which such patches (up to rotational symmetry) do occur. Furthermore we do not have to look far in any part of the tiling to find such examples. This is the feature that allows us to distinguish the coset $q_1 + 2Q$ from the remaining cosets of $Q \bmod 2Q$.

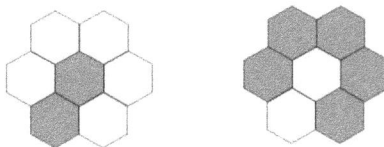

**Figure 29.** Patches of (parity) tiles in which the central tile has 5 of its surrounding tiles of the same color.

The actual proof goes in three steps. In the first we show the "no five hexagons" rule for hexagons centered on the vertices of triangles of level 1. Next we do the same for all the vertices of triangles of level $\geq 2$. This deals with all hexagons centered on $q_1 + 2Q$. Finally we show that for each of the other cosets of $Q \bmod 2Q$ the "no five hexagons" rule fails at least somewhere.

1.  **The vertices of 1-level triangles**: We wish to show that around each hexagon centered on a point of $q_1 + 2Q$ there are at least two different pairs of tiles with mismatched colors amongst its six surrounding tiles (and hence the no five-hexagons rule is true). This is explained in the text below and in Figure 30.

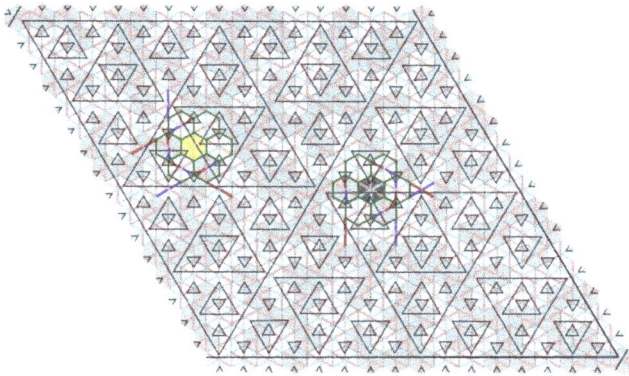

**Figure 30.** Examples are shown that demonstrate how we can see that around each hexagon centered on a level 1 vertex of $q_1 + 2Q$ there are at least two different pairs of tiles with mismatched colors amongst its six surrounding tiles. The two cases correspond to a level 1 vertex at the mid-point of a level 2 triangle and a level 1 vertex a non-midpoint of a higher level triangle. As explained in the text, one pair is found in a uniform way in both cases. The other pair is found by one method in the first case and another method in the other.

There are two different situations for a hexagon $H$ centered on the vertex $v$ of a 1-level triangle. One is the case that $v$ is at the midpoint of a 2-level triangle. This is indicated on the left side of Figure 30. The other is the case that $v$ is on the edge of a $n$-level triangle where $n \geq 3$. This is indicated on the right side of Figure 30. In both cases, we note that the two tiles on opposite sides of $H$ which share the long edge of a triangle of level $\geq 2$ have different colors. The reason is the following. Apart from the red-blue diameters, the long red and blue diameters and black stripes are same for the both the Taylor–Socolar tile and the reflected Taylor–Socolar tile. However the red and blue diameters of the middle tiles of 2-level triangles determine different red-blue diameters for the two tiles, and so they have different parity.

Now we wish to find another pair of tiles with opposite colors for each of the cases. Let us look at the first case (see the left side of Figure 30). Consider the corner of the level 3 triangle $T$ that is defined by $H$, and consider the two edges of $T$ that bound this corner. The red and blue diameters of the tiles at the mid-points of these two edges of $T$ determine different red-blue diameters for two neighbouring tiles in the surrounding tiles of $H$. This again results in different parities.

Finally consider the second case (see the right side of Figure 30). Notice that the long black stripe of $H$ is the part of the long edge of 3 or higher level triangle. If this long edge is from a level 4 or higher triangle, we consider just the part of it that is the level 3 triangle $T$ whose edge coincides with the stripe of $H$. The red and blue diameters of the tile centered on the mid-point of this edge of $T$ determine different red-blue diameters for two of the ring of tiles around $H$, as shown.

2. **The vertices of level $\geq 2$ triangles**: Next we look at the six tiles surrounding the corner tiles of level $\geq 2$ triangles in a Taylor–Socolar tiling. Notice that the pattern of the colored diameters of six surrounding tiles is same for every corner tile of a level 2 or higher triangle, Figure 31. So what determines the basic patches of 7 tiles around the corner tiles of level 2 or higher triangles is the pattern of black stripes on it. Furthermore, there is a one-to-one correspondence between these basic patches of 7 tiles and the basic patches of 7 tiles of white and gray colors. This is shown in Figure 32. The key point is that this means that the basic patches of 7 tiles around the vertices of the level 2 or higher triangles already determine the coloring.

At this point we know that all the vertices of level $\geq 1$ have at least two tiles of each color in the ring of any basic patch of 7 tiles.

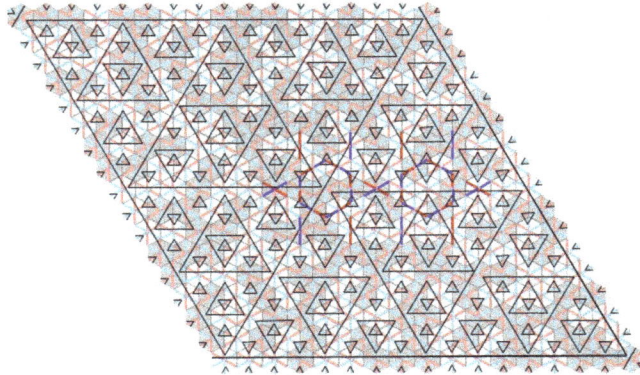

**Figure 31.** A figure showing that the pattern of the colored diameters of six surrounding tiles is same for every corner tile of a level 2 or higher triangle.

3. **Seeing how the other cosets of $Q$ mod $2Q$ violate the "two tiles of each color in the ring of any basic patch of 7 tiles" rule.** From a given parity tiling, there are four choices in determining the 1-level triangles. One of these is the coset $q_1 + 2Q$, and we know that the 7 tile patches around each of these points satisfy the "no five-hexagon" rule. However the three choices, corresponding to the other three cosets of $Q$ mod $2Q$ all have some 7 tile patches that violate the rule. We can see violations to the rule for each of the other three cosets in Figure 33, which is a small piece from the lower right corner of Figure 1. Since any parity tilings in the hull are repetitive, we observe the patches frequently over the parity tiling. Furthermore, since there is only one way that is allowed to determine 1-level triangles, it does not depend on where one starts to find the 1-level triangles. They will all match in the end.

This finishes the discussion of how the coset $q_1 + 2Q$ is identified in the parity tiling $\mathcal{T}_P$. The scaling argument shows that we can continue the process to identify $q_2 + 4Q, q_3 + 8Q$ and so on. This finally identifies $\mathbf{q}$. In fact, even stronger, once we identify $q_2 + 4Q$, we can use the one-to-one correspondence between these basic patches of 7 tiles and the basic patches of 7 tiles of white and gray colors as it is shown in Figure 32 to determine the entire triangularization.

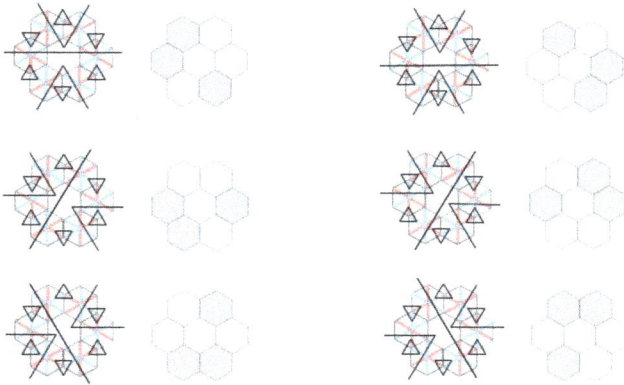

**Figure 32.** The edge and color patterns and corresponding parity patterns that can occur in the hexagons surrounding the corners of triangles of level $\geq 2$.

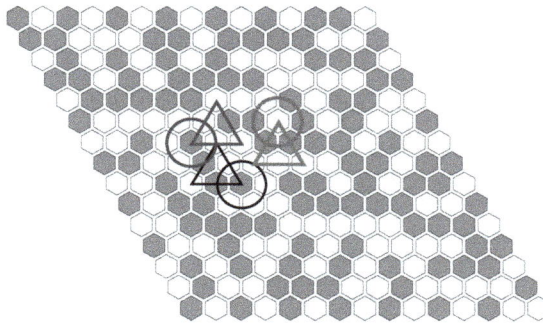

**Figure 33.** Here we see three violations to the five-hexagons rule, the rings of hexagons being indicated by the circles. The shaded triangles show how the cosets determined by the center look, and show that the violating hexagons are from three different cosets. The hexagonal pattern comes from the lower right corner of Figure 1.

We note two things here: That this argument of this subsection has not required that the tiling be generic, and that the process of reconstruction of the triangularization from a parity tiling is local in character.

We conclude that the parity hull loses no color information, and hence the factor mapping from the Taylor–Socolar hull to the parity hull that simply forgets all information about each tile except for its parity, is in fact an isomorphism. In fact the two hulls are mutually locally derivable (MLD) in the sense that each tiling in one is derivable from a corresponding tiling in the other, using only local information around each tile (see [7]).

**Corollary 9.1** *Each Taylor–Socolar tiling and its corresponding parity tiling are mutually locally derivable.*

## 10. Concluding Remarks

The paper has developed an algebraic setting for the Taylor–Socolar hexagonal tilings. This approach leads naturally to a cut and project scheme with a compact $Q$-adic internal space $\overline{Q}$. We have determined the structure of the tiling hull and in particular the way in which it lies over compact group $Q$. Each tiling is a model set from this cut and project formalism. The corresponding parity tilings are in fact completely equivalent to the fully decorated tilings. We have also reproved the parity formula of [2].

The tiling is both remarkably simple and remarkably subtle. The parity tilings remain fascinating and inviting of further study.

**Acknowledgments:** We are grateful to Michael Baake and Franz Gähler for their interest and helpful suggestions in the writing of this paper, and to Joan Taylor who provided the idea behind Section 9. The first author is grateful for the support of Basic Research Program through the National Research Foundation of Korea (NRF) funded by the Ministry of Education, Science and Technology (2010-0011150) as well as the support of KIAS.

## References

1.  Taylor, J. Aperiodicity of a functional monotile. 2010. Avaliable online: http://www.math.uni-bielefeld.de/sfb701/files/preprints/sfb10015.png (accessed on 12 December 2012).
2.  Socolar, J.; Taylor, J. An aperiodic hexagonal tile. *J. Comb. Theory Ser. A* **2011**, *118*, 2207–2231.
3.  Baake, M.; Grimm, U.; Open University, Milton Keynes, UK. A hexagonal monotile for the Euclidean plane, talk at the KIAS Workshop on The Mathematics of Aperiodic Order. 2010.
4.  Robinson, R.M. Undecidability and nonperiodicity for tilings of the plane. *Invent. Math.* **1971**, *12*, 177–209.
5.  Gähler, F.; Julien, A.; Savinien, J. Combinatorics and topology of the Robinson tiling. *Comptes Rendus Math.* **2012**, *350*, 627–631.
6.  Frettlöh, D. Nichtperiodische Pflasterungen Mit Ganzzahligem Inflationsfaktor. Ph.D. Thesis, Universität Dortmund, Dortmund, Germany, 2002.
7.  Frettlöh, D.; Harriss, E. Tilings Encyclopedia. Avaliable online: http://tilings.math.uni-bielefeld.de/ (accessed on 12 December 2012).
8.  Grünbaum, B.; Shephard, G.C. *Tilings and Patterns*; W.H. Freeman and Company: New York, NY, USA, 1987.
9.  Penrose, R. Remarks on a tiling: Details of a $(1 + \epsilon + \epsilon^2)$-aperiodic set. In *The Mathematics of Long-Range Aperiodic Order*; Moody, R.V., Ed.; NATO ASI Series C: 489; Kluwer Academic Publishers: Dordrecht, the Netherlands, 1997; pp. 467–497.
10. Penrose, R. Twistor newsletter. In *Roger Penrose: Collected Works*; Oxford University Press: New York, NY, USA, 2010; Volume 6, pp. 1997–2003.
11. Baake, M.; Gähler, F.; Grimm, U. Hexagonal inflation tilings and planar monotiles. *Symmetry* **2012**, *4*, 581–602.
12. Akiyama, S.; Lee, J.-Y. The computation of overlap coincidence in Taylor–Socolar substitution tilings. 2012; Preprint arXiv:12124209v1.
13. Nearest neighbours in $Q$ are distance 1 apart and the short diameters of the hexagons are of length 1 while the edges of the hexagons are of length $r = 1/\sqrt{3}$. The main diagonals of the hexagons are of length $2r$ in the directions of $\pm w_1$, $\pm w_2$, $\pm (w_2 - w_1)$. One notes that each of these vectors of $P$ is also of length $r$.
14. We shall introduce levels for a number of objects that appear in this paper: Points, lines, edges, triangles.
15. Lee, J.-Y.; Moody, R.V.; Solomyak, B. Pure point dynamical and diffraction spectra. *Ann. Henri Poincaré* **2002**, *3*, 1003–1018.
16. Baake, M.; Lenz, D.; Moody, R.V. Characterizations of model sets by dynamical systems. *Ergod. Theory Dyn. Syst.* **2007**, *27*, 341–382. [CrossRef]
17. Schlottmann, M. Generalized model sets and dynamical systems. In *Directions in Mathematical Quasicrystals*; Baake, M., Moody, R.V., Eds.; CRM Monograph Series 13; AMS: Providence, RI, USA, 2000; pp. 143–159.
18. Baake, M.; Moody, R.V. Weighted Dirac combs with pure point diffraction. *J. Reine Angew. Math.* **2004**, *573*, 61–94. [CrossRef]
19. Lee, J.-Y.; Moody, R.V. Lattice substitution systems and model sets. *Discrete Comput. Geom.* **2001**, *25*, 173–201. [CrossRef]
20. Lee, J.-Y.; Moody, R.V.; Solomyak, B. Consequences of pure point diffraction spectra for multiset substitution systems. *Discrete Comp. Geom.* **2003**, *29*, 525–560. [CrossRef]
21. Moody, R.V. Model sets: A survey. In *From Quasicrystals to More Complex Systems*; Les Editions de Physique; Axel, F., Gazeau, J.-P., Eds.; Springer-Verlag: Berlin, Germany, 2000; pp. 145–166.

22. In our original analysis of the Taylor–Socolar tilings, we chose the triangle vertices to be in $Q$, whereas higher level triangles have vertices in cosets of various $2^k Q$, not necessarily in $2^k Q$ itself. This initial choice of elements in $Q$ was convenient to keep all the triangle vertices in $Q$ itself. In the halving process we are doing here, we may choose any of the cosets of $(\frac{1}{2}Q)/Q$. The canonical choice might be to choose 0, but this is not necessary to get a tiling.

23. Baake, M.; Schlottmann, M.; Jarvis, P.D. Quasiperiodic tilings with tenfold symmetry and equivalence with respect to local derivability. *J. Phys. A* **1991**, *19*, 4637–4654. [CrossRef]

*symmetry*

MDPI

*Article*

# The Roundest Polyhedra with Symmetry Constraints

## András Lengyel *,†, Zsolt Gáspár † and Tibor Tarnai †

Department of Structural Mechanics, Budapest University of Technology and Economics,
H-1111 Budapest, Hungary; gaspar@ep-mech.me.bme.hu (Z.G.); tarnai@ep-mech.me.bme.hu (T.T.)
* Correspondence: lengyel.andras@epito.bme.hu; Tel.: +36-1-463-4044
† These authors contributed equally to this work.

Academic Editor: Egon Schulte
Received: 5 December 2016; Accepted: 8 March 2017; Published: 15 March 2017

**Abstract:** Amongst the convex polyhedra with $n$ faces circumscribed about the unit sphere, which has the minimum surface area? This is the isoperimetric problem in discrete geometry which is addressed in this study. The solution of this problem represents the closest approximation of the sphere, i.e., the roundest polyhedra. A new numerical optimization method developed previously by the authors has been applied to optimize polyhedra to best approximate a sphere if tetrahedral, octahedral, or icosahedral symmetry constraints are applied. In addition to evidence provided for various cases of face numbers, potentially optimal polyhedra are also shown for $n$ up to 132.

**Keywords:** polyhedra; isoperimetric problem; point group symmetry

## 1. Introduction

The so-called isoperimetric problem in mathematics is concerned with the determination of the shape of spatial (or planar) objects which have the largest possible volume (or area) enclosed with given surface area (or circumference). The isoperimetric problem for polyhedra can be reflected in a question as follows: What polyhedron maximizes the volume if the surface area and the number of faces $n$ are given? The problem can be quantified by the so-called Steinitz number [1] $S = A^3/V^2$, a dimensionless quantity in terms of the surface area $A$ and volume $V$ of the polyhedron, such that solutions of the isoperimetric problem minimize $S$. Lindelöf proved that in order to reach the maximum volume, the polyhedron must have an insphere such that all faces are tangent to this sphere at their centroid [2,3]. If the radius of such an inscribed sphere is taken to be unity, one can easily find that $S = 27V = 9A$. This way, the problem is reformulated so as to find polyhedra with $n$ faces circumscribed about the unit sphere which have the minimum surface area (or volume).

An alternative measure of roundness of solids applied in various fields of science is the so-called isoperimetric quotient ($IQ$) introduced by Pólya [4], which is a normed inverse of the Steinitz number:

$$IQ = 36\pi \frac{V^2}{A^3}. \tag{1}$$

The isoperimetric quotient is a positive dimensionless number that takes unity for the sphere and less for other solids. The closer the $IQ$ is to unity, the more spherical the polyhedron is. Goldberg proposed a lower bound [5] for the Steinitz number, from which a lower bound $A_G(n)$ for the surface area can be easily obtained if Lindelöf's necessary condition is satisfied:

$$A_G(n) = 6(n-2)\tan E(4\sin^2 E - 1) \tag{2}$$

where,

$$E = \frac{n\pi}{6(n-2)}. \tag{3}$$

Furthermore, on the one hand, as the isoperimetric quotient is inversely proportional to the Steinitz number, for a given $n$ an upper bound $4\pi/A_G$ for the maximum value of $IQ$ can be simply derived as $IQ = 36\pi/S = 4\pi/A$. On the other hand, any polyhedron constructed with $n$ faces constitutes a lower bound for the maximum value of $IQ$ since it has a surface area larger than or equal to that of the optimal polyhedron. A polyhedron here is referred to as optimal if it has the largest possible $IQ$ among all possible polyhedra of any topology for the given $n$. Any candidate polyhedron construction is called hereafter conjectural unless rigorous mathematical proof is provided for its optimality.

For this isoperimetric problem where no symmetry constraint is enforced, mathematical proof of optimality exists only in certain cases of a small number of faces. Fejes Tóth [6] proved that Goldberg's formula provides a lower bound of the minimum value of the Steinitz number for any $n$, and it is exact for $n = 4, 6, 12$. In this way he proved that the regular tetrahedron, hexahedron, and dodecahedron are the optimal polyhedra in the cases of 4, 6, and 12 faces, respectively. The case of 5 faces is also proven [2,3].

Conjectured solutions with numerically optimized geometry were shown by Goldberg [5] for $n = 4$ to $10, 12, 14$ to $16, 20, 32, 42$, preceding Fejes Tóth's evidence, and later by Schoen [7,8] for cases up to $n = 43$. Mutoh [9] also dealt with polyhedra with minimum volume circumscribed about the unit sphere, and by using a computer-aided search, provided a series of conjectured optimal polyhedra with the number of faces ranging between 4 and 30. In an earlier paper [10] the authors pointed out that in some cases the isoperimetric problem for $n$ faces and the problem of the minimum covering of a sphere by $n$ equal circles have the same proven or conjectured solution (the points of tangency of the faces and the centres of the circles are identical). In some cases, the obtained polyhedra are only topologically identical. Sometimes, the topology of the optimal polyhedron matches that of a suboptimal (nearly optimal) circle covering, and not the optimal one. We developed an iterative numerical method for determining a locally optimal solution of the isoperimetric problem, where the starting point for the topology of the polyhedron is a conjectured optimal circle covering on a sphere. Using this method and starting with the conjectured best coverings with 50 [11] and 72 [12] equal circles, conjectural solutions to the isoperimetric problem for $n = 50$ and 72 were obtained [10].

Some of the above-mentioned proven or conjectured solutions have tetrahedral, octahedral or icosahedral symmetry even though such symmetry constraints were not enforced. The solutions for $n = 4, 6$, and 12 investigated by Fejes Tóth [6] have tetrahedral, octahedral, and icosahedral symmetry, respectively. The conjectured solution suggested by Goldberg [5] for $n = 16$ has tetrahedral symmetry, and for $n = 32$ and 42 has icosahedral symmetry. The conjectured solution for $n = 72$ provided in [10] has icosahedral symmetry.

It turns out that in some artistic or practical applications it is not the solutions to the unconstrained isoperimetric problem which play an important role, but rather those with tetrahedral, octahedral or icosahedral symmetry constraints. For instance, a turned ivory piece from Germany, and a wooden die from Korea are exact representations of the conjectured roundest polyhedron with octahedral symmetry constraints for $n = 14$ (Figure 1). Similarly, another turned ivory object in Germany and a modern soccer ball show the conjectured roundest polyhedron with icosahedral symmetry constraint for $n = 32$ (Figure 2). More details on these artifacts can be found in [13]. Multi-symmetric roundest polyhedra for $n = 14, 26, 32, 62$ can be particularly important in practical applications, e.g., soccer ball design [14,15]. The conjectured solution for the case of 14 (32) faces is an *isodistant* truncation of the octahedron (icosahedron) where the centroids of the faces lie at the vertices and the face centres of a spherical octahedron (icosahedron). A truncation of a Platonic polyhedron is said to be *isodistant* if the distances from the centre of the Platonic polyhedron to the truncating planes and to the faces of the Platonic polyhedron are equal, that is, if the truncated polyhedron has an insphere. Polyhedra with 26 and 62 faces are obtained by double isodistant truncations of the octahedron and the icosahedron, respectively [14,15].

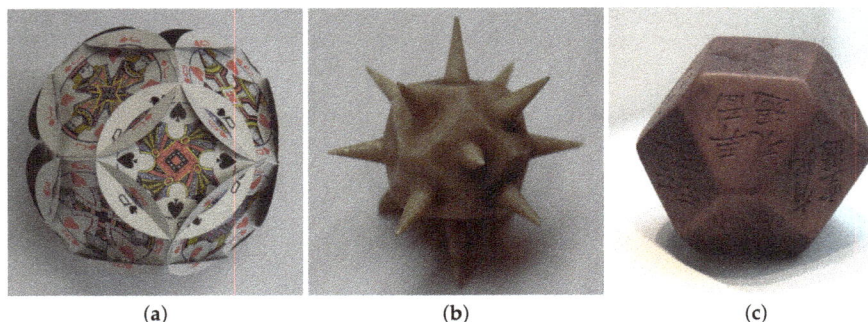

**Figure 1.** The roundest polyhedron with 14 faces and with octahedral symmetry constraint. (a) Minimum covering of a sphere by 14 equal circles (card model; photo: A. Lengyel); (b) core of a broken turned ivory sphere, 17th century. (Grünes Gewölbe, Dresden, Germany; photo: T. Tarnai); (c) Wooden die, 7th–9th centuries (Gyeongju National Museum, Korea; photo: K. Hincz).

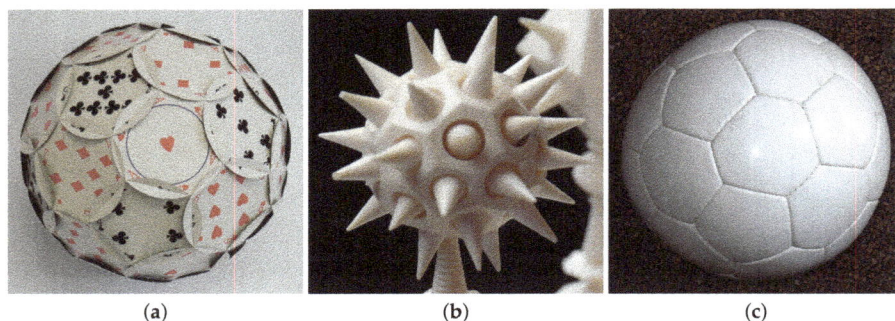

**Figure 2.** The roundest polyhedron with 32 faces and with icosahedral symmetry constraint. (a) Minimum covering of a sphere by 32 equal circles (card model; photo: A. Lengyel); (b) part of a turned ivory object, around 1600 (Grünes Gewölbe, Dresden, Germany, inv. no. 255; photo: T. Tarnai); (c) the Hyperball designed by P. Huybers, the underlying polyhedron is somewhat different from that in (a) (photo: T. Tarnai).

The artistic and practical interest in finding the roundest object as shown above is an important motivation to study the isoperimetric problem for polyhedra with symmetry constraints. The primary aim of this paper is to prove the solutions for small values of $n$ ($n = 8, 10, 14$) and to present conjectured solutions for $n = 18, 20, 22$. Recalling that some solutions for small even numbers of polygonal faces have been proven by Fejes Tóth [6], our new results complete the sequence of proven and conjectured solutions for even $n$ up to 22. Goldberg [16] has shown how to construct multi-symmetric polyhedra by using a triangular lattice on the tetrahedron, octahedron, and icosahedron (the details will be discussed later in the paper). After his name, these polyhedra now are called Goldberg polyhedra. The secondary aim of this paper is to present octahedrally symmetric Goldberg polyhedra for $n = 30$ and 38 as conjectural solutions, and icosahedrally symmetric Goldberg polyhedra for $n = 92, 122, 132$. In this way it is possible to extend the list of icosahedral Goldberg polyhedra up to $n = 132$.

## 2. Multi-Symmetric Point Arrangements on the Sphere

If a plane is tangent to a sphere, then the position of the plane is uniquely determined by the position of the point of tangency. In this way we can investigate the faces of polyhedra having an insphere via investigating the points of tangency on spherical tetrahedra, octahedra, and icosahedra. Identifying and exploring the degrees of freedom of tangent placement on these spherical polyhedra allows the generation of polyhedra with varying face numbers and set symmetries, among which our numerical method can identify the locally optimal (roundest) solution.

Consider the Platonic polyhedra $\{3, q\}$ with triangular faces and $q$-valent vertices where $q = 3, 4, 5$ (that is the regular tetrahedron, octahedron, icosahedron). Let $V$, $E$, and $F$ denote their number of vertices, edges, and faces, respectively. According to the relationships derived in §10.3 of Coxeter's book [17]:

$$V = \frac{12}{6 - q}, \quad E = \frac{6q}{6 - q}, \quad F = \frac{4q}{6 - q}. \tag{4}$$

The symmetry of these polyhedra is characterized by the fact that $q$-fold, 2-fold and 3-fold axes of rotation go through their vertices, edge midpoints and face centres, respectively. If we take one face of the spherical version of these polyhedra, then we can put one point of tangency or none to a vertex, any number of points of tangency on an edge, and a number of points of tangency on a face, divisible by three (points arranged in threefold rotational symmetry) or the remainder after division by three is one (points arranged in threefold rotational symmetry and a point in the face centre). Let $v$, $e$, and $f$ denote the number of points of tangency at a vertex, on an edge, and on a face, respectively. Then these non-negative integers can be expressed as follows:

$$v = 0 \text{ or } 1, \quad e = 0, 1, 2, 3, \ldots, \quad f = 3k \text{ or } 1 + 3k, \quad k = 0, 1, 2, 3, \ldots \tag{5}$$

such that $v + e + f \neq 0$. In this way the number of all points of tangency on the sphere is:

$$n = vV + eE + fF. \tag{6}$$

Since $V$, $E$, and $F$ are even, it is obvious that $n$ is even, too. Additionally, $n \geq 4$.

**Proposition 1.** *Any even number of points, not less than 4, can be arranged on the sphere in tetrahedral symmetry.*

**Proof.** For tetrahedral symmetry, Equation (6) yields $n = 4v + 6e + 4f = 2(2v + 3e + 2f)$. We prove the proposition if we see that $2v + 3e + 2f$ is an arbitrary natural number $\geq 2$. First select $v = 1$, $f = 0$, $e = 0, 1, 2, 3, \ldots$ In this case $2v + 3e + 2f = 2 + 3k$, $k = 0, 1, 2, 3, \ldots$ Then select $v = 0$, $f = 0$, $e = 1, 2, 3, \ldots$ In this case $2v + 3e + 2f = 3 + 3k$, $k = 0, 1, 2, 3, \ldots$ Finally, select $v = 1$, $f = 1$, $e = 0, 1, 2, 3, \ldots$ In this case $2v + 3e + 2f = 4 + 3k$, $k = 0, 1, 2, 3, \ldots$ The union of these three sets of numbers contains all natural numbers $\geq 2$, consequently, the set of numbers $n$ contains all even numbers $n \geq 4$. $\square$

**Remark 1.** *From the proposition it follows that for any even number $n \geq 4$ there exists a polyhedron with $n$ faces and tetrahedral symmetry. In the proof, for $f$ we used only two values out of the infinitely many. Therefore, for many values of $n$, there exist not only one but more polyhedra with different topology in tetrahedral symmetry.*

In the case of octahedral symmetry, the number of points of tangency takes the form $n = 6v + 12e + 8f$. It can be verified quickly that there exist polyhedra with an even number of faces and octahedral symmetry if $n \geq 6$ except if $n = 4 + 6k$, $k = 1, 2, 3, \ldots$

In the case of icosahedral symmetry, the number of points of tangency takes the form $n = 12v + 30e + 20f$. Here, it can be verified quickly that there exist polyhedra with $n$ faces and icosahedral symmetry only for the following values of $n$: $n = 12 + 30k$, $n = 20 + 30k$, $n = 30 + 30k$, $n = 32 + 30k$, where $k = 0, 1, 2, 3, \ldots$

The vertices, edge midpoints, and face centres are the points of a face triangle of the tetrahedron, octahedron, and icosahedron, through which rotation axes pass. If a point of tangency is put to such a point, then the position of the point of tangency is fixed. It cannot be slightly moved without destroying the symmetry. We say that the degree of freedom of the point of tangency is zero in this position. If we put one point of tangency on an edge and another one symmetrically to the edge midpoint, then the point and its symmetrical counterpart can be moved along the edge without breaking the symmetry. We say that the degree of freedom of the point of tangency is one. If we put a point of tangency inside the triangle and two others according to threefold symmetry with respect to the face centre, then we can move the point and its symmetrical counterparts simultaneously in the triangle so that the symmetry is maintained. We say that the degree of freedom of the point of tangency is two.

Consider the simplest cases where $v$, $e$, and $f$ in Equation (5) take the values 1 or 0. Since here the points of tangency have zero degree of freedom, they uniquely determine the respective polyhedra. Since the number of 3-tuples of a 2-set is $2^3$, we have eight possibilities, but the case $v = e = f = 0$, as meaningless, is left out. Thus in each of the tetrahedral, octahedral, icosahedral symmetries, we have seven polyhedra. The face numbers $n$ of these polyhedra (the numbers of points of tangency) are collected in Table 1.

**Table 1.** Number of points of tangency $n$ if they lie only on $q$-fold and/or 2-fold and/or 3-fold rotation axes.

| $v$ | $e$ | $f$ | $n$ | | |
|-----|-----|-----|-----|-----|-----|
| | | | $q = 3$ | $q = 4$ | $q = 5$ |
| 1 | 0 | 0 | 4 | 6 | 12 |
| 0 | 1 | 0 | 6 | 12 | 30 |
| 0 | 0 | 1 | 4 | 8 | 20 |
| 0 | 1 | 1 | 10 | 20 | 50 |
| 1 | 0 | 1 | 8 | 14 | 32 |
| 1 | 1 | 0 | 10 | 18 | 42 |
| 1 | 1 | 1 | 14 | 26 | 62 |

Polyhedra determined by the data of Table 1 have also planes of symmetry. Thus, their symmetry groups are the full tetrahedral, octahedral, and icosahedral groups, respectively. Among these 21 polyhedra some are identical. It is trivial to see, for instance, that in the case $n = 4$, each of the two polyhedra is the same regular tetrahedron. This kind of coincidence in some other cases will be shown later.

For $n > 14$ the tetrahedral arrangements of points of tangency surely will also contain points with one and/or two degrees of freedom. That means that the configuration with the highest $IQ$ can be found only by an optimization process.

For larger values of $n$, it is expected that trivalent polyhedra with mostly hexagonal faces provide the best results, especially in icosahedral symmetry. Trivalent vertices are found to be visually less "pointed", consider for example the classic 32-panel soccer ball. Such polyhedra are the Goldberg polyhedra [16] explained below.

In virus research, Caspar and Klug [18] discovered a regular tessellation on the regular triangle-faced polyhedra $\{3, q\}$, $q = 3, 4, 5$, which consist of small equal equilateral triangles. Coxeter [19] denoted this tessellation by $\{3, q+\}_{b,c}$, where $q+$ indicates that $q$ and more than $q$ (i.e., six) small triangles meet at each vertex, some of which are folded, such that they tessellate the surface of the regular polyhedron $\{3, q\}$. The subscripts $b$ and $c$ show that a vertex of the regular polyhedron $\{3, q\}$ can be arrived at from an adjacent one along the edges of the tessellation by $b$ steps on the vertices in one direction, then proceeding by $c$ steps after a change in direction by $60°$. Figure 3 shows a part of this tessellation.

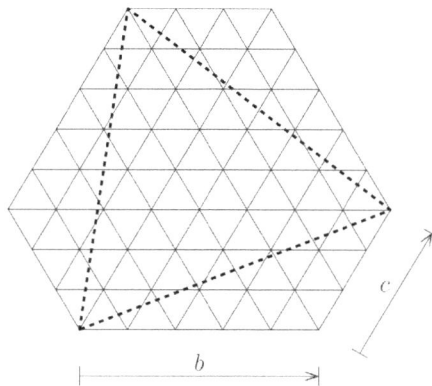

**Figure 3.** The meaning of the Goldberg–Coxeter parameters $b$ and $c$. The large equilateral triangle drawn with dashed lines is a face of the regular tetrahedron, octahedron or icosahedron.

The pair of non-negative integers $b$, $c$, called Goldberg–Coxeter parameters, generate the tessellation and determine the so-called triangulation number $T$.

$$T = b^2 + bc + c^2. \tag{7}$$

By "blowing up" the tessellation, we obtain a triangular lattice on the sphere. This lattice determines a polyhedron whose all vertices lie on the sphere and all faces are triangles. Its numbers of vertices $V_T$, edges $E_T$, and faces $F_T$ can be expressed with $T$, by using the expressions for $V$, $E$, and $F$, respectively in (4), in the form:

$$V_T = T\frac{2q}{6-q} + 2, \quad E_T = T\frac{6q}{6-q}, \quad F_T = T\frac{4q}{6-q}, \quad q = 3, 4, 5. \tag{8}$$

The reciprocal (dual) of this triangle-faced polyhedron is called Goldberg polyhedron, which is a trivalent polyhedron which has $12/(6-q)$ $q$-gonal faces, and $(T-1)2q/(6-q)$ hexagonal faces. For $n = V_T$, the Goldberg polyhedra are good candidates for the roundest polyhedra with $n$ faces, especially with octahedral and icosahedral symmetry where,

$$n = 4T + 2 \tag{9}$$

and,

$$n = 10T + 2, \tag{10}$$

respectively.

When the above-mentioned spherical triangular lattice is applied in the case of small values of $n$, sometimes not all vertices are considered as points of tangency. Sometimes extra points of tangency, which are not vertices of the lattice, are added to the point system. In Figure 4, in a schematic view, we present a part of the triangular lattice—lying on a face of the spherical tetrahedron, octahedron or icosahedron—together with the points of tangency, for the polyhedra shown in Figure 5. The degree of freedom of the points of tangency is also indicated. The subfigures of Figure 4 are in correspondence with those of Figure 5.

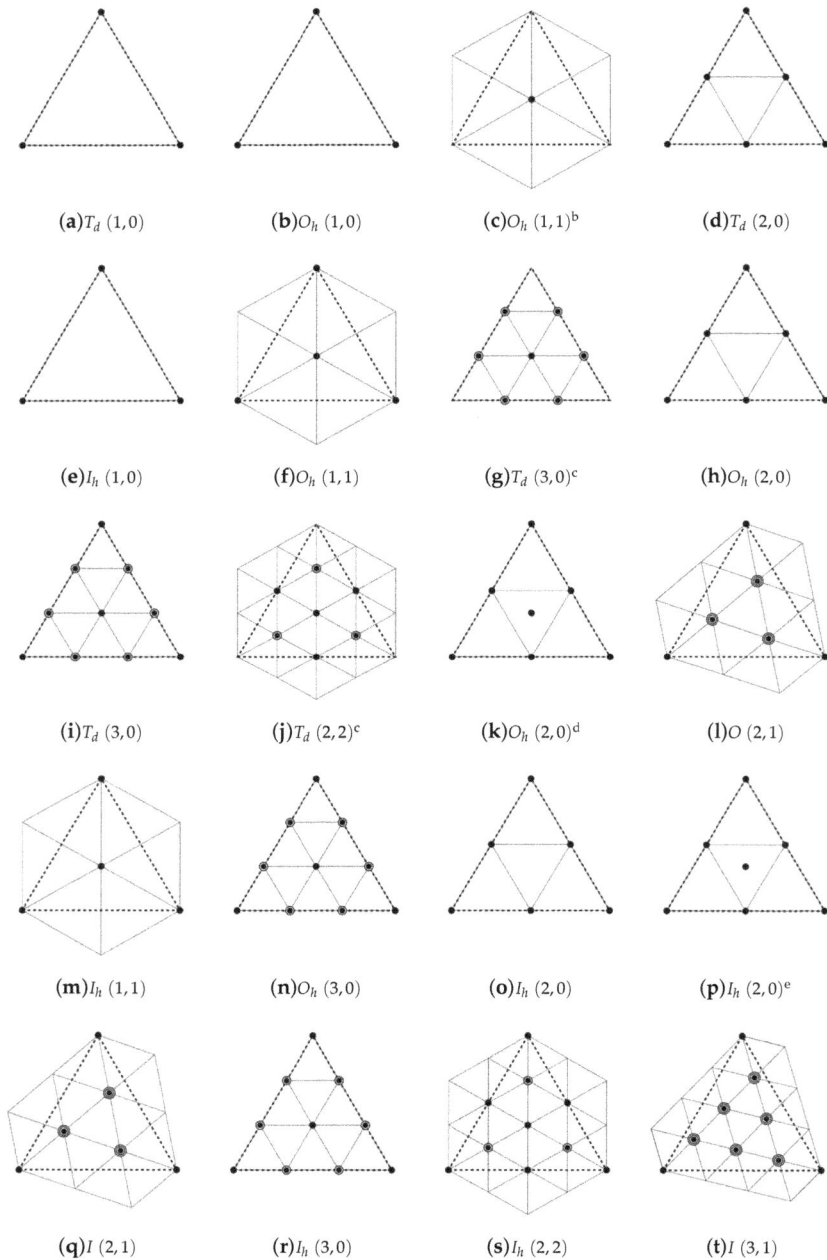

**Figure 4.** Schematic view of the points where the faces of the proven and conjectured roundest polyhedra are tangent to a sphere. Triangular surface lattice on a face of the spherical tetrahedron, octahedron or icosahedron with the degrees of freedom of the points of tangency. In the legends of the subfigures, the symmetry and the Goldberg–Coxeter parameters are given. Superscripts (**b–e**) are explained in Table 2. Symbols ● ◉ ◎ denote points of tangency with zero, one, and two degrees of freedom, respectively.

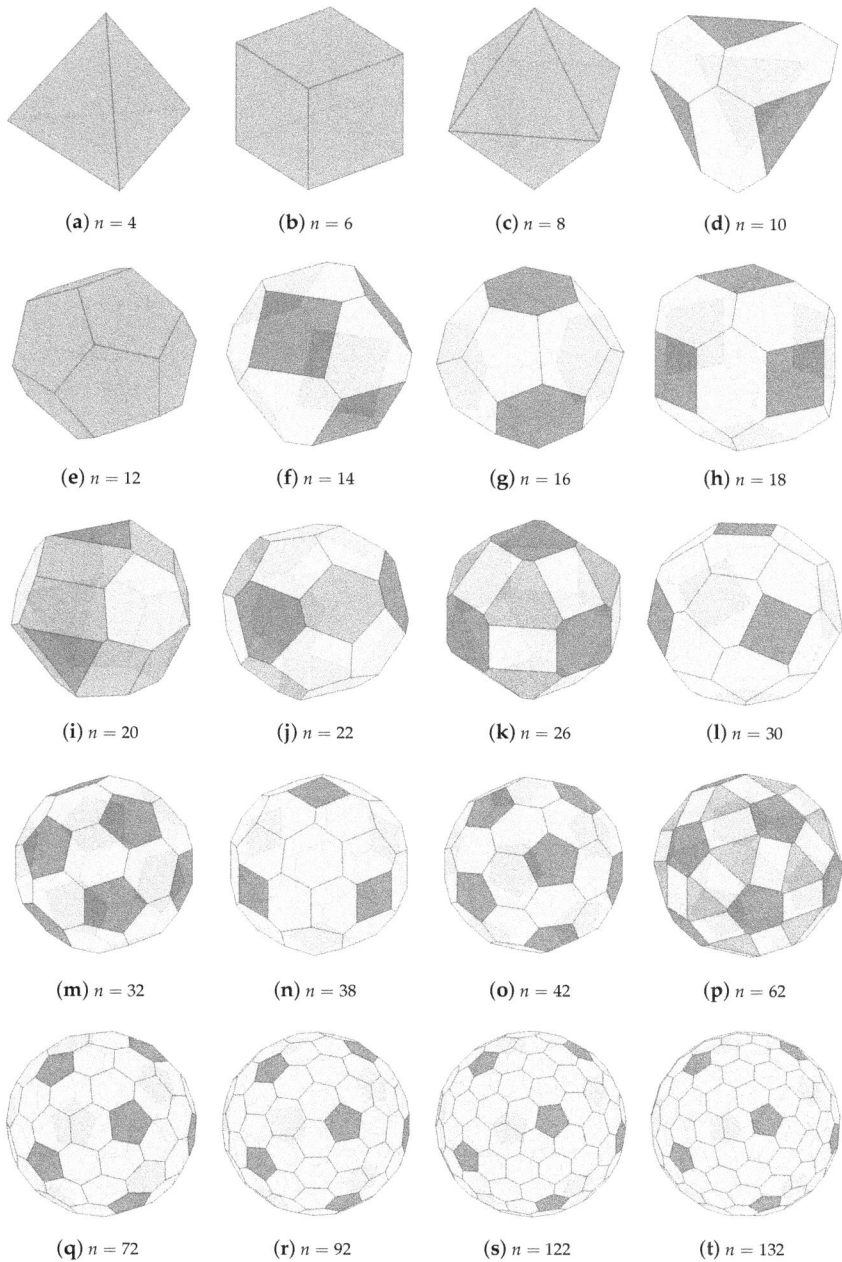

**Figure 5.** Polyhedra with *n* faces that maximize the isoperimetric quotient under tetrahedral, octahedral, or icosahedral symmetry constraints. See main text for description.

### 3. Method

In a recent paper [10] the authors presented a numerical iterative method for the surface minimization of convex polyhedra circumscribed about the unit sphere with given number of faces and topology (edge graph). The main point of the method is to regard the surface area of a trivalent polyhedron with $n$ faces as the potential energy of a mechanical system to minimize and to create the associated (dual) polyhedron whose faces are considered as triangle elements. Edge forces of the triangle elements are defined as derivatives of the surface area with respect to the lengths, and the resultant of these forces meeting at each vertex can be computed. The set of vertices reaches an equilibrium position (i.e., the surface area is minimized locally) if the resultants at all vertices pass through the centre of the sphere, which implies that the components of the forces lying in the tangent plane (i.e., the plane of the respective face of the parent polyhedron) are in equilibrium at each vertex. We usually start the iteration with the polyhedron determined by the conjectured optimal or suboptimal covering of the sphere by $n$ equal circles. A detailed description and mathematical evaluation of the method is given in [10].

Since we determine the required optimum numerically, despite the care in calculation, we are not able to guarantee that we found the global optimum. What we surely found is a local optimum, and our solution is a conjectured solution.

### 4. Results

*4.1. Small Values of n*

*4.1.1. $n = 8$*

**Proposition 2.** *The roundest polyhedron with 8 faces and with tetrahedral or octahedral symmetry is the regular octahedron.*

**Proof.** The proof is simple. According to Table 1 there exist two configurations. The first has tetrahedral symmetry, and the points of tangency are at the vertices and the face centres of the spherical tetrahedron. These points on the sphere are identical to the vertices of a cube. The second has octahedral symmetry, and the points of tangency are at the face centres of the spherical octahedron (Figure 4c). These eight points are identical to the vertices of a cube. Consequently, the two configurations of the points of tangency are congruent, and we have only one polyhedron, which is the reciprocal (dual) of the cube, that is the regular octahedron (Figure 5c). □

Its symmetry is $O_h$, and $IQ = 0.604599788\ldots$

*4.1.2. $n = 10$*

**Proposition 3.** *The roundest polyhedron with 10 faces and with tetrahedral symmetry is a cube whose four non-adjacent vertices are truncated.*

**Proof.** According to Table 1 there exist two configurations, both with tetrahedral symmetry. In the first, say configuration A, the points of tangency are at the vertices and the edge midpoints of the spherical tetrahedron (Figure 4d). In the second, say configuration B, the points of tangency are at the face centres and the edge midpoints of the same spherical tetrahedron. Since the vertices and the face centres of the spherical tetrahedron coincide with the vertices of a spherical cube, and the edge midpoints of the spherical tetrahedron coincide with the face centres of the same spherical cube, configuration A can be taken into coincidence with configuration B by a rotation of 90° about an axis passing through two opposite face midpoints of the spherical cube. Consequently, configurations A and B are congruent, and we have only one polyhedron. The planes tangent to the vertices and edge midpoints of the spherical tetrahedron determine a cube whose four non-adjacent vertices are truncated, or what is the same, a tetrahedron whose edges are chamfered (Figure 5d). □

Its symmetry is $T_d$, and $IQ = 0.630745372\ldots$

### 4.1.3. $n = 14$

**Proposition 4.** *The roundest polyhedron with 14 faces and with octahedral symmetry is a truncated octahedron (truncated octahedron here does not mean Archimedean-truncated octahedron).*

**Proof.** According to Table 1 there exist two configurations. The first has tetrahedral symmetry, and the points of tangency are at the vertices, the edge midpoints and the face centres of the spherical tetrahedron. These points on the sphere are identical to the vertices and the face centres of a spherical cube. The second has octahedral symmetry, and the points of tangency are at the vertices and the face centres of the spherical octahedron (Figure 4f). Because the cube and the regular octahedron are mutually reciprocal to each other, the two configurations of points of tangency on the sphere are congruent, and we have only one polyhedron, that is a truncated octahedron (Figure 5f). □

Its symmetry is $O_h$, and $IQ = 0.781638893\ldots$

### 4.1.4. $n = 18$

According to Table 1 there exists one configuration with octahedral symmetry, where the points of tangency are at the vertices and the edge midpoints of the spherical octahedron (Figure 4h). There are two additional configurations with tetrahedral symmetry, the first for $v = 0, e = 3, f = 0$, the second for $v = 0, e = 1, f = 3$. If only octahedral symmetry is considered then it is obvious, if the tetrahedral symmetry is also considered then it is conjectured that the best configuration is obtained with octahedral symmetry $O_h$, and the roundest polyhedron is the rhombic dodecahedron whose four-valent vertices are truncated, or what is the same, the cube whose edges are chamfered (Figure 5h). For this $IQ = 0.823218074\ldots$

### 4.1.5. $n = 20$

According to Table 1 there exist two configurations. The first has octahedral symmetry, and the points of tangency are at the edge midpoints and the face centres of the spherical octahedron. The second has icosahedral symmetry, and the points of tangency are at the face centres of the spherical icosahedron. There are two additional configurations with tetrahedral symmetry, the first for $v = 1, e = 0, f = 4$, the second for $v = 1, e = 2, f = 1$ (Figure 4i). The polyhedron with icosahedral symmetry is the regular icosahedron itself, for which $IQ = 0.828797719\ldots$ The octahedral configuration and the first tetrahedral configuration determine the same polyhedron, if the tetrahedral configuration has planes of symmetry. This common polyhedron is the rhombic dodecahedron whose three-valent vertices are truncated, for which $IQ = 0.784085714\ldots$ The second tetrahedral configuration, where the points of tangency on the edges can concertedly move with one degree of freedom, is optimized. That resulted in the polyhedron with symmetry $T_d$ in Figure 5i, for which $IQ = 0.830222439\ldots$ which is the conjectured best for $n = 20$.

### 4.1.6. $n = 22$

There are four configurations, all with tetrahedral symmetry: (1) for $v = 0, e = 3, f = 1$; (2) for $v = 1, e = 1, f = 3$; (3) for $v = 0, e = 1, f = 4$, (4) for $v = 1, e = 3, f = 0$. Configuration (4) is not a good candidate since all the points of tangency are concentrated along the edges instead of being more or less uniformly distributed on the sphere. Configuration (1) is not promising either since here apart from rectangular and pentagonal, even nonagonal faces are expected, which is not advantageous. Because of self-duality of the tetrahedron it can be established that configurations (2) and (3) are identical. In this way, only one candidate remained: configuration (3) (Figure 4j). Although the edge graph of the polyhedron related to this configuration is uniquely determined by the data, since $f > 1$, the polyhedron itself is not uniquely determined. On a face of the spherical tetrahedron, there is one point of tangency at the face centre, and there are three points of tangency in threefold rotational symmetry with respect to the face centre. In theory, these three points can concertedly move on the face with two degrees of freedom. However, we suppose that the arrangement of the points of tangency has also planes of symmetry, and so we have symmetry $T_d$ instead of $T$. The three points lie on the three altitudes of the face triangle, where they can concertedly move with one degree of freedom. Consequently, finding the maximum of $IQ$ required optimization in one variable. It resulted in $IQ = 0.862408738\ldots$ The polyhedron obtained is shown in Figure 5j.

### 4.2. Octahedral Goldberg Polyhedra

#### 4.2.1. $n = 30$

Let the pair of Goldberg–Coxeter parameters be $(2, 1)$ for which $T = 7$. From Equation (9), $n = 30$ is obtained. Here $v = 1, e = 0, f = 3$ (Figure 4l). The three points of tangency can concertedly move with two degrees of freedom on the face triangle of the spherical octahedron, while maintaining threefold symmetry with respect to the face centre. In this way, finding the optimal polyhedron, for which $IQ$ is the maximum, required optimization in two variables. The executed numerical optimization resulted in $IQ = 0.896930384\ldots$ The obtained polyhedron is presented in Figure 5l. Since the polyhedron has no planes of symmetry, the symmetry of the polyhedron is $O$.

#### 4.2.2. $n = 38$

Let the pair of Goldberg–Coxeter parameters be $(3, 0)$ for which $T = 9$. The relationship from Equation (9) yields $n = 38$. Here $v = 1, e = 2, f = 1$ (Figure 4n). Since $e > 1$, the Goldberg polyhedron is not uniquely determined. We have one one-degree-of-freedom point of tangency on one of the edges of the underlying spherical octahedron, and the positions of all the other points on the edges are given by symmetry which is $O_h$, because the arrangement of points of tangency has also planes of symmetry. The value of $IQ$ is a function of the distance between the considered point and the nearest vertex of the octahedron. Finding the polyhedron for which the surface area is the minimum, and so $IQ$ is the maximum, requires optimization in one variable. The obtained maximum is $IQ = 0.917445003\ldots$ The obtained polyhedron is displayed in Figure 5n.

### 4.3. Icosahedral Goldberg Polyhedra

#### 4.3.1. $n = 92$

Here, the configuration on a face of the underlying spherical polyhedron is the same as that in the case of $n = 38$, that is, the pair of Goldberg–Coxeter parameters is $(3, 0)$ for which $T = 9$, and $v = 1, e = 2, f = 1$ (Figure 4r). However, the underlying polyhedron is not the octahedron but the icosahedron for which the relationship in Equation (10) yields $n = 92$. Because of this coincidence, finding the polyhedron for which $IQ$ is a maximum needs optimization in one variable. The obtained maximum $IQ$ is $0.966957236\ldots$ The obtained polyhedron having symmetry $I_h$ is shown in Figure 5r.

#### 4.3.2. $n = 122$

Let the pair of Goldberg–Coxeter parameters be $(2, 2)$ for which $T = 12$. From Equation (10), $n = 122$ is obtained. Here $v = 1, e = 1, f = 4$ (Figure 4s). On the face of the spherical icosahedron, there is one point of tangency at the face centre, and there are three points of tangency in threefold rotational symmetry with respect to the face centre. In theory, these three points can concertedly move on the face with two degrees of freedom. However, as the configuration of the related spherical covering with 122 equal circles has planes of symmetry [12], we suppose that the arrangement of the points of tangency also has planes of symmetry, and so we have symmetry $I_h$ instead of $I$. The three points lie on the three altitudes of the face triangle, where they can concertedly move with one degree of freedom.

Consequently, finding the maximum of $IQ$ required optimization in one variable. The executed numerical optimization resulted in $IQ = 0.975117622\ldots$ and in the polyhedron in Figure 5s.

### 4.3.3. $n = 132$

Let the pair of Goldberg–Coxeter parameters be $(3, 1)$ for which $T = 13$. From Equation (10), $n = 132$ is obtained. Here $v = 1, e = 0, f = 6$ (Figure 4t). The six points of tangency form two sets of threes, where the three points in each set can concertedly move with two degrees of freedom on the face triangle of the spherical icosahedron, while maintaining threefold symmetry with respect to the face centre. Since the two sets of points can move independently of each other, the whole arrangement has four degrees of freedom. In this way, finding the optimal polyhedron, for which $IQ$ is the maximum, required optimization in four variables. The numerical optimization process started with a configuration where the centres of the 132 equal circles forming the best known covering of the sphere [20] were considered as the points of tangency of the 132 faces of the polyhedron. The process eventually resulted in $IQ = 0.976993221\ldots$ The polyhedron shown in Figure 5t. Since the polyhedron has no planes of symmetry, the symmetry of the polyhedron is $I$.

The roundest multi-symmetric polyhedra proven or conjectured by us together with those previously published by other authors are shown in Figure 5, and the data characterizing these polyhedra are collected in Table 2.

**Table 2.** Polyhedra with $n$ faces that maximize the isoperimetric quotient $IQ$ under tetrahedral, octahedral, or icosahedral symmetry constraints. Polyhedra are characterized by $n$, point group symmetry $G$, Goldberg–Coxeter parameters $(b, c)$, isoperimetric quotient $IQ$. Upper bound of $IQ$ is given via Goldberg's formula. Particular properties are discussed in footnotes.

| $n$ | $G$ | $(b, c)$ | $IQ$ | Upper Bound | Remarks |
|---|---|---|---|---|---|
| 4 | $T_d$ | $(1, 0)$ | 0.302299894 [a] | 0.302299894 | Proven, Fejes Tóth [6] |
| 6 | $O_h$ | $(1, 0)$ | 0.523598775 [a] | 0.523598775 | Proven, Fejes Tóth [6] |
| 8 | $O_h$ | $(1, 1)$ [b] | 0.604599788 | 0.637349714 | Proven, this work |
| 10 | $T_d$ | $(2, 0)$ | 0.630745372 | 0.707318712 | Proven, this work |
| 12 | $I_h$ | $(1, 0)$ | 0.754697399 [a] | 0.754697399 | Proven, Fejes Tóth [6] |
| 14 | $O_h$ | $(1, 1)$ | 0.781638893 | 0.788894402 | Huybers [14]; proven, this work |
| 16 | $T_d$ | $(3, 0)$ [c] | 0.812189098 | 0.814733609 | Goldberg [5] |
| 18 | $O_h$ | $(2, 0)$ | 0.823218074 | 0.834942754 | This work |
| 20 | $T_d$ | $(3, 0)$ | 0.830222439 | 0.851179828 | This work |
| 22 | $T_d$ | $(2, 2)$ [c] | 0.862408738 | 0.864510388 | This work |
| 26 | $O_h$ | $(2, 0)$ [d] | 0.876811431 | 0.885098414 | Huybers [15] |
| 30 | $O$ | $(2, 1)$ | 0.896930384 | 0.900256896 | This work |
| 32 | $I_h$ | $(1, 1)$ | 0.905798260 | 0.906429544 | Goldberg [5] |
| 38 | $O_h$ | $(3, 0)$ | 0.917445003 | 0.921082160 | This work |
| 42 | $I_h$ | $(2, 0)$ | 0.927651905 | 0.928542518 | Goldberg [5] |
| 62 | $I_h$ | $(2, 0)$ [e] | 0.945021022 | 0.951478663 | Huybers [15] |
| 72 | $I$ | $(2, 1)$ | 0.957881213 | 0.958189143 | Tarnai et al. [10] |
| 92 | $I_h$ | $(3, 0)$ | 0.966957236 | 0.967248411 | This work |
| 122 | $I_h$ | $(2, 2)$ | 0.975117622 | 0.975282102 | This work |
| 132 | $I$ | $(3, 1)$ | 0.976993221 | 0.977150391 | This work |

[a] Mathematically proven to be the best without enforcing symmetry; [b] The vertices of the regular octahedron are not points of tangency; [c] The vertices of the regular tetrahedron are not points of tangency; [d] The system of the lattice points is supplemented with the face midpoints of the regular octahedron; [e] The system of the lattice points is supplemented with the face midpoints of the regular icosahedron.

## 5. Discussion

A numerical iterative method developed recently by the authors has now been applied to produce conjectured solutions for the isoperimetric problem of polyhedra if tetrahedral, octahedral, or icosahedral symmetry constraints are prescribed. The algorithm ensures that the polyhedra are local optima, and they are either identical or close to global optima if initial face arrangements are chosen appropriately. In some of the cases such initial arrangements for the conjectured solutions proposed by the authors originated from optimal circle coverings on a sphere. The small differences between the actual $IQ$ values and the corresponding upper bounds (e.g., for $n = 72, 92, 122, 132$) render the polyhedra likely to be global optima.

Using our numerical investigations, we have found (proven or conjectured) the roundest Goldberg polyhedra with icosahedral symmetry for the Goldberg–Coxeter parameters $(1,0)$, $(1,1)$, $(2,0)$, $(2,1)$, $(2,2)$, $(3,0)$, $(3,1)$; that is for face numbers $n = 12, 32, 42, 72, 122, 92, 132$. This list is complete up to $(3,1)$. We also have octahedrally symmetric Goldberg polyhedra for the Goldberg–Coxeter parameters $(1,0)$, $(1,1)$, $(2,0)$, $(2,1)$ and $(3,0)$, that is, for face numbers $n = 6, 14, 18, 30$ and $38$, as well as tetrahedrally symmetric Goldberg polyhedra for the Goldberg–Coxeter parameters $(1,0)$, $(1,1)$, $(2,0)$ and $(3,0)$, that is for face numbers $n = 4, 8, 10$ and $20$. Here the tetrahedral case $(1,1)$ is special because the two trivalent vertices on each of the edges of the underlying tetrahedron coincide forming four-valent vertices, and the hexagons become triangles. In this way, the truncated tetrahedron becomes a regular octahedron. Recently, intense research was conducted into constructing "equilateral" Goldberg polyhedra [21]. This raised the question as to whether these polyhedra could be used to extend the range of the conjectured roundest Goldberg polyhedra. Unfortunately, the answer is in the negative. The "equilateral" Goldberg polyhedra are only "nearly" spherical, and so they have no insphere.

We note that Schoen [22] mentioned that recently Deeter numerically determined the solution for $n = 122$ without enforcing any symmetry. The optimal polyhedron he obtained has icosahedral symmetry $I_h$. Unfortunately, the numerical data of this polyhedron are not provided, so we could not decide whether his polyhedron is identical to ours.

Four out of the five Platonic polyhedra are the roundest multi-symmetric polyhedra for $n = 4, 6, 8, 12$. A natural question is whether among the semi-regular polyhedra there are some which are the roundest for $n > 12$. The semi-regular polyhedra have tetrahedral, octahedral or icosahedral symmetry, look quite spherical, and additionally their $IQ$ values are known [23]. The Archimedean polyhedra are out of question because they have no insphere. However, it is possible to modify them according to isodistant truncation, as happened, for instance, to the truncated octahedron and truncated icosahedron ($n = 14$ and $32$). The Archimedean duals, that is, the Catalan polyhedra are worth investigating because they do have an insphere. Among the Archimedean duals there are two polyhedra for $n = 12$, four polyhedra for $n = 24$, one polyhedron for $n = 30$, one polyhedron for $n = 48$, four polyhedra for $n = 60$, and one polyhedron for $n = 120$. Polyhedra for $n = 12$ are not interesting because the roundest is the regular dodecahedron. Among the four polyhedra for $n = 24$, the dual of the snub cube is the best, for which $IQ = 0.872628\ldots$ However, the points of tangency have two degrees of freedom here, therefore by optimization this $IQ$ value can be increased. In the case of $n = 30$, the rhombic triacontahedron has $IQ = 0.887200\ldots$, which is smaller than that we found for the polyhedron in Figure 5l. In the case of $n = 48$, the dual of the great rhombicuboctahedron has $IQ = 0.910066\ldots$, which is smaller than that for $n = 42$ given by Goldberg, therefore it is probably not optimal. Among the four polyhedra for $n = 60$, the dual of the snub dodecahedron is the best, for which $IQ = 0.945897\ldots$ However, because of an argument similar to that in the case of $n = 24$, the polyhedron is not optimal. In the case of $n = 120$, the dual of the great rhombicosidodecahedron has $IQ = 0.957765\ldots$, which is much smaller than that we have for $n = 92$, therefore, probably it cannot be optimal.

Figures 1 and 2 show some applications for $n = 14$ and 32. After having some new results we discovered that the conjectured roundest polyhedron obtained for $n = 18$ (Figure 5h) decorates the memorial to Thomas Bodley in the chapel of Merton College, Oxford, which was erected in 1615 [24,25].

**Acknowledgments:** This work was supported by NKFI under Grants K81146 and K119440.

**Author Contributions:** Tibor Tarnai worked on the theoretical backgorund. Zsolt Gáspár and András Lengyel contributed equally to the numerical computations.

**Conflicts of Interest:** The authors declare no conflict of interest. The founding sponsors had no role in the design of the study; in the collection, analyses, or interpretation of data; in the writing of the manuscript, and in the decision to publish the results.

# References

1.  Deza, A.; Deza, M.; Grishukhin, V. Fullerenes and coordination polyhedra versus half-cubes embeddings. *Discret. Math.* **1998**, *192*, 41–80.
2.  Lindelöf, L. Propriétés générales des polyedres qui, sous une étendue superficielle donnée, renferment le plus grand volume. *Bull. Acad. Sci. St. Pétersb.* **1869**, *14*, 257–269.
3.  Lindelöf, L. *Recherches sur les Polyèdres Maxima*; Officina Typographica Societatis Litterariae Fennicae: Helsingfors, Finland, 1899; Volume 24.
4.  Pólya, G. *Mathematics and Plausible Reasoning. Vol. I, Induction and Analogy in Mathematics*; Princeton University Press: Princeton, NJ, USA, 1954.
5.  Goldberg, M. The isoperimetric problem for polyhedra. *Tôhoku Math. J.* **1935**, *40*, 226–236.
6.  Fejes Tóth, L. The isepiphan problem for *n*-hedra. *Am. J. Math.* **1948**, *70*, 174–180.
7.  Schoen, A. A defect-correction algorithm for minimizing the volume of a simple polyhedron which circumscribes a sphere. In Proceedings of the 2nd Annual ACM Symposium on Computational Geometry, Yorktown Heights, NY, USA, 2–4 June 1986; ACM Press: New York, NY, USA, 1986; pp. 159–168.
8.  Schoen, A. *Supplement to a 'Defect-Correction Algorithm for Minimizing the Volume of a Simple Polyhedron Which Circumscribes a Sphere'*; Technical Report No 86-01*; Department of Computer Science, Southern Illinois University: Carbondale, IL, USA, 1986.
9.  Mutoh, N. The polyhedra of maximal volume inscribed in the unit sphere and of maximal volume circumscribed about the unit sphere. In *Discrete and Computational Geometry*; Akiyama, J., Kano, M., Eds.; Volume 2866 of the Series Lecture Notes in Computer Science; Springer: New York, NY, USA, 2003; pp. 204–214.
10. Tarnai, T.; Gáspár, Z.; Lengyel, A. From spherical circle coverings to the roundest polyhedra. *Philos. Mag.* **2013**, *93*, 3970–3982.
11. Tarnai, T.; Gáspár, Z. Covering the sphere by equal circles, and the rigidity of its graph. *Math. Proc. Camb. Philos. Soc.* **1991**, *110*, 71–89.
12. Tarnai, T.; Wenninger, M. Spherical circle-coverings and geodesic domes. *Struct. Topol.* **1990**, *16*, 5–21.
13. Tarnai, T. Hidden geometrical treasures. *Math. Intell.* **2013**, *35*, 76–80.
14. Huybers, P. Isodistant polyhedra or 'isohedra'. *J. Int. Assoc. Shell Spat. Struct.* **2013**, *54*, 15–25.
15. Huybers, P. Soccer ball geometry, a matter of morphology. *Int. J. Space Struct.* **2007**, *22*, 151–160.
16. Goldberg, M. A class of multi-symmetric polyhedra. *Tôhoku Math. J.* **1937**, *43*, 104–108.
17. Coxeter, H.S.M. *Introduction to Geometry*; Wiley: New York, NY, USA, 1961.
18. Caspar, D.L.D.; Klug, A. Physical principles in the construction of regular viruses. *Cold Spring Harb. Symp. Quant. Biol.* **1962**, *27*, 1–24.
19. Coxeter, H.S.M. Virus macromolecules and geodesic domes. In *A Spectrum of Mathematics*; Butcher, J.C., Ed.; Auckland University Press and Oxford University Press: Auckland, New Zealand, 1971; pp. 98–107.
20. Gáspár, Z.; Tarnai, T. Cable nets and circle-coverings on a sphere. *Z. Angew. Math. Mech.* **1990**, *70*, T741–T742.
21. Schein, S.; Gayed, J.M. Fourth class of convex equilateral polyhedron with polyhedral symmetry related to fullerenes and viruses. *Proc. Natl. Acad. Sci. USA* **2014**, *111*, 2920–2925.
22. Schoen, A.H. Roundest Polyhedra. Available online: http://schoengeometry.com/a_poly.html (accessed on 1 December 2016).
23. Aravind, P.K. How spherical are the Archimedean solids and their duals? *Coll. Math. J.* **2011**, *42*, 98–107.

24. Wilson, J. The memorial by Nicholas Stone to Sir Thomas Bodley. *Church Monum.* **1993**, *8*, 57–62.
25. Tarnai, T.; Krähling, J. Polyhedra in churches. In Proceedings of the IASS-SLTE 2008 Symposium, Acapulco, Mexico, 27–31 October 2008; Oliva Salinas, J.G., Ed.; Universidad Nacional Autonoma de Mexico: Mexico City, Mexico, 2008.

*symmetry*

MDPI

*Article*
# Barrel Pseudotilings

**Undine Leopold**

Department of Mathematics, Northeastern University, Boston, MA 02115, USA;
E-Mail: leopold.u@husky.neu.edu; Tel.: +1-617-373-5655; Fax: +1-617-373-5658

Received: 29 June 2012; in revised form: 13 August 2012 / Accepted: 19 August 2012 /
Published: 30 August 2012

**Abstract:** This paper describes 4-valent tiling-like structures, called pseudotilings, composed of barrel tiles and apeirogonal pseudotiles in Euclidean 3-space. These (frequently face-to-face) pseudotilings naturally rise in columns above 3-valent plane tilings by convex polygons, such that each column is occupied by stacked congruent barrel tiles or congruent apeirogonal pseudotiles. No physical space is occupied by the apeirogonal pseudotiles. Many interesting pseudotilings arise from plane tilings with high symmetry. As combinatorial structures, these are abstract polytopes of rank 4 with both finite and infinite 2-faces and facets.

**Keywords:** tiling; pseudotiling; abstract polytope; realization

---

## 1. Introduction

An *n-gonal barrel* is a simple polyhedron whose outer shell is easily constructed. Take two *n*-gons, surround each by a ring of *n* pentagons, then glue the two halves together at their respective boundaries. For example, a hexagonal barrel is shown in Figure 1a, and a pentagonal barrel is a (not necessarily regular) pentagonal dodecahedron.

This article defines and investigates structures in Euclidean 3-space which are closely related to both normal simple tilings by barrels and normal simple tilings in the plane. The deviation from classical tiling theory occurs when we introduce *apeirobarrels*, barrels over relatively tame apeirogons (infinite simple polygons in 3-space), which are not tiles in the typical sense as they contain no physical volume. For this reason, we call our constructed objects *barrel pseudotilings*.

Allowing infinite regular polygons as faces, Grünbaum [1] discovered many new regular polyhedra in Euclidean 3-space besides the already well-known Platonic solids, planar tessellations, Kepler–Poinsot polyhedra, and Petrie–Coxeter polyhedra (see also [2]). The list of 47 regular polyhedra presented in [1] was just one polyhedron short of being complete, which was remedied by Dress ([3,4]) a few years later, who also proved the completeness of the list. Later, Leytem [5] found an alternative way to obtain the elusive 48th Grünbaum–Dress polyhedron.

Over the last few decades, the development of the theory of abstract polytopes and their realizations provided a common framework for the investigation of regular and other polytopes. This generalization of polyhedra and polytopes to combinatorial objects seemed only natural after a long evolution of the terms *polygon* and *polyhedron*. While the Grünbaum–Dress polyhedra are regular (abstract) 3-polytopes, realized in 3-space, McMullen and Schulte [6] investigated faithful and discrete realizations of regular 4-polytopes in 3-space, in addition to providing a new proof of Dress' completeness result. Their complete list encompasses 8 regular 4-apeirotopes (infinite polytopes) in $\mathbb{E}^3$. The subject of abstract polytopes and their realizations has developed greatly since and is still evolving (see, e.g., [7,8]).

After giving a first example of a barrel pseudotiling in Section 2, we review the terminology in Section 3, and then proceed to Section 4 for the proper definitions. We create each barrel pseudotiling from a normal, simple, plane tiling by convex *polygons*, by a process outlined in Section 5.

After analyzing the properties of barrel pseudotilings in Section 6, we arrive at an equivalent definition for certain classes in the form of decorated plane triangulations (Section 7). More examples and some remarks follow in Sections 8 and 9.

What makes the pseudotilings interesting, among other things, is that they are faithful realizations of certain abstract rank 4 polytopes (see Sections 3.4 and 4). For faces of rank 0, 1, and 2, we have vertices, edges, and *planar* finite faces, as well as non-planar apeirogons. The tiles (barrels) and apeirogonal pseudotiles (apeirogonal barrels) form the set of 3-faces (facets). These polytopes are simple, as well as both infinite in all inner ranks and with (infinitely many) infinite rank 2 faces, but they can be realized faithfully in low-dimensional Euclidean 3-space with a translational symmetry in one direction. While realizations of *regular* and *chiral* abstract 4-polytopes in $\mathbb{E}^3$ have received the most attention (e.g. [6,7], Section 7F of [8]), it still seems worthwhile to investigate other, less symmetrical, classes as well. The construction process connects the barrel pseudotilings to (partially) directed infinite graphs stemming from triangulations, many of which are visually appealing. It may be worth exploring these connections in the future.

As a note of caution to the reader, in deviating from the original usage in [1], the term *apeirogon* denotes in this paper *any* infinite, simple polygon. The prefix *apeir-* is used to emphasize the presence of infinite faces or infinitely many faces of some kind (similar to the usage in [6,8]), however it is *not* used to denote regularity or even a particular symmetry.

## 2. A First Example

Start with the regular hexagonal tiling $(6^3)$ of the plane. Color the tiles in the tiling properly (*i.e.*, such that no two adjacent tiles have the same color) with three colors, see Figure 2a. For reasons that will soon become clear, the color labels have been chosen from the set $\{0, 0.5, \infty\}$. Consider a hexagonal barrel stemming from a hexagonal *right* prism of height 1, such as in Figure 1b, with the prism's mantle altered as shown "unwrapped" in Figure 1c. We can construct an infinite stack of pairwise congruent hexagonal barrels of this kind whose common faces are hexagons. Call this object in space a hex-stack. Now, consider our planar hexagonal tiling as situated on a horizontal plane in Euclidean 3-space. Each hexagon gives rise to a hexagonal column which extends indefinitely in either direction perpendicular to the plane. The columns associated to colors 0 and 0.5 of our base tiling can be filled up with directly congruent hex-stacks. It is possible to do that in such a way that all the inclined (*i.e.*, neither horizontal nor vertical) edges in the barrels' sides face a column associated to color $\infty$. Then, in order to obtain a partial face-to-face tiling (see Section 3 for terminology) of 3-space from all these hexagonal barrels, we have to match up the pentagons which, as points sets, appear as rectangles. For this we shift the columns over hexagonal tiles with color 0.5 vertically by half a step relative to those over hexagonal tiles with color 0 (see Figure 2b), which explains the choice of labels. Note that the planar tiling $(6^3)$ functions merely as a guide in the construction of the columns. The plane in which $(6^3)$ is situated serves as a reference plane for applying shifts when positioning adjacent barrel-filled columns relative to each other in order to match the pentagons.

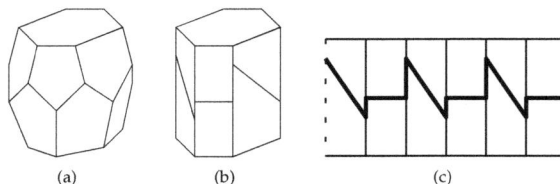

(a)    (b)    (c)

**Figure 1.** (**a**) A hexagonal barrel; (**b**) A hexagonal barrel stemming from an hexagonal *right* prism; (**c**) The mantle of a hexagonal barrel stemming from an hexagonal *right* prism.

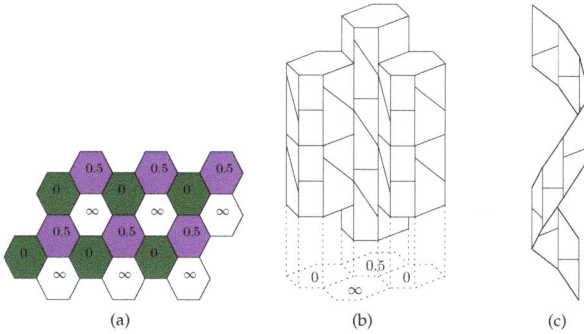

**Figure 2.** (**a**) A proper 3-coloring of $(6^3)$; (**b**) Part of the constructed barrel pseudotiling. Note the emerging face structure of the column associated to color ∞, particularly the apeirogons (spirals); (**c**) Part of a single apeirobarrel.

The remaining part of 3-space separates into infinite columns associated to tiles of color ∞. Let these columns inherit the boundary structure from the hexagonal barrels in adjacent columns. As a result, the boundary of each column colored ∞ falls apart into three "intertwined" or "stacked" apeirobarrels (see Figure 2b,c), where an *apeirobarrel* is defined as two parallel apeirogons joined by two infinite "rings" of pentagons. Even though no physical volume is occupied by these *pseudotiles*, the last set of columns can now also be regarded as "tiled", preserving the face-to-face property and simplicity. Another way to say this is that apeirobarrels can be stacked so that the union of their mantles covers the mantle of an infinite prismatic column completely without overlaps, which is consistent with the way finite barrels stack. This can be done in a left-handed and a right-handed version in the obvious way.

Moreover, a similar construction is possible for several other plane tilings, including the uniform tilings $(4.6.12)$ and $(4.8^2)$, in which the regular convex polygons surrounding each vertex have four, six, and twelve, respectively four, eight, and eight vertices. Naturally, the question comes up which kinds of *barrel pseudotilings* arise this way. This is explored in this article.

Related kinds of tiling-like structures with apeirogons, but in the Euclidean plane instead of 3-space, have been investigated by Grünbaum, Miller, and Shephard in [9]. The similarity with the structures considered in this article is perhaps greatest in some of the planar structures (called strip tilings) depicted in Figure 6 of [9].

## 3. Basic Notions

In order to keep the exposition as self-contained as possible, we will now review basic notions from several areas of relevance.

### 3.1. Tilings and Barrel Polyhedra

A *tiling* of Euclidean *d*-space $\mathbb{E}^d$ is a locally finite, countable collection of closed topological cells (*tiles*), which satisfy the following three conditions. First, each tile is a *topological d-polytope*, a homeomorphic image of a convex *d*-polytope. Second, the union of all tiles is $\mathbb{E}^d$. Third, the intersection of any two distinct tiles is either empty or a proper face of each; in particular, if this intersection is a *k*-face of each tile, then it is a *k-face* of the tiling. In other places in the literature, such a tiling may be called *face-to-face* (compare [10,11]). A *normal* tiling is a tiling in which the tiles are uniformly bounded in size (*i.e.*, each tile contains a *d*-ball of some small radius, and is contained in a *d*-ball of some larger radius). A tiling of $\mathbb{E}^d$ is *simple* if all vertices are $(d+1)$-valent. However, this latter notion is only applicable when "vertices" and "edges" are well-defined, such as in the context of plane tilings or face-to-face tilings (for any *d*). The face poset of a (face-to-face) tiling, when amended by a unique largest and a unique smallest (empty) face, can be interpreted as an abstract polytope, see Section 3.4.

A *barrel over a topological n-gon* is a simple topological 3-polytope with two disjoint distinguished *n*-gonal 2-faces, such that all other 2-faces are pentagonal and adjacent to exactly one of the distinguished faces. Thus, the distinguished faces—also called *top* and *bottom* (*base*) of the barrel—have *n* pentagonal adjacents each, giving a total of 2*n* pentagonal faces (if $n = 5$, all $12 = 2n + 2$ faces are pentagons). In short, the sides of the barrel are formed by two rings of pentagons. In an extreme case, such a barrel can be obtained (but not as a convex polytope in the strict sense!) from an *n*-gonal right prism by partitioning the rectangles in the prism's mantle accordingly (as indicated in Figure 1b). In this article, we will explicitly allow faces in polytopes to be coplanar, even when they are adjacent.

## 3.2. Plane Graphs

All graphs considered in this paper will be *simple*, which means that no multiple edges or loops occur, but they will not necessarily be finite. In fact, most graphs we encounter will be infinite.

Let $G = (V, E)$ be such a (simple) graph. A *dominating set* (compare [12]) is a subset of vertices $V' \subseteq V$, such that each vertex $v \in V$ is either in $V'$ itself, or adjacent (via an edge) to a vertex in $V'$, or both. By contrast, a subset $V' \subseteq V$ is *independent* if no two distinct vertices $v, w$ of $V'$ are adjacent in $G$. A *maximal independent set* is an independent set $V'$ of vertices with the property that there is no independent superset ($V'$ cannot be enlarged without becoming dependent, *i.e.*, without introducing a pair of distinct adjacent vertices). As such, maximal independent sets are equivalent to dominating independent sets ([12], p. 117, the result for finite graphs can be extended to infinite graphs).

## 3.3. Simplicial Complexes

A *(geometric) simplicial complex* $\Delta$, in the classical sense, is a collection of geometric *simplices* (the convex hull of $(d + 1)$ affinely independent points, where *d* denotes the dimension of the simplex) in some Euclidean space, with the following two properties. Faces of simplices in $\Delta$ are again in $\Delta$, and the intersection of any two simplices from $\Delta$ is a face of each (and therefore in $\Delta$). If $\Delta$ is not an empty collection, then these properties imply that it contains at least the empty face (as a $(-1)$-dimensional simplex). As usual, and compatible with the notions from graph theory or tiling theory, the zero-dimensional (one-dimensional, two-dimensional) simplices are called *vertices* (*edges*, *triangles*).

The (open) star of a vertex $v$, $St_\Delta(v)$, is the collection of all simplices incident to $v$. By contrast, the *simplicial neighborhood*, or *closed star*, denoted $Nb_\Delta(v)$, is defined as the collection consisting of all simplices incident to $v$ (*i.e.*, containing $v$ as a face), and all their (simplicial) faces. Therefore, the simplicial neighborhood is a simplicial complex. The *link* of a vertex $v$ in $\Delta$, denoted $Lk_\Delta(v)$, is defined as $Lk_\Delta(v) := Nb_\Delta(v) \setminus St_\Delta(v)$. (For further reference, see also [13], pp. 31–42.)

For the purpose of this article, we will use the term *geometric simplicial complex* even when the simplices are only homeomorphic images of the standard simplex in the appropriate dimension, as long as all other properties are retained. For example, a (topological) triangulation of the plane is a two-dimensional geometric simplicial complex.

The *k-skeleton* of a simplicial complex ignores all faces of dimension greater than $k$, and is, in fact, a simplicial complex. We will use the notation $G(\Delta)$ to refer to the 1-skeleton, or induced *edge graph*, of a simplicial complex $\Delta$.

## 3.4. Abstract Polytopes

We will only need some very basic ideas about abstract polytopes, which are combinatorial objects generalizing the previously existing notions of (convex or non-convex) polyhedra and polytopes. For a thorough discussion of abstract polytopes it is recommended to consult the standard reference by McMullen and Schulte [8]. Briefly, an *abstract n-polytope* $\mathcal{P}$ is a certain kind of graded poset of *faces* (including a unique smallest face of rank $-1$, and a unique largest face of rank *n*), with the partial order being *incidence*. The rank of a face corresponds to the notion of *dimension* for traditional polytopes (such as the convex polytopes), and so the traditional terms *vertex* and *edge* are used for the rank 0 and rank 1 faces, respectively. The poset $\mathcal{P}$ needs to satisfy the following two additional combinatorial properties taken from standard polytopes. First, $\mathcal{P}$ is *strongly flag-connected*, and second, if two incident faces $F < G$ are exactly *two* ranks apart, then there exist precisely *two* faces $H$ such that $F < H < G$ (compare the definitions in [8], pp. 22–25).

In order to understand the first condition, we need to know that a *section* of $\mathcal{P}$ is a graded poset defined for any two incident faces $F \le G$ as $\{H \in \mathcal{P} | F \le H \le G\}$, again with incidence as partial order. A *flag* is a maximal chain in a poset, and for graded posets of finite rank all flags have the same length (for example, all flags in $\mathcal{P}$ have length $n + 2$). $\mathcal{P}$ is said to be *strongly flag-connected* if each section of $\mathcal{P}$ (including $\mathcal{P}$ itself) is *flag-connected*, that is, if each flag can be joined to any other flag (within the same section) via a sequence of flags, changing only one element (face) in the flag at a time.

The second condition is commonly called the *diamond condition*, as $F$, $G$, and the two faces $H$ in between form the shape of a diamond in the Hasse diagram $\mathcal{P}$. For example, in an abstract 4-polytope (which is the kind of polytope that we will encounter in this article), the diamond condition stipulates the following:

(1) Each edge is incident to precisely two vertices;

(2) For each 2-face and each of its incident vertices, there are precisely two edges which are incident with both the 2-face and the vertex;

(3) For each 3-face (*facet*) and each of its incident edges, there are precisely two 2-faces incident with both;

(4) There are precisely two facets incident with each 2-face.

A *faithful realization* of an abstract polytope $\mathcal{P}$ is an injective mapping of its vertices into a suitable Euclidean space, along with a suitable interpretation of its combinatorial structure in the geometric setting. When considering realizations of abstract *regular* polytopes (see [6–8], Section 7F), one usually requires that the symmetry is retained, *i.e.*, that the flag-transitive automorphism group of the abstract polytope carries over to a group of isometries which acts transitively on the flags of the realization. In this paper, however, we drop this requirement in order to obtain more realizations. To give an example of a realization of a 4-polytope that is not necessarily regular, any face-to-face, normal tiling of $\mathbb{E}^3$ is a realization of a so-called abstract 4-*apeirotope* (polytope containing either infinitely many finite faces, or infinite faces, or both).

## 4. Definition of Barrel Pseudotilings

This section is devoted to the proper, but rather technical, definition of a *barrel pseudotiling*. All the barrels in the considered pseudotilings stem from (right) prisms in one form or another, so we begin by explaining the concept of a *stack of apeiroprisms*.

**Definition 4.1.** *Let $D$ be a plane polygonal disc in a horizontal plane of $\mathbb{E}^3$ with unit normal vector $u = (0, 0, 1)$. A STACK OF APEIROPRISMS, or an $\infty$-PRISM STACK, $\Phi$ associated with $D$, is a doubly infinite right prism, or cylinder, $Z = D + \mathbb{R}u$ together with a family $\mathcal{A} = \{A_0, \ldots, A_{k-1}\}$ of $k > 1$ mutually non-intersecting apeirogons in $\partial Z = \partial D + \mathbb{R}u$ (the mantle of the cylinder) with the following properties:*

*(1) $\mathcal{A}$ is strictly monotone in the direction of $u$, meaning that every section of $Z$ by a plane parallel to $\mathrm{aff}(D)$ meets every apeirogon in $\mathcal{A}$ in exactly one point;*

*(2) Any two adjacent edges of an apeirogon in $\mathcal{A}$ lie in adjacent bounding walls of $Z$. Thus, each apeirogon $A$ in $\mathcal{A}$ spirals around $Z$, in both directions, such that its vertices lie above the vertices of $D$;*

*(3) For each $i = 0, \ldots, k-1$ and each $x$ in $A_i$, we have $x + ru \in A_l$ for some $l$ if and only if $r \in \mathbb{Z}$; and in particular, $x + ju \in A_{i+j}$ for each $j \in \mathbb{Z}$, with subscripts taken modulo $k$. (Thus, more informally, the points directly above $x$ on apeirogons occur at integer distances from $x$, with the apeirogons repeating periodically modulo $ku$.)*

Any $A_i + [0,1]u$ (for $i = 0, 1, \ldots, k-1$) is the mantle of a single *apeiroprism* in a faithful realization (the apeiroprism itself, however, is an abstract polytope whose face poset contains all faces of the mantle, *i.e.*, the parallelograms and their faces, as well as the apeirogons $A_i$ and $A_{i+1}$, and a unique three-dimensional largest face). Subsequently, each apeiroprism gives rise to an *apeirobarrel* via a subdivision of the mantle parallelograms into pentagons.

**Definition 4.2.** *A* STACK OF APEIROBARRELS, *or an* ∞-STACK, $\Omega$, *is a stack of apeiroprisms as in Definition 4.1, together with a partition of its mantle into (topological) pentagon faces obtained in the following way. All edges of the apeirogons in $\mathcal{A}$ are retained as edges of pentagons; each vertical edge (parallel to $u$) of the mantle is split into three edges by two new vertices, called* PARTITION VERTICES, *in its relative interior; and each parallelogram on the mantle $A_i + [0,1]u$ of an apeiroprism (for $i = 0, 1, \ldots, k-1$) is subdivided into two coplanar pentagon faces by a new edge that connects a pair of new vertices on opposite vertical edges of the parallelogram. These new edges are called* PARTITION EDGES.

A *staircase* is any infinite polygon in the mantle of $\Omega$ whose successive edges consist alternately of partition edges and vertical edges, such that all traversed vertices are partition vertices (see Figure 3a for an example). Intuitively, the partition edges and vertical edges in these mathematical staircases correspond to the treads and risers of ordinary staircases. Note that staircases do not cross the apeirogonal bases of apeiroprisms (which, in extending the analogy, would take the role of stringers). The *winding orientation* of an ∞-stack is the orientation (clockwise or counterclockwise) in which the apeirogonal bases wind upwards (*i.e.*, in the direction of $u = (0,0,1)$) and is already prescribed by the apeirogons in $\Phi$.

**Definition 4.3.** *Let B be a finite barrel stemming from an n-gonal right prism of height 1 whose polygonal disc base D is in a horizontal plane with unit normal vector $u = (0,0,1)$. Call the added vertices in the vertical edges of the prism* PARTITION VERTICES, *and the added non-vertical edges separating a rectangle into pentagons* PARTITION EDGES. *Then, in this finite case, a* STAIRCASE *is a simple closed polygon consisting of edges of the barrel and passing only through partition vertices. A* STACK OF n-GONAL BARRELS *or n-STACK, $\Omega$, *is the set of barrels $\{B + lu \mid l \in \mathbb{Z}\}$, whose union fills the doubly infinite right prism (cylinder) $Z = D + \mathbb{R}u$. A* FINITE STACK *is an n-stack for some finite n.*

Thus, every barrel in a finite stack has an "impossible staircase" (familiar, e.g., from Dutch artist M.C. Escher's print *Ascending and Descending* from 1960, see [14], p. 146): Partition edges (treads) and vertical edges linking partition edges (risers) close up to give a finite, non-planar polygon.

**Definition 4.4.** *A* BARREL PSEUDOTILING $\mathcal{B}$ *is a countable collection of vertical ∞-stacks and vertical finite stacks with the following two properties. First, the underlying family of doubly infinite prisms (cyclinders) gives a locally finite tiling of $\mathbb{E}^3$ which, when cut by a horizontal plane, determines a locally finite, simple, face-to-face tiling in this plane. Second, any two barrels associated to distinct stacks in $\mathcal{B}$ intersect, if at all, in a common finite face (vertex, edge, or pentagon). Note that the construction of the stacks implies that distinct barrels in the same finite stack also intersect either in a common face (finite polygon), or not at all.*

Observe that translation by $u$ is a built-in symmetry of the pseudotiling $\mathcal{B}$. The finite barrels by themselves form a tessellation of a noncompact 3-manifold with infinitely many boundary components. In a way, the boundary components get sewn together by the stacks of apeirobarrels.

Combinatorially, the faces of the constructed object, together with a unique largest face of rank 4 and a unique smallest (empty) face of rank $-1$, form a graded poset as required for an abstract 4-polytope. The faces of rank 0 are the vertices, the faces of rank 1 are the edges of the pseudotiling, the faces of rank 2 are the planar polygons and non-planar apeirogons, and the faces of rank 3 are the tiles (barrels) and pseudotiles (apeirobarrels). There are infinitely many faces of each of these kinds. Additionally, we have *infinite 2-faces*, the bases of the apeirobarrels (*i.e.*, the apeirogons). Strong flag-connectedness is easily verified. The diamond condition holds because the criteria for a 4-polytope as listed in Section 3.4 are fulfilled. In particular, each 2-face is incident to precisely two 3-faces; it is here where the condition of $k > 1$ apeirogons (and thus $k > 1$ apeirobarrels) per $\infty$-stack is required. Thus, the abstract polytope underlying $\mathcal{B}$ is a (non-regular) 4-apeirotope.

This 4-apeirotope is realized faithfully in Euclidean 3-space, such that all finite faces lie in an affine subspace of the appropriate dimension and coincide with the convex hull of their vertices (straight edges, plane polygons, finite barrels stemming from right prisms). The size of the finite barrels is uniformly bounded, so a property very close to normality is retained. All vertices are 4-valent, so, in many ways, barrel pseudotilings are similar to simple normal tilings of 3-space.

We conclude this section with a note on the face-to-face property.

**Remark 4.1.** $\mathcal{B}$ is FACE-TO-FACE, *in the sense that any two distinct barrels (finite or infinite) intersect, if at all, in a common face, precisely when each $\infty$-stack consists of at least three apeirobarrels.*

## 5. Outline of the Construction

A rough outline of constructing a pseudotiling by polygonal barrels and $\infty$-barrels is provided by the following five steps.

First of all, select a plane, simple, normal tiling $\mathcal{T}$ by (convex, simple) polygons, situated in the horizontal plane through the origin of $\mathbb{E}^3$, where the unit normal vector $u = (0, 0, 1)$ denotes the direction "up". All plane tilings mentioned in this paper are assumed to be of this type (recall that our definition of tiling includes the face-to-face property). We call $\mathcal{T}$ a *base tiling*. The faces of $\mathcal{T}$ will not (or not necessarily) be used in the completed barrel pseudotiling; rather, its tiles serve as guides for the infinite columns in $\mathbb{E}^3$ which will extend in the directions $\pm u$. Second, for each $n$-gonal tile ($n \geq 3$) of $\mathcal{T}$ assign either a color in $[0, 1)$, or $\infty$, according as the associated column is to be used for a stack of finite barrels, or apeirobarrels. We obtain a *coloring* or *shift function* $f: T(\mathcal{T}) \to [0, 1) \cup \{\infty\}$, where $T(\mathcal{T})$ denotes the set of tiles of $\mathcal{T}$. Third, for each finite-colored polygon $P$, let $P + \mathbb{R}u$ be the underlying cylinder for a stack of right *prisms*, with the (disjoint) union of prism bases being $P + \mathbb{Z}u$.

The fourth step deserves a more detailed explanation and will be carefully analyzed in the next section: Shift each stack of $n$-prisms associated to a finite-colored tile $P$ of $\mathcal{T}$ by the amount $\alpha = f(P)$ in $[0, 1)$ in the direction of $u$, as prescribed by the function $f$ (*i.e.*, shifting by $(0, 0, \alpha)$). The disjoint union of prism bases is now at $P + (\mathbb{Z} + \cancel{\geq})u$. The aim of our construction is to obtain a new face structure on each prism, turning each $n$-prism into an $n$-barrel. This is done by respecting topological adjacencies, in fact, by taking the final facial structure of the constructed object (up to rank 2) to be completely determined by the topology.

In order to assure that each rectangle in a prism's mantle splits into precisely two 2-faces, it is necessary and sufficient to shift adjacent stacks (*i.e.*, stacks over adjacent tiles in the underlying plane tiling $\mathcal{T}$) by different amounts. Since $\mathcal{T}$ is locally finite and consists of countably many tiles, it is clearly possible to find a shift function with this property. However, we need to be more restrictive in order to assure that the partition edges and vertices produce only pentagons. Therefore, consider any tile in $\mathcal{T}$. When cyclically traversing the adjacent tiles in *counterclockwise* order, and listing the corresponding function values of $f$ (which will be the shift lengths applied to the stacks of finite prisms), using asterisks $*$ as placeholders for tiles colored $\infty$, we will obtain a symbol like $(*, \alpha_0, \alpha_1, \ldots, *, \ldots)$. Let us call such a symbol a *neighborhood symbol* for the respective tile. A neighborhood symbol for a tile is not unique, since the neighbors can be traversed from different starting points, although we do adopt the convention to list the values of $f$ in counterclockwise order around the tile. All neighborhood symbols for the same tile (using the same coloring function $f$) differ only by a cyclic permutation, so we need to keep in mind that the neighborhood symbol is a cyclic symbol.

The requirement of splitting the rectangles in the mantle of an $n$-prism into pentagons necessitates a condition on the neighborhood symbols, which is explained subsequently. Let $\alpha$ be the amount of shift applied to the stack associated to the *finite*-colored $n$-gonal tile $P$ with neighborhood symbol

$(*, \alpha_0, \alpha_1, \ldots, *, \ldots)$. Let $(*, (\alpha_0 - \alpha) \bmod 1, (\alpha_1 - \alpha) \bmod 1, \ldots, *, \ldots)$ be the *augmented neighborhood symbol* for $P$, with the convention that "$x \bmod 1$" for $x \in (-1, 0)$ equals $x + 1 \in (0, 1)$. Then we require that, for finite-colored tiles, the (contiguous) subsequences of (cyclically) consecutive numbers in the augmented neighborhood symbol—as delimited by asterisks—are either all increasing or all decreasing. Violating this condition results in the introduction of hexagons and quadrilaterals on the mantle. An obvious consequence is that each tile which has *not* been colored by $\infty$ itself must be adjacent to an $\infty$-colored tile (so the neighborhood symbol for finite-colored tiles may be written down starting with $*$), otherwise we would have no delimiter, and the cyclic symbol would have to consist of ever increasing entries.

It may not be possible to carry out step four consistently, depending on earlier decisions, *i.e.*, selection and coloring of the plane tiling. However, if step four can be carried out successfully, then step five is to fit a stack of apeirogonal barrels ($\infty$-stack) on each infinite column $P + \mathbb{R}u$ perpendicular to an $\infty$-colored tile $P$ of the plane tiling. Again, depending on earlier decisions, this may or may not be possible (recall that the condition in step four was only a necessary one). The next section investigates the conditions on $f$ which must be satisfied so that the last two steps of the outline can be carried out.

## 6. Barrel Pseudotilings with Isolated Apeirostacks

Recall that for the remainder of this article, we consider our plane base tiling $\mathcal{T}$ as situated in the horizontal plane $\pi$ through the origin of $\mathbb{E}^3$, always looking at the plane tiling from "above", where $u = (0, 0, 1)$ denotes the direction "up". Recall further that $\mathcal{T}$ functions as an aid in the construction of a barrel pseudotiling $\mathcal{B}$, and that its faces (vertices, edges, tiles) need not be present in $\mathcal{B}$.

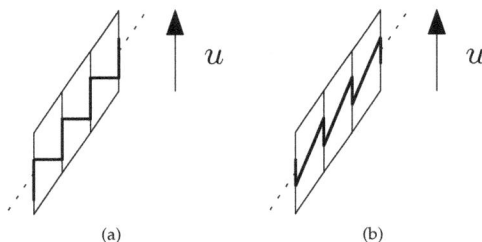

(a)  (b)

**Figure 3.** Unwrapped: An apeirobarrel winding up upstairs (**a**); and an apeirobarrel winding up downstairs (**b**). The staircases are highlighted.

To each barrel and apeirobarrel in a barrel pseudotiling $\mathcal{B}$ based on $\mathcal{T}$ there is an associated *upstairs orientation* (clockwise or counterclockwise): This is the orientation in which the staircase in the mantle leads "upstairs", if one were to step on the partition edges (treads), so this is either counterclockwise (up) or clockwise (up). For example, in Figure 1c we go upstairs from left to right, so if what is depicted corresponds to the unwrapped view of the mantle when walking around the outside of the barrel, then the upstairs orientation is counterclockwise. By contrast, remember that the *winding orientation* of an $\infty$-stack is the orientation (clockwise or counterclockwise) in which the apeirogonal bases wind upwards. Note here that, for apeirobarrels, the upstairs orientation does *not generally* correspond to the winding orientation, compare the examples of unwrapped apeirobarrels in Figure 3. However, in pseudotilings with only isolated $\infty$-stacks there is indeed such a correspondence, see Lemma 6.3. Before concentrating on pseudotilings with only isolated $\infty$-stacks, let us establish two general facts.

**Lemma 6.1.** *All barrels (both finite and infinite) in a barrel pseudotiling $\mathcal{B}$ have the same upstairs orientation. This is the* GLOBAL UPSTAIRS ORIENTATION *of $\mathcal{B}$.*

**Proof.** Consider any vertex $v$ of the plane base tiling $\mathcal{T}$ in the plane $\pi$, and assume that tiles $P_1$, $P_2$, and $P_3$ meet there as pictured in Figure 4 (in particular, $(P_1, P_2, P_3)$ is the listing of tiles in counterclockwise order around $v$). In the pseudotiling, $v$ corresponds to a line $v + \mathbb{R}u$ perpendicular to the plane. We move upwards on this line (in direction $u = (0,0,1)$) starting at $v$, and we mark all vertices of the pseudotiling (note that $v$ may not be one of them). In addition, we label a vertex on this line 1, 2, or 3, depending on whether the vertex in question is incident to a base of a (finite or infinite) barrel of the stack associated to $P_1$, $P_2$, or $P_3$. We obtain a repeating pattern since translation by the vertical unit vector $u$ is a built-in symmetry of the pseudotiling. If the repeating pattern is $(1,2,3)$ (*i.e.*, coherent with the right-hand rule), then the upstairs orientation of each barrel meeting a point on this line is counterclockwise, as can easily be seen. Otherwise, if the repeating pattern is $(1,3,2)$, then the upstairs orientation of each barrel meeting a point of the line is clockwise. Moreover, if the upstairs orientation for barrels in one stack is set (e.g., in a construction process), then this determines the orientations for all barrels in adjacent stacks. These stacks, in turn, determine the orientations for all stacks adjacent to them, and so forth. Thus, the upstairs orientations of barrels must be globally compatible, *i.e.*, the same. □

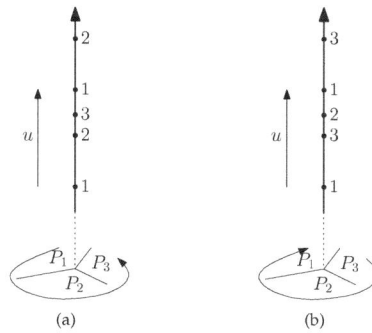

**Figure 4.** Determining upstairs orientations: (**a**) Counterclockwise; (**b**) clockwise.

**Remark 6.1.** *From the proof of Lemma 6.1 follows that if $P_1$, $P_2$, $P_3$ are assigned* FINITE *colors $\alpha_1$, $\alpha_2$, $\alpha_3$, respectively, of which w.l.o.g. $\alpha_1$ is the smallest, then the global upstairs orientation is counterclockwise if $\alpha_1 < \alpha_2 < \alpha_3$, and clockwise if $\alpha_1 < \alpha_3 < \alpha_2$. Incidentally, any shift function $f$ which assigns its finite values around such vertices of $\mathcal{T}$ coherently with a particular global upstairs orientation automatically satisfies the necessary condition on the augmented neighborhood symbols for finite-colored tiles (see step four in Section 5). Thus, we have a new and more practical necessary condition on $f$.*

**Lemma 6.2.** *If a colored plane tiling $(\mathcal{T}, f)$ gives rise to a barrel pseudotiling in one upstairs orientation, then the same base tiling—with altered coloring—also gives rise to a barrel pseudotiling in the reverse upstairs orientation.*

**Proof.** One way to construct a compatible barrel pseudotiling is to reflect the existing barrel pseudotiling in the plane $\pi$, which clearly yields a barrel pseudotiling of opposite upstairs orientation. Note that, in this case, the winding orientation in each apeirostack changes as well, whereas the altered coloring $f'$ of the base tiling $\mathcal{T}$ is given by:

$$f'(T) = \begin{cases} f(T) & \text{if } f(T) \in \{0, \infty\} \\ 1 - f(T) & \text{otherwise} \end{cases} \qquad \text{for each tile } T \text{ of } \mathcal{T}.$$

□

This fact allows us to adopt the convention that, unless otherwise stated, all subsequent barrel pseudotilings are assumed to have a counterclockwise upstairs orientation. We now focus on the properties of barrel pseudotilings with only *isolated* ∞-stacks, *i.e.*, barrel pseudotilings where apeirostacks are not directly adjacent.

**Lemma 6.3.** *In a barrel pseudotiling with only isolated ∞-stacks the winding orientation of any ∞-stack is coherent with (i.e., the same as) the global upstairs orientation.*

**Proof.** Select any ∞-stack $\Omega$. Because $\Omega$ has no faces in common with other ∞-stacks, the infinite staircases in the mantle of $\Omega$ have horizontal steps. Consequently, when unwinding the mantle as pictured in Figure 5, all staircases will be monotonically increasing when following the global upstairs orientation.

Let $A$ be any apeirogonal face (base apeirogon of an apeirobarrel) in the ∞-stack. Suppose the winding orientation of $A$ is opposed to the global upstairs orientation. Then $A$ must intersect a staircase, which is impossible. Therefore, the winding orientation of $A$, and thus of any apeirogonal face, is coherent with the global upstairs orientation. □

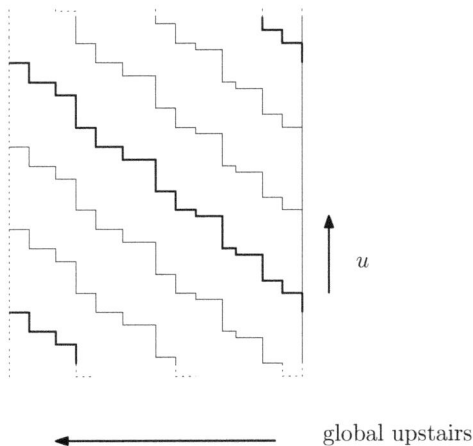

**Figure 5.** Staircase lines on unwrapped mantle of ∞-stack (schematic). Other edges have been omitted for clarity.

It is not known if there is any such correspondence for pseudotilings with adjacent ∞-stacks, which makes it harder to explore them. Therefore, for the remainder of this article, all considered barrel pseudotilings have only isolated apeirostacks.

The following Lemma characterizes when *isolated* ∞-stacks are compatible with a given shift function $f$ and a counterclockwise global upstairs orientation. Recall that the neighborhood symbol is taken in counterclockwise orientation, so it corresponds to our default choice of upstairs orientation.

**Lemma 6.4.** *Let $T$ be an $n$-gonal tile in the base tiling $\mathcal{T}$ with $f(T) = \infty$ and a neighborhood symbol $(\alpha_0, \alpha_1, \ldots, \alpha_{n-1})$ with finite entries. Assume the global upstairs orientation is counterclockwise. An isolated ∞-stack $\Omega$ with $k \geq 2$ distinct apeirogonal faces (spirals), which consequently consists of $k$ distinct apeirogonal barrels, can be constructed over $T$ if and only if we have $\alpha_i > \alpha_{i+1}$ for precisely $k$ mutually distinct indices $i$ (indices modulo $n$).*

**Remark 6.2.** *The case $k = 1$ would lead to a degenerate $\infty$-stack and cannot happen in a pseudotiling (compare Definitions 4.1 and 4.2).*

**Proof.** Let us first show that a constructed $\infty$-stack $\Omega$ with $k$ distinct apeirobarrels implies the condition on the neighborhood symbol. In $\Omega$, any of the $k$ apeirogonal bases climbs $k$ units before making a complete turn around the stack, and consequently, so do the $k$ staircases. By definition, in an isolated $\infty$-stack all entries in the neighborhood symbol are finite. Consequently, any staircase consists of vertical and horizontal segments only, is monotone in the direction of $u$, and the height of two partition vertices stemming from *any* base in the $i$-th adjacent stack determines the corresponding shift entry $\alpha_i$ in the neighborhood symbol (by taking the remainder of the height modulo 1, *i.e.*, disregarding full units of distance and reducing to an entry in $[0, 1)$). Fixing an arbitrary staircase in $\Omega$ and starting at a step (tread) on height $r$ which reduces to the smallest $\alpha_i$, w.l.o.g. $\alpha_1$, the corresponding contiguous entries in the neighborhood symbol increase as long as the staircase is trapped between $r$ and $\lfloor r \rfloor + 1$. Only when the staircase reaches a step at a height of $\lfloor r \rfloor + 1$ or greater will there be a jump from a larger to a smaller entry in the neighborhood symbol. Subsequently, the next jump will occur when a height of $\lfloor r \rfloor + 2$ or greater is reached and so forth. Thus, if the staircase climbs $k$ units when following the global upstairs orientation once around the stack, we obtain $k$ jumps from an $\alpha_i$ to a strictly smaller $\alpha_{i+1}$ in the neighborhood symbol.

Conversely, in constructing an $\infty$-stack $\Omega$, any apeirogonal faces are constructed as polygonal lines not intersecting the staircases. Any time there is a jump in the entries of the neighborhood symbol from an $\alpha_i$ to a strictly smaller $\alpha_{i+1}$, the staircase has climbed another unit. Consequently, if there are $k$ such jumps, the staircase climbs $k$ units before making a complete turn around the stack. An apeirogonal base must therefore also climb $k$ units. Translations in the direction of $u$ by integer multiples are prescribed symmetries of the pseudotiling. A translation by $ku$ is a symmetry that transforms the staircase under consideration into itself, but this is not true for translation by less than $k$ units in the direction of $u$. Thus, we must have $k$ staircases in total in $\Omega$, and therefore $k$ apeirogonal faces and *barrels*. □

Observe that the condition formulated in Remark 6.1 assures that step four of the construction outline can be carried out, whereas the condition in Lemma 6.4 assures that $\infty$-stacks can be fitted in step five (as long as $f$ isolates all $\infty$-stacks). This, however, constructs not only the apeirostacks, but also the missing vertices and edges on the finite stacks, so we are done. We are now in the position to summarize the necessary and sufficient conditions for when a pair $(\mathcal{T}, f)$ of a base tiling and coloring (shift) function gives rise to a barrel pseudotiling with only isolated $\infty$-stacks.

**Theorem 6.1.** *A pair $(\mathcal{T}, f)$ of a normal, simple, plane base tiling $\mathcal{T}$ and a coloring (shift) function $f : T(\mathcal{T}) \to [0, 1) \cup \{\infty\}$ gives rise to a barrel pseudotiling with only isolated $\infty$-stacks and counterclockwise global upstairs orientation if and only if the following two conditions hold.*
*(1) For each vertex $v$ of $\mathcal{T}$ where tiles $P_1$, $P_2$, $P_3$ meet in counterclockwise order around $v$, such that their colors $\alpha_1 = f(P_1)$, $\alpha_2 = f(P_2)$, $\alpha_3 = f(P_3)$ are finite, and of which w.l.o.g. $\alpha_1$ is the smallest, the inequality $\alpha_1 < \alpha_2 < \alpha_3$ holds;*
*(2) For each n-gonal tile $P$ of $\mathcal{T}$ with $f(P) = \infty$, all entries in a neighborhood symbol $(\alpha_0, \alpha_1, \ldots, \alpha_{n-1})$ are finite, and there are at least two distinct indices $i$ such that $\alpha_i > \alpha_{i+1}$ (indices taken modulo n).*

## 7. Barrel Pseudotilings and Decorated Plane Triangulations

Any normal simple tiling of the plane has a normal simple dual, which is a (topological) triangulation of the plane (*cf.* [10], Chapter 4.2). We can thus formulate properties of any simple, normal, plane base tiling $\mathcal{T}$ with associated coloring of tiles, which gives rise to a barrel pseudotiling with isolated $\infty$-stacks, in terms of a (normal) dual triangulation $\mathcal{T}^*$. We do this not only for the sake of having dual statements, but also to draw connections to plane graphs (specifically, edge graphs of triangulations of the plane). Note that we do not make the assumption of convexity in the dual, since it is not generally known whether any tiling by convex polygons has a dual which also consists of convex polygons (*cf.* [10], p. 174).

Assume that $\mathcal{T}$ is a normal, simple, plane tiling by convex polygons with an appropriate coloring $f$ on the tiles, which gives rise to a barrel pseudotiling with only isolated $\infty$-stacks in counterclockwise global upstairs orientation (as usual). Then let $\mathcal{T}^*$ be a dual to $\mathcal{T}$, which is a (normal, topological) triangulation of the plane with vertex set $V(\mathcal{T}^*)$. By slight abuse of notation we also denote the associated infinite geometric simplicial complex with $\mathcal{T}^*$. The vertices of $\mathcal{T}^*$ inherit the coloring or shift function $f^*$ : $V = V(\mathcal{T}^*) \rightarrow [0,1) \cup \{\infty\}$ from the function $f$ on the tiles of $\mathcal{T}$. This adds decoration to the triangulation, but we will also add further decoration by directing some of the edges. Let $V^0 = (f^*)^{-1}([0,1))$ denote the set of vertices of $\mathcal{T}^*$ with finite value under $f^*$, and let $\mathcal{T}^0$ denote the simplicial subcomplex of $\mathcal{T}^*$ induced by $V^0$ (it contains all simplices which have only vertices in $V^0$, and the empty simplex).

**Note 7.1.** *The set $V^\infty := V \setminus V^0$ is a dominating independent set in the edge graph of $\mathcal{T}^*$ (see [12] or Section 3 of this article for terminology).*

This is simply the dual version of the statement that all tiles are either colored $\infty$, or adjacent to an $\infty$-colored tile, but not both. As a consequence, $\mathcal{T}^0$ contains no complete simplicial neighborhood of a vertex in $\mathcal{T}^*$, *i.e.*, there are no vertices of $\mathcal{T}^0$ which lie in its interior (we do not use the term *relative interior* here because $\mathcal{T}^0$ may be a one-dimensional subcomplex of $\mathcal{T}$). We have:

**Note 7.2.** *The underlying topological space $|\mathcal{T}^0|$ of $\mathcal{T}^0$ is (path-)connected. In fact, $|\mathcal{T}^0|$ is obtained from $\mathbb{E}^2$ by deleting the mutually disjoint open discs given by the open stars of the vertices in $V^\infty$.*

Furthermore, the function $f^*$ naturally induces directions on the edges in $\mathcal{T}^0$, where $\{x, y\}$ turns into the directed edge, or *arc*, $(x, y)$ with *tail* $x$ and *head* $y$ if $f^*(x) < f^*(y)$ (these values are finite, by definition of $\mathcal{T}^0$). Furthermore, every arc is *oriented* in mathematically positive or negative direction *with respect to an incident triangle*, depending on whether it runs counterclockwise or clockwise around a point in the triangle's interior. Since the global upstairs orientation is assumed as mathematically positive (for the pseudotiling based on $\mathcal{T}$), Remark 6.1 implies that every triangle in $\mathcal{T}^0$ has two mathematically positive and one mathematically negative oriented arcs. Note that the resulting directed edge graph of $\mathcal{T}^0$, which we denote by $\overrightarrow{G^0}$, must not have directed cycles, as the value of $f^*$ increases for successive vertices along directed paths.

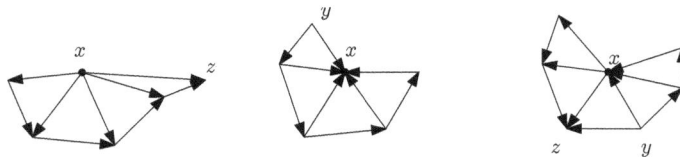

**Figure 6.** The neighborhood of a regular vertex $v$ in $\mathcal{T}^0$.

A *cut-vertex*—for the purpose of this paper—is a vertex $x$ of $\mathcal{T}^0$ such that the underlying topological space $|Lk_{\mathcal{T}^0}(x)|$ of its link in $\mathcal{T}^0$ is disconnected. A vertex of $\mathcal{T}^0$ which is not a cut-vertex is called a *regular vertex*. Note that the simplicial neighborhood $Nb_{\mathcal{T}^0}(x)$ of any regular vertex $x$ in $\mathcal{T}^0$ has as underlying topological space a closed disc whose boundary is a simple cycle containing $x$ (as $x$ cannot be an interior vertex). With directed arcs, there are three types of neighborhoods for regular vertices $x$, see Figure 6, as classified by the digraph of edges in that neighborhood (closed star), denoted $\overrightarrow{G(Nb_{\mathcal{T}^0}(x))}$:

- $x$ is the tail of all incident arcs, *i.e.*, a *source* in $\overrightarrow{G(Nb_{\mathcal{T}^0}(x))}$. In this case, $x$ is the only source in $\overrightarrow{G(Nb_{\mathcal{T}^0}(x))}$. The boundary cycle has a single edge, $(x, z)$, which is oriented negatively (with respect to (wrt) an interior point of the neighborhood), and $z$ is the only sink in $\overrightarrow{G(Nb_{\mathcal{T}^0}(x))}$;
- $x$ is the head of all incident arcs, *i.e.*, a *sink* in $\overrightarrow{G(Nb_{\mathcal{T}^0}(x))}$. In this case, $x$ is the only sink in $\overrightarrow{G(Nb_{\mathcal{T}^0}(x))}$, the boundary cycle of $Nb_{\mathcal{T}^0}(x)$ has a single edge, $(y, x)$, which is oriented negatively, and $y$ is the only source in $\overrightarrow{G(Nb_{\mathcal{T}^0}(x))}$;
- $x$ is neither a source nor a sink in $\overrightarrow{G(Nb_{\mathcal{T}^0}(x))}$. In this case, $x$ is incident to a unique triangle $yxz$ (labelled counterclockwise) for which it is a *transition vertex*, that is $x$ is the head of $(y, x)$ and the tail of $(x, z)$. The only source and sink in $\overrightarrow{G(Nb_{\mathcal{T}^0}(x))}$ are $y$ and $z$, respectively. The boundary cycle of $Nb_{\mathcal{T}^0}(x)$ has a single negatively oriented edge, $(y, z)$.

For cut-vertices $x$, the neighborhood splits into components for which $x$ is a regular vertex, or which are single arcs containing $x$.

We call two triangles $s$, $t$ of $\mathcal{T}^0$ *triangle-connected* if there is a finite sequence $s = t_0, t_1, \ldots, t_n = t$ of triangles in $\mathcal{T}^0$, such that for $i = 1, \ldots, n$ the triangles $t_{i-1}$ and $t_i$ are adjacent (*i.e.*, they share an edge). We call a simplicial subcomplex of $\mathcal{T}^0$ induced by a set of mutually triangle-connected triangles of $\mathcal{T}^0$ a *triangle component*. By definition, the underlying topological space of any triangle component is connected. However, we can also prove the following:

**Lemma 7.1.** *If $S$ is a (maximal) triangle component of $\mathcal{T}^0$, then its underlying topological space $|S|$ is simply connected. In particular, for finite triangle components $S$, the space $|S|$ is a closed disc.*

**Proof.** Assume the contrary, *i.e.*, that there exists a noncontractible (topological) cycle $C$ in $|S|$. Then there exists a *finite* triangle-connected subcomponent $S'$ of $S$ whose underlying space still contains $C$. We prove inductively that $|S'|$ is a closed topological disc (the second statement), and thus contains no nontrivial cycle, which implies that $|S|$ is simply connected.

First of all, observe that, by definition, any (finite) triangle component $S'$ of $S$ can be built up by successively attaching triangles along at least one edge (arc) to smaller triangle components. We now make use of the directed edges and their orientation with respect to incident triangles, in particular on the boundary of the component. Since $S'$ is finite, $|S'|$ is bounded and has finitely many boundary arcs (after assigning directions). We show that, topologically, the boundary is a simple closed curve, and thus encloses a disc.

A single triangle, as the base case, has two positively and exactly one negatively oriented arc (wrt an interior point). Suppose now that after $k$ steps we have built up a component $S_k$ which has exactly one negatively oriented arc on its boundary (wrt an interior point of the incident boundary triangle). We know that directed cycles cannot occur in the directed graph of our constructed component. When adding a triangle to the component in order to obtain $S_{k+1}$ the following cases need to be considered:

- The new triangle is glued on at a single positive boundary arc only. In this case this positive boundary arc (which now is no longer on the boundary) is replaced by two positive boundary arcs in $S_{k+1}$, and $S_{k+1}$ retains exactly one negative arc on its boundary;

- The new triangle is glued on at a single negative boundary arc only. In this case this negative boundary arc is replaced by a positive and a negative boundary arc in $S_{k+1}$, and $S_{k+1}$ still has exactly one negative arc on its boundary;
- The new triangle is glued to *two* boundary arcs. Since by the induction hypothesis there is only one negative arc in the boundary of $S_k$, and the new triangle cannot be glued to two previously positive arcs (because they would have to match up with two negative arcs in the triangle), exactly one of those arcs must be positive, and one must be negative. These boundary arcs are replaced by a single positive boundary arc, leading to $S_{k+1}$ having only positive arcs in its boundary. This is impossible because it would lead to an oriented cycle in the boundary, *i.e.*, to cyclically ever increasing values of $f^*$ on the traversed vertices;
- The new triangle is glued to *three* boundary arcs. This is also impossible, because as in the previous case, two positive and one negative arc would be removed from the boundary, leading to an oriented cycle in the boundary;
- The new triangle is glued on at a single positive (negative) boundary arc, and its third vertex is also glued. However, then the new boundary would (topologically) not be a simple closed curve, as *four* boundary arcs meet at that (glued) vertex, creating more than one cycle (regardless of edge direction) in the boundary. As there is still only one negatively oriented arc, this would lead to an oriented cycle in the boundary, which is forbidden.

Thus, in passing from $S_k$ to $S_{k+1}$, the new triangle can only be attached along a single edge on the boundary of $S_k$, as described in the first two cases. The boundary must form a simple closed curve (dangling edges *etc.* do not occur by definition). In particular, $|S_{k+1}|$ is a closed topological disc because it is also a bounded region. The component $S_{k+1}$ then has one more arc on the boundary, which is positively oriented, and retains one negatively oriented arc. This satisfies the induction hypothesis, so we have established that $|S'|$ is a closed topological disc.

Note that if a finite triangle component $S$ consists of $n$ triangles, then the boundary contains $(n+1)$ positive arcs and one negative arc. $\square$

We can now give a dual version of Theorem 6.1, which is not formulated in terms of a precise coloring $f^*$, but in terms of directed and undirected edges. This seems easier to handle in any kind of construction attempt, as the images in the next section show.

**Theorem 7.1.** *A normal simple tiling $\mathcal{T}$ of $\mathbb{E}^2$ by convex polygons gives rise to a barrel pseudotiling with isolated $\infty$-stacks and counterclockwise upstairs orientation if and only if some edges of the (any) dual triangulation $\mathcal{T}^*$ can be directed such that the following conditions are all satisfied:*

*(1) The set of vertices which are only incident to* UNDIRECTED *edges, denoted $V^\infty$, is a dominating independent set in the edge graph $G(\mathcal{T}^*)$;*

*(2) For each $x$ in $V^\infty$, the boundary of $\overrightarrow{G(Nb_{\mathcal{T}^*}(x))}$ has at least two directed arcs going in* CLOCKWISE ORIENTATION *around an interior point;*

*(3) Each triangle in $\mathcal{T}^*$ which has all its boundary edges oriented contains precisely one clockwise arc and two counterclockwise arcs.*

**Remark 7.1.** *We note without proof that condition (3) can be rephrased as follows. Each maximal triangle component of $\mathcal{T}^0$ has, when edges are directed, exactly one arc on its boundary which is negatively oriented with respect to an interior point. It is not entirely obvious, and therefore remarkable, that conditions (1), (2), (3) prevent directed cycles in the partially directed edge graph of $T^*$ (note that a directed cycle should not contain undirected edges).*

## 8. A Small Zoo of Examples

For all examples shown in this section, we still implicitly assume that the global upstairs direction is mathematically positive (*i.e.*, counterclockwise).

A class of base tilings which reproduce the examples mentioned in Section 2 are the simple normal plane tilings which are properly 3-colorable. In this case, we can assume the colors (shifts) to be 0, 0.5, and $\infty$. Figure 7a,b show (in the dual base tiling) how barrel pseudotilings can arise from the uniform tilings $(6^3)$ and $(4.8^2)$. Recall that the uniform tilings of the plane are precisely the vertex-transitive tilings by regular convex polygons. Recall further that $(n_1, n_2, \ldots, n_k)$ is the standard notation for a uniform tiling of the plane where each vertex is cyclically surrounded by an $n_1$-gon, $n_2$-gon, and so forth, in that order (compare [10], chapter 2.1). Further 3-colorable tilings are also shown in parts Figure 7c–h. The ease of finding barrel pseudotilings for these lies in the consequences of 3-colorability for the conditions in Theorem 6.1. The first condition is automatically fulfilled with an $\infty$-colored tile at each vertex; and $\infty$-colored tiles have at least four adjacents, so their neighborhood symbols fall apart into at least two cyclically increasing subsequences, thus satisfying the second condition.

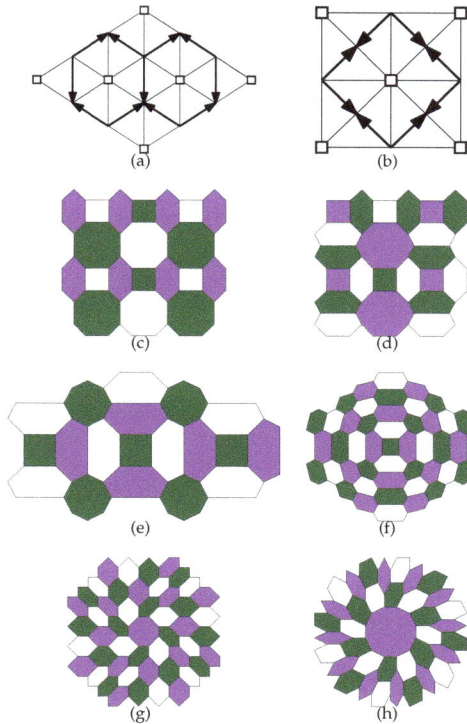

**Figure 7.** Decorated duals of the uniform base tilings $(6^3)$ and $(4.8^2)$ (**a**,**b**). The small boxes □ indicate $\infty$-colored vertices, and arrows indicate an increase in the function $f^*$. Patches of properly 3-colorable normal simple plane tilings, (**c**)–(**e**) periodic, (**f**)–(**h**) non-periodic.

More interesting things happen when the base tiling is not 3-colorable, or if one deliberately uses more than three colors or shift lengths (possibly infinitely many). In this case, the conditions on the neighborhood symbol have significant implications, unlike in the previous case. In the decorated dual base tiling we can now see longer paths and possibly nonempty triangle components forming. Figure 8 shows some examples, all of which yield periodic pseudotilings, *i.e.*, barrel pseudotilings with translational symmetry in three linearly independent directions, almost all of which are face-to-face. Observe that the absolute magnitude of the (finite) shift lengths is of little relevance, but rather is the shift difference among neighboring tiles (or vertices, in the dual). Finitely many colors (shift lengths) are sufficient if and only if all oriented paths in the decorated dual base tiling are of finite length, and the minimal number of colors needed is the same as the number of vertices on the longest directed path.

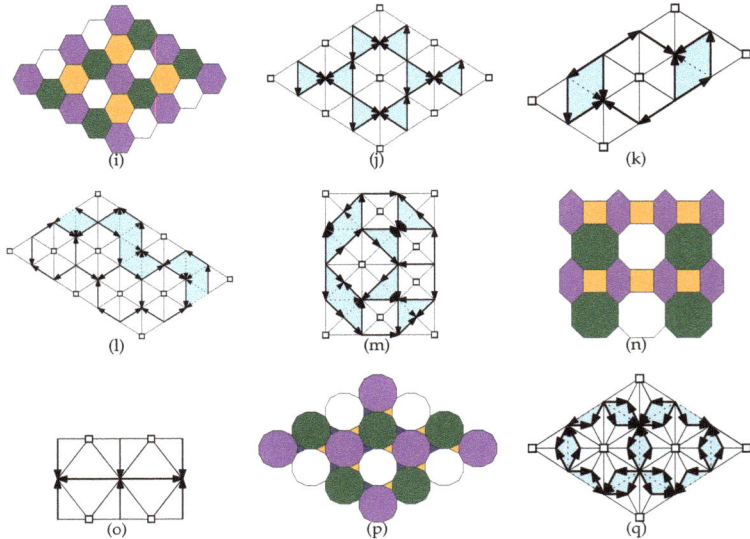

**Figure 8.** Barrel pseudotilings arise from the information encoded in a coloring of $(6^3)$ with four (**a**) and (**b**), five (**c**), and nine (**d**) colors. More examples stem from $(4.8^2)$ with eleven colors (**e**), from a 4-colored tiling (**f**) and (**g**), and a 5-colored version of $(3.12^2)$ (**h**) and (**i**). All except (**e**) give face-to-face barrel pseudotilings.

## 9. Remarks

We conclude the paper with some remarks about barrel pseudotilings and suggestions for further investigation in more general settings. In Section 8, many periodic examples of barrel pseudotilings were shown. It would be nice to have an algorithm to construct periodic barrel pseudotilings from periodic base tilings. As we have seen, not all suitable base tilings are periodic, but periodic base tilings seem to be a more interesting case because of their additional symmetry. Furthermore, the terrain is wide open when it comes to allowing adjacent ∞-stacks; it is unclear if there is a connection between winding orientation and upstairs orientation along the lines of Lemma 6.3, as well as how a decorated planar triangulation could capture all of the essential information in this case.

We have made many assumptions and chosen the definitions in this article in a certain way. First, we require that (finite) barrels stem from a *right* prism and that planar base tilings consist of *convex* polygons. However, it is conceivable that this is only a technical assumption to simplify the description of statements and procedures. It may be possible—but it has not been investigated—to deform the pseudotilings so that no coplanar adjacent 2-faces occur. For some pseudotilings, it may even be possible to have strictly convex polytopes emerge as the finite cells.

Second, we selected the definitions so as to produce structures which are realizations of abstract rank 4 polytopes. Several very similarly arising structures have been discarded on the grounds of not being associated with an abstract rank 4 polytope. Nevertheless, it may be of interest to chemists or crystallographers to identify the corresponding 4-valent atomic networks. Figure 9 shows an example patch of a structure, which has translational symmetry in three directions (and additional symmetries)

*Symmetry* **2012**, *4*, 545–565

yet fails to be classified as a barrel pseudotiling because only one apeirogon appears per stack (the degenerate case mentioned in Remark 6.2).

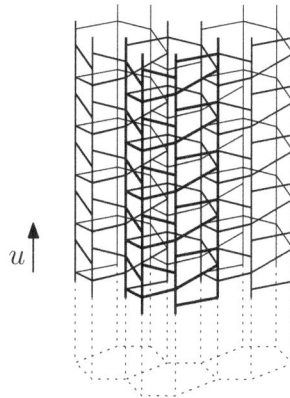

**Figure 9.** This patch with base $(6^3)$ is not part of a barrel pseudotiling, as the face structure of the object does not correspond to an abstract 4-polytope.

In order to further explore the connection with abstract polytopes, one could use a similar construction not only on tilings of the plane, but also on polyhedral maps. This would not fit as nicely into $\mathbb{E}^3$, but perhaps it would still produce interesting results. Finally, it should be possible to quotient by a multiple of $u$ in order to produce and study finite objects.

**Acknowledgments:** The author would like to thank Egon Schulte for his careful reading and many helpful suggestions. Also, the author would like to thank the two anonymous referees for their insightful comments which led to the improvement of this paper.

# References

1. Grünbaum, B. Regular polyhedra—old and new. *Aequ. Math.* **1977**, *16*, 1–20.
2. Schulte, E. Symmetry of polytopes and polyhedra. In *Handbook of Discrete and Computational Geometry*, 2nd ed.; Goodman, J.E., O'Rourke, J., Eds.; Chapman & Hall/CRC: Boca Raton, FL, USA, 2004; pp. 431–454.
3. Dress, A.W.M. A combinatorial theory of Grünbaums's new regular polyhedra, Part I: Grünbaum's new regular polyhedra and their automorphism group. *Aequ. Math.* **1981**, *23*, 252–265.
4. Dress, A.W.M. A combinatorial theory of Grünbaums's new regular polyhedra, Part II: Complete enumeration. *Aequ. Math.* **1985**, *29*, 222–243.
5. Leytem, C. Pseudo-Petrie operators on Grünbaum polyhedra. *Math. Slovaca* **1997**, *47*, 175–188.
6. McMullen, P.; Schulte, E. Regular polytopes in ordinary space. *Discrete Comput. Geom.* **1997**, *17*, 449–478.
7. McMullen, P.; Schulte, E. Regular and chiral polytopes in low dimensions. In *The Coxeter Legacy: Reflections and Projections*; Davis, C., Ellers, E.W., Eds.; AMS: Providence, RI, USA, 2005; pp. 87–106.
8. McMullen, P.; Schulte, E. *Abstract Regular Polytopes (Encyclopedia of Mathematics and its Applications)*; Cambridge University Press: Cambridge, UK, 2002; Volume 92.
9. Grünbaum, B.; Shephard, G.C.; Miller, J.C.P. Uniform tilings with hollow tiles. In *The Geometric Vein: The Coxeter Festschrift*; Davis, C., Grünbaum, B., Sherk, F.A., Eds.; Springer: New York, NY, USA, 1981; pp. 17–64.
10. Grünbaum, B.; Shephard, G.C. *Tilings and Patterns*; W.H. Freeman and Company: New York, NY, USA, 1987.
11. Schattschneider, D.; Senechal, D. Tilings. In *Handbook of Discrete and Computational Geometry*, 2nd ed.; Goodman, J.E., O'Rourke, J., Eds.; Chapman & Hall/CRC: Boca Raton, FL, USA, 2004; pp. 53–72.
12. West, D.B. *Introduction to Graph Theory*; Prentice Hall Inc.: Upper Saddle River, NJ, USA, 1996.

13. Maunder, C.R.F. *Algebraic Topology*; Cambridge University Press: Cambridge, UK, 1980.
14. Locher, J.L. *The Magic of M.C. Escher*; Harry N. Abrams: New York, NY, USA, 2000.

# symmetry

MDPI

*Article*

# Superspheres: Intermediate Shapes between Spheres and Polyhedra

Susumu Onaka

Department of Materials Science and Engineering, Tokyo Institute of Technology, 4259-J2-63 Nagatsuta, Yokohama 226-8502, Japan; onaka.s.aa@m.titech.ac.jp; Tel.: +81-45-924-5564; Fax: +81-45-924-5566

Received: 16 May 2012; in revised form: 20 June 2012; Accepted: 25 June 2012; Published: 3 July 2012

**Abstract:** Using an $x$-$y$-$z$ coordinate system, the equations of the superspheres have been extended to describe intermediate shapes between a sphere and various convex polyhedra. Near-polyhedral shapes composed of {100}, {111} and {110} surfaces with round edges are treated in the present study, where {100}, {111} and {110} are the Miller indices of crystals with cubic structures. The three parameters $p$, $a$ and $b$ are included to describe the {100}-{111}-{110} near-polyhedral shapes, where $p$ describes the degree to which the shape is a polyhedron and $a$ and $b$ determine the ratios of the {100}, {111} and {110} surfaces.

**Keywords:** supersphere; particle; precipitate; materials science; crystallography

## 1. Introduction

Small crystalline precipitates often form in alloys and have near-polyhedral shapes with round edges. Figure 1 is a transmission electron micrograph showing an example of this where the dark regions, which have shapes between a circle and a square, are Co-Cr alloy particles precipitated in a Cu matrix [1,2]. Why such precipitate shapes form has been explained by the anisotropies of physical properties of metals and alloys originating from the crystal structures [2,3]. Both the Co-Cr alloy particles and Cu matrix have cubic structures. The three-dimensional shapes of the particles shown in Figure 1 are intermediate between a sphere and a cube composed of crystallographic planes {100} as indicated by the Miller indices.

Even if the alloy system such as the Co-Cr alloy particles in the Cu matrix is fixed, the precipitate shapes change as a function of the precipitate size [1,2]. In the case of the Co-Cr alloy precipitates, the spherical to cubical shape transition occurs as the precipitate size increases [2,3]. The size dependence of the precipitate's equilibrium shape determines the shape transitions [2,3]. When we discuss such physical phenomenon, it is convenient to use simple equations that can approximate the precipitate shapes [2–5]. In the present study, we discuss a simple equation that gives shapes intermediate between a sphere and various polyhedra.

Symmetry **2012**, 4, 336–343

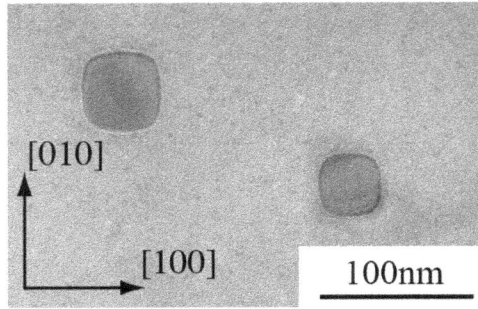

**Figure 1.** Transmission electron micrograph showing the Co-Cr alloy precipitates in a Cu matrix [1,2].

## 2. Cubic Superspheres

The solid figure described by

$$|x / R|^{p} + |y / R|^{p} + |z / R|^{p} = 1 \quad (R > 0, \; p \geq 2) \tag{1}$$

expresses a sphere with radius $R$ when $p = 2$ and a cube with edges $2R$ as $p \to \infty$ [2–4]. It is reported in [6] that the 19th century French mathematician Gabriel Lamé first presented this equation. Intermediate shapes between these two limits can be represented by choosing the appropriate value of $p > 2$. In [2–4], such shapes are called superspheres, and Figure 2 shows the shapes given by (1) for (a) $p = 2$, (b) $p = 4$ and (c) $p = 20$. The parameter $R$ determines the size and $p$ determines the polyhedrality, *i.e.*, the degree to which the supersphere is polyhedron. If $|x| > |y|$ and $|x| > |z|$, $|x/R|^{p} + |y/R|^{p} + |z/R|^{p} = 1$ as $p \to \infty$ means $|x/R| = 1$. This describes the limit for (1) as $p \to \infty$ which gives a cube surrounded by three sets of parallel planes, $x = \pm R$, $y = \pm R$ and $z = \pm R$.

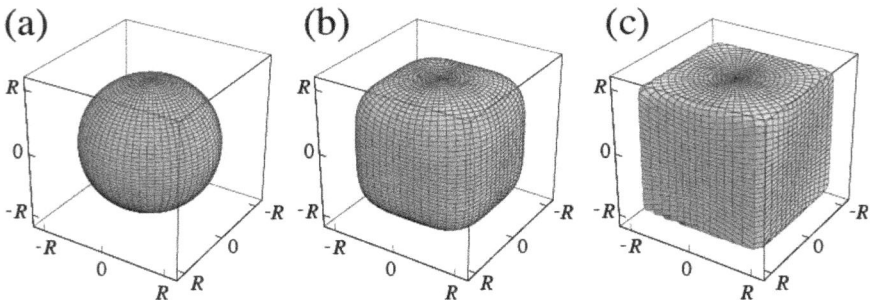

**Figure 2.** Shapes of the cubic superspheres given by (1); (a) $p = 2$; (b) $p = 4$ and (c) $p = 20$.

## 3. {111} Regular-Octahedral and {110} Rhombic-Dodecahedral Superspheres

Equation (1) can be rewritten as

$$\left[ h_{cube}(x,y,z) \right]^{1/p} = R \quad \text{where} \quad h_{cube}(x,y,z) = |x|^{p} + |y|^{p} + |z|^{p} \tag{2}$$

This expression has been extended to describe other convex polyhedra [7]. Although the original superspheres discussed in [2–4] are intermediate shapes between a sphere and a cube, now the superspheres can refer to shapes intermediate between various convex polyhedra and a sphere [8].

Superspheres have been used to discuss the shapes of small crystalline particles and precipitates [2, 3,5,8,9]. The planes of crystal facets are indicated by their Miller indices. We use this notation in the present study. The cube given by (2) as $p \to \infty$ is the {100} cube composed of six {100} faces. Assuming crystals with cubic structures, the regular octahedron is the {111} octahedron and the rhombic dodecahedron is the {110} dodecahedron [7].

The {111} octahedral superspheres are given by the following equation:

$$\left[ h_{octa}\left(x,y,z\right)\right]^{1/p} = R \tag{3a}$$

where

$$h_{octa}\left(x,y,z\right) = \left|x+y+z\right|^{p} + \left|-x+y+z\right|^{p} + \left|x-y+z\right|^{p} + \left|x+y-z\right|^{p} \tag{3b}$$

The shapes given by (3) are shown in Figure 3.

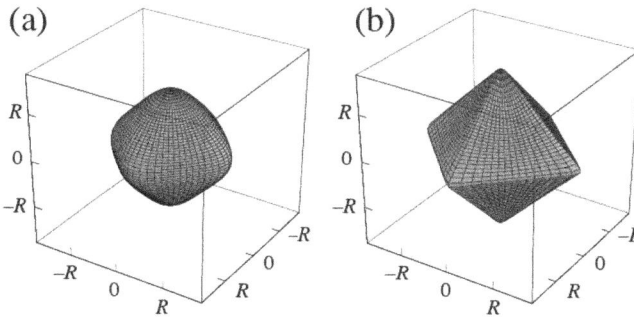

**Figure 3.** Shapes of the {111} regular-octahedral superspheres given by (3); (a) $p = 4$ and (b) $p = 40$.

On the other hand, the {110} dodecahedral superspheres are given by

$$\left[ h_{dodeca}\left(x,y,z\right)\right]^{1/p} = R \tag{4a}$$

where

$$h_{dodeca}\left(x,y,z\right) = \left|x+y\right|^{p} + \left|x-y\right|^{p} + \left|y+z\right|^{p} + \left|y-z\right|^{p} + \left|x+z\right|^{p} + \left|x-z\right|^{p} \tag{4b}$$

The shapes given by (4) are shown in Figure 4. Equations (2–4) expressed by the spherical coordinate system are shown in [7].

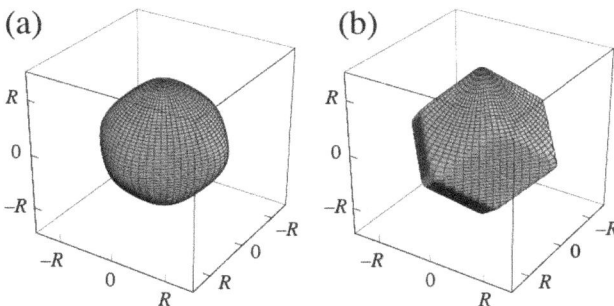

**Figure 4.** Shapes of the {110} rhombic-dodecahedral superspheres given by (4); (a) $p = 6$ and (b) $p = 40$.

*Symmetry* **2012**, *4*, 336–343

## 4. {100}-{111}-{110} Polyhedral Superspheres

Combined superspheres can be expressed by combining the equations of each supersphere. Combining (2), (3) and (4), we get

$$\left[ h_{cube}(x,y,z) + \frac{1}{a^p} h_{octa}(x,y,z) + \frac{1}{b^p} h_{dodeca}(x,y,z) \right]^{1/p} = R.$$

(5)

The parameters $a > 0$ and $b > 0$ are those for determining the ratios of the {100}, {110} and {111} surfaces. The shapes of the supersphere given by (5) are shown in Figure 5 when

$$a = \sqrt{3}$$

'

$$b = \sqrt{2}$$

for two values of $p$.

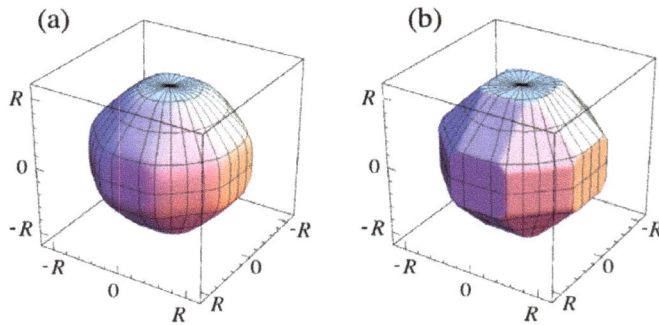

**Figure 5.** Shapes of the {100}-{111}-{110} polyhedral superspheres given by (5); (a) $p = 20$ and (b) $p = 100$.

The $a$ and $b$ dependences of the shapes given by (5) are understood by examining the polyhedral shapes as $p \to \infty$. Among the three polyhedra given by $[h_{cube}(x,y,z)]^{1/p} = R$, $[h_{octa}(x,y,z)]^{1/p} = aR$ and $[h_{dodeca}(x,y,z)]^{1/p} = bR$, the innermost surfaces of the polyhedra are retained to form the combined polyhedron. Figure 6 shows the effect of $a$ and $b$ on the shapes given by (5) as $p \to \infty$. The shape is determined by their location in the quadrilateral surrounded by the points $P(a,b) = (3,2)$, $Q(2,2)$, $R(1,1)$ and $S(3/2,1)$. Various shapes in and around the quadrilateral are shown by the insets in Figure 6 can be summarized as follows:

1. Three basic polyhedra

   (a) {100} cube at point $P$.
   (b) {111} octahedron at point $R$.
   (c) {110} dodecahedron at point $S$.

2. Combination of two basic polyhedra

   (a) {100}-{111} polyhedra changing from the {100} cube to the {111} octahedron along the line from $P$ to $R$ via $Q$, by truncating the eight vertices of the cube (The shape at point $Q$ is {100}-{111} cuboctahedron).

(b) {111}-{110} polyhedra changing from the {111} octahedron to the {110} dodecahedron along the line from R to S, by chamfering the 12 edges of the octahedron.

(c) {110}-{100} polyhedra changing from the {110} dodecahedron to the {100} cube along the line from S to P, by truncating six of the 14 vertices of the dodecahedron.

3. Combinations of all three basic polyhedra

(a) {100}-{111}-{110} polyhedra with mutually non-connected {110} surfaces in Region 1 (R-1).
(b) {100}-{111}-{110} polyhedra with mutually connected {110} surfaces in Region 2 (R-2).

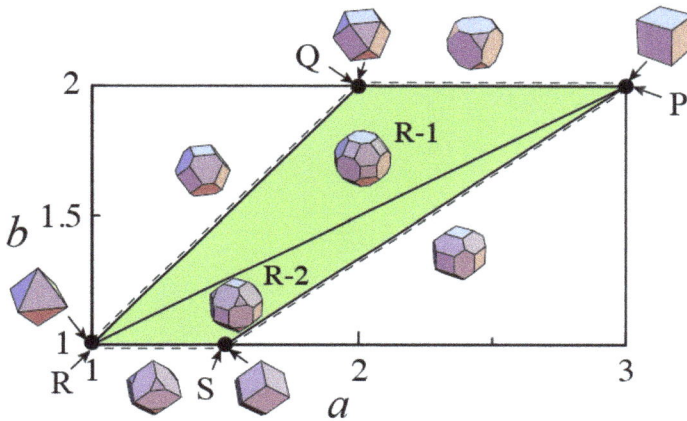

**Figure 6.** Diagram showing the variation in the shapes of the {100}-{111}-{110} polyhedral superspheres given by (5) as $p \to \infty$.

The boundary between Regions 1 and 2, expressed by the line from P to R, is written as:

$$b = (a+1)/2 \qquad (6)$$

Figure 6 is essentially the same as Figure 3 in [7,8] where the parameters $\alpha = 1/a$ and $\beta = 1/b$ are used instead of $a$ and $b$. In the appendix, the volume and surface area of the polyhedra shown in Figure 6 are written as a function of $a$ and $b$. The use of the parameters $a$ and $b$ gives a more intuitive diagram (Figure 6), compared with the diagram given by $\alpha$ and $\beta$.

## 5. Discussion

### 5.1. Shape Transitions of Superspheres from a Sphere to Various Polyhedra

Shape transitions of superspheres from a sphere to a polyhedron are characterized by the change in the normalized surface area $N = S/V^{2/3}$, where $S$ is the surface area and $V$ the volume of the supersphere. For a sphere, $N = 6^{2/3}\pi^{1/3} \approx 4.84$. Figure 7 shows the variations in $N$ as a function of $p$ for the following the superspheres as indicated by the insets:

(i) the {100} cube type given by (2),
(ii) the {111} regular-octahedral type given by (3),
(iii) the {110} rhombic-dodecahedral type given by (4) and
(iv) the {100}-{111}-{110} polyhedral type given by (5) with

$$a = \sqrt{3}$$

and

$$b = \sqrt{2}$$

.

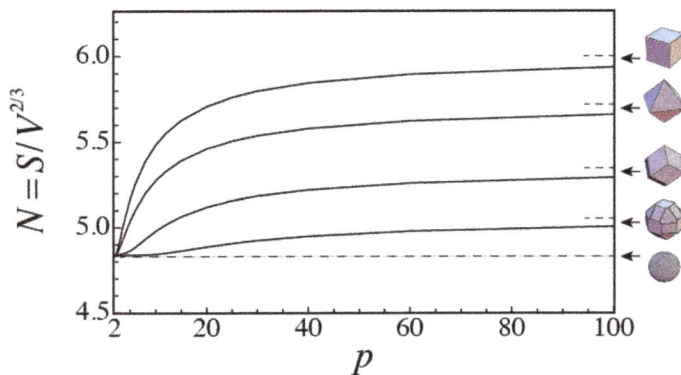

**Figure 7.** Dependence of the normalized surface area $N = S/V^{2/3}$ on $p$, where $S$ is the surface area and $V$ the volume for various superspheres: (i) the {100} cube type given by (2); (ii) the {111} octahedral type given by (3); (iii) the {110} dodecahedral type given by (4) and (iv) the {100}-{111}-{110} polyhedral type given by (5) with $a = \sqrt{3}$ and $b = \sqrt{2}$.

The broken lines at the right show the values of $N$ for the polyhedra as $p \to \infty$.

As shown in Figure 7, the change in $N$ with increasing $p$ becomes smaller as the number of faces of polyhedra increases from the {100} cube with 6 to the {100}-{111}-{110} polyhedron with 26. Among the various polyhedra shown in Figure 3, the polyhedron given by

$$a = \sqrt{3}$$

and

$$b = \sqrt{2}$$

in Region 1 with $N = S/V^{2/3} \approx 5.05$ has the minimum total surface area $S$ for the same $V$ [8,10]. The $a$ and $b$ dependence of $N$ can be calculated easily using the results shown in the appendix.

### 5.2. Shape of Small Metal Particles

The shapes of small metal particles observed in previous studies have been discussed previously using the superspherical approximation [8]. Menon and Martin reported the production of ultrafine Ni particles by vapor condensation in an inert gas plasma reactor [11]. They have also reported the crystallographic characterization of these particles by transmission electron microscopy [11]. Near-polyhedral shapes of nanoparticles have been observed to discuss their properties [12–15]. The superspherical approximation is a useful geometrical tool to describe the near-polyhedral shapes.

**Acknowledgments:** This research was supported by a Grand-in-Aid for Scientific Research C (22560657) by the Japan Society for the Promotion of Science.

appendix

**6.**

The volume and surface area of the polyhedra shown in Figure 3.

The volume $V$ and the {100}, {111} and {110}

surface area, $S_{100}$, $S_{111}$ and $S_{110}$ of the polyhedra shown in Figure 6 are written as a function of $a$ and $b$. In Region 1, these are given by

$$V = 4\left[\frac{a^3}{3} - (a-1)^3 - (a-b)^2 (6-a-2b)\right]R^3 \tag{A1}$$

$$S_{100} = 12\left[(a-1)^2 - 2(a-b)^2\right]R^2 \tag{A2}$$

$$S_{111} = 4\sqrt{3}\left[(3-a)^2 - 3(2-b)^2\right]R^2 \tag{A3}$$

and

$$S_{110} = 24\sqrt{2}\,(a-b)(2-b)R^2 \tag{A4}$$

In Region 2, these are

$$V = 2\left[b^3 - \frac{1}{3}(3b-2a)^3 - 4(b-1)^3\right]R^3 \tag{A5}$$

$$S_{100} = 24(b-1)^2\,R^2 \tag{A6}$$

$$S_{111} = 4\sqrt{3}(3b-2a)^2\,R^2 \tag{A7}$$

and

$$S_{110} = 6\sqrt{2}\left[b^2 - (3b-2a)^2 - 4(b-1)^2\right]R^2 \tag{A8}$$

when $a = 1$ and $b = 1$, the shape given by (5) as $p \to \infty$ is the {111} regular-octahedron as shown by Figure 6. Since the {111} regular-octahedron belongs to both Regions 1 and 2, from both (A1) to (A4) and (A5) to (A8), we get $V = (4/3)R^3$, $S_{100} = 0$, $S_{111} = 4\sqrt{3}R^2$ and $S_{110} = 0$ as it should be.

**References**

1. Fujii, T.; Tamura, T.; Kato, M.; Onaka, S. Size-Dependent equilibrium shape of Co-Cr particles in Cu. *Microsc. Microanal.* **2002**, *8*, 1434–1435.
2. Onaka, S.; Kobayashi, N.; Fujii, T.; Kato, M. Energy analysis with a superspherical shape approximation on the spherical to cubical shape transitions of coherent precipitates in cubic materials. *Mat. Sci. Eng.* **2002**, *A347*, 42–49.
3. Onaka, S.; Kobayashi, N.; Fujii, T.; Kato, M. Simplified energy analysis on the equilibrium shape of coherent $\gamma'$ precipitates in gamma matrix with a superspherical shape approximation. *Intermetallics* **2002**, *10*, 343–346. [CrossRef]
4. Onaka, S. Averaged Eshelby tensor and elastic strain energy of a superspherical inclusion with uniform eigenstrains. *Phil. Mag. Let.* **2001**, *81*, 265–272. [CrossRef]
5. Onaka, S. Geometrical analysis of near polyhedral shapes with round edges in small crystalline particles or precipitates. *J. Mat. Sci.* **2008**, *43*, 2680–2685. [CrossRef]
6. Jaklic, A.; Leonardis, A.; Solina, F. *Segmentation and Recovery of Superquadrics (Computational Imaging and Vision)*; Kluwer Academic: Dordrecht, The Netherlands, 2000; Volume 20, pp. 13–39.

7. Onaka, S. Simple equations giving shapes of various convex polyhedra: The regular polyhedra and polyhedra composed of crystallographically low-index planes. *Phil. Mag. Let.* **2006**, *86*, 175–183. [CrossRef]

8. Miyazawa, T.; Aratake, M.; Onaka, S. Superspherical-shape approximation to describe the morphology of small crystalline particles having near-polyhedral shapes with round edges. *J. Math. Chem.* **2012**, *50*, 249–260. [CrossRef]

9. Onaka, S.; Fujii, T.; Kato, M. Elastic strain energy due to misfit strains in a polyhedral precipitate composed of low-index planes. *Acta Mater.* **2007**, *55*, 669–673. [CrossRef]

10. Suárez, J.; Gancedo, E.; Manuel Álvarez, J.; Morán, A. Optimum compactness structures derived from the regular octahedron. *Eng. Struct.* **2008**, *30*, 3396–3398. [CrossRef]

11. Menon, S.K.; Martin, P.L. Determination of the anisotropy of surface free energy of fine metal particles. *Ultramicroscopy* **1986**, *20*, 93–98. [CrossRef]

12. Wang, G.L. Transmission electron microscopy of shape-controlled nanocrystals and their assemblies. *J. Phys. Chem.* **2000**, *104*, 1153–1175. [CrossRef]

13. Niu, W.; Zheng, S.; Wang, D.; Liu, X.; Li, H.; Han, S.; Chen, J.; Tang, Z.; Xu, G. Selective synthesis of single-crystalline rhombic dodecahedral, octahedral, and cubic gold nanocrystals. *J. Am. Chem. Soc.* **2009**, *131*, 697–703.

14. Jeong, G.W.; Kim, M.; Lee, Y.W.; Choi, W.; Oh, W.T.; Park, Q.-H.; Han, S.W. Polyhedral Au Nanocrystals exclusively bound by {110} facts: The rhombic dodecahedron. *J. Am. Chem. Soc.* **2009**, *131*, 1672–1673.

15. Ribis, J.; de Carlan, Y. Interfacial strained structure and orientation relationships of the nanosized oxide particles deduced from elasticity-driven morphology in oxide dispersion strengthened materials. *Acta Mater.* **2012**, *60*, 238–252. [CrossRef]

*symmetry*

MDPI

*Article*

# Regular and Chiral Polyhedra in Euclidean Nets

Daniel Pellicer

Centro de Ciencias Matemáticas, UNAM-Morelia, Antigua Carretera a Patzcuaro 8701, 58087 Morelia, Mexico; pellicer@matmor.unam.mx; Tel.: +52-443-322-2783

Academic Editor: Egon Schulte
Received: 14 September 2016; Accepted: 23 October 2016; Published: 28 October 2016

**Abstract:** We enumerate the regular and chiral polyhedra (in the sense of Grünbaum's skeletal approach) whose vertex and edge sets are a subset of those of the primitive cubic lattice, the face-centred cubic lattice, or the body-centred cubic lattice.

**Keywords:** regular polyhedron; chiral polyhedron; net

## 1. Introduction

The primitive cubic lattice, face-centred cubic lattice, and body-centred cubic lattice are well-known geometric objects widely studied in mathematics, and also used in other fields as models of various concepts. There are several interesting topics closely related to these lattices; for example, packings of spheres in Euclidean space [1] and crystal systems [2]. These lattices are a useful tool for the study of the affinely-irreducible discrete groups of isometries of Euclidean space [3], and are the ambient space for other interesting mathematical objects [4–6].

According to [7], a polyhedron is a connected and discrete collection of polygons in Euclidean space where every edge belongs to two cycles (faces). Polyhedra are not required to be convex, are not required to be finite, and their faces are not required to be contained in a plane. Regular and chiral polyhedra admit combinatorial rotations along all their faces and around all their vertices. Regular polyhedra admit combinatorial reflections, while chiral polyhedra do not. Regular polyhedra were classified in [8,9]; chiral polyhedra were classified in [10,11].

In this paper, we prove the following theorem.

**Theorem 1.** *The only chiral polyhedra in Euclidean space whose vertex and edge sets are subsets of* **pcu**, **fcu**, *or* **bcu** *are* $P(1,0)$ *and* $P_1(1,0)$ *in* **pcu**, *and* $Q(1,1)$ *and* $P_2(1,-1)$ *in* **bcu**.

The polyhedra $P(1,0)$, $P_1(1,0)$, $Q(1,1)$, and $P_2(1,-1)$ are among the chiral polyhedra described in [10,11]. We recall some of their main aspects and provide some local pictures in Section 4.

In Section 2, we provide background on the primitive cubic lattice **pcu**, face-centred cubic lattice **fcu**, and body-centred cubic lattice **bcu** in the context of nets of Euclidean space. Rotary, chiral, and regular polyhedra are defined and described in Section 3. Finally, in Section 4, we prove Theorem 1 by enumerating the rotary polyhedra whose vertex and edge sets can be taken from those of one of the three lattices mentioned above.

## 2. Nets

By a net, we mean a connected graph embedded with straight edges in Euclidean space $\mathbb{R}^3$, invariant under translations by three linearly independent vectors. The vertex set of the net must be discrete. Nets arise naturally when modelling periodic structures in chemistry. They are commonly denoted by three letters **abc** that often carry information about chemical compounds whose links can be represented by that net. They are also natural structures for mathematicians to study.

A symmetry of a net is an isometry of $\mathbb{R}^3$ that preserves the vertex and edge sets. A wealth of highly-symmetric nets can be found in the Reticular Chemistry Structure Resource database [12]. In what follows, we describe some of these nets that are relevant for this work. They are all highly related with the cubic tessellation $\mathcal{T}$ of Euclidean space. For convenience, we assume that the vertex set of $\mathcal{T}$ is the set of points with integer coordinates.

The primitive cubic lattice **pcu** consists of all vertices and edges of $\mathcal{T}$ (see Figure 1a). Every vertex is 6-valent and has edges with three distinct direction vectors; namely, those of the coordinate axes.

The face-centred cubic lattice **fcu** can be constructed from $\mathcal{T}$ by taking as vertices all vertices of **pcu** whose sum of coordinates is even (one part of the natural bipartition of **pcu**) and as edges all diagonals of squares of $\mathcal{T}$ with endpoints in the vertex set (see Figure 1b). The vertices are 12-valent, and there are six distinct directions of the edges. There are precisely four vertices of each cube of $\mathcal{T}$ in the vertex set of **fcu**; their convex hull is a regular tetrahedron. Given any vertex of **pcu** that is not a vertex of **fcu**, its six neighbours in **pcu** all belong to **fcu**, and are the vertices of an octahedron. These tetrahedra and octahedra are the cells of tessellation #1 in [13]. Each triangle of that tessellation can be extended to a plane tessellation by equilateral triangles, where all triangles are also triangles of tessellation #1 in [13]. By gluing sets of six triangles together, we can obtain the vertex and edge set of a tessellation by regular hexagons as a subset of the net **fcu**.

The body-centred cubic lattice **bcu** has as vertex set all vertices of **pcu** whose coordinates are either all odd or all even; two vertices are adjacent whenever they are endpoints of a diagonal of a cube of $\mathcal{T}$ (see Figure 1c, where the thin gray lines represent only those lines containing edges of $\mathcal{T}$ where two of the coordinates are even). The vertices are 8-valent, and there are four distinct directions of the edges.

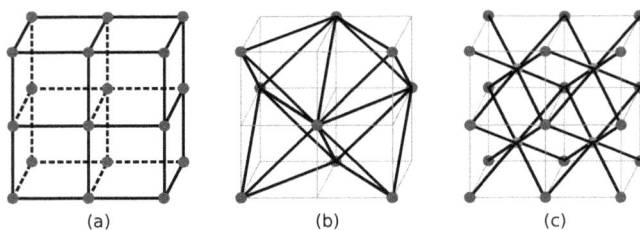

**Figure 1.** Nets **pcu** (a); **fcu** (b); and **bcu** (c).

The three nets just described are the only nets whose symmetry groups induce only one kind of vertex and whose vertex-stabilizers are isomorphic to the symmetry group [3, 4] of the octahedron.

The symmetry group of any net is a crystallographic group, and hence it contains no rotations of order 5 (for example, see [10], Lemma 4.1).

In this paper, we shall think of the nets as rigid objects in Euclidean space, although the same combinatorial structures could be embedded in a less symmetric way. Except in the third column of Tables 2–4, all nets should be understood as geometric objects and not only as combinatorial ones.

We shall need the following straightforward result.

**Lemma 1.** *The angle between two edges incident to the same vertex is*

- *either $\pi$ or $\pi/2$ if the edges are in* **pcu***;*
- *either $\pi/3$, $\pi/2$, $2\pi/3$, or $\pi$ if the edges are in* **fcu***;*
- *and either $\cos^{-1}(1/3)$, $\cos^{-1}(-1/3)$, or $\pi$ if the edges are in* **bcu***.*

The angles listed in Lemma 1 are highlighted in Figure 2.

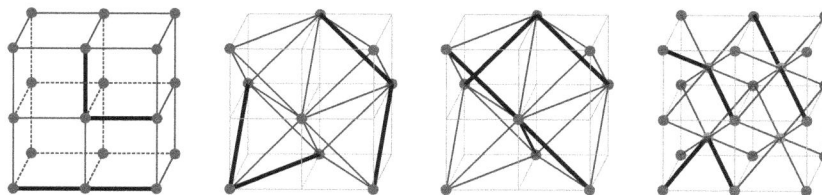

**Figure 2.** The angles formed by edges of the nets **pcu**, **fcu** and **bcu**.

## 3. Regular and Chiral Polyhedra

Highly symmetric convex polyhedra have been studied for centuries. It is well-known that there are only five convex regular polyhedra and 13 Archimedean solids, which, together with the infinite families of prisms and antiprisms, are the only convex polyhedra whose faces are regular polygons and whose symmetry groups induce only one kind of vertex.

In order to get a richer theory, we admit other combinatorial structures in $\mathbb{R}^3$ as polyhedra. In particular, we allow infinite polyhedra and even infinite polygons, and we abandon the idea of polygons being spanned by a 2-dimensional membrane.

### 3.1. Definitions

For us, a polygon is an embedding to $\mathbb{R}^3$ of a connected 2-regular graph; that is, of a cycle or of a two-sided infinite path. Polygons are explicitly allowed to be • skew (non-planar) and infinite, but we require the vertex set to be discrete.

A polyhedron is a (finite or infinite) collection of polygons (also called faces) with the properties that

- the set of vertices is discrete,
- the graph determined by all vertices and edges is connected,
- every edge belongs to exactly two faces,
- the vertex-figure at every vertex is a finite polygon. (The *vertex-figure* at a vertex $v$ is the graph whose vertices are the neighbours of $v$, two of them joined by an edge whenever they are the neighbours of $v$ in some face of the polyhedron.)

Convex polyhedra clearly satisfy the previous definition, as also do face-to-face tilings of the Euclidean plane (embedded in $\mathbb{R}^3$) and many more interesting structures.

A symmetry of a polyhedron $\mathcal{P}$ is an isometry of $\mathbb{R}^3$ that preserves $\mathcal{P}$. The group of all isometries of $\mathcal{P}$ is denoted $G(\mathcal{P})$.

Whenever there is a symmetry of $\mathcal{P}$ that cyclically permutes the vertices of a face $F$, we say that $\mathcal{P}$ has *abstract rotations along* $F$. Similarly, if there is a symmetry of $\mathcal{P}$ that cyclically permutes the neighbours of a given vertex $v$, we say that $\mathcal{P}$ has *abstract rotations around* $v$. An *abstract reflection* of $\mathcal{P}$ is a symmetry that, for some triple of mutually incident vertex, edge, and face, it preserves two of the elements while moving the third.

When $\mathcal{P}$ is a convex polyhedron, abstract rotations and abstract reflections are indeed rotations around some axes and reflections with respect to planes. However, if the faces are not planar, or they are not finite, then the abstract rotations about the faces are determined by the nature of the faces, and cannot be rotations about lines.

Some polygons admitting abstract rotations are shown in Figure 3. An abstract rotation of a finite polygon that has all vertices on a plane may be either a geometric rotation or a rotatory reflection (composition of a rotation about a line $l$ and a reflection with respect to a plane perpendicular to $l$). If the polygon is finite but skew (no plane contains all vertices), then an abstract rotation mapping a vertex to an adjacent vertex is necessarily a rotatory reflection. Some examples of finite polygons can be seen in the left side of Figure 3. The abstract rotation that maps a vertex of a planar zigzag to a neighbouring vertex is either a twist (composition of a translation and a rotation with axis in the direction of the translation) or a glide reflection (composition of a translation and a reflection about the plane containing the direction of the translation). If the polygon is a helix then such an abstract rotation is necessarily a twist. A helix and a zigzag are shown in the right side of Figure 3. In addition to the polygons mentioned above, polygons having all edges in the same line also admit abstract rotations, but they are not relevant for this work.

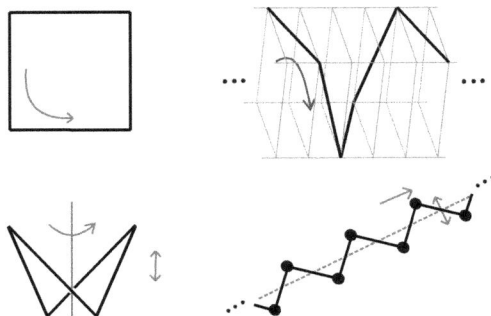

**Figure 3.** Abstract rotations of polygons.

We say that a polyhedron is *rotary* whenever it admits all possible abstract rotations around all its faces and around all its vertices. A rotary polyhedron is *regular* if in addition it admits an abstract reflection, and *chiral* otherwise. The Platonic solids are the only convex polyhedra that are regular under this definition. Furthermore, there are no convex chiral polyhedra.

The definitions of polyhedron, regular, and chiral in this section are equivalent to those in [10,11,14], and differ mildly only on the condition on the vertex-figure with those in [7]. The above use of the term "chiral" is widely accepted in the community studying abstract polytopes and related topics; to avoid confusion, the reader should bear in mind that in chemistry this word is used to denote a substantially different property.

All faces of a regular or chiral polyhedron have the same number $p$ of edges, and all vertices have the same degree $q$. The pair $\{p,q\}$ is called the *Schläfli type* (or just *type*) of the polytope. When studying rotary polyhedra, $p$ is allowed to be $\infty$, but $q$ must be finite to prevent the vertex set from being non-discrete.

A *Petrie polygon* of a polyhedron $\mathcal{P}$ is a closed walk on the vertex and edge sets of $\mathcal{P}$, where any two consecutive edges—but not three—belong to the same face. The structure obtained from the vertex and edge set of $\mathcal{P}$, but considering the Petrie polygons as faces is often a polyhedron. When this occurs, it is called the *Petrial* of $\mathcal{P}$.

The Petrial of the cube is outlined in Figure 4. The four faces are the hexagonal Petrie polygons in thick lines. Note that every edge belongs to precisely two such polygons.

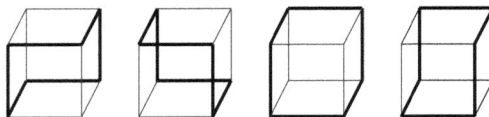

**Figure 4.** Petrial of the cube.

### 3.2. Regular Polyhedra

In [7], Grünbaum introduced the idea of polyhedron used in this and many other papers. There, he also described 47 regular polyhedra. The classification of the 48 regular polyhedra was achieved by Dress in [8,9]. A shorter proof of the completeness of the classification can be found in [14]. Throughout, we shall use the names of the polyhedra given in [14].

There are 18 finite regular polyhedra. They are the five Platonic solids, the four Kepler–Poinsot polyhedra, and the Petrials of the previous nine (see [7,15] for further details).

Six of the infinite regular polyhedra are in fact planar. Three of them are the regular tessellations of $\mathbb{R}^2$ by squares, equilateral triangles, and regular hexagons, denoted by $\{4,4\}$, $\{3,6\}$, and $\{6,3\}$, respectively. The remaining three are the Petrials $\{\infty,4\}_4$, $\{\infty,6\}_3$, and $\{\infty,3\}_6$ of these tessellations. Figure 5 shows two Petrie polygons of $\{\infty,4\}_4$, three of $\{\infty,6\}_3$, and three of $\{\infty,3\}_6$; all other Petrie polygons are translates of these.

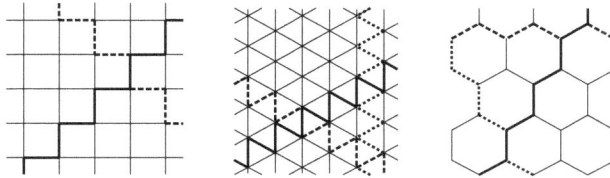

**Figure 5.** Petrie polygons of the planar polyhedra.

The remaining 24 infinite regular polyhedra live properly in $\mathbb{R}^3$, and can be evenly divided into those that are *blended* (their automorphism groups permute the translates of some plane) and those that are *pure* (not blended).

Each blended polyhedron $\mathcal{P}$ has a regular planar polyhedron $\mathcal{Q}$ as its image under the orthogonal projection to some plane $\Pi$, with the property that edges are mapped to edges and faces to faces. The orthogonal projection of $\mathcal{P}$ to the line $\Pi^\perp$ perpendicular to $\Pi$ is either a line segment $\{\}$ (the only regular polytope of rank 1) or a tessellation $\{\infty\}$ of $\Pi^\perp$ by equal segments (the only regular polygon on the line), and we shall denote it by $\mathcal{R}$ in either case. The polyhedron is then denoted by $\mathcal{Q}\#\mathcal{R}$.

The vertices of the polyhedron $\mathcal{P}\#\{\}$ are contained in two parallel planes, and every edge joins a vertex in one plane to a vertex in the other. The polyhedron $\{4,4\}\#\{\}$ is illustrated in Figure 6a; the faces are skew quadrillaterals that project to the lower (or upper) plane into squares. The faces of $\mathcal{P}\#\{\infty\}$ are helices over the faces of $\mathcal{P}$. If two such faces share an edge, then one is obtained from the other by the reflection about a wall of the helix. Figure 6c shows three faces of $\{4,4\}\#\{\infty\}$, one in solid lines, one in dotted lines and one in dashed lines. One zigzag of $\{\infty,4\}_4\#\{\}$ and one zigzag of $\{\infty,4\}_4\#\{\infty\}$ are shown in Figure 6b,d, respectively.

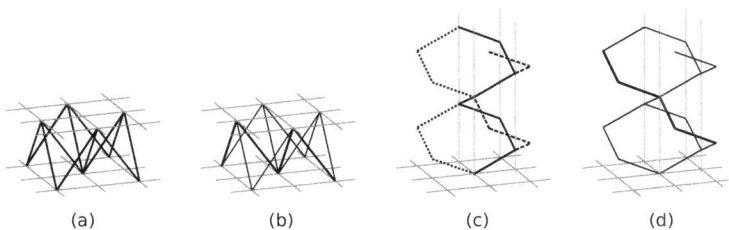

|  (a) | (b) | (c) | (d) |

**Figure 6.** Blended polyhedra $\{4,4\}\#\{\}$ (**a**); $\{\infty,4\}_4\#\{\}$ (**b**); $\{4,4\}\#\{\infty\}$ (**c**) and $\{\infty,4\}_4\#\{\infty\}$ (**d**).

To each of the 12 blended polyhedra, we may associate a real positive parameter $\beta$ corresponding to the ratio between the lengths of an edge of $\mathcal{Q}$ and a line segment of $\mathcal{R}$. The parameter $\beta$ determines the angle $\alpha$ between consecutive edges of a face. Assuming that $\theta$ is the angle between two consecutive edges of a face of the planar polyhedron $\mathcal{Q}$, the parameter $\alpha$ satisfies that $0 < \alpha < \theta$ when $\mathcal{P}$ is $\mathcal{Q}\#\{\}$, whereas $\theta < \alpha < \pi$ if $\mathcal{P}$ is $\mathcal{Q}\#\{\infty\}$. The parameter $\alpha$ completely determines $\mathcal{P}$ up to similarity (see the polyhedra in Class 6 of [7]).

Three of the pure polyhedra have finite planar faces and skew vertex-figures. Two of them were discovered by Petrie, and the remaining by Coxeter (see [16]). The faces of $\{4, 6|4\}$ are squares of the cubic tessellation, while the faces of $\{6, 4|4\}$ and of $\{6, 6|3\}$ are hexagons in the lattice **fcu**. Partial views of the polyhedra $\{4, 3|4\}$ and $\{6, 4|4\}$ are shown in Figure 7.

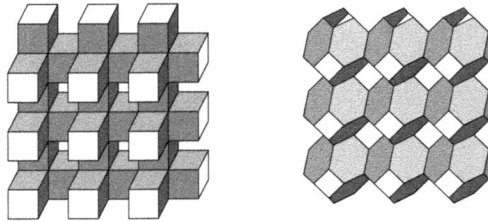

**Figure 7.** The polyhedra $\{4, 3|4\}$ and $\{6, 4|4\}$. Squares and hexagons in the same shade of gray represent polygons in parallel planes

Three infinite pure polyhedra have finite skew faces and planar vertex figures. The faces of the polyhedron $\{6, 4\}_6$ consist of one Petrie polygon of each cube in the cubic tessellation, suitably chosen; this polyhedron is self-Petrial. The faces of the polyhedron $\{4, 6\}_6$ are Petrie polygons of tetrahedra of the tiling of tetrahedra and octahedra; the faces of the polyhedron $\{6, 6\}_4$ are Petrie polygons of octahedra of the same tiling. These two polyhedra are Petrials of each other.

The remaining six pure regular polyhedra have helical faces; three have skew vertex-figures and three planar vertex-figures. The polyhedra $\{\infty, 6\}_{4,4}$, $\{\infty, 4\}_{6,4}$, and $\{\infty, 3\}_{6,3}$ are the Petrials of $\{4, 6|4\}$, $\{6, 4|4\}$, and $\{6, 6|3\}$, respectively, and therefore have skew vertex-figures. The faces of $\{\infty, 6\}_{4,4}$ and $\{\infty, 4\}_{6,4}$ are helices over triangles, whereas those of $\{\infty, 3\}_{6,3}$ are helices over squares. The polyhedra $\{\infty, 3\}^{(a)}$ and $\{\infty, 3\}^{(b)}$ are Petrials of each other; the faces of the former are helices over triangles, and those of the latter are helices over squares. The polyhedron $\{\infty, 4\}_{\cdot,*3}$ is self-Petrial, and its facets are helices over triangles.

In contrast to the blended polyhedra, the pure polyhedra are unique up to similarity. More details on their geometry can be found in [7].

### 3.3. Chiral Polyhedra

In 2005, all chiral polyhedra in $\mathbb{R}^3$ were described by Schulte in [10,11]. Here, we briefly summarise that description. They are all infinite and pure, and can be classified into six infinite families.

The chiral polyhedra in the families $P(a, b)$, $Q(c, d)$, and $Q(c, d)^*$ have finite skew faces and skew vertex-figures, whereas those in families $P_1(a, b)$, $P_2(c, d)$, and $P_3(c, d)$ have helical faces and planar vertex-figures. The parameters take real values, not both 0, and a polyhedron with parameters $(a, b)$ or $(c, d)$ is similar to that with parameters $(ka, kb)$ or $(kc, kd)$, respectively, for any $k \neq 0$. This makes it possible to consider the polyhedra in each family to be parametrised by only one real parameter $a/b$ or $c/d$.

The parameters for polyhedra in the families $P(a, b)$, $Q(c, d)$, and $Q(c, d)^*$ must be rational multiples of each other (or one of them 0), since otherwise the vertex set is not discrete. There is no such restriction for the parameters of polyhedra in the remaining three families. Each of the six families has two distinguished parameters for which the corresponding polyhedra are regular; the polyhedra determined by the remaining parameters are chiral.

The polyhedra in the family $P(a,b)$ have type $\{6,6\}$. Two of them, $P(a,b)$ and $P(a',b')$ (say), are combinatorially isomorphic if and only if $(a,b) \in \{(ka',kb'),(kb',ka')\}$ for some $k \neq 0$, and they are congruent if and only if $(a,b) \in \{\pm(a',b'),\pm(b',a')\}$. The polyhedra $P(1,1)$ and $P(1,-1)$ are the regular polyhedra $\{6,6|3\}$ and $\{6,6\}_4$, respectively.

The polyhedra in the family $Q(c,d)$ have type $\{4,6\}$. Two of them, $Q(c,d)$ and $Q(c',d')$, are combinatorially isomorphic if and only if $(c,d) \in \{(kc',kd'),(-kc',kd')\}$ for some $k \neq 0$, and they are congruent if and only if $(c,d) \in \{\pm(c',d'),\pm(-c',d')\}$. The polyhedra $Q(0,1)$ and $Q(1,0)$ are the regular polyhedra $\{4,6|4\}$ and $\{4,6\}_6$, respectively. When $c$ and $d$ are relatively prime, with $c$ odd and $d \equiv 2$ modulo 4, then the vertex-figure at every vertex is the union of two cycles, and thus in that case, $Q(c,d)$ is not a polyhedron.

The polyhedron $Q(c,d)^*$ is the dual of $Q(c,d)$, meaning that its vertices are at the centres of the faces of $Q(c,d)$, and each of its faces can be constructed around some vertex of $Q(c,d)$. Hence, $Q(c,d)^*$ has type $\{6,4\}$. Furthermore, $Q(c,d)^*$ and $Q(c',d')^*$ are combinatorially isomorphic if $(c,d) \in \{(kc',kd'),(-kc',kd')\}$ for some $k \neq 0$, and they are congruent if and only if $(c,d) \in \{\pm(c',d'),\pm(-c',d')\}$. The polyhedra $Q(0,1)^*$ and $Q(1,0)^*$ are the regular polyhedra $\{6,4|4\}$ and $\{6,4\}_6$, respectively.

If a polyhedron in one of the families $P(a,b)$, $Q(c,d)$, and $Q^*(c,d)$ described above is not combinatorially isomorphic to any of the two regular members of the family, it is geometrically chiral and also chiral as a combinatorial structure.

The polyhedra in the family $P_1(a,b)$ have type $\{\infty,3\}$, and their faces are helices over triangles with the exception of $P_1(1,1)$. The regular instances of this family are $P_1(1,1) = \{3,3\}$ (the tetrahedron) and $P_1(1,-1) = \{\infty,3\}^{(a)}$. Any other member of the family is geometrically chiral but combinatorially isomorphic to $\{\infty,3\}^{(a)}$. Two polyhedra $P_1(a,b)$ and $P_1(a',b')$ are similar when $(a,b) = k(a',b')$ or $(a,b) = k(b',a')$ for some $k \neq 0$; and they are congruent if and only if $(a,b) = \pm(a',b')$ or $(a,b) = \pm(b',a')$.

The polyhedra in the family $P_2(c,d)$ also have type $\{\infty,3\}$, but their faces are helices over squares with the exception of $P_2(0,1)$. The regular instances of this family are $P_2(0,1) = \{4,3\}$ (the cube) and $P_2(1,0) = \{\infty,3\}^{(b)}$. Any other member of the family is geometrically chiral but combinatorially isomorphic to $\{\infty,3\}^{(b)}$. Two polyhedra $P_2(c,d)$ and $P_2(c',d')$ are similar when $(c,d) = k(c',d')$ or $(c,d) = k(-c',d')$ for some $k \neq 0$; and they are congruent if and only if $(c,d) = (\pm c,\pm d)$.

The two regular members of the family $P_3(c,d)$ are $P_3(1,0) = \{3,4\}$ (the octahedron) and $P_3(0,1) = \{\infty,4\}_{\cdot,*3}$. The remaining polyhedra of the family have type $\{\infty,4\}$, and their faces are helices over triangles; they are all geometrically chiral but combinatorially isomorphic to the regular double cover $\{\infty,4\}_{\cdot,*6}$ of $\{\infty,4\}_{\cdot,*3}$. Two polyhedra $P_3(c,d)$ and $P_3(c',d')$ are similar when $(c,d) = k(c',d')$ or $(c,d) = k(-c',d')$ for some $k \neq 0$; and they are congruent if and only if $(c,d) = (\pm c,\pm d)$.

The continuous movement of the parameters of the polyhedra in the last three families can be understood as a continuous movement of the polyhedra $\{\infty,3\}^{(a)}$, $\{\infty,3\}^{(b)}$, and $\{\infty,4\}_{\cdot,*6}$ that preserves at all times the index 2 subgroup of the symmetry group generated by the abstract rotations. More details about the chiral polyhedra described here can be found in [10,11,17].

## 4. Polyhedra in Euclidean Lattices

In this section, we prove Theorem 1 by listing the regular and chiral polyhedra whose underlying graph is contained in one of the three lattices defined in Section 2. In other words, we want to find all possible sets $\mathcal{S}$ of polygons (faces) in **pcu**, **fcu**, and **bcu** such that

- the union of the polygons yields a connected graph,
- every edge of the lattice belongs to precisely two polygons, or to none of them,
- every vertex-figure is a finite polygon,
- there are abstract rotations preserving $\mathcal{S}$ along every face,

135

- there are abstract rotations preserving $S$ around every vertex.

A large part of this work was done in [18] in a slightly different context, where the nets of the regular infinite polyhedra are studied. Here, we also study polyhedra whose vertex and edge sets are proper subsets of the nets **pcu**, **fcu**, and **bcu**, including the finite polyhedra; we still mention all regular polyhedra for the sake of completeness. The main contribution of this paper, then, is the study of the chiral polyhedra that admit an embedding into the nets **pcu**, **fcu**, and **bcu**.

The vertex and edge sets of the cube and of its Petrial can be easily seen as subsets of **pcu**. As explained in Section 2, **fcu** contains subsets of vertices and edges isometric to those of tetrahedra, octahedra, and hence also of their Petrials.

Among the finite regular polyhedra, only the six mentioned above have no 5-fold rotation in their symmetry groups. The underlying graphs of the remaining twelve cannot be embedded in any of the three nets while preserving their symmetries.

The vertex and edge sets of the polyhedra $\{4,4\}$ and $\{\infty,4\}_4$ can be found as subsets of **pcu** in the obvious way. They can also be found in **fcu**, for example, by considering only the vertices and edges of the net whose third coordinates equal to 0. As mentioned in Section 2, the vertices and edges of each of the remaining four planar polyhedra—$\{3,6\}$, $\{6,3\}$, $\{\infty,3\}_6$, and $\{\infty,6\}_3$—can be seen as subsets of those of **fcu**.

In Table 1, we summarise the lattices containing finite and planar regular polyhedra.

**Table 1.** Finite and planar polyhedra and the nets where they can be embedded.

| Polyhedra | Net | Remarks | Polyhedra | Net | Remarks |
|-----------|-----|---------|-----------|-----|---------|
| $\{3,3\}, \{4,3\}_3$ | fcu | finite | $\{3,6\}, \{\infty,6\}_3$ | fcu | planar |
| $\{3,4\}, \{6,4\}_3$ | fcu | finite | $\{6,3\}, \{\infty,3\}_6$ | fcu | planar |
| $\{4,3\}, \{6,3\}_4$ | pcu | finite | $\{4,4\}, \{\infty,4\}_4$ | pcu, fcu | planar |

A blended polyhedron may be embedded in different nets for different values of its parameter $\alpha$. Several possibilities can be discarded by noting the angles between two edges incident to the same vertex in **pcu**, **fcu**, and **bcu** (see Lemma 1).

The angle between two consecutive edges in a face of $\{4,4\}\#\{\}$ or $\{\infty,4\}_4\#\{\}$ is strictly less than $\pi/2$, and so these polyhedra cannot be found as a subset of **pcu**. A sample square of one embedding of $\{4,4\}\#\{\}$ in **fcu** has vertices

$$(0,0,0), (1,0,1), (1,1,0), (0,1,1),$$

and one in **bcu** has vertices

$$(0,0,0), (1,1,1), (2,0,0), (1,-1,1).$$

In both cases, the polyhedron can be embedded in such a way that the vertices all have third coordinates equal to 0 or to 1.

Consecutive edges on a face of $\{3,6\}\#\{\}$ make an angle smaller than $\pi/3$, and hence the vertex and edge sets of this polyhedron (and of its Petrial) are not subsets of any of the lattices **pcu**, **fcu**, and **bcu**.

The vertex and edge sets of $\{6,3\}\#\{\}$ and of $\{\infty,3\}_6\#\{\}$ can be found as subsets of any of **pcu**, **fcu**, and **bcu**. A sample hexagon of $\{6,3\}\#\{\}$ in each of these lattices has vertex set

$$(0,0,0), (1,0,0), (1,-1,0), (1,-1,1), (0,-1,1), (0,0,1),$$
$$(0,0,0), (1,1,0), (1,0,-1), (2,0,0), (1,-1,0), (1,0,1),$$
$$(0,0,0), (1,1,-1), (2,0,-2), (3,-1,-1), (2,-2,0), (1,-1,1),$$

respectively. When extending these hexagons to the entire polyhedron, half of the vertices are in the plane $x + y + z = 0$, and half in the plane $x + y + z = a$, where $a = 2$ for **fcu** and $a = 1$ for the remaining two nets.

The angle between two consecutive edges in a face of $\{4,4\}\#\{\infty\}$ or $\{\infty,4\}_4\#\{\infty\}$ is strictly greater than $\pi/2$ but less than $\pi$, and so these polyhedra cannot be found as a subset of **pcu**. A sample helix of one embedding of $\{4,4\}\#\{\infty\}$ in **fcu** has vertices

$$\ldots, (1,1,-2), (0,1,-1), (0,0,0), (1,0,1), (1,1,2), (0,1,3), \ldots$$

and one in **bcu** has vertices

$$\ldots, (2,0,-2), (1,-1,-1), (0,0,0), (1,1,1), (2,0,2), (1,-1,3), \ldots$$

In both nets, the axes of the helices are parallel to a coordinate axis (the $z$-axis in the case of the embeddings containing the two helices above).

The vertex and edge sets of the polyhedron $\{3,6\}\#\{\infty\}$ and of $\{\infty,6\}_3\#\{\infty\}$ can be found as subsets of any of **pcu**, **fcu**, and **bcu**. A sample hexagonal helix of $\{3,6\}\#\{\infty\}$ in each of these lattices has vertex set

$$\ldots, (0,0,0), (1,0,0), (1,1,0), (1,1,1), (2,1,1), (2,2,1), \ldots$$
$$\ldots, (0,0,0), (1,1,0), (1,2,1), (2,2,2), (3,3,2), (3,4,3), \ldots$$
$$\ldots, (0,0,0), (1,1,-1), (2,0,0), (1,1,1), (2,2,0), (3,1,1), \ldots$$

respectively. In all these helices, the axis has direction vector $(1,1,1)$. In general, the direction axes of all helices are parallel to exactly one diagonal of a cube of the cubic tessellation.

Consecutive edges on a face of $\{6,3\}\#\{\infty\}$ make an angle greater than $2\pi/3$, and hence the vertex and edge sets of this polyhedron (and of its Petrial) are not subsets of any of the lattices **pcu**, **fcu**, or **bcu**.

The nets containing blended regular polyhedra are summarised in Table 2. The polyhedra blended with $\{\}$ are combinatorially isomorphic to the planar polyhedra, and their nets are intrinsically planar; the name of these planar nets according to the Reticular Chemistry Structure Resource database appear in the column "Net". The nets of the polyhedra blended with $\{\infty\}$ admit several embeddings in Euclidean space. In the column "Net", we indicate the name of the most symmetric such embedding according to Reticular Chemistry Structure Resource database. The nets **pcu** and **dia** have more symmetries than the blended poyhedra they carry.

**Table 2.** Blended polyhedra and the nets where they can be embedded.

| Polyhedra | Ambient Net | Net |
|---|---|---|
| $\{3,6\}\#\{\}, \{\infty,6\}_3\#\{\}$ | none | hxl |
| $\{6,3\}\#\{\}, \{\infty,3\}_6\#\{\}$ | pcu, fcu, bcu | hcb |
| $\{4,4\}\#\{\}, \{\infty,4\}_4\#\{\}$ | fcu, bcu | sql |
| $\{3,6\}\#\{\infty\}, \{\infty,6\}_3\#\{\infty\}$ | pcu, fcu, bcu | pcu |
| $\{6,3\}\#\{\infty\}, \{\infty,3\}_6\#\{\infty\}$ | none | acs |
| $\{4,4\}\#\{\infty\}, \{\infty,4\}_4\#\{\infty\}$ | fcu, bcu | dia |

In Table 3, we list the nets where the pure polyhedra can be embedded. In the column "Net", we indicate the name of the net consisting of the vertex and edge sets of each polyhedron. This table has a large intersection with Table 1 in [18].

We now turn our attention to the chiral polyhedra. The procedure we will follow consists of first determining two consecutive edges at a face of the polyhedra in each family. For simplicity, we choose the common vertex to be the origin, except for the polyhedra $Q(c,d)^*$, where we consider them as the duals of the polyhedra $Q(c,d)$. To determine whether these two edges at the origin can be embedded in **pcu**, **fcu**, or **bcu**, we use the standard inner product to take the cosine of the angle between them and compare with the cosine of the angles described in Lemma 1. That is, the cosine must equal 0 or $-1$ if the edges are in **pcu**; $1/2$, $0$, $-1/2$, or $-1$ if the edges are in **fcu**; and $1/3$, $-1/3$, or $-1$ if the edges are in **bcu**. It will then remain to determine if the parameters yield a polyhedron; in particular, if the polyhedron in question has finite faces, we still have to verify if the obtained parameters are rational multiples of each other, or if one of them is 0.

**Table 3.** Pure polyhedra and the nets where they can be embedded.

| Polyhedra | Ambient Net | Net |
|---|---|---|
| $\{4,6|4\}$, $\{\infty,6\}_{4,4}$ | pcu | pcu |
| $\{6,4|4\}$, $\{\infty,4\}_{6,4}$ | fcu | sod |
| $\{6,6|3\}$, $\{\infty,6\}_{6,3}$ | fcu | crs |
| $\{4,6\}_6$, $\{6,6\}_4$ | fcu | hxg |
| $\{6,4\}_6$, $\{\infty,4\}_{\cdot,*3}$ | pcu | nbo |
| $\{\infty,3\}^{(a)}$, $\{\infty,3\}^{(b)}$ | fcu | srs |

## 4.1. Polyhedra $P(a,b)$

According to Section 5 of [10], the neighbours of the origin in the base face of $P(a,b)$ are $(a,0,b)$ and $(0,-b,-a)$; that is, the image of the origin under $S_1$ and $S_1^{-1}$, where $S_1$ is given by

$$(x,y,z) \mapsto (-y, z-b, x-a).$$

They form an angle $\alpha$ with the origin, given by

$$\cos(\alpha) = \frac{(a,0,b)\cdot(0,-b,-a)}{|(a,0,b)||(0,-b,-a)|} = \frac{-ab}{a^2+b^2}.$$

We use the fact that the polyhedra $P(ka,kb)$ and $P(kb,ka)$ are congruent for any $k \neq 0$ to assume without loss of generality that $a=1$. Then, $\cos(\alpha) = -b/(1+b^2)$. This equals 0 if and only if $b=0$ and the polyhedron is $P(1,0)$. This polyhedron has the same vertex and edge sets as **pcu**; the faces are some Petrie polygons of the cubes in the cubic tiling. The six faces around the origin are illustrated in the left of Figure 8. Two of the six faces are in solid lines, two in dotted lines and two in dashed lines.

On the other hand, $\cos(\alpha) = 1/z$ if and only if $b^2 + bz + 1 = 0$. When $z \in \{2,-2\}$, then the polyhedra are $P(1,1)$ and $P(1,-1)$, which are regular. If $z \in \{3,-3\}$, then $b$ is not rational, and so $P(1,b)$ is not a (discrete) polyhedron. The equation has no solution when $z=-1$, and hence $P(1,0)$ is the only chiral polyhedron in the family whose vertex and edge sets are subsets of **pcu**, **fcu**, or **bcu**.

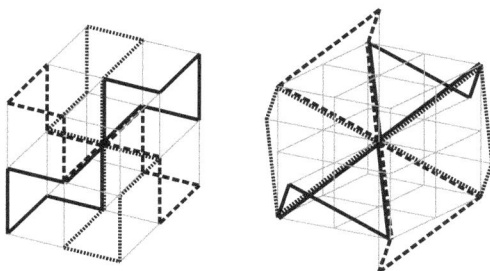

**Figure 8.** The polyhedra $P(1,0)$ and $Q(1,1)$.

*4.2. Polyhedra $Q(c,d)$*

The neighbours of the origin in the base face of $Q(c,d)$ are its images of $S_1$ and $S_1^{-1}$, where $S_1$ maps $(x,y,z)$ to $(-x+c, z-d, -y-c)$, as in Section 6 of [10]. Thus, these neighbours are $(c,-c,d)$ and $(c,-d,-c)$, while the fourth vertex of the base face is $(0,-c-d,d-c)$. The neighbours form an angle $\alpha$ with the origin whose cosine is equal to $c^2/(2c^2+d^2)$.

Recall that $Q(0,1)$ is regular, and by similarity of the polyhedra $Q(c,d)$ and $Q(kc,kd)$, we may assume that $c=1$. Then, $\cos(\alpha) = 1/(2+d^2)$. This number is always strictly greater than 0, and it equals $1/z$ if and only if $d^2 - z + 2 = 0$. If $z = 2$, then the polyhedron is $Q(1,0)$, which is also regular. Finally, if $z = 3$, then we may assume that the polyhedron is $Q(1,1)$, since $Q(1,-1)$ is congruent to $Q(1,1)$.

To describe the polyhedron $Q(1,1)$, we first observe that the vertices of **bcu** are the union of 8 disjoint copies of the vertices of **2bcu**, the net similar to **bcu** whose edges are twice as long. The following list contains a representative in each of these copies:

$$\{(0,0,0), (1,1,1), (1,1,-1), (1,-1,1), (-1,1,1), (2,0,0), (0,2,0), (0,0,2)\}.$$

The vertices of $Q(1,1)$ are those of **bcu**; the edges are those of **bcu** after removing:

- all edges with direction vector $(1,1,1)$ at vertices in **2bcu**;
- all edges with direction vector $(1,-1,1)$ at vertices in $(-1,1,1)+$**2bcu**;
- all edges with direction vector $(1,1,-1)$ at vertices in $(1,-1,1)+$**2bcu**;
- all edges with direction vector $(-1,1,1)$ at vertices in $(1,1,-1)+$**2bcu**.

This removes two edges from every vertex of **bcu**, and hence the vertices of $Q(1,1)$ are 6-valent. The faces are skew quadrilaterals congruent to the base quadrilateral with vertices $(0,0,0)$, $(1,-1,-1)$, $(0,-2,0)$, and $(1,-1,1)$. The corresponding net then has only one kind of vertex and one kind of edge under $G(Q(1,1))$, it is bipartite (as a subnet of **bcu**), and its smallest rings have 4-edges. The author does not know if this net already has a name. The six faces of this polyhedron at the origin are shown in Figure 8. Two of the six faces are in solid lines, two in dotted lines and two in dashed lines.

*4.3. Polyhedra $Q(c,d)^*$*

The polyhedron $Q(c,d)^*$ is the geometric dual of $Q(c,d)$, and so the vertices are in the geometric centres of the faces of $Q(c,d)$. The centre of the base face of $Q(c,d)$, and hence the base vertex of $Q(c,d)^*$ is $v_0 = 1/2(c, -d-c, d-c)$. The base face of $Q(c,d)$ shares consecutive edges with its images under the isometries $S_2$ and $S_2^{-1}$, where $S_2$ maps $(x,y,z)$ to $(-z,-x,-y)$. Therefore, the two neighbours of $v_0$ in the base face of $Q(c,d)^*$ are $v_0 S_2 = 1/2(c-d, -c, c+d)$ and $v_0 S_2^{-1} = 1/2(c+d, c-d, -c)$. By translating by $-v_0$, we get that the cosine of the angle $\alpha$ formed by the two neighbours of $v_0$ with $v_0$ is $-d^2/(2d^2+4c^2)$.

The polyhedron $Q(0,1)^*$ is regular, and therefore we may assume that $c = 1$. Then, $\cos(\alpha) = -d^2/(4 + 2d^2)$. This number is in the interval $(-1/2, 0]$, and it is 0 only when $d = 0$. Since $Q(1,0)^*$ is regular, we only need to explore the possibility of $\cos(\alpha) = -1/3$.

If $\cos(\alpha) = -1/3$, then $d = \pm 2$. As for the polyhedra $Q(c,d)$ with $c$ odd and $d \equiv 2$ modulo 4, the structure $Q(1,2)^*$ is not a polyhedron. Here, every edge belongs to more than one hexagon. Hence, none of the polyhedra $Q(c,d)^*$ live in **pcu**, **fcu**, or **bcu**.

### 4.4. Polyhedra $P_1(a,b)$

According to Section 4 of [11], the neighbours of $(0,0,0)$ in the base helix of the polyhedron $P_1(a,b)$ are $(b,a,0)$ and $(a,0,b)$; that is, the images of $(0,0,0)$ under $S_1$ and $S_1^{-1}$, where $S_1$ maps $(x,y,z)$ to $(-z+b, -x+a, y)$. The cosine of the angle $\alpha$ formed by these two neighbours with $(0,0,0)$ equals $ab/(a^2 + b^2)$.

Since $P_1(a,b)$ is similar to $P_1(b,a)$ and to $P(ka, kb)$ for every $k \neq 0$, we may assume that $a = 1$. Hence, $\cos(\alpha) = b/(1 + b^2)$. If $b \neq 0$ and $\cos(\alpha) = 1/z$, then $b^2 - bz + 1 = 0$. If $z = -1$, then the equation has no solution. If $z \in \{2, -2\}$, then we obtain one of the regular polyhedra $P(1,1)$ or $P(1,-1)$. On the other hand, if $z \in \{3, -3\}$, then $b \notin \{0, 1, -1\}$. We claim that these choices of $b$ do not yield polyhedra having their vertex and edge sets on **bcu**, although the angles between consecutive edges of a face suggest that they could. To see this, we recall that the three neighbours of $(0,0,0)$ in $P_1(1,b)$ are $(b,1,0)$, $(1,0,b)$, and $(0,b,1)$, and note that the neighbours of $(b,1,0)$ are $(0,0,0)$, $(b-1,1,b)$, and $(b, 1-b, 1)$ (see ([11], Page 198)). This implies that the directions of the edges at $(0,0,0)$ are $(b,1,0)$, $(1,0,b)$, and $(0,b,1)$; and that at $(b,1,0)$, there are edges in the directions of $(-1,0,b)$ and $(0,-b,1)$. Therefore $P_1(1,b)$ has edges with at least five different directions. Since **bcu** has edges in only four different directions (the main diagonals of a cube of the cubic tiling), there is no chiral polyhedron $P_1(a,b)$ with $a, b \neq 0$ whose vertex and edge sets are subsets of **pcu**, **fcu**, or **bcu**.

The polyhedron $P_1(1,0)$ is described in detail in [19]. Its faces are helices over triangles embedded in **pcu**. The three edges at every vertex in $P_1(1,0)$ are in the directions of the canonical axes. The axes of the helices are in the directions of the diagonals of a cube of the cubic tiling. The 1-skeleton of $P_1(1,0)$ is illustrated in the left of Figure 9. The three helical faces at some point are shown in the right part of the same figure.

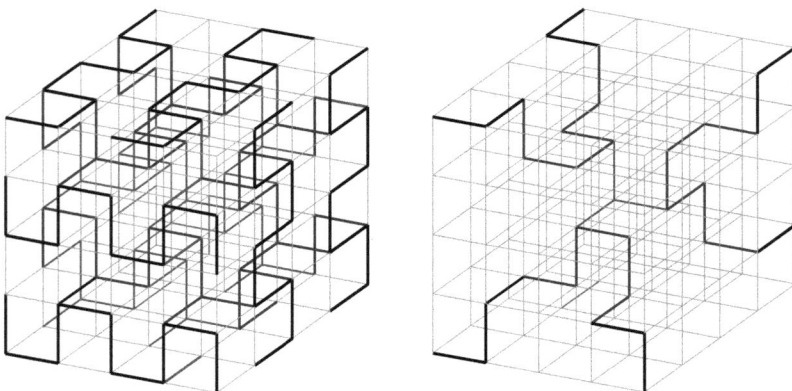

**Figure 9.** The polyhedron $P_1(1,0)$.

### 4.5. Polyhedra $P_2(c,d)$

The neighbours of $(0,0,0)$ in the base helix of the polyhedron $P_2(c,d)$ are $(d,c,-c)$ and $(c,-c,d)$; that is, the images of $(0,0,0)$ under $S_1$ and $S_1^{-1}$, where $S_1$ maps $(x,y,z)$ to $(-z+d, y+c, x-c)$

as in Section 5 of [11]. The cosine of the angle $\alpha$ formed by these two neighbours with $(0,0,0)$ equals $-c^2/(d^2 + 2c^2)$.

Taking on account that $P_2(0,1)$ and $P_2(1,0)$ are regular, and that $\cos(\alpha) \in (-1/2, 0)$ if $c, d \neq 0$, we only need to consider the possibility of $\cos(\alpha) = -1/3$. This gives the parameters $c = d = 1$ (recall here that $P_2(c,d)$ is isomeric to $P_2(c,-d)$). The vertex and edge sets of this polyhedron are indeed subsets of **bcu**. This can be seen by noting that the direction of the three edges at $(0,0,0)$ of this polyhedron are $(1,-1,-1)$, $(-1,1,-1)$, and $(-1,-1,1)$; and that the isometries $S_1$ and $S_2$ (the latter mapping $(x,y,z)$ to $(y,z,x)$) that generate the symmetry group of the polyhedron preserve the set of directions $\{(1,-1,-1),(-1,1,-1),(-1,-1,1),(1,1,1)\}$, all directions of edges of **bcu**. In fact, this polyhedron has its vertices and edges in the diamond net **dia**, which is contained in **bcu**. In the left of Figure 10, we show a portion of the 1-skeleton of $P_2(1,1)$; the three helices at a point are illustrated in the right of the same figure.

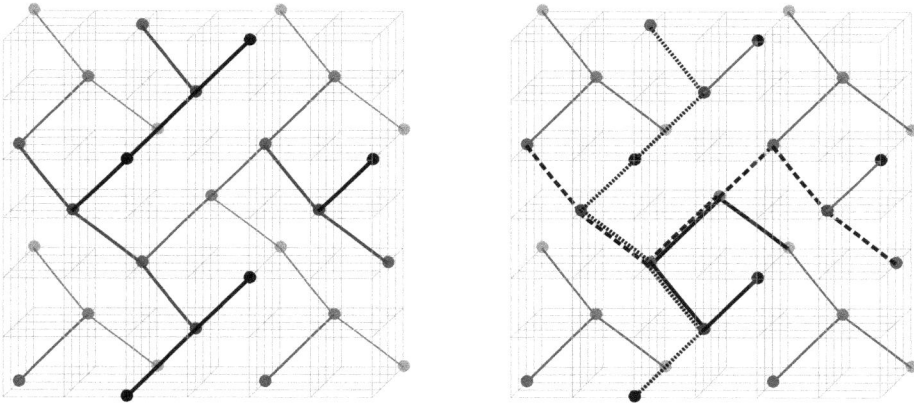

**Figure 10.** The polyhedron $P_2(1,1)$.

*4.6. Polyhedra $P_3(c,d)$*

According to Section 6 of [11], the neighbours of $(0,0,0)$ in the base helix of the polyhedron $P_3(c,d)$ are $(-d,-c,c)$ and $(c,-c,d)$; that is, the images of $(0,0,0)$ under $S_1$ and $S_1^{-1}$, where $S_1$ maps $(x,y,z)$ to $(z-d, x-c, y+c)$. The cosine of the angle $\alpha$ formed by these two neighbours with $(0,0,0)$ equals $c^2/(d^2 + 2c^2)$.

If $c = 0$ or $d = 0$, then $P_3(c,d)$ is regular, so we may assume that $c = 1$ and $d \neq 0$. In this situation, $\cos(\alpha) \in (0, 1/2)$, and it equals $1/3$ whenever $d \in \{1,-1\}$. However, Lemma 6.3 of [11] states that if $c/d$ is a non-zero integer, then $P(c,d)$ is not a geometric polyhedron. In fact, every edge of $P_3(1,\pm 1)$ belongs to three helical faces.

With this, we conclude the proof of Theorem 1. The previous discussion in summarised in the following table.

**Table 4.** Pure polyhedra and the nets where they can be embedded.

| Polyhedra | Ambient Net | Net |
|-----------|-------------|---------|
| $P(1,0)$ | pcu | pcu |
| $Q(1,1)$ | bcu | unknown |
| $P_1(1,0)$ | pcu | srs |
| $P_2(1,-1)$ | bcu | srs |

*Symmetry* **2016**, *8*, 115

**Acknowledgments:** The author thanks the annonymous referees for helpful suggestions that substantially improved this paper. The author was supported by PAPIIT-UNAM under project grant IN101615. This work was elaborated while the author was on sabbatical at the Faculty of Education of the University of Ljubljana. The author thanks that institution and also the program PASPA-DGAPA and UNAM for their support to this sabbatical stay.

**Conflicts of Interest:** The author declares no conflicts of interest.

## References

1. Rogers, C.A. The packing of equal spheres. *Proc. Lond. Math. Soc.* **1958**, *8*, 609–620.
2. *International Tables for Crystallography*; Hahn, T., Ed.; Published for International Union of Crystallography, Chester; D. Reidel Publishing Co.: Dordrecht, The Netherlands, 1983; Volume A, p. 869.
3. Conway, J.H.; Burgiel, H.; Goodman-Strauss, C. *The Symmetries of Things*; A K Peters, Ltd.: Wellesley, MA, USA, 2008; p. 444.
4. Pellicer, D.; Schulte, E. Regular polygonal complexes in space, I. *Trans. Am. Math. Soc.* **2010**, *362*, 6679–6714.
5. Pellicer, D.; Schulte, E. Regular polygonal complexes in space, II. *Trans. Am. Math. Soc.* **2013**, *365*, 2031–2061.
6. Schulte, E. Polyhedra, complexes, nets and symmetry. *Acta Crystallogr. Sect. A* **2014**, *7*, 203–216.
7. Grünbaum, B. Regular polyhedra—Old and new. *Aequ. Math.* **1977**, *16*, 1–20.
8. Dress, A.W.M. A combinatorial theory of Grünbaum's new regular polyhedra. I. Grünbaum's new regular polyhedra and their automorphism group. *Aequ. Math.* **1981**, *23*, 252–265.
9. Dress, A.W.M. A combinatorial theory of Grünbaum's new regular polyhedra. II. Complete enumeration. *Aequ. Math.* **1985**, *29*, 222–243.
10. Schulte, E. Chiral polyhedra in ordinary space. I. *Discrete Comput. Geom.* **2004**, *32*, 55–99.
11. Schulte, E. Chiral polyhedra in ordinary space. II. *Discrete Comput. Geom.* **2005**, *34*, 181–229.
12. O'Keeffe, M.; Peskov, M.A.; Ramsden, S.J.; Yaghi, O.M. Reticular Chemistry Structure Resource. *Accts. Chem. Res.* **2008**, *41*, 1782–1789.
13. Grünbaum, B. Uniform tilings of 3-space. *Geombinatorics* **1994**, *4*, 49–56.
14. McMullen, P.; Schulte, E. Regular polytopes in ordinary space. *Discrete Comput. Geom.* **1997**, *17*, 449–478.
15. Coxeter, H.S.M. *Regular Polytopes*, 3rd ed.; Dover Publications, Inc.: Mineola, NY, USA, 1973; p. 335.
16. Coxeter, H.S.M. Regular skew polyhedra in three and four dimensions, and their topological analogues. *Proc. Lond. Math. Soc.* **1937**, *43*, 33–62.
17. Pellicer, D.; Ivić Weiss, A. Combinatorial structure of Schulte's chiral polyhedra. *Discrete Comput. Geom.* **2010**, *44*, 167–194.
18. O'Keeffe, M. Three-periodic nets and tilings: Regular and related infinite polyhedra. *Acta Crystallogr. Sect. A* **2008**, *64*, 425–429.
19. Pellicer, D. A chiral 4-polytope in $\mathbb{R}^3$. *Ars Math. Contemp.* **2016**, in press.

symmetry

MDPI

*Article*

# Non-Crystallographic Symmetry in Packing Spaces

**Valery G. Rau [1,*], Leonty A. Lomtev [2] and Tamara F. Rau [1]**

[1]  Vladimir State University, Gorkogo St., 87, Vladimir 600000, Russia; vgrautf@mail.ru
[2]  Joint Stock Company "Magneton", Kuibyshev St., 26, Vladimir 600026, Russia; leontijlomtev@rambler.ru
[*]  Author to whom correspondence should be addressed; vgrau@mail.ru; Tel.: +79-066-103-496.

Received: 25 November 2012; Accepted: 5 December 2012; Published: 9 January 2013

**Abstract:**  In the following, isomorphism of an arbitrary finite group of symmetry, non-crystallographic symmetry (quaternion groups, Pauli matrices groups, and other abstract subgroups), in addition to the permutation group, are considered. Application of finite groups of permutations to the packing space determines space tilings by policubes (polyominoes) and forms a structure. Such an approach establishes the computer design of abstract groups of symmetry. Every finite discrete model of the real structure is an element of symmetry groups, including non-crystallographic ones. The set packing spaces of the same order $N$ characterizes discrete deformation transformations of the structure.

**Keywords:** tilings; finite groups of permutations; packing spaces; polyominoes; quaternion group; cayley tables; Pauli matrices; dirac matrices

## 1. Introduction

One-dimensional integer partition of periodic space into segments (of *mod N*) containing k-points-atoms was proposed by Patterson [1] and researched [2,3] to solve a homometrical theory problem that leads to the ambiguity of crystal structure interpretation. In particular, it was demonstrated [4] that the vector system point disposition along the nodal directions of the orthogonal cell of $N \times N$ nodes allows us to have a set of two-dimensional (2D) homometric structures for each direction. In that case, the vector system for each of these structures contains all points of the lattice constant $N$. In combinatorics (Table 3 [5]), the coordinates of the structure of the complete vector system of points form a block diagram and $(v, k, \lambda)$- difference set, where $v = N$-, $k$- is the minimum number of points with integer coordinates and $\lambda$- is the multiplicity of points in the vector system. When building a complete system of vector points in the two-dimensional lattice and the centered cells of the lattice, its versions form the basis of the discrete two-dimensional model, and the $n$-dimensional packing space is formed $N$ times. The choice of direction in the lattice containing $N \times N$ nodes defines one of the sublattices, and its cell is composed of $N$ lattice nodes.

Enumerating the options of homometric structures with a given number $k$ $(k + 1) = N$ of vector system points in the cell of a two-dimensional structure showed that this number coincides with $(N + 1)$, but only for simple $N$. A solution of the general problem of enumerating the number of two-dimensional (2D) and three-dimensional (3D) lattices with $N$-nodes in the translational lattice was obtained after the introduction of the concept of packing space (PS) and its formalization.

Report [6] had the goal of using PS for the search, generation and prediction of structures based on the partition of space into polyominoes. Some practical steps were used to establish the method of discrete modeling of partitions and packages [7] to predict variations of molecular crystal structure (see, for example, [8]). The basis of the method was the partition of periodic space into policubes (3D-polyomino). Figure 1 represents molecules in the form of policubes.

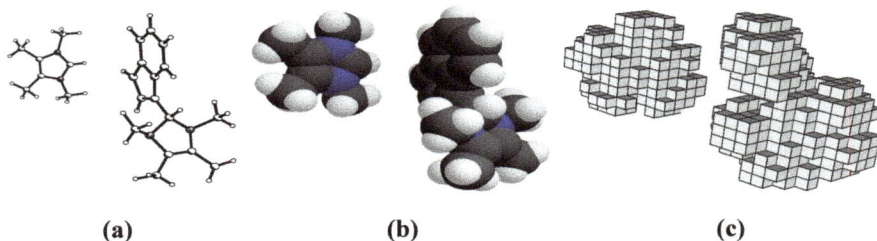

Figure 1. (a) The rod model of molecules; (b) geometrical (spherical) model; and (c) discrete model (policube).

Generally, a growing number of publications have been devoted to the problems of the space partition (see for example [9] or a collection dedicated to M. Gardner [10]). Recent works published in *Symmetry* [11] were also concerned with the problems of space partition into polyominoes and polyiamonds.

The results represented in article [11] show that the interest in partitions is not decreasing and that many problems have not yet been solved. Research carried out by a group of scientists from Vladimir State University (VISU, Vladimir, Russia), described elsewhere [7,12], continues. To the best of our knowledge, this paper represents another way to partition periodic space into 2D-, 3D-polyomino with the help of final permutation groups never described before.

The presentation of a generalized packaging $N \times N$ space containing all two-dimensional lattices of packing spaces of $N$th order was introduced in [12]. Generalized criterion of the space partition into the polyominoes and policubes based on the Patterson model of structure as a vector system of points with integer values of the coordinates (in fractions of the cell) has been proposed. Algorithms and programs of the space partition into the polyominoes (2D), policubes (3D) and prediction of the model and actual molecular structures were partially presented in [13,14].

These investigations have proceeded in another direction in the description of the mechanism of the layered growth of real and model crystal structures [15,16].

The ideas of the approach are based on the research by Conway, Sloane, *et al.* [17]. The properties of the resulting 3D-packaging polyhedron (Figure 2) in a mathematical model of the growing connection graph of the polyominoes in the partition have been studied and described in [18,19]. Article [20] presented the results of the packing polyhedra calculation of growth in crystal nuclei of real structures (nanoclusters) of rhombic and monoclinic sulfur. The possibilities of using the Cambridge Structural Database (CSD) for a great number of nanoclusters of different molecular compounds were presented in [20]. Anthracene, pentacene, coronene and halite nanoclusters were also researched. Based on the calculation of nanoclusters, a multicenter computer model of random nucleation and growth of crystal nuclei in the material has been developed. The output for the two-phase polycrystalline samples of monoclinic and rhombic sulfur $S_8$ is presented in Figure 3. The sizes of the nuclei in the figure range from 1 nm to 8 nm.

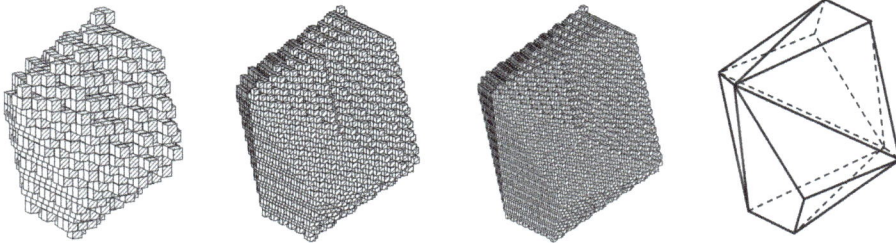

**Figure 2.** Formation of a grown polyhedron by tiling of the plane into policubes. The number of layers *n* [15] is consistently indicated (*n* = 5; *n* = 10; *n* = 15).

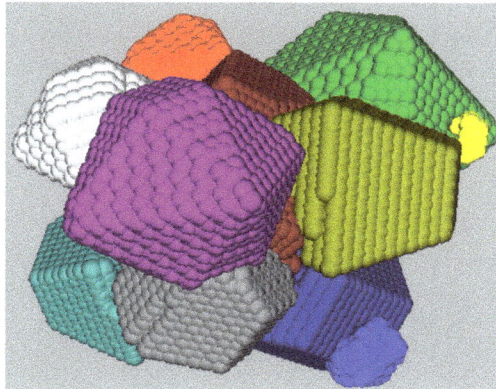

**Figure 3.** A multicenter model of the growth of monoclinic and rhombic sulfur $S_8$ nanoclusters presented in "the Nanoscopic Computer". Color of the clusters identifies various moments of the random nucleation of the future crystalline phases.

The concept of *n*-dimensional packing space was introduced in [21]. In [22], it was proposed to use a mathematical permutation group for the analysis of the symmetry of the packing space (PS). The crystallographic symmetry PS$^{2D}$ based on classical symmetry transformations in the Weyl group was presented in [23]. Broadened options to use the method for the description of fractal non-periodic partitions at present were shown in [24].

In this paper, we describe new results of a discrete model of packing space usage for the visualization of abstract finite groups of substitutions, quaternions, Pauli matrices, Dirac matrices, and a description of the real two-dimensional layer structure.

## 2. Basic Concepts and Definitions

In crystallography, figures built on the basis of joining squares are called polyomino, on the basis of triangles are called "polyiamonds" and on the basis of hexagon are called polyhexes (see for example, [11]). The shape of the square and cube has been chosen to describe the space partition for simplicity of algorithms and software partition implementation and continuity of these figures in the transition from two- to three-dimensional, and then to *n*-dimensional space [21]. Three-dimensional analogs of the triangle (tetrahedron) and hexagon (hexagonal bipyramid or prism) do not give simple partitions.

The figures compiled from the *N*-unicellular-squares or cubes in E$^3$ so that each square (or cube) is adjacent to at least one neighbor that has a common side, are called polyominoes [11] (3D-polyomino, or policube). Figure 6 shows an example of figures "polyomino-cross" with $N = 5$, which splits the space, that is, fills it without voids or overlays. It is evident that any $N$-cellular polyomino can be fully presented on the plane in the area of the square containing the $N \times N$ cells, each of which coincides with the size of the cells polyomino (Figure 4, where $N = 6$). Centers $N^2$ elementary cell of the square form $N^2$ nodes of the lattice translations of all the space. If the basic directions of space are used to select the horizontal and vertical lines, the nodes in this basis will have integer values (Figure 4). The beginning of the reference in the chosen basis accept node is located in the upper left hand corner of the square. In the general case of not-primitive sublattice, this lattice, which has $N$ nodes, has a fundamental area of $N$ of the corresponding cells polyomino. The number of $N$ in this case is called the index of a sublattice or the order. Red (Figure 4) highlights a nodal direction, determined by the translation of $0 \rightarrow 8$ that will be permutated in the cycles:

(0 8 16 18 26 34)(1 9 17 19 27 35)(2 10 12 20 28 30)(3 11 13 21 29 31)(4 6 14 22 24 32)(5 7 15 23 25 33).

This is a permutation of highlights in the lattice of the series of parallel nodal lines (Figure 4). Cell sublattice defines the coordinates of nodes: (6,0) and (2,1), but the fundamental area of the cell sublattice contains $N = 6$ nodes, or 6 cells polyomino. The coordinates of the selected nodes can be represented in the form of vector-matrix columns:

$$\begin{bmatrix} 6 & 2 \\ 0 & 1 \end{bmatrix}$$

and all the information about the selected cell sublattice is recorded as follows: PS-P 6 $1_2$, where the designation PS is a brief record of the "Pack Space", and $P$–plane. To partition, the policubes will use the symbol PS S, where S is the space—"Space".

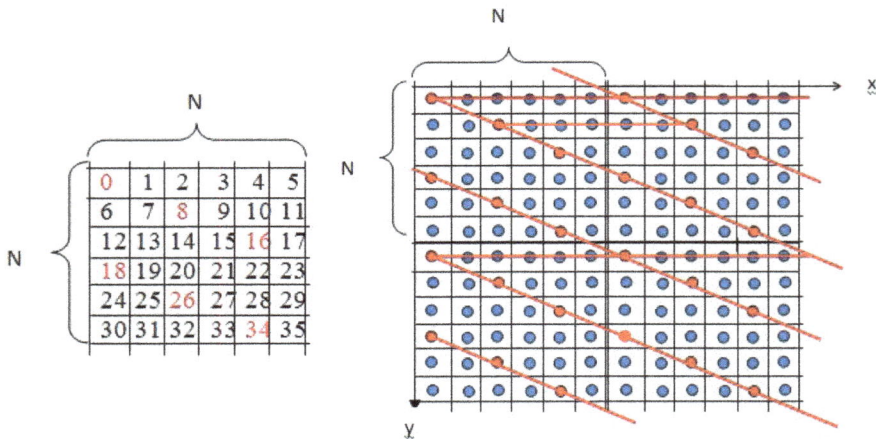

Figure 4. A series of parallel nodal lines.

As sublattice in two-dimensional space is determined by two independent directions coming from the origin, the result was the following theorem:

**Theorem 1.** A beam translational independent nodal line conducted through the origin and every node of the square $N \times N$ is a subgroup "$N^2$" of the permutations of the order of $N^2$. By forming a group of some of the directions in the lattice, it is simple, with the help of operations of multiplication, to find all of the subgroup "$N^2$" and its table Cayley. The result of this calculation for a simple case of $N = 4$ is presented in Table 1.

The splitting of the elements of the group on the equivalence classes that are defined as independent lines, and irreducible representations of the group "$N^2$", which have the same color, is presented in the Table 2. The number of directions is 16. The group is divided into 5 subgroups of 4 elements each, including the identity transformation. In each subgroup, operations are divided into cycles. Each cycle determines the number of equivalent nodes that form the sublattice translations. The number of cycles determines the number of independent nodes in the cell of the sublattice, which is the index of a sublattice.

From the geometry of the nodal lines in the space $4 \times 4$ (Figure 5) and Table 2, the operations of $g(2)$, $g(8)$ and $g(10)$ are distinguished by one cell with an index of 8. Simple division of these cells into two independent parts leaves only two of them ($g(10)$ and $g(8)$), each of which will be characterized by dividing the index, equal to 4. The remaining operations are divided into each subgroup to the equivalent of a pair, the number of which is equal to 5. Thus, the number of independent nodal areas is 7 for lattices with the index $N = 4$, which coincides with the sum of the divisors of the number 4, as shown in the simple brute force of previously published works [13].

**Table 1.** Cayley table for $N = 4$.

| | g[0] | g[1] | g[2] | g[3] | g[4] | g[5] | g[6] | g[7] | g[8] | g[9] | g[10] | g[11] | g[12] | g[13] | g[14] | g[15] |
|---|---|---|---|---|---|---|---|---|---|---|---|---|---|---|---|---|
| g[0]=(0 1 2 3 4 5 6 7 8 9 10 11 12 13 14 15); | g[0] | g[1] | g[2] | g[3] | g[4] | g[5] | g[6] | g[7] | g[8] | g[9] | g[10] | g[11] | g[12] | g[13] | g[14] | g[15] |
| g[1]=(5 6 7 4 9 10 11 8 13 14 15 12 1 2 3 0); | g[1] | g[2] | g[3] | g[0] | g[5] | g[6] | g[7] | g[4] | g[9] | g[10] | g[11] | g[8] | g[13] | g[14] | g[15] | g[12] |
| g[2]=(10 11 8 9 14 15 12 13 2 3 0 1 6 7 4 5); | g[2] | g[3] | g[0] | g[1] | g[6] | g[7] | g[4] | g[5] | g[10] | g[11] | g[8] | g[9] | g[14] | g[15] | g[12] | g[13] |
| g[3]=(15 12 13 14 3 0 1 2 7 4 5 6 11 8 9 10); | g[3] | g[0] | g[1] | g[2] | g[7] | g[4] | g[5] | g[6] | g[11] | g[8] | g[9] | g[10] | g[15] | g[12] | g[13] | g[14] |
| g[4]=(6 7 4 5 10 11 8 9 14 15 12 13 2 3 0 1); | g[4] | g[5] | g[6] | g[7] | g[8] | g[9] | g[10] | g[11] | g[12] | g[13] | g[14] | g[15] | g[0] | g[1] | g[2] | g[3] |
| g[5]=(11 8 9 10 15 12 13 14 3 0 1 2 7 4 5 6); | g[5] | g[6] | g[7] | g[4] | g[9] | g[10] | g[11] | g[8] | g[13] | g[14] | g[15] | g[12] | g[1] | g[2] | g[3] | g[0] |
| g[6]=(12 13 14 15 0 1 2 3 4 5 6 7 8 9 10 11); | g[6] | g[7] | g[4] | g[5] | g[10] | g[11] | g[8] | g[9] | g[14] | g[15] | g[12] | g[13] | g[2] | g[3] | g[0] | g[1] |
| g[7]=(1 2 3 0 5 6 7 4 9 10 11 8 13 14 15 12); | g[7] | g[4] | g[5] | g[6] | g[11] | g[8] | g[9] | g[10] | g[15] | g[12] | g[13] | g[14] | g[3] | g[0] | g[1] | g[2] |
| g[8]=(8 9 10 11 12 13 14 15 0 1 2 3 4 5 6 7); | g[8] | g[9] | g[10] | g[11] | g[12] | g[13] | g[14] | g[15] | g[0] | g[1] | g[2] | g[3] | g[4] | g[5] | g[6] | g[7] |
| g[9]=(13 14 15 12 1 2 3 0 5 6 7 4 9 10 11 8); | g[9] | g[10] | g[11] | g[8] | g[13] | g[14] | g[15] | g[12] | g[1] | g[2] | g[3] | g[0] | g[5] | g[6] | g[7] | g[4] |
| g[10]=(2 3 0 1 6 7 4 5 10 11 8 9 14 15 12 13); | g[10] | g[11] | g[8] | g[9] | g[14] | g[15] | g[12] | g[13] | g[2] | g[3] | g[0] | g[1] | g[6] | g[7] | g[4] | g[5] |
| g[11]=(7 4 5 6 11 8 9 10 15 12 13 14 3 0 1 2); | g[11] | g[8] | g[9] | g[10] | g[15] | g[12] | g[13] | g[14] | g[3] | g[0] | g[1] | g[2] | g[7] | g[4] | g[5] | g[6] |
| g[12]=(14 15 12 13 2 3 0 1 6 7 4 5 10 11 8 9); | g[12] | g[13] | g[14] | g[15] | g[0] | g[1] | g[2] | g[3] | g[4] | g[5] | g[6] | g[7] | g[8] | g[9] | g[10] | g[11] |
| g[13]=(3 0 1 2 7 4 5 6 11 8 9 10 15 12 13 14); | g[13] | g[14] | g[15] | g[12] | g[1] | g[2] | g[3] | g[0] | g[5] | g[6] | g[7] | g[4] | g[9] | g[10] | g[11] | g[8] |
| g[14]=(4 5 6 7 8 9 10 11 12 13 14 15 0 1 2 3); | g[14] | g[15] | g[12] | g[13] | g[2] | g[3] | g[0] | g[1] | g[6] | g[7] | g[4] | g[5] | g[10] | g[11] | g[8] | g[9] |
| g[15]=(9 10 11 8 13 14 15 12 1 2 3 0 5 6 7 4); | g[15] | g[12] | g[13] | g[14] | g[3] | g[0] | g[1] | g[2] | g[7] | g[4] | g[5] | g[6] | g[11] | g[8] | g[9] | g[10] |

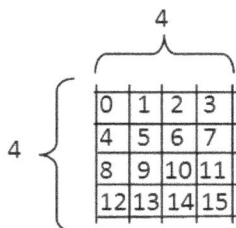

**Figure 5.** A series of parallel nodal lines.

**Table 2.** The splitting of the elements of the group on the equivalence classes.

g[0] = (0 1 2 3 4 5 6 7 8 9 10 11 12 13 14 15 ) = (0) (1) (2) (3) (4) (5) (6) (7) (8) (9) (10) (11) (12) (13) (14) (15)
g[1] = (5 6 7 4 9 10 11 8 13 14 15 12 1 2 3 0 ) = (0 5 10 15) (1 6 11 12) ( 2 7 8 13) (3 4 9 14)
g[2] = (10 11 8 9 14 15 12 13 2 3 0 1 6 7 4 5 ) = (0 10) (1 11) (2 8) (3 9) (4 14) (5 15) ( 6 12) (7 13)
g[3] = (15 12 13 14 3 0 1 2 7 4 5 6 11 8 9 10 ) = (0 15 10 5) (1 12 11 6) (2 13 8 7) (3 14 9 4)
g[4] = (6 7 4 5 10 11 8 9 14 15 12 13 2 3 0 1 ) = (0 6 8 14) (1 7 9 15) (2 4 10 12) (3 5 11 13)
g[5] = (11 8 9 10 15 12 13 14 3 0 1 2 7 4 5 6 ) = (0 11 2 9) (1 8 3 10) (4 15 6 13) (5 12 7 14)
g[6] = (12 13 14 15 0 1 2 3 4 5 6 7 8 9 10 11 ) = (0 12 8 4 ) (1 13 9 5) (2 14 10 6) (3 15 11 7)
g[7] = (1 2 3 0 5 6 7 4 9 10 11 8 13 14 15 12 ) = (0 1 2 3) (4 5 6 7) (8 9 10 11) (12 13 14 15)
g[8] = (8 9 10 11 12 13 14 15 0 1 2 3 4 5 6 7 ) = (0 8) (1 9) (2 10) (3 11) (4 12) (5 13) (6 14) (7 15)
g[9] = (13 14 1 5 12 1 2 3 0 5 6 7 4 9 10 11 8 ) = (0 13 10 7) (1 14 11 4) (2 15 8 5) (3 12 9 6)
g[10] = (2 3 0 1 6 7 4 5 10 11 8 9 14 15 12 13 ) = (0 2) (1 3) (4 6) (5 7) (8 10) (9 11) (12 14) (13 15)
g[11] = (7 4 5 6 11 8 9 10 15 12 13 14 3 0 1 2 ) = (0 7 10 13) (1 4 11 14) (2 5 8 15) (3 6 9 12)
g[12] = (14 15 12 13 2 3 0 1 6 7 4 5 10 11 8 9 ) = (0 14 8 6) (1 15 9 7) (2 12 10 4) (3 13 11 5)
g[13] = (3 0 1 2 7 4 5 6 11 8 9 10 15 12 13 14 ) = (0 3 2 1) (4 7 6 5) (8 11 10 9) (12 15 14 13)
g[14] = (4 5 6 7 8 9 10 11 12 13 14 15 0 1 2 3 ) = (0 4 8 12) (1 5 9 13) (2 6 10 14) (3 7 11 15)
g[15] = (9 10 11 8 13 14 15 12 1 2 3 0 5 6 7 4 ) = (0 9 2 11) (1 10 3 8) (4 13 6 15) (5 14 7 12)

Packing space (PS) or partition space is a lattice in which each node is assigned weight (symbol or color) so that any set of lattice nodes with the same weight makes the same (up to shift) sublattice of translations of the original lattice.

The columns of the vectors' coordinates (based on the original lattice) of one of the sublattice bases of this lattice form an integral matrix:

$$Y = \begin{bmatrix} x_1 & x_2 & x_3 \\ 0 & y_2 & y_3 \\ 0 & 0 & z_3 \end{bmatrix}$$

where $0 \le x_2 < x_1, 0 \le x_3 < x_1, 0 \le y_3 < y_2$, and $z_3 > 0$.

Matrix $Y$ is a matrix of three-dimensional packing space. The order of packing space coincides with the index of the sublattice and is the multiplication of the diagonal elements of the packing matrix $N = x_1 y_2 z_3$.

For a two-dimensional lattice in the matrix, only numbers $x_1$, $x_2$, and $y_2$ will remain. The order of the packing space in this case is $N = x_1 y_2$.

For each node of the packing space, we construct a partition of the Dirichlet domains by dividing the interstitial distances by the median plane, and we obtain a partition of space into minimum areas that do not contain nodes, except one. These are cubes in three dimensions, and squares in the two dimensions (Figure 6). In solid state physics, such lattice domains are called Wigner–Seitz cells. In [25], the question of the structural organization of the complex compound with carbamide on the basis of the division of space into Dirichlet-polihedras was considered.

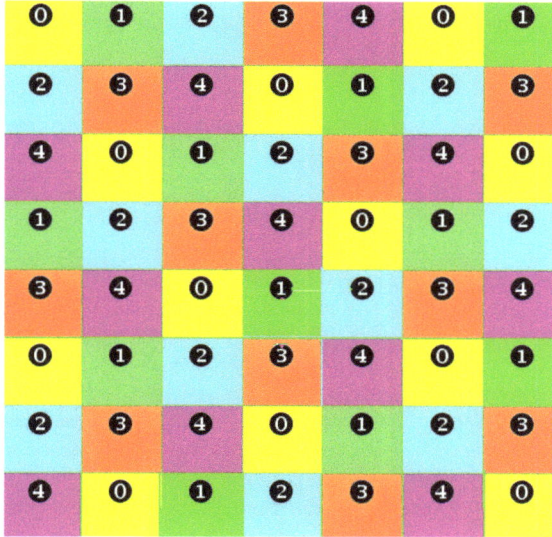

**Figure 6.** Definition of the packing space. PS P $51_3$.

Let the size of the square coincide with the size of cells that make up the $N$-minoes of arbitrary shape. Then, we can address the problem of investing polyomino into the fundamental domain of two-dimensional packing space. If this can be conducted, the selected polyomino will tile space.

For example, we choose a polyomino of 5 cells in the form of a cross and put its middle cell with any of the packing spaces of $P51_2$ (Figure 7). We assume the following statement (criterion of partition): if all the cells with the given orientation of the polyomino coincide with cells that have different weights in the packing space, the chosen polyomino breaks space. This conclusion is general and answers the question: Which polyomino (policube) or its assembly may become a fundamental domain of space partition?

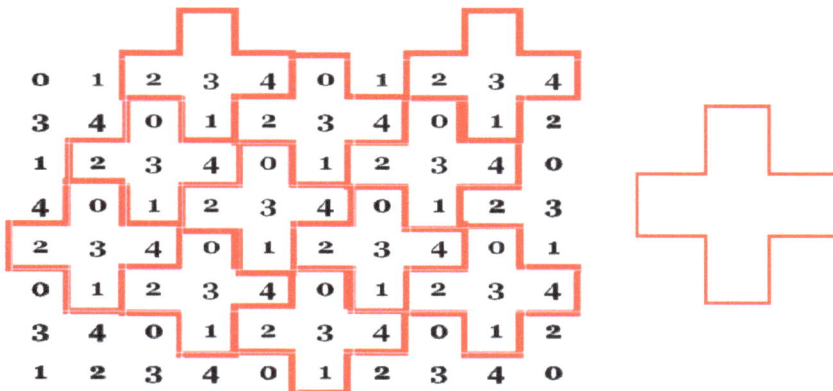

**Figure 7.** PS $P51_2$. One of the cells, corresponding to the range of PS, is selected by vectors. The criterion of the space partition is tested by the polyomino shape—the "cross". The center of the figure coincides with the "weight 3" of PS.

The formalization of the basic concepts of the $n$-dimensional packing space discrete model, policube and the partition criterion are given in [21].

For the space presented in Figure 6, we have $x_1 = 5$, $x_2 = 3$, and $y_2 = 1$. The two-dimensional packing space, presented in Figure 3, is denoted as P5 $1_3$.

The number $N$ of packing spaces with a given order is finite and is determined by formula (1) [7]:

$$N_p \text{ (two-dimensional packing space)} = \sum_{d/N} d \text{ ; } N_S \text{ (three-dimensional packing space)} = \sum_{d_1 d_2 | N} d_1^2 d_2$$

(1)

For the image splitting, in addition to specifying the packaging space, each cell in the program of images in succession is encoded with the help of 4 characters: 0, 1, 2, 3. The first character is associated with the absence of the cells of the borders in the division. The following three are responsible for the existence of the border only on the left, only from the top and the left and top, respectively. The boundaries from the bottom and on the right are not taken into account because these limits are defined as cells, located below and to the right. The encoding of the cell structure, shown in Figure 7, is recorded as follows: 3 1 3 0 2. On the encoding in the given packing space, all partitioning is restored. Procedure busting encoding characters and allocation in the division of areas with closed borders allow one to try all possible variants of partition space sublattice with the index $N$. The code of the cell-neighbors is agreed upon so that the program can consistently surround each polyomino in the division of neighboring and find the structure of the layer-by-layer growth. In the computer experiment, it was found, and in [18], strictly proved, that in the periodic partitioning the existence of a polygon growth (or growth of the polyhedron in three-dimensional space) is obligatory. Therefore, all of the crystals under ideal conditions grow up in the form of polyhedra.

Traditionally, analyzing a particular crystal structure, in accordance with the theorem (Cayley) about the isomorphism of an arbitrary finite group and the permutation group, for each type of crystallographic symmetry we can find an alternative compared to the transformation of space and the subgroup of permutations (Example 1.1 in this article), which is of academic interest. We proceed differently and apply the subgroup of permutations of the PS elements, and then, we perform the analysis of the resulting structure. Now, we have the first practical result immediately: If transformation is applied to the same PS, the resulting structures are geometrically conjugate to the periods and angles of the lattice in accordance with the concept of PS. This conjugation, in particular, is important for micro- and nano-electronics. Then, as each transformation of substitution can be represented in the form of cycles, each of them in the PS will match some domain that consists of equivalent points, and if we assign each set of equivalent points the same color, then we obtain space partitioned into separate fields. This is the second practical result. Partition models are important in the analysis of coordination packages in crystal chemistry. For every space partition, we can calculate packing polyhedron (polygon) of structure growth, the construction of which (by prescription [15,16]) helps us to determine how the number of dots on each layer of the growing shape varies (magic numbers) and how structural motives are arranged relating to faces (or edges) of the growth shape. This is the third practical result, which is used to construct models of nanoclusters [20,26].

To achieve these results, we compiled a program for sorting permutations with a given number of elements $N$, tiling into separate subgroups and construction of the Cayley table for each subgroup and presentation of the group operations in the form of cycles. An algorithm and program for each of the substitution operations to the packing space and the space partition into the colored areas were developed.

## 3. Examples of Permutation Groups on Packing Spaces

### 3.1. Arbitrary Permutation Groups

As an example, we analyze the structure of the partition on the symmetry plane of P4mm (Figure 8). If you select symmetrically identical points of the structure and paint them one color (pick out point of the orbit), and then write the transformation of symmetry group with substitutions operations, you obtain eight operations, making up a group. A Cayley table of that group is presented in Table 3. The group also includes the translation operation g(0), leaving all the points in their places. A column of Table 3, delivered to the right of the Cayley table, presents the elements of points symmetry determined by the operations g(*N*), which is written in the corresponding row.

**Table 3.** Cayley table for the crystallographic group P4mm (Figure 7).

| g[0] = (0 1 2 3 4 5 6 7); | g[0] | g[1] | g[2] | g[3] | g[4] | g[5] | g[6] | g[7] | e |
|---|---|---|---|---|---|---|---|---|---|
| g[1] = (0 4 2 6 3 7 1 5); | g[1] | g[2] | g[3] | g[0] | g[5] | g[6] | g[7] | g[4] | 4 |
| g[2] = (0 3 2 1 6 5 4 7); | g[2] | g[3] | g[0] | g[1] | g[6] | g[7] | g[4] | g[5] | 2 |
| g[3] = (0 6 2 4 1 7 3 5); | g[3] | g[0] | g[1] | g[2] | g[7] | g[4] | g[5] | g[6] | 4 |
| g[4] = (0 1 2 3 6 7 4 5); | g[4] | g[7] | g[6] | g[5] | g[0] | g[3] | g[2] | g[1] | mh |
| g[5] = (0 6 2 4 3 5 1 7); | g[5] | g[4] | g[7] | g[6] | g[1] | g[0] | g[3] | g[2] | md1 |
| g[6] = (0 3 2 1 4 7 6 5); | g[6] | g[5] | g[4] | g[7] | g[2] | g[1] | g[0] | g[3] | mv |
| g[7] = (0 4 2 6 1 5 3 7); | g[7] | g[6] | g[5] | g[4] | g[3] | g[2] | g[1] | g[0] | md2 |

Another result with the other Cayley table is obtained when the cells belonging to each polyomino in the structure are painted the same color. In accordance with this approach, Figure 8a shows the tri-colored structure, and its growth structure is shown in Figure 8.

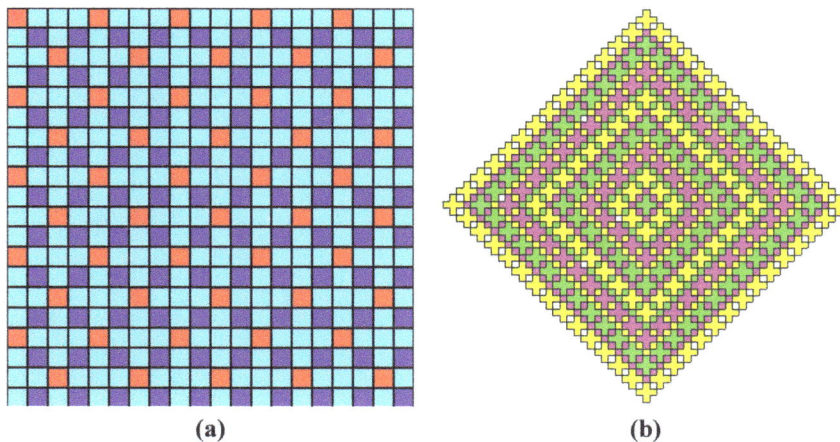

**(a)** **(b)**

**Figure 8.** (**a**) An example of P42$_2$ space partition into the polyomino and (**b**) the form of layered structure growth.

The corresponding 3-cycle transformation cannot be obtained in the group with the crystallographic operations presented in Table 3. In the first case (Table 3), we address the crystallographic operating space. Table 4, in which the color is determined by the structure of the space tiling into the polyominoes, gives another representation of packing space. In this case, we assume that the space is the structural space. There is a variant, when these spaces are isomorphic, but it is obvious that the structural space is more general. Any crystallographic operational space defines a tiling of PS into polyominoes, but not every tiling is determined by the transformations of the operational space with the crystallographic symmetry of the plane (or space) group.

Table 4. Cayley table for transformations PS P42$_2$ (for Figure 8a).

| g[0] = (0 1 2 3 4 5 6 7) | g[0] | g[1] | g[2] | g[3] | g[4] | g[5] | g[6] | g[7] | g[8] | g[9] | 8-colored |
|---|---|---|---|---|---|---|---|---|---|---|---|
| g[1] = (0 2 3 4 6 7 1 5) | g[1] | g[2] | g[3] | g[4] | g[5] | g[6] | g[7] | g[8] | g[9] | g[0] | 3-colored |
| g[2] = (0 3 4 6 1 5 2 7) | g[2] | g[3] | g[4] | g[5] | g[6] | g[7] | g[8] | g[9] | g[0] | g[1] | 4-colored |
| g[3] = (0 4 6 1 2 7 3 5) | g[3] | g[4] | g[5] | g[6] | g[7] | g[8] | g[9] | g[0] | g[1] | g[2] | 3-colored |
| g[4] = (0 6 1 2 3 5 4 7) | g[4] | g[5] | g[6] | g[7] | g[8] | g[9] | g[0] | g[1] | g[2] | g[3] | 4-colored |
| g[5] = (0 1 2 3 4 7 6 5) | g[5] | g[6] | g[7] | g[8] | g[9] | g[0] | g[1] | g[2] | g[3] | g[4] | 7-colored |
| g[6] = (0 2 3 4 6 5 1 7) | g[6] | g[7] | g[8] | g[9] | g[0] | g[1] | g[2] | g[3] | g[4] | g[5] | 4-colored |
| g[7] = (0 3 4 6 1 7 2 5) | g[7] | g[8] | g[9] | g[0] | g[1] | g[2] | g[3] | g[4] | g[5] | g[6] | 3-colored |
| g[8] = (0 4 6 1 2 5 3 7) | g[8] | g[9] | g[0] | g[1] | g[2] | g[3] | g[4] | g[5] | g[6] | g[7] | 4-colored |
| g[9] = (0 6 1 2 3 7 4 5) | g[9] | g[0] | g[1] | g[2] | g[3] | g[4] | g[5] | g[6] | g[7] | g[8] | 3-colored |

In general, the second approach is characterized by non-crystallographic transformations of permutations.

To demonstrate the proposed method of analysis of any structure, including one with non-crystallographic symmetry, we consider a specific subgroup of permutations of the twelfth order. A Cayley table of the group is presented in Table 5.

Table 5. Cayley table for the subgroup of permutations of the twelfth order.

| g[0] | g[1] | g[2] | g[3] | g[4] | g[5] | g[6] | g[7] | g[8] | g[9] | g[10] | g[11] |
|---|---|---|---|---|---|---|---|---|---|---|---|
| g[1] | g[0] | g[3] | g[2] | g[5] | g[4] | g[7] | g[6] | g[9] | g[8] | g[11] | g[10] |
| g[2] | g[4] | g[0] | g[5] | g[1] | g[3] | g[8] | g[10] | g[6] | g[11] | g[7] | g[9] |
| g[3] | g[5] | g[1] | g[4] | g[0] | g[2] | g[9] | g[11] | g[7] | g[10] | g[6] | g[8] |
| g[4] | g[2] | g[5] | g[0] | g[3] | g[1] | g[10] | g[8] | g[11] | g[6] | g[9] | g[7] |
| g[5] | g[3] | g[4] | g[1] | g[2] | g[0] | g[11] | g[9] | g[10] | g[7] | g[8] | g[6] |
| g[6] | g[7] | g[8] | g[9] | g[10] | g[11] | g[0] | g[1] | g[2] | g[3] | g[4] | g[5] |
| g[7] | g[6] | g[9] | g[8] | g[11] | g[10] | g[1] | g[0] | g[3] | g[2] | g[5] | g[4] |
| g[8] | g[10] | g[6] | g[11] | g[7] | g[9] | g[2] | g[4] | g[0] | g[5] | g[1] | g[3] |
| g[9] | g[11] | g[7] | g[10] | g[6] | g[8] | g[3] | g[5] | g[1] | g[4] | g[0] | g[2] |
| g[10] | g[8] | g[11] | g[6] | g[9] | g[7] | g[4] | g[2] | g[5] | g[0] | g[3] | g[1] |
| g[11] | g[9] | g[10] | g[7] | g[8] | g[6] | g[5] | g[3] | g[4] | g[1] | g[2] | g[0] |

Multiplication of permutations allows the group to be split into classes of conjugate elements by iterating over the operations of multiplication, determining the conjugate elements by the rule:

$$g(i) = g(\alpha)g(k)g^{-1}(\alpha), \alpha \in (0,11)$$

As a result, we have 6 classes taken in braces:
{g(0)}; {g(1), g(2), g(5)}; {g(3), g(4)}; {g(6)}; {g(7), g(8) g(11)}; and {g(9), g(10)}.

The operations of substitutions and their cyclic representation, the number of colored areas in the PS P61$_1$ and distribution of the color designated by Latin letters in the layered structure of the packing space P61$_1$, are presented in Table 6.

**Table 6.** The operations of substitutions and their cyclic representation and the number of colored areas in the PS P61$_1$.

| Operations of substitutions | Cyclic representation | Number of colored areas | The alternation of layers in the PS P61$_1$ (Figure 9) |
|---|---|---|---|
| g[0] = ( 0 1 2 3 4 5 ) | (0)(1)(2)(3)(4)(5) | 6-colored | i j k l m n (Figure 9a) |
| g[1] = ( 0 1 4 5 2 3 ) | (0)(1)(24)(35) | 4-colored | i j i j k l (Figure 9b) |
| g[2] = ( 4 5 2 3 0 1 ) | (04)(15)(2)(3) | 4-colored | i j i j k l (Figure 9b) |
| g[3] = ( 4 5 0 1 2 3 ) | (042)(153) | 2-colored | i j i j i j (Figure 9c) |
| g[4] = ( 2 3 4 5 0 1 ) | (024)(135) | 2-colored | i j i j i j (Figure 9c) |
| g[5] = ( 2 3 0 1 4 5 ) | (02)(13)(4)(5) | 4-colored | i j i j k l (Figure 9b) |
| g[6] = ( 1 0 3 2 5 4 ) | (01)(23)(45) | 3-colored | i i j j k k (Figure 9d) |
| g[7] = ( 1 0 5 4 3 2 ) | (01)(25)(34) | 3-colored | i i j k k j (Figure 9e) |
| g[8] = ( 5 4 3 2 1 0 ) | (05)(14)(03) | 3-colored | i i j k k j (Figure 9e) |
| g[9] = ( 5 4 1 0 3 2 ) | (052143) | 1-colored | i i i i i i (Figure 9f) |
| g[10] = ( 3 2 5 4 1 0 ) | (034125) | 1-colored | i i i i i i (Figure 9f) |
| g[11] = ( 3 2 1 0 5 4 ) | (03)(12)(45) | 3-colored | i i j k k j (Figure 9e) |

Below (in Figure 9 for P61$_1$), all independent structures are given. Under each of the colored variants, there are marked operations of the permutation groups applied to the PS in accordance with Table 4, which replaces the numbering of figures.

Thus, there are only six independent colored PS P61$_1$ in the subgroup of permutations of the 12th order. Each variant matches its own class of conjugate group elements.

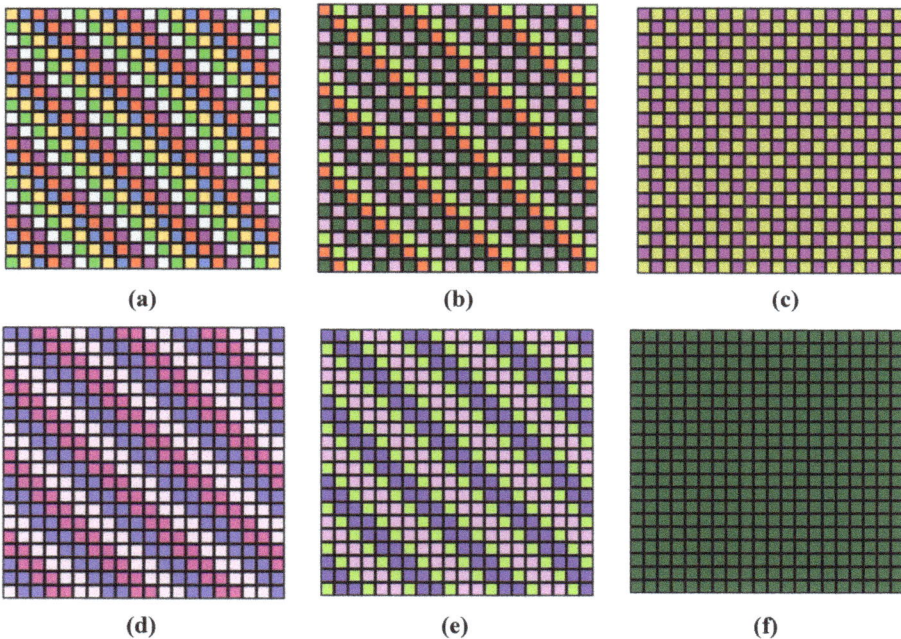

(a)       (b)       (c)

(d)       (e)       (f)

**Figure 9.** Colored PS P61$_1$, obtained by partition of space by a subgroup of permutations of the 12th permutation order. (**a**) g(0); (**b**) g(1) = g(2) = g(5); (**c**) g(3) = g(4); (**d**) g(6); (**e**) g(7) = g(8) = g(11); (**f**) g(9) = g(10).

### 3.2. Quaternion Group $D_4{}^2$

If the basis of the complex space are selected vectors:

$$0 = (1, i),\ 1 = (-1,-i),\ 2 = (i, 1),\ 3 = (-i, -1),\ 4 = (1,-i),\ 5 = (-1, i),\ 6 = (-i, 1),\ \text{and } 7 = (i, -1)$$

as it is well known in mathematics, they will match the operational set, forming a two-dimensional representation:

$$e = (0)(1)(2)(3)(4)(5)(6)(7) = \begin{pmatrix} 1 & 0 \\ 0 & 1 \end{pmatrix};\ -e = (0\,1)(2\,3)(4\,5)(6\,7) = \begin{pmatrix} -1 & 0 \\ 0 & -1 \end{pmatrix}$$

$$i = (0\,2\,1\,3)(4\,7\,5\,6) = \begin{pmatrix} i & 0 \\ 0 & -i \end{pmatrix};\ -i = (0\,3\,1\,2)(4\,6\,5\,7) = \begin{pmatrix} -i & 0 \\ 0 & i \end{pmatrix}$$

$$j = (0\,4\,1\,5)(2\,6\,3\,7) = \begin{pmatrix} 0 & -i \\ -i & 0 \end{pmatrix};\ -j = (0\,5\,1\,4)(2\,7\,3\,6) = \begin{pmatrix} 0 & i \\ i & 0 \end{pmatrix}$$

$$k = (0\,6\,1\,7)(2\,5\,3\,4) = \begin{pmatrix} 0 & -1 \\ 1 & 0 \end{pmatrix};\ -k = (0\,7\,1\,6)(2\,4\,3\,5) = \begin{pmatrix} 0 & 1 \\ -1 & 0 \end{pmatrix}$$

Expressing the elements of representation set in terms of substitution, we have eight operations:

$g(0) = 1 = e = (0\,1\,2\,3\,4\,5\,6\,7) = (0)(1)(2)(3)(4)(5)(6)(7)$
$g(1) = I = (2\,3\,1\,0\,7\,6\,4\,5) = (0213)(4756)$
$g(2) = j = (4\,5\,6\,7\,1\,0\,3\,2) = (0415)(2637)$
$g(3) = k = (6\,7\,5\,4\,2\,3\,1\,0) = (0617)(2534)$
$g(4) = -1 = (1\,0\,3\,2\,5\,4\,7\,6) = (01)(23)(45)(67)$
$g(5) = -I = (3\,2\,0\,1\,6\,7\,5\,4) = (0312)(4657)$
$g(6) = -j = (5\,4\,7\,6\,0\,1\,2\,3) = (0514)(2736)$
$g(7) = -k = (7\,6\,4\,5\,3\,2\,0\,1) = (0716)(2435)$

by simply composing permutation matrices, we can construct the following group Cayley table (Table 7).

**Table 7.** Cayley table for the quaternion group.

|   | 0 | 1 | 2 | 3 | 4 | 5 | 6 | 7 |
|---|---|---|---|---|---|---|---|---|
| 0 | 0 | 1 | 2 | 3 | 4 | 5 | 6 | 7 |
| 1 | 1 | 4 | 3 | 6 | 5 | 0 | 7 | 2 |
| 2 | 2 | 7 | 4 | 1 | 6 | 3 | 0 | 5 |
| 3 | 3 | 2 | 5 | 4 | 7 | 6 | 1 | 0 |
| 4 | 4 | 5 | 6 | 7 | 0 | 1 | 2 | 3 |
| 5 | 5 | 0 | 7 | 2 | 1 | 4 | 3 | 6 |
| 6 | 6 | 3 | 0 | 5 | 2 | 7 | 4 | 1 |
| 7 | 7 | 6 | 1 | 0 | 3 | 2 | 5 | 4 |

The symbols in the rows and columns of Table 7 for each transformation of the group are replaced by the number corresponding to this transformation of permutation operation.

We will consider each row of Table 7 as a new operation of substitution. It is easy to verify that the resulting new operating set will generate a group isomorphic to the quaternion group. Table 8, derived from elements of rows 5, will have the properties of the original Table 7.

**Table 8.** Cayley table for the quaternion group.

| g[0] = (0 1 2 3 4 5 6 7) | g[0] | g[1] | g[2] | g[3] | g[4] | g[5] | g[6] | g[7] |
|---|---|---|---|---|---|---|---|---|
| g[1] = (2 7 4 1 6 3 0 5) | g[1] | g[2] | g[3] | g[0] | g[5] | g[6] | g[7] | g[4] |
| g[2] = (4 5 6 7 0 1 2 3) | g[2] | g[3] | g[0] | g[1] | g[6] | g[7] | g[4] | g[5] |
| g[3] = (6 3 0 5 2 7 4 1) | g[3] | g[0] | g[1] | g[2] | g[7] | g[4] | g[5] | g[6] |
| g[4] = (7 6 1 0 3 2 5 4) | g[4] | g[7] | g[6] | g[5] | g[2] | g[1] | g[0] | g[3] |
| g[5] = (1 4 3 6 5 0 7 2) | g[5] | g[4] | g[7] | g[6] | g[3] | g[2] | g[1] | g[0] |
| g[6] = (3 2 5 4 7 6 1 0) | g[6] | g[5] | g[4] | g[7] | g[0] | g[3] | g[2] | g[1] |
| g[7] = (5 0 7 2 1 4 3 6) | g[7] | g[6] | g[5] | g[4] | g[1] | g[0] | g[3] | g[2] |

We apply the same approach to the resulting Cayley Table 8 again. However, despite the changed permutation operation workflow, the table view remains the same (Table 9). Therefore, all operating sets obtained from the original group of quaternions are isomorphic, and any of them can be used to represent corresponding structures in the packing space.

**Table 9.** Cayley table for the quaternion group.

| g[0] = (0 1 2 3 4 5 6 7) | g[0] | g[1] | g[2] | g[3] | g[4] | g[5] | g[6] | g[7] | (0)(1)(2)(3)(4)(5)(6)(7) |
|---|---|---|---|---|---|---|---|---|---|
| g[1] = (1 2 3 0 5 6 7 4) | g[1] | g[2] | g[3] | g[0] | g[5] | g[6] | g[7] | g[4] | (0123)(4567) |
| g[2] = (2 3 0 1 6 7 4 5) | g[2] | g[3] | g[0] | g[1] | g[6] | g[7] | g[4] | g[5] | (02)(13)(46)(57) |
| g[3] = (3 0 1 2 7 4 5 6) | g[3] | g[0] | g[1] | g[2] | g[7] | g[4] | g[5] | g[6] | (0321)(4765) |
| g[4] = (6 5 4 7 0 3 2 1) | g[4] | g[7] | g[6] | g[5] | g[2] | g[1] | g[0] | g[3] | (0426)(1735) |
| g[5] = (5 4 7 6 3 2 1 0) | g[5] | g[4] | g[7] | g[6] | g[3] | g[2] | g[1] | g[0] | (0527)(1436) |
| g[6] = (4 7 6 5 2 1 0 3) | g[6] | g[5] | g[4] | g[7] | g[0] | g[3] | g[2] | g[1] | (0624)(1537) |
| g[7] = (7 6 5 4 1 0 3 2) | g[7] | g[6] | g[5] | g[4] | g[1] | g[0] | g[3] | g[2] | (0725)(1634). |

Obviously, when we visualize the quaternion group, the order of the appropriate packing space should be a chosen multiple of the order of the permutations subgroup, that is, $N = 8k$. In this case, such as in Example 1.1, every cycle in an operation of the permutation subgroup selects in PS identical points (elementary squares, cubes), which are assigned a certain color. The whole space is divided into a number of one-colored, but differing from each other and in general unconnected colored areas, or into a number of corresponding substitution operation cycles. Thus, cycles determine the internal symmetry of the colored field.

We demonstrate this approach with examples of the PS $P81_1$ partition (Figures 6–13) by the operations of quaternion permutation group in accordance with Cayley Table 5 and the PS $P42_1$ partition (Figure 14) operations in accordance with Table 10.

**Table 10.** Cayley table for the quaternion group.

| | | g[1] | g[2] | g[3] | g[4] | g[5] | g[6] | g[7] |
|---|---|---|---|---|---|---|---|---|
| g[0] = (0 1 2 3 4 5 6 7) | g[0] | g[1] | g[2] | g[3] | g[4] | g[5] | g[6] | g[7] |
| g[1] = (1 6 3 4 5 2 7 0) | g[1] | g[2] | g[3] | g[0] | g[5] | g[6] | g[7] | g[4] |
| g[2] = (6 7 4 5 2 3 0 1) | g[2] | g[3] | g[0] | g[1] | g[6] | g[7] | g[4] | g[5] |
| g[3] = (7 0 5 2 3 4 1 6) | g[3] | g[0] | g[1] | g[2] | g[7] | g[4] | g[5] | g[6] |
| g[4] = (2 5 6 1 0 7 4 3) | g[4] | g[7] | g[6] | g[5] | g[2] | g[1] | g[0] | g[3] |
| g[5] = (5 4 1 0 7 6 3 2) | g[5] | g[4] | g[7] | g[6] | g[3] | g[2] | g[1] | g[0] |
| g[6] = (4 3 0 7 6 1 2 5) | g[6] | g[5] | g[4] | g[7] | g[0] | g[3] | g[2] | g[1] |
| g[7] = (3 2 7 6 1 0 5 4) | g[7] | g[6] | g[5] | g[4] | g[1] | g[0] | g[3] | g[2] |

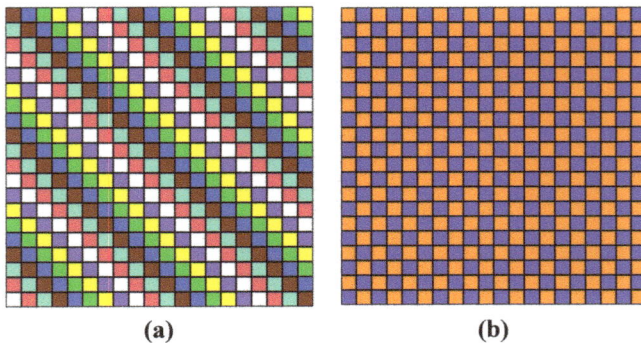

**Figure 10.** (**a**) PS $81_1$, operation (01234567); (**b**) PS $81_1$, operation (65470321).

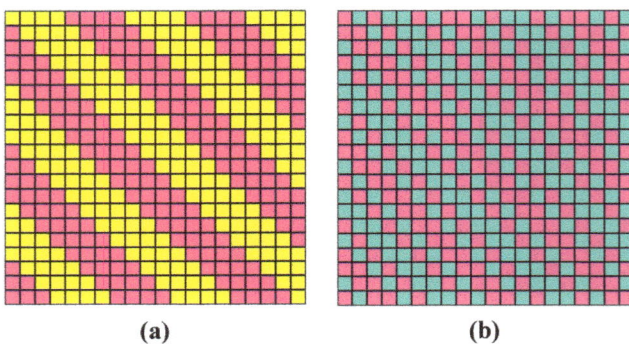

**Figure 11.** (**a**) PS $81_1$, operation (12305674); (**b**) PS $81_1$, operation (54763210).

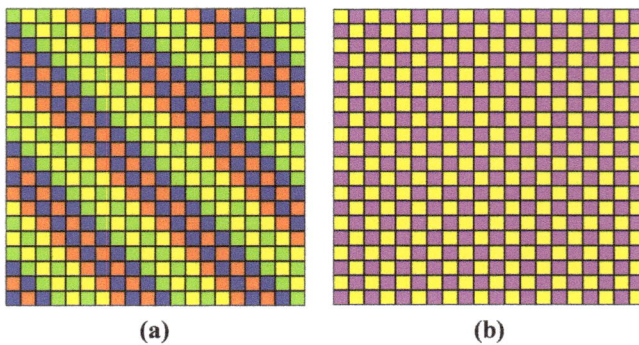

**Figure 12.** (**a**) PS $81_1$, operation (23016745); (**b**) PS $81_1$, operation (47652103).

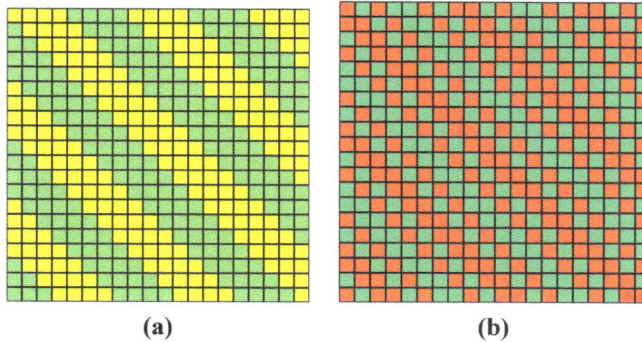

**Figure 13.** (a) PS 81$_1$, operation (30127456); (b) PS 81$_1$, operation (76541032).

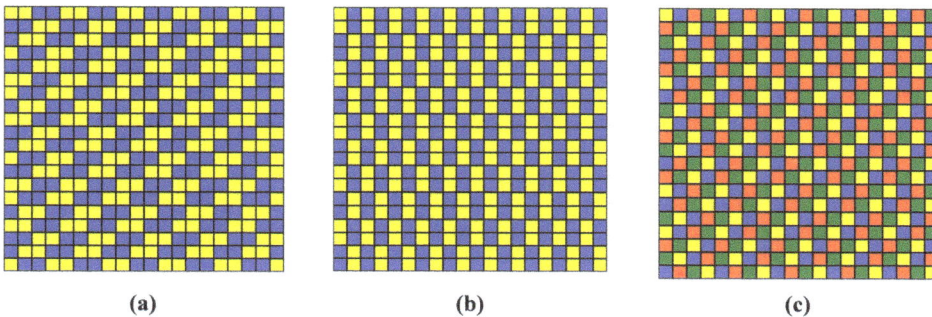

**Figure 14.** Quaternion group of P42$_1$ in the packing space. (a) The equivalent transformations of the group: (1 6 3 4 5 2 7 0) = (3 2 7 6 1 0 5 4) = (5 4 1 0 7 6 3 2) = (7 0 5 2 3 4 1 6); (b) the equivalent transformations of the group (2 5 6 1 0 7 4 3) = (4 3 0 7 6 1 2 5); (c) (6 7 4 5 2 3 0 1).

In packing space, the P81$_1$ quaternion group of symmetry is represented by a set of 8 structures. Among them, there is one representation of the operation of identical 8-colored transformation g(0) = (0) (1) (2) (3) (4) (5) (6) (7) (Figure 10a), one 4-colored transformation operation g(1) = (02) (13) (46) (57) (Figure 12a) and six two-colored transformation operations, dividing space into three pairs of equivalent structures. Thus, there are only five independent structures that correspond to the five classes of quaternions:

C$_0$ (Figure 10a) = {g(0)}, C$_1$ (Figure 12a) = {g(2)}, C$_2$ (Figure 11a, Figure 13a) = {g(1),g(3)}
C$_3$ (Figure 10b, Figure 12b) = {g(4),g(6)}, C$_4$ (Figure 11b, Figure 13b) = {g(5),g(7)}

and consequently, the group has five irreducible representations. From a geometrical point of view, the quaternion group splits PS P81$_1$ into heterolayers (see, for example, Figures 7–11) and P42$_1$ into regions in the form of polyomino. There is a "degeneration" of permutation transformations in PS P42$_1$: (1 6 3 4 5 2 7 0), (3 2 7 6 1 0 5), (5 4 1 0 7 6 3 2 4), (7 0 5 2 3 4 1 6), which gives the same result. Thus, the structure of the packing space affects the number of independent partitions.

The transition between structures that are determined by one fixed operation of the permutation group, belonging to different packing spaces of the same order *N*, is discrete due to changes in the angular parameters of the sublattices. More information will be shown below in Example 1.6. Therefore, we can assume that the lattice is deformed in such a transition, and matrices of packing space function as a discrete deformation transformation (DDT). The number *N* of these transformations is finite and determined by (1).

## 3.3. Quaternion Group at the Extended Packing Space

Quaternion groups in the packing spaces of higher order can be used. We consider this possibility in the 16th order PS. We match two (or more) cells of packing space to each number in the record of quaternion group permutation transformations. As the number of characters that determine the quaternion group is 8, the order of packing space becomes equal to $N = 16$. A correspondence table of old and new characters can be selected on the basis of a simple ordered set of numbers:

$0 = (0,1)$; $1 = (2,3)$; $2 = (4,5)$; $3 = (6,7)$; $4 = (8,9)$; $5 = (10,11)$; $6 = (12,13)$; and $7 = (14,15)$

The subsequent note of new permutation operations and their cyclic partitions will appear as follows:

1H(0 1 2 3 4 5 6 7 8 9 10 11 12 13 14 15) = (0)(1)(2)(3)(4)(5)(6)(7)(8)(9)(10)(11)(12)(13)(14)(15);
*i*H (4 5 6 7 2 3 0 1 14 15 12 13 8 9 10 11) = (0 4 2 6)(1 5 3 7)(8 14 10 12)(9 15 11 13);
*j*H(8 9 10 11 12 13 14 15 2 3 0 1 6 7 4 5) = (0 8 2 10)(1 9 3 11)(4 12 6 14)(5 13 7 15);
*k*H(12 13 14 15 10 11 8 9 4 5 6 7 2 3 0 1) = (0 12 2 14)(1 13 3 15)(4 10 6 8)(5 11 7 9);
−1H(2 3 0 1 6 7 4 5 10 11 8 9 14 15 12 13) = (0 2)(1 3)(4 6)(5 7)(8 10)(9 11)(12 14)(13 15);
−*i*H(6 7 4 5 0 1 2 3 12 13 14 15 10 11 8 9) = (0 6 2 4)(1 7 3 5)(8 12 10 14)(9 13 11 15);
−*j*H(10 11 8 9 14 15 12 13 0 1 2 3 4 5 6 7) = (0 10 2 8)(1 11 3 9)(4 14 6 12)(5 15 7 13);
−*k*H(14 15 12 13 8 9 10 11 6 7 4 5 0 1 2 3) = (0 14 2 12)(1 15 3 13)(4 8 6 10)(5 9 7 11).

You can check that by simple multiplication of these operations, constructed in such a way that a new table will be identical to the original Cayley table of the quaternion group; the symmetry groups of packing spaces of the 8th and 16th order are isomorphic.

To compare colored structures PS P161$_1$ and PS P81$_1$, we choose operations $i_H$ and i, and then, we present the results of their application to selected packing spaces (Figure 15).

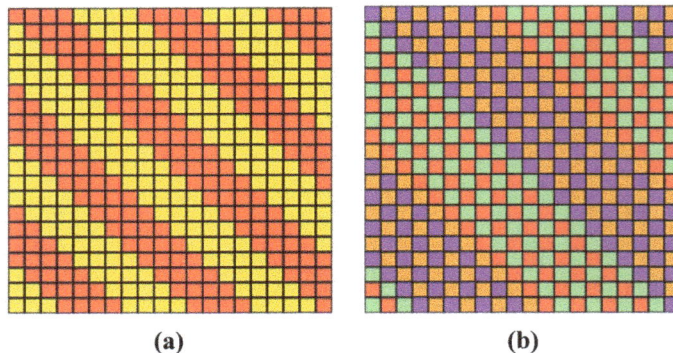

**(a)**                                        **(b)**

**Figure 15.** Compliance of the quaternion symmetry group and dual quaternions in PS: (a) P81$_1$ (2 3 1 0 7 6 4 5); (b) PS P16 1$_1$ (4 5 6 7 2 3 0 1 14 15 12 13 89 10 11).

The heterolayers symmetry of selected structures is clear in the figures. This example shows that it is possible to increase the size of the structure while maintaining symmetry.

## 3.4. The Pauli Matrices Group

We apply the technique for imaging the symmetry group with the help of packing spaces to Pauli matrices, which are well known in physics. For this purpose, we will expand their number to the total subgroup, using the operation of matrices multiplication. The operation of the Pauli matrices multiplication in quantum mechanics is defined and leads to the commutation relations.

The Pauli matrices group is a subgroup of the order 16:

$$e = \begin{pmatrix} 1 & 0 \\ 0 & 1 \end{pmatrix} (0); \quad -e = \begin{pmatrix} -1 & 0 \\ 0 & -1 \end{pmatrix} (1); \quad \sigma_x = x = \begin{pmatrix} 0 & 1 \\ 1 & 0 \end{pmatrix} (2);$$

$$\sigma_y = y = \begin{pmatrix} 0 & -i \\ i & 0 \end{pmatrix} (3); \quad \sigma_z = z = \begin{pmatrix} 1 & 0 \\ 0 & -1 \end{pmatrix} (4); \quad \sigma_{-x} = -x = \begin{pmatrix} 0 & -1 \\ -1 & 0 \end{pmatrix} (5);$$

$$\sigma_{-y} = -y = \begin{pmatrix} 0 & i \\ -i & 0 \end{pmatrix} (6); \quad \sigma_{-z} = -z = \begin{pmatrix} -1 & 0 \\ 0 & 1 \end{pmatrix} (7); \quad \sigma_{xy} = xy = \begin{pmatrix} i & 0 \\ 0 & -i \end{pmatrix} (8);$$

$$\sigma_{yx} = yx = \begin{pmatrix} -i & 0 \\ 0 & i \end{pmatrix} (9); \quad \sigma_{yz} = yz = \begin{pmatrix} 0 & i \\ i & 0 \end{pmatrix} (10); \quad \sigma_{zy} = zy = \begin{pmatrix} 0 & -i \\ -i & 0 \end{pmatrix} (11);$$

$$\sigma_{xz} = xz = \begin{pmatrix} 0 & -1 \\ 1 & 0 \end{pmatrix} (12); \quad \sigma_{zx} = zx = \begin{pmatrix} 0 & 1 \\ -1 & 0 \end{pmatrix} (13); \quad \sigma_{xyz} = xyz = \begin{pmatrix} -i & 0 \\ 0 & -i \end{pmatrix} (14);$$

As a result, we obtain the following Cayley table for the Pauli matrices group (Table 11).

**Table 11.** Cayley table for the Pauli matrices group.

| g(i) | [0] | [1] | [2] | [3] | [4] | [5] | [6] | [7] | [8] | [9] | [10] | [11] | [12] | [13] | [14] | [15] |
|---|---|---|---|---|---|---|---|---|---|---|---|---|---|---|---|---|
| [0] | [0] | [1] | [2] | [3] | [4] | [5] | [6] | [7] | [8] | [9] | [10] | [11] | [12] | [13] | [14] | [15] |
| [1] | [1] | [0] | [5] | [6] | [7] | [2] | [3] | [4] | [9] | [8] | [11] | [10] | [13] | [12] | [15] | [14] |
| [2] | [2] | [5] | [0] | [8] | [12] | [1] | [9] | [13] | [3] | [6] | [15] | [14] | [4] | [7] | [11] | [10] |
| [3] | [3] | [6] | [9] | [0] | [10] | [8] | [1] | [11] | [5] | [2] | [4] | [7] | [14] | [15] | [12] | [13] |
| [4] | [4] | [7] | [13] | [11] | [0] | [12] | [10] | [1] | [15] | [14] | [6] | [3] | [5] | [2] | [9] | [8] |
| [5] | [5] | [2] | [1] | [9] | [13] | [0] | [8] | [12] | [6] | [3] | [14] | [15] | [7] | [4] | [10] | [11] |
| [6] | [6] | [3] | [8] | [1] | [11] | [9] | [0] | [10] | [2] | [5] | [7] | [4] | [15] | [14] | [13] | [12] |
| [7] | [7] | [4] | [12] | [10] | [1] | [13] | [11] | [0] | [14] | [15] | [3] | [6] | [2] | [5] | [8] | [9] |
| [8] | [8] | [9] | [6] | [2] | [15] | [3] | [5] | [14] | [1] | [0] | [12] | [13] | [11] | [10] | [4] | [7] |
| [9] | [9] | [8] | [3] | [5] | [14] | [6] | [2] | [15] | [0] | [1] | [13] | [12] | [10] | [11] | [7] | [4] |
| [10] | [10] | [11] | [15] | [7] | [3] | [14] | [4] | [6] | [13] | [12] | [1] | [0] | [8] | [9] | [2] | [5] |
| [11] | [11] | [10] | [14] | [4] | [6] | [15] | [7] | [3] | [12] | [13] | [0] | [1] | [9] | [8] | [5] | [2] |
| [12] | [12] | [13] | [7] | [14] | [2] | [4] | [15] | [5] | [10] | [11] | [9] | [8] | [1] | [0] | [6] | [3] |
| [13] | [13] | [12] | [4] | [15] | [5] | [7] | [14] | [2] | [11] | [10] | [8] | [9] | [0] | [1] | [3] | [6] |
| [14] | [14] | [15] | [11] | [12] | [9] | [10] | [13] | [8] | [4] | [7] | [2] | [5] | [6] | [3] | [1] | [0] |

We choose to visualize a 4-color transformation g(8) = (8 9 6 2 15 3 5 14 1 0 12 13 11 10 4 7) = (0819) (2653) (4 15 7 14) (10 12 11 13) in the packing space P82$_2$. The result is presented in Figure 16.

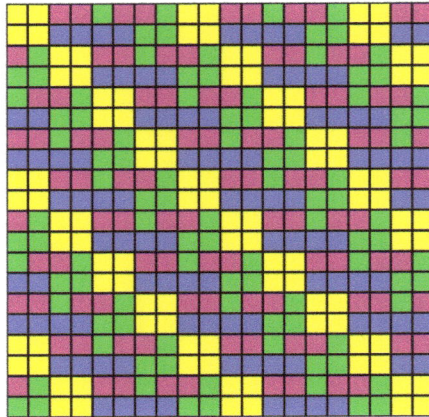

**Figure 16.** Partition of PS P82$_2$ by element g(8) of the Pauli matrices permutation group.

Package code in the program partition into polyomino (at [7]): 3232132310322230.

Layer growth leads to the growth of shape in the form of a parallelogram, which is also characterized by an element of the Pauli matrices group (Figure 17).

The sequence of growth is demonstrated on the diagram in Figure 18. Numerical information is placed on the three linear functions, which correspond to the layer-growth studies of partition into polyomino and other pieces of plane periodic mosaics [16,17].

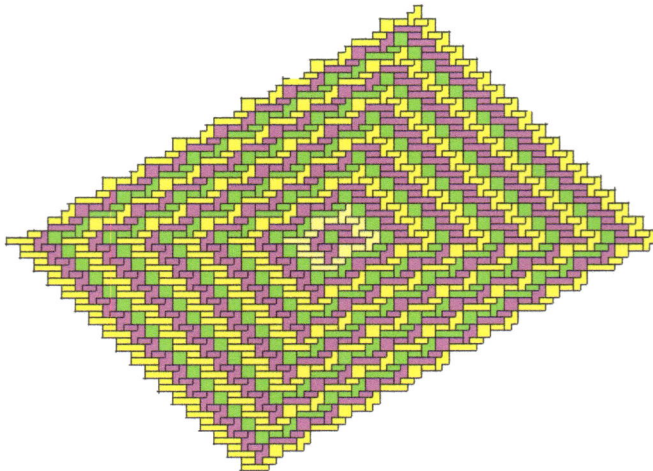

**Figure 17.** The shape of the growth of the partition PS P82$_2$ by element of g(8) of the permutation group of the Pauli matrices.

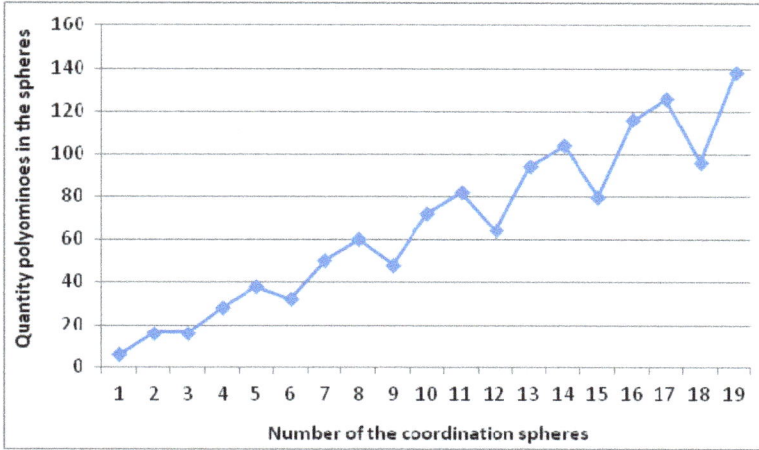

**Figure 18.** Coordination (magic) numbers of layer growth structure.

*3.5. The Dirac Matrices Group*

Similarly, we proceed with the Dirac matrices. The number of elements in the full group was found to be 32 with the Cayley table shown in Table 12.

The packing space P84$_1$ 8-color permutation operation g(13) = (0 13 7 14) (1 11 6 12) (2 10 5 9) (3 15 4 8) (16 30 23 31)(17 28 18 29) (19 27 20 26) (21 25 22 24) of the Cayley table of the Dirac matrices gives rise to partition (Figure 19). The growth form of the partition structure is shown in Figure 20.

**Table 12.** Cayley table for the Dirac matrices group.

| g[i] | [0] | [1] | [2] | [3] | [4] | [5] | [6] | [7] | [8] | [9] | [10] | [11] | [12] | [13] | [14] | [15] | [16] | [17] | [18] | [19] | [20] | [21] | [22] | [23] | [24] | [25] | [26] | [27] | [28] | [29] | [30] | [31] |
|---|---|---|---|---|---|---|---|---|---|---|---|---|---|---|---|---|---|---|---|---|---|---|---|---|---|---|---|---|---|---|---|---|
| [0] | [0] | [1] | [2] | [3] | [4] | [5] | [5] | [7] | [8] | [9] | [10] | [11] | [12] | [13] | [14] | [15] | [16] | [17] | [18] | [19] | [20] | [21] | [22] | [23] | [24] | [25] | [26] | [27] | [28] | [29] | [30] | [31] |
| [1] | [1] | [0] | [3] | [2] | [5] | [4] | [7] | [6] | [9] | [8] | [15] | [13] | [14] | [11] | [12] | [10] | [17] | [16] | [23] | [21] | [22] | [19] | [20] | [18] | [26] | [27] | [24] | [25] | [30] | [31] | [28] | [29] |
| [2] | [2] | [4] | [0] | [6] | [1] | [7] | [3] | [5] | [11] | [14] | [13] | [8] | [15] | [10] | [9] | [12] | [19] | [22] | [21] | [16] | [23] | [18] | [17] | [20] | [28] | [29] | [31] | [30] | [24] | [25] | [27] | [26] |
| [3] | [3] | [5] | [1] | [7] | [0] | [6] | [2] | [4] | [13] | [12] | [11] | [9] | [10] | [15] | [8] | [14] | [21] | [20] | [19] | [17] | [18] | [23] | [16] | [22] | [30] | [31] | [29] | [28] | [26] | [27] | [25] | [24] |
| [4] | [4] | [2] | [6] | [0] | [7] | [1] | [5] | [3] | [14] | [11] | [12] | [10] | [9] | [8] | [15] | [13] | [22] | [19] | [20] | [18] | [17] | [16] | [23] | [21] | [31] | [30] | [28] | [29] | [27] | [26] | [24] | [25] |
| [5] | [5] | [3] | [7] | [1] | [6] | [0] | [4] | [2] | [12] | [13] | [14] | [15] | [8] | [9] | [10] | [11] | [20] | [21] | [22] | [23] | [16] | [17] | [18] | [19] | [29] | [28] | [30] | [31] | [25] | [24] | [26] | [27] |
| [6] | [6] | [7] | [4] | [5] | [2] | [3] | [0] | [1] | [10] | [15] | [8] | [14] | [13] | [12] | [11] | [9] | [18] | [23] | [16] | [22] | [21] | [20] | [19] | [17] | [27] | [26] | [25] | [24] | [31] | [30] | [29] | [28] |
| [7] | [7] | [6] | [5] | [4] | [3] | [2] | [1] | [0] | [15] | [10] | [9] | [12] | [11] | [14] | [13] | [8] | [23] | [18] | [17] | [20] | [19] | [22] | [21] | [16] | [25] | [24] | [27] | [26] | [29] | [28] | [31] | [30] |
| [8] | [8] | [10] | [12] | [13] | [14] | [11] | [9] | [15] | [0] | [6] | [1] | [5] | [2] | [3] | [4] | [7] | [24] | [27] | [26] | [29] | [28] | [30] | [31] | [25] | [16] | [23] | [18] | [17] | [20] | [19] | [21] | [22] |
| [9] | [9] | [15] | [14] | [11] | [12] | [13] | [8] | [10] | [1] | [7] | [0] | [4] | [3] | [2] | [5] | [6] | [26] | [25] | [24] | [31] | [30] | [28] | [29] | [27] | [17] | [18] | [23] | [16] | [22] | [21] | [19] | [20] |
| [10] | [10] | [8] | [13] | [12] | [11] | [14] | [15] | [9] | [6] | [0] | [7] | [3] | [4] | [5] | [2] | [1] | [27] | [24] | [25] | [30] | [31] | [29] | [28] | [26] | [18] | [17] | [16] | [23] | [21] | [22] | [20] | [19] |
| [11] | [11] | [13] | [15] | [10] | [9] | [8] | [14] | [12] | [2] | [3] | [4] | [7] | [0] | [6] | [1] | [5] | [28] | [30] | [31] | [25] | [24] | [27] | [26] | [29] | [19] | [20] | [21] | [22] | [23] | [16] | [18] | [17] |
| [12] | [12] | [14] | [8] | [9] | [10] | [15] | [13] | [11] | [5] | [4] | [3] | [0] | [7] | [1] | [6] | [2] | [29] | [31] | [30] | [24] | [25] | [26] | [27] | [28] | [20] | [19] | [22] | [21] | [16] | [23] | [17] | [18] |
| [13] | [13] | [11] | [10] | [15] | [8] | [9] | [12] | [14] | [3] | [2] | [5] | [6] | [1] | [7] | [0] | [4] | [30] | [28] | [29] | [27] | [26] | [25] | [24] | [31] | [21] | [22] | [19] | [20] | [18] | [17] | [23] | [16] |
| [14] | [14] | [12] | [9] | [8] | [15] | [10] | [11] | [13] | [4] | [5] | [2] | [1] | [6] | [0] | [7] | [3] | [31] | [29] | [28] | [26] | [27] | [24] | [25] | [30] | [22] | [21] | [20] | [19] | [17] | [18] | [16] | [23] |
| [15] | [15] | [9] | [11] | [14] | [13] | [12] | [10] | [8] | [7] | [1] | [6] | [2] | [5] | [4] | [3] | [0] | [25] | [26] | [27] | [28] | [29] | [31] | [30] | [24] | [23] | [16] | [17] | [18] | [19] | [20] | [22] | [21] |
| [16] | [16] | [18] | [20] | [21] | [22] | [19] | [17] | [23] | [25] | [26] | [27] | [28] | [29] | [31] | [30] | [24] | [0] | [6] | [1] | [5] | [2] | [3] | [4] | [7] | [15] | [8] | [9] | [10] | [11] | [12] | [14] | [13] |
| [17] | [17] | [23] | [22] | [19] | [20] | [21] | [16] | [18] | [27] | [24] | [25] | [30] | [31] | [29] | [28] | [26] | [1] | [7] | [0] | [4] | [3] | [2] | [5] | [6] | [10] | [9] | [8] | [15] | [13] | [14] | [12] | [11] |
| [18] | [18] | [16] | [21] | [20] | [19] | [22] | [23] | [17] | [26] | [25] | [24] | [31] | [30] | [28] | [29] | [27] | [6] | [0] | [7] | [3] | [4] | [5] | [2] | [1] | [9] | [10] | [15] | [8] | [14] | [13] | [11] | [12] |
| [19] | [19] | [21] | [23] | [18] | [17] | [16] | [22] | [20] | [29] | [31] | [30] | [24] | [25] | [26] | [27] | [28] | [2] | [3] | [4] | [7] | [0] | [6] | [1] | [5] | [12] | [11] | [14] | [13] | [8] | [15] | [9] | [10] |
| [20] | [20] | [22] | [16] | [17] | [18] | [23] | [21] | [19] | [28] | [30] | [31] | [25] | [24] | [27] | [26] | [29] | [5] | [4] | [3] | [0] | [7] | [1] | [6] | [2] | [11] | [12] | [13] | [14] | [15] | [8] | [10] | [9] |
| [21] | [21] | [19] | [18] | [23] | [16] | [17] | [20] | [22] | [31] | [29] | [28] | [26] | [27] | [24] | [25] | [30] | [3] | [2] | [5] | [6] | [1] | [7] | [0] | [4] | [14] | [13] | [12] | [11] | [9] | [10] | [8] | [15] |
| [22] | [22] | [20] | [17] | [16] | [23] | [18] | [19] | [21] | [30] | [28] | [29] | [27] | [26] | [25] | [24] | [31] | [4] | [5] | [2] | [1] | [6] | [0] | [7] | [3] | [13] | [14] | [11] | [12] | [10] | [9] | [15] | [8] |
| [23] | [23] | [17] | [19] | [22] | [21] | [20] | [18] | [16] | [24] | [27] | [26] | [29] | [28] | [30] | [31] | [25] | [7] | [1] | [6] | [2] | [5] | [4] | [3] | [0] | [8] | [15] | [10] | [9] | [12] | [11] | [13] | [14] |
| [24] | [24] | [26] | [28] | [30] | [31] | [29] | [27] | [25] | [23] | [18] | [17] | [20] | [19] | [22] | [21] | [16] | [8] | [9] | [10] | [11] | [12] | [13] | [14] | [15] | [7] | [0] | [6] | [1] | [5] | [2] | [4] | [3] |
| [25] | [25] | [27] | [29] | [31] | [30] | [28] | [26] | [24] | [16] | [17] | [18] | [19] | [20] | [21] | [22] | [23] | [15] | [10] | [9] | [12] | [11] | [14] | [13] | [8] | [0] | [7] | [1] | [6] | [2] | [5] | [3] | [4] |
| [26] | [26] | [24] | [30] | [28] | [29] | [31] | [25] | [27] | [18] | [23] | [16] | [22] | [21] | [20] | [19] | [17] | [9] | [8] | [15] | [13] | [14] | [11] | [12] | [10] | [6] | [1] | [7] | [0] | [4] | [3] | [5] | [2] |
| [27] | [27] | [25] | [31] | [29] | [28] | [30] | [24] | [26] | [17] | [16] | [23] | [21] | [22] | [19] | [20] | [18] | [10] | [15] | [8] | [14] | [13] | [12] | [11] | [9] | [1] | [6] | [2] | [5] | [3] | [4] | [0] | [7] |
| [28] | [28] | [31] | [24] | [27] | [26] | [25] | [30] | [29] | [20] | [21] | [22] | [23] | [16] | [17] | [18] | [19] | [11] | [14] | [13] | [8] | [15] | [10] | [9] | [12] | [5] | [2] | [3] | [4] | [7] | [0] | [1] | [6] |
| [29] | [29] | [30] | [25] | [26] | [27] | [24] | [31] | [28] | [19] | [22] | [21] | [16] | [23] | [18] | [17] | [20] | [12] | [13] | [14] | [15] | [8] | [9] | [10] | [11] | [2] | [5] | [4] | [3] | [0] | [7] | [6] | [1] |
| [30] | [30] | [29] | [26] | [25] | [24] | [27] | [28] | [31] | [22] | [19] | [20] | [18] | [17] | [16] | [23] | [21] | [13] | [12] | [11] | [9] | [10] | [15] | [8] | [14] | [4] | [3] | [2] | [5] | [6] | [1] | [0] | [7] |
| [31] | [31] | [28] | [27] | [24] | [25] | [26] | [29] | [30] | [21] | [20] | [19] | [17] | [18] | [23] | [16] | [22] | [14] | [11] | [12] | [10] | [9] | [8] | [15] | [13] | [3] | [4] | [5] | [2] | [1] | [6] | [7] | [0] |

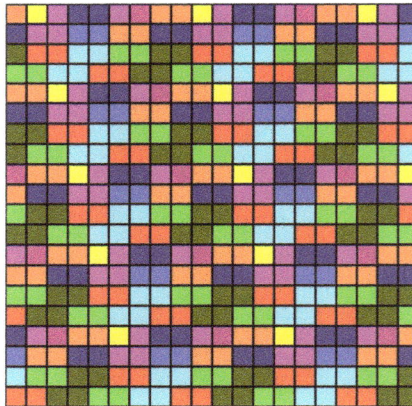

**Figure 19.** Tiling PS P84₁ by permutation operation g (13) (Table 12) of the Dirac matrices group.

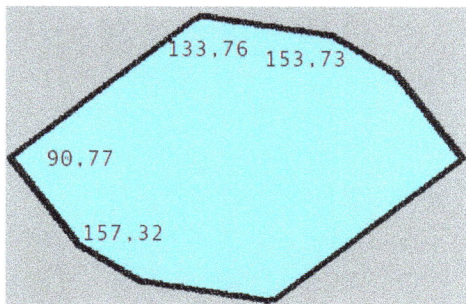

**Figure 20.** Octagon of the growth structure partition (Figure 19) on the PS P84$_1$.

The numerical sequence of filling the coordination spheres of the growing figure in the PS 84$_1$ depends on the choice of "seed polyomino", but the limiting form of the growth, which is the choice of the different growth beginning, is always the same, as presented in [18]. The only linear dependence of polyominoes quantity in the layer is on the layer number "on average" (Figure 21), which is uncommon for two-dimensional partitions [16].

In this article, we will not analyze the use of information received on the Pauli and Dirac matrices and the structure of the groups, although there is a connection of physical theory with the design obtained by partition of packing spaces. A search for the connection can be conducted in two ways: First, to identify what group properties determine the structure of the partition, and, second, to determine if there is a real structure, which is subordinate to discrete representation of these groups.

To answer the arising questions, it is necessary to consider the transition procedure from the real structure to its discrete model based on permutation operations.

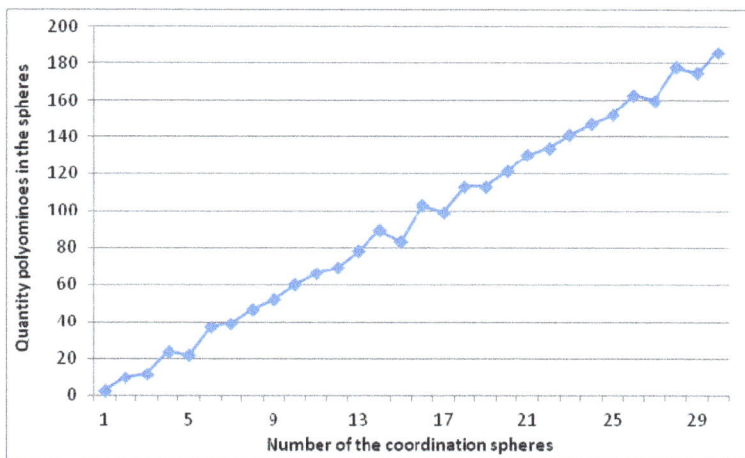

**Figure 21.** Coordination (magic) numbers of growth structure (Figure 19).

*3.6. An Example of a Real Structure with Non-Crystallographic Permutation Group Symmetry*

For the partition of three-dimensional structures, you can use the methodology for constructing molecular domains of Voronoi–Dirichlet by means of Fisher and Koch (see for example, reference [25]). However, as the analysis of three-dimensional partitions of a periodic lattice is hard to perceive, we choose the real structure of $[(C_2H_5)^4N^+]_2 \cdot C_4O_4^{2-} \cdot 2(C_2H_5)_4N + HCO_3^- \cdot 4(NH_2)_2CO \cdot 6H_2O$ [27], based on a molecular layer (Figure 22). We use a qualitative approach: select the layer of the two types of dimer: $2(HCO_3)^-$, $2CO(NH_2)_2$ and two pairs of water molecules per unit cell, bound by hydrogen bonds. The result is a three-colored partition "map" into non-overlapping substructures (Figure 22b).

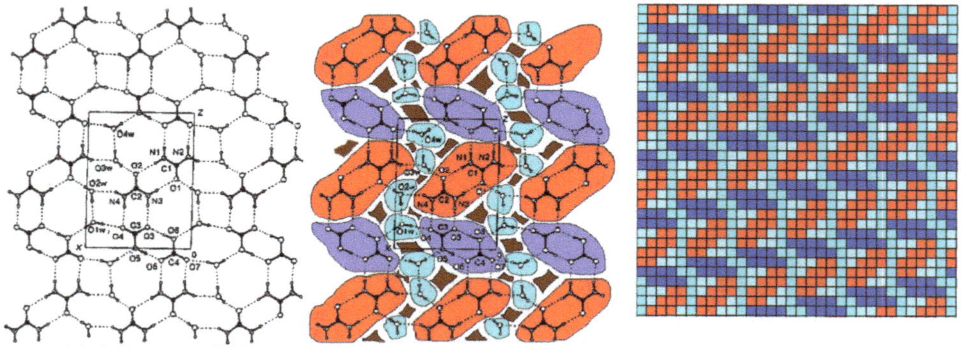

**Figure 22.** The molecular layer (**a**) in the structure of the compound $[(C_2H_5)^4N^+]_2 \cdot C_4O_4^{2-} \cdot 2(C_2H_5)_4N + HCO_3^- \cdot 4(NH_2)_2CO \cdot 6H_2O$, "map" of its partition; (**b**) into disjoint regions and (**c**) the discrete layer structure model.

Without changing the coordination environment of each region, a polyomino-like presentation of this structure's layer (Figure 22c) is obtained for the packing space P31 $1_6$. The sizes of cells are selected in such way that the order of the cell of packing space is a prime number $N = 31$. The number of independent types of lattices in this case is minimal, which is important when iterating the options in the prediction structures task. The image of the PS P31$_6$ structure, rotated to locate the yellow-red layer consisting of water molecules horizontally, is presented in Figure 23. The form of layered structure growth is presented in [16].

**Figure 23.** A discrete model of the structure layer [27] in PS P31 $1_6$.

In this example, the formed four-colored operation of permutation is written as follows:
G(1) = (1 2 3 28 10 11 5 8 14 15 9 18 6 12 17 16 22 23 25 13 19 27 21 24 7 26 20 4 29 30 0)

= (0 1 2 3 28 29 30) (4 10 9 15 16 22 21 27 ) (5 11 18 25 26 20 19 13 12 6) (7 8 14 17 23 24).

In this article, we do not give a Cayley matrix of the cyclic group $(G (1))^m$, which has an order greater than $2^N$. In terms of group properties, it is interesting that when the order of $m = p$, where $p$ is a prime number >10, the structure is invariant. Rearrangements occur within the number of cycles, and we use this fact to calculate the statistical probability and statistical entropy:

$W_T = 31!/6!7!8!10! = 10^{13}$; $\sigma = \ln W_T = 13 \ln 10 = 29{,}9$.

The formula of Boltzmann's entropy of the structure is $S = k \ln W_T = 41.3 \times 10^{-23}$ (J/K).

In other orders of m, the number of cycles (color set in the PS P31 $1_6$ partition) changes, as well as their structure and entropy.

Keeping the original structure unchanged, that is, without changing the operation $G(1)$, we consider gradual deformation of the structure when there is a transition from one packing space with $N = 31k$ to another, the number of which, as for the primes, is equal to $N + 1 = 32$ (Figure 24).

The first and last PS P31 $1_0$ and P31 $1_{30}$ are not completely presented in Figure 24. It is easily observed from the slope of the layers in the partition that the deformation transformations of the lattice shift result in similar pairs of partitions associated with each other as "left" and "right" forms. These phase structures are obtained by a mirror image.

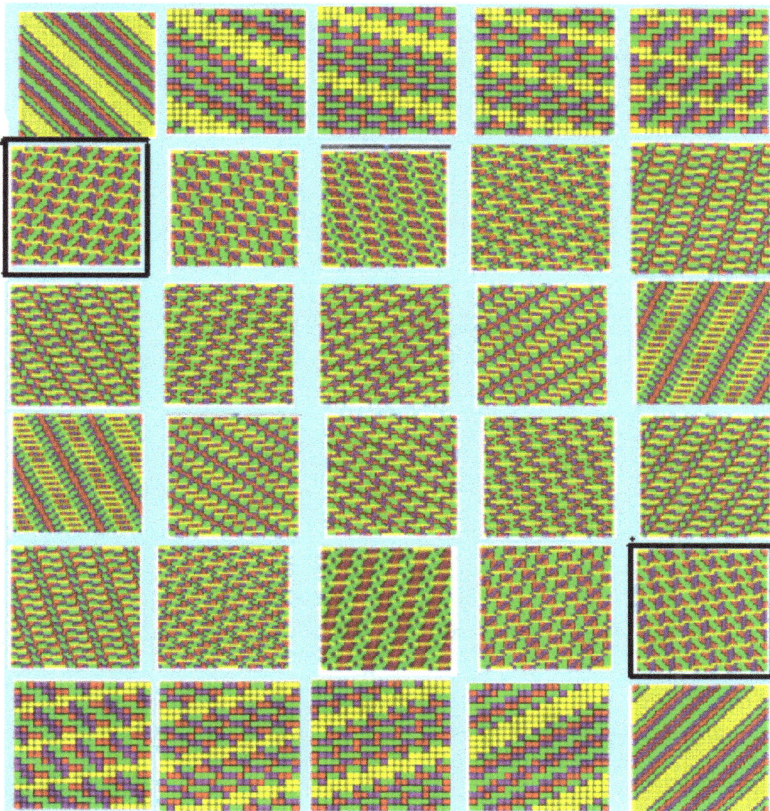

**Figure 24.** Discrete deformation transformations of the layer structure [27] in the PS P31 $1_6$ partition model (in Figure 22). The frames mark "left" and "right" forms of the structure.

## 4. Brief Summary

1. Discrete modeling of tilings and packages, as well as the introduction of packing space, allow us to make space partitions into any fragments in a periodic lattice. In this article, we present for the first time the method of the exhaustive search option sublattices with index $N$ based on the analysis of irreducible representations of a mathematical group of permutations of the $N^2$ elements.

2. The developed partition algorithms and software packages (not covered in this article) make it possible to partition and predict the structure of molecular compounds using the criterion of partition presented in this article. The criterion answers the question: What polyomino or policube may be a fundamental area in a periodic lattice?

3. Application of the layer growth model and growth of the graph in the space partition allows us to calculate structures of real crystalline nuclei and nanoclusters model. Algorithms and programs for the calculations in the article are not presented as they were released previously. The results of calculating the polyhedra growth of structures formed by partitioning of the space with the use of groups of permutations are considered for the first time in this work, and the figures of growth can be one of the characteristics of group 4. Application of finite groups of permutations to the packing space determines space partition into policubes (polyominoes), and forms a structure. This allows you to expand the capabilities of the description of symmetry of periodic structures and is not limited to 230 space groups.

5. Any finite groups, expressed through a group of permutations and applied to the packing space, are structurally visualized.

6. The totality of packaging spaces of the same order $N$ characterizes the discrete deformation transformation of the corresponding periodic structure.

7. Application of the methodology, based on the PS formalism, will contribute to the transfer to the numeric "language", and it is essential that a significant amount of information about the structures that do work be obtained with computers that operate only with numbers.

## References

1. Patterson, A.L. A Fourier series method for the determination of the components of interatomic distances in crystals. *Phys. Rev.* **1934**, *46*, 372–376. [CrossRef]
2. Buerger, M.J. *Vector Space and Its Application in Crystal Structure Investigation*; John Wiley & Sons, Inc.: New York, NY, USA, 1959; p. 384.
3. Rau, V.G.; Parkhomov, L.G.; Ilyukhin, V.V.; Belov, N.V. Calculation pattersonovskiy cyclotomic sets. *Sov. Phys. Dokl.* **1980**, *25*, 960–963.
4. Rau, V.G.; Ilyukhin, V.V.; Belov, N.V. The isovector structures of pseudosimmetric crystals. *Dokl. Akad. Nauk SSSR* **1979**, *249*, 611–613.
5. Hall, M. *Combinatorial Theory*; Blaisdell Publishing Company: Waltham, UK, 1967; p. 424.
6. Rau, V.G.; Parkhomov, L.G.; Kotov, N.A.; Maleev, A.V. The problem of "structure seeker" creation. In Proceedings of XII European Crystallographic Meeting; 1989; Volume 2, p. 48.
7. Maleev, A.V.; Rau, V.G.; Potekhin, K.A.; Parkhomov, L.G.; Rau, T.F.; Stepanov, S.V.; Struchov, Yu.T. The method of discrete modeling of packings in molecular crystals. *Sov. Phys. Dokl.* **1990**, *35*, 997–1000.
8. Maleev, A.V.; Zhitkov, I.K.; Rau, V.G. Generation of crystal structures of heteromolecular compounds by the method of discrete modeling of packings. *Crystallogr. Rep.* **2005**, *50*, 727–734. [CrossRef]
9. Grünbaum, B.; Shephard, G.C. The ninety one types of isogonal tiling in the plane. *Trans. Am. Math. Soc.* **1978**, *242*, 335–353. [CrossRef]
10. Klarner, D.A. *The Mathematical Gardner*; Prindle, Weber & Schmidt: Boston, MA, USA, 1981; p. 494.
11. Fukuda, H.; Kanomata, C.; Mutoh, N.; Nakamura, G.; Schattschneider, D. Polyominoes and polyiamonds as fundamental domains of isohedral tiling with rotational symmetry. *Symmetry* **2011**, *3*, 828–851. [CrossRef]
12. Rau, V.G. Simple criterion of division of two-dimensional space of a periodic lattice into arbitrarily shaped poyominoes. *Crystallogr. Rep.* **2000**, *45*, 199–202. [CrossRef]

*Symmetry* **2013**, *5*, 54–80

13. Maleev, A.V. An algorithm and program of exhaustive search for possible tiling of a plane with polyominoes. *Crystallogr. Rep.* **2000**, *46*, 154–156. [CrossRef]

14. Maleev, A.V. Generation of the structures of molecular crystals with two molecules related by the center of inversion in a primitive unit cell. *Crystallogr. Rep.* **2002**, *47*, 731–735. [CrossRef]

15. Rau, V.G.; Zhuravlev, V.G.; Rau, T.F.; Maleev, A.V. Morfogenesis of crystal structures in the discrete modeling of packings. *Crystallogr. Rep.* **2002**, *47*, 727–730. [CrossRef]

16. Rau, V.G.; Pugaev, A.A.; Rau, T.F. Coordination sequeces and coordination waves in matter. *Crystallogr. Rep.* **2006**, *51*, 2–10. [CrossRef]

17. Conway, J.H.; Sloan, N.J.A. Low-dimensional lattices. VII. Coordination sequences. *Proc. R. Soc. Lond. A* **1997**, *453*, 2369–2389. [CrossRef]

18. Zhuravlev, V.G. Self-similar growth of periodic partitions and graphs. *St. Petersb. Math. J.* **2001**, *13*, 201–220.

19. Zhuravlev, V.G.; Maleev, A.V.; Rau, V.G.; Shutov, A.V. Growth of planar random graphs and packaging. *Crystallogr. Rep.* **2002**, *47*, 907–912.

20. Rau, V.G.; Skvortzov, K.V.; Potekhin, K.A.; Maleev, A.V. Geometric analysis of the models of sulfur ($S_8$). Molecular nanoclusters in computer experiment. *J. Struct. Chem.* **2011**, *52*, 781–786. [CrossRef]

21. Maleev, A.V. *n*-Dimensional packing spaces. *Crystallogr. Rep.* **1995**, *40*, 354–358.

22. Rau, V.G.; Rau, T.F. Tiling a space with Dirichlet polyominoes. *Crystallogr. Rep.* **1995**, *40*, 154–157.

23. Maleev, A.V.; Lisov, A.E.; Potekhin, K.A. Symmetry of the *n*-divensional packing spaces. *Kristallografiya* **1998**, *43*, 775–781.

24. Maleev, A.V.; Shutov, A.V.; Zhuravlev, V.G. 2D quasi-periodic rauzy tiling as a section of 3D Periodic tiling. *Crystallogr. Rep.* **2010**, *55*, 723–733. [CrossRef]

25. Rau, V.G.; Rau, T.F.; Lebedev, G.O.; Kurkutova, E.N. Geometrical model of the Structure of the Heterocomplex Compound $[Cr(OCN_2H_4)_6]^{3+}[Co(DH)_2(NO_2)_2]_3^-·2H_2O$. *Crystallogr. Rep.* **2000**, *45*, 595–600. [CrossRef]

26. Rau, V.G.; Pugaev, A.A.; Rau, T.F.; Maleev, A.V. Geometrical aspect of solving the problem of real structure growth on the model of alkali metal halides of the NaCl type. *Crystallogr. Rep.* **2009**, *54*, 28–34. [CrossRef]

27. Lam, Ch.K.; Mak, T.K.U. The new layered anionic host lattice is formed by molecules of urea, skvarata, bicarbonate and water. *J. Struct. Chem.* **1999**, *40*, 883–891. [CrossRef]

*symmetry*

MDPI

*Article*

# Decoration of the Truncated Tetrahedron—An Archimedean Polyhedron—To Produce a New Class of Convex Equilateral Polyhedra with Tetrahedral Symmetry

**Stan Schein** [1,2,3,*], **Alexander J. Yeh** [4], **Kris Coolsaet** [5] **and James M. Gayed** [6]

[1]  California NanoSystems Institute, Mailcode 722710, University of California, Los Angeles (UCLA), Los Angeles, CA 90095, USA

[2]  Brain Research Institute, Mailcode 951761, University of California, Los Angeles (UCLA), Los Angeles, CA 90095, USA

[3]  Department of Psychology, Mailcode 951563, University of California, Los Angeles (UCLA), Los Angeles, CA 90095, USA

[4]  Department of Chemistry and Biochemistry, Mailcode 951569, University of California, Los Angeles (UCLA), Los Angeles, CA 90095, USA; alexanderjyeh@gmail.com

[5]  Department of Applied Mathematics, Computer Science and Statistics, Ghent University, Krijgslaan 281-S9, B-9000 Gent, Belgium; kris.coolsaet@ugent.be

[6]  Department of Psychology, Mailcode 951563, University of California, Los Angeles (UCLA), Los Angeles, CA 90095, USA; gayed@ucla.edu

*   Correspondence: stan.schein@gmail.com; Tel.: +1-310-855-4769

Academic Editor: Egon Schulte
Received: 27 July 2016; Accepted: 16 August 2016; Published: 20 August 2016

**Abstract:** The Goldberg construction of symmetric cages involves pasting a patch cut out of a regular tiling onto the faces of a Platonic host polyhedron, resulting in a cage with the same symmetry as the host. For example, cutting equilateral triangular patches from a 6.6.6 tiling of hexagons and pasting them onto the full triangular faces of an icosahedron produces icosahedral fullerene cages. Here we show that pasting cutouts from a 6.6.6 tiling onto the full hexagonal and triangular faces of an Archimedean host polyhedron, the truncated tetrahedron, produces two series of tetrahedral ($T_d$) fullerene cages. Cages in the first series have $28n^2$ vertices ($n \geq 1$). Cages in the second (leapfrog) series have $3 \times 28n^2$. We can transform all of the cages of the first series and the smallest cage of the second series into geometrically convex equilateral polyhedra. With tetrahedral ($T_d$) symmetry, these new polyhedra constitute a new class of "convex equilateral polyhedra with polyhedral symmetry". We also show that none of the other Archimedean polyhedra, six with octahedral symmetry and six with icosahedral, can host full-face cutouts from regular tilings to produce cages with the host's polyhedral symmetry.

**Keywords:** Goldberg polyhedra; cages; fullerenes; tilings; cutouts

---

## 1. Introduction

Known to the ancient Greeks, the five Platonic polyhedra and 13 Archimedean polyhedra are the first two classes of convex equilateral polyhedra with polyhedral symmetry (icosahedral, octahedral or tetrahedral) [1]. In 1611, Johannes Kepler added a third class, the rhombic polyhedra [2]. A fourth class, the "Goldberg polyhedra", was recently described [3]. These are primarily icosahedral fullerene cages transformed into geometrically convex equilateral polyhedra—which necessarily have convex planar faces [4,5].

Preceding the work on icosahedral viruses by Caspar and Klug [6], Goldberg's construction of icosahedral fullerene cages employed decoration of the full equilateral triangular faces of a host icosahedron (Figure 1a, left) with equilateral triangular cutouts from a tiling of regular hexagons (Figure 1b, left) [7–9]. The resulting cages have the same symmetry as the host icosahedron. They also have 3-valent vertices, hexagons and 12 pentagons (Figure 1c, left). Of the other of the five Platonic (or regular) polyhedra (Figure 2a), the octahedron (Figure 1a, middle) and the tetrahedron (Figure 1a, right) also have equilateral triangular faces and can serve as hosts that can be decorated similarly (Figure 1b) to produce octahedral cages with hexagons and six squares (Figure 1c, middle) and tetrahedral cages with hexagons and four triangles (Figure 1c, right) [7].

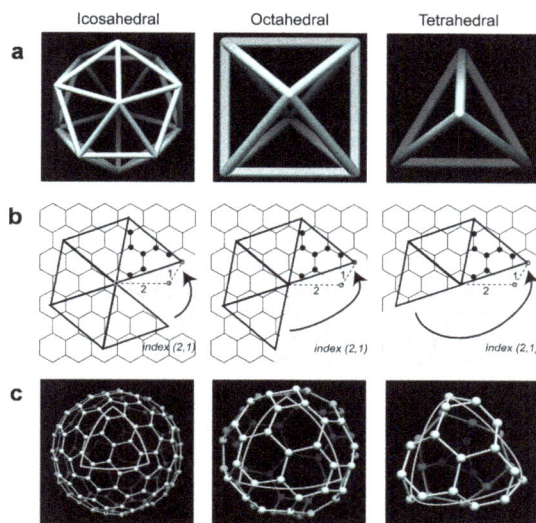

**Figure 1.** Decoration of Platonic polyhedra with a tiling of hexagons. (**a**) The icosahedron, the octahedron and the tetrahedron, Platonic polyhedra with equilateral triangular faces; (**b**) a tiling of hexagons with cutouts with 5 triangular sectors (**left**), suitable for decorating the full triangular faces of an icosahedron with 5-valent vertices, 4 triangular sectors (**middle**) for the octahedron with 4-valent vertices, and 3 triangular sectors (**right**) for the tetrahedron with 3-valent vertices. The dashed arrows show some of the edges that anneal to one another in the cage. The edge of each triangle can be described by indices ($h,k$), here (2,1) in all three cases, corresponding to two steps in one direction in the tiling and one step after a turn of 60°, and containing $T = 7$ vertices; (**c**) icosahedral, octahedral and tetrahedral equilateral cages with indices (2,1) and nonplanar faces. The triangle from the tiling, which contains 7 vertices, becomes a spherical triangle with 7 vertices on the cage. These geometrically icosahedral, octahedral or tetrahedral cages are equilateral, with regular small faces (5-gons, 4-gons or 3-gons) but nonplanar 6-gons.

The triangular cutouts of the tiling of hexagons can be different sizes and orientations, with the pattern still neatly continuing across the borders of adjacent triangular faces and across the gaps that span 1, 2 or 3 triangles (Figure 1b). These size and orientation variants can be expressed by indices ($h,k$) that characterize one side of the equilateral triangular cutout, where $h$ indicates steps (from the center of one hexagon to the center of next) along one axis of the tiling of hexagons and $k$ indicates steps along a second axis at 60 degrees to the first [6–9]. For example, the indices in Figure 1b are (2,1). These indices also make it easy to calculate the number $T = h^2 + hk + k^2$ of vertices in each equilateral triangle [6–9]. In Figure 1, there are thus $2^2 + 1 \times 1 + 1^2 = 7$ vertices in each triangle. Similarly, square cutouts in a variety of sizes and orientations from a tiling of squares can decorate the full square faces of a cube to produce octahedral cages with 4-gons and eight triangles [10].

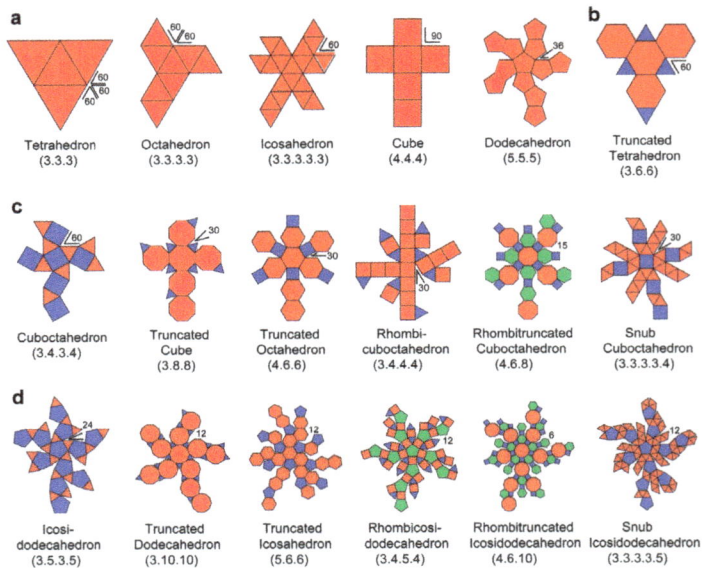

**Figure 2.** Platonic and Archimedean polyhedra and their angle deficits. Five Platonic polyhedra (**a**); one tetrahedral Archimedean polyhedron (**b**); six octahedral Archimedean polyhedra (**c**) and six icosahedral Archimedean polyhedra (**d**). Polyhedra with angle deficits of 60°, 120° and 180° are compatible with decoration by a 6.6.6 tiling, the one with 90° by a 4.4.4.4 tiling. However, a 6.6.6 tiling of the square faces in the cuboctahedron—with an angle deficit of 60°—cannot knit across the boundaries of the squares.

Here, we show that just one of the Archimedean polyhedra (Figure 2b–d), the truncated tetrahedron (Figure 2b), can be similarly decorated. Specifically, we paste regular hexagonal and triangular patches cut out of a 2D tiling of hexagons onto the full hexagonal and triangular faces of a truncated tetrahedron. This "re-tiling" of the truncated tetrahedron produces 3D cages with the same symmetry as the host polyhedron, in this case tetrahedral ($T_d$) symmetry. For a subset of these new $T_d$ cages we can produce geometrically convex equilateral polyhedra with $T_d$ symmetry, thus creating another class of convex equilateral polyhedra with polyhedral symmetry.

## 2. Materials and Methods

We used Carbon Generator (CaGe) software [11] (https://caagt.ugent.be/CaGe/) to produce cages, specifically Schlegel diagrams, w3d files and spinput files. We used the Equi program (http://caagt.ugent.be/equi/) to read the w3d files and make the cages equilateral (with edges of 5 units) and with planar faces, producing w3d and spinput files for the resulting equilateral polyhedra. We used Spartan software (Wavefunction, Inc., Irvine, CA, USA) [12] to read spinput files and produce the data in the supplementary tables and the pdb files that can by read by UCSF (University of California, San Francisco) Chimera [13].

## 3. Results

### 3.1. Decoration of an Archimedean Polyhedron to Produce New Cages

We ask if we can produce cages by decorating any of the 13 Archimedean polyhedra (Figure 2b–d), each with more than one type of regular face, with cutouts of a regular tiling. Although the truncated icosahedron—or soccer ball—is the most famous of the Archimedean polyhedra [14], the first one we consider is the truncated tetrahedron, which has four regular hexagons and four equilateral triangles (Figures 2b and 3a). We can decorate its eight full faces with contiguous hexagonal and triangular cutouts from a 2D tiling of hexagons (top of Figure 3b, labeled (3,0 and 1,1)), with the pattern of the tiling neatly annealing across the borders of all the faces, as indicated by the dashed arrows. Only two orientations and certain sizes of hexagonal and triangular cutouts neatly fit the full faces of the truncated tetrahedron, one group (e.g., (3,0 and 1,1) and (6,0 and 2,2)) illustrated in Figure 3b, the other group (e.g., (1,1 and 1,0), (2,2 and 2,0) and (3,3 and 3,0)) in Figure 3c. By contrast, cutouts from a tiling of squares would not fit neatly into hexagons or triangles (Figure 4).

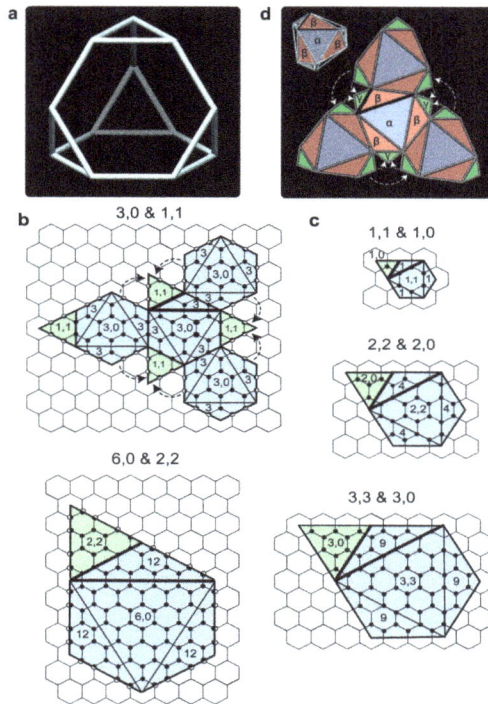

**Figure 3.** Decoration of the full hexagonal and triangular faces of a truncated tetrahedron with a tiling of hexagons. (**a**) The truncated tetrahedron, an Archimedean polyhedron; (**b**) patterns of cutouts for the new $T_d$ cages in the leapfrog series. For (3,0 and 1,1), the cutout consists of four regular hexagons (blue) and four equilateral triangles (green). The dashed arrows show a few of the edges that anneal to one another when the cutout is folded to create the cage. The regular hexagon may be subdivided into a large equilateral triangle and three isosceles triangles. The index (3,0) characterizes one (bolded) edge of the large equilateral triangle that contains 9 vertices, and the index (1,1) characterizes one (bolded) edge of the small equilateral triangle that contains 3 vertices. The isosceles triangles have the same number of internal vertices, 3, as the small equilateral triangle. For the other cage in this series, (6,0 and 2,2), only one regular hexagon (containing a large equilateral triangle and three isosceles triangles) and one small equilateral triangle are shown; (**c**) patterns of cutouts for the new $T_d$ cages in the first series. Only one hexagon (blue) and one small equilateral triangle (green) are shown; (**d**) the construction of a general tetrahedral fullerene [15] includes 20 triangles, composed of 4 sets of 5 triangles, with one set shown in the inset. Each set contains a large equilateral triangle ($\alpha$), three scalene triangles ($\beta$) and three thirds (=one whole) of one small equilateral triangle ($\gamma$)—in thirds to illustrate the symmetry. The dashed arrows show a few of the edges that anneal to one another when the cutout is folded to create the cage.

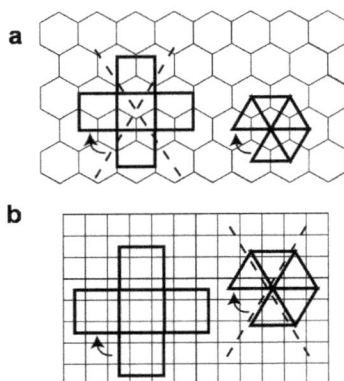

**Figure 4.** Compatibility between regular tilings and face type. (**a**) The 6.6.6 tiling has 6-fold symmetry about the centers of hexagonal faces, with the pattern repeated every 60°. The edges of an icosahedron neatly anneal because of its 60°-angle deficit at each vertex, whereas the edges of a cube do not anneal, as indicated by the X, because of the cube's 90°-angle deficit at each vertex; (**b**) the 4.4.4.4 tiling has 4-fold symmetry about the centers of square faces, with the pattern repeated every 90°. The edges of a cube neatly anneal because of the cube's 90°-angle deficit at each vertex, whereas the edges of an icosahedron do not anneal, as indicated by the X, because of its 60°-angle deficit at each vertex.

The new cages based on decoration of the full faces of the truncated tetrahedron have tetrahedral symmetry, specifically the point group $T_d$ of the host, 3-valent vertices, hexagons and 12 pentagons. They are therefore a subset of the tetrahedral ($T_h$, $T_d$ and T) fullerene cages, even a subset of the $T_d$ fullerene cages, the construction of which was described in 1988 [15]. The latter construction created tetrahedral fullerenes from four sets of five triangles (Figure 3d), each set containing a large equilateral triangle ($\alpha$), three scalene triangles ($\beta$) and a small equilateral triangle ($\gamma$), all decorated with a tiling of hexagons. (To show the 3-fold axis centered on the large equilateral triangle ($\alpha$) in Figure 3d, each small equilateral triangle ($\gamma$) is divided into three thirds.) Our new $T_d$ cages, with four hexagons and four triangles, can be similarly described, with each hexagonal cutout providing the large equilateral triangle ($\alpha$) and the three scalene—isosceles in this case—triangles ($\beta$), and each triangular cutout providing the small equilateral triangle ($\gamma$) (Figure 3b,c).

The equilateral triangles assembled to produce icosahedral fullerene cages can be described by Goldberg indices ($h,k$) that characterize one edge of the triangle (e.g., 2,1 in Figure 1b). Likewise, for the tetrahedral fullerenes, the large equilateral triangle in the construction can be described by indices ($i,j$) (containing $i^2 + ij + j^2$ vertices), and the small equilateral triangle can be described by its own indices ($k,l$) (containing $k^2 + kl + l^2$ vertices) [15]. The isosceles triangles contain the same number of vertices as the small equilateral triangle (Figure 3b,c). Therefore, for most of the cutouts in Figure 3b,c, we show just one regular hexagon (with its large equilateral triangle and three isosceles triangles) and one equilateral triangle (that becomes the small equilateral triangle).

### 3.2. Two Series of the New $T_d$ Cages

One series of the new cages have large and small equilateral triangles with indices (1,1 and 1,0), (2,2 and 2,0), (3,3 and 3,0), etc. (Figure 3c) The other series are leapfrogs [9] of the first, with indices (3,0 and 1,1), (6,0 and 2,2), (9,0 and 3,3), etc. (Figure 3b). Arranging the indices and calculating the total numbers of vertices (Table 1) shows that the cages in the first series have $28n^2$ vertices ($n \geq 1$), whereas the cages in the second series have $3 \times 28n^2$, the triplication expected for leapfrogs. These new $T_d$ fullerene cages represent only a few of the possible $T_d$ fullerene cages [15].

**Table 1.** Application of a 6.6.6 tiling to the truncated tetrahedron [1].

| Large Equilateral | | | Small Equilateral | | Scalene | | | | Equilateral |
| --- | --- | --- | --- | --- | --- | --- | --- | --- | --- |
| Triangle | | | Triangle | | Triangle | Total Vertices | | | Polyhedron |
| $i$ | $j$ | Vertices | $k$ | $l$ | Vertices | Vertices | | | + or − |
| 1 | 1 | 3 | 1 | 0 | 1 | 1 | 28 | 1 × 28 | + |
| 2 | 2 | 12 | 2 | 0 | 4 | 4 | 112 | 4 × 28 | + |
| 3 | 3 | 27 | 3 | 0 | 9 | 9 | 252 | 9 × 28 | + |
| 4 | 4 | 48 | 4 | 0 | 16 | 16 | 448 | 16 × 28 | + |
| 5 | 5 | 75 | 5 | 0 | 25 | 25 | 700 | 25 × 28 | + |
| 6 | 6 | 108 | 6 | 0 | 36 | 36 | 1008 | 36 × 28 | + |
| 7 | 7 | 147 | 7 | 0 | 49 | 49 | 1372 | 49 × 28 | + |
| 8 | 8 | 192 | 8 | 0 | 64 | 64 | 1792 | 64 × 28 | + |
| 9 | 9 | 243 | 9 | 0 | 81 | 81 | 2268 | 81 × 28 | + |
| 10 | 10 | 300 | 10 | 0 | 100 | 100 | 2800 | 100 × 28 | too large |
| 3 | 0 | 9 | 1 | 1 | 3 | 3 | 84 | 1 × 3 × 28 | + |
| 6 | 0 | 36 | 2 | 2 | 12 | 12 | 336 | 4 × 3 × 28 | − |
| 9 | 0 | 81 | 3 | 3 | 27 | 27 | 756 | 9 × 3 × 28 | − |
| 12 | 0 | 144 | 4 | 4 | 48 | 48 | 1344 | 16 × 3 × 28 | − |
| 15 | 0 | 225 | 5 | 5 | 75 | 75 | 2100 | 25 × 3 × 28 | − |
| 18 | 0 | 324 | 6 | 6 | 108 | 108 | 3024 | 36 × 3 × 28 | too large |
| 21 | 0 | 441 | 7 | 7 | 147 | 147 | 4116 | 49 × 3 × 28 | too large |
| 24 | 0 | 576 | 8 | 8 | 192 | 192 | 5376 | 64 × 3 × 28 | too large |
| 27 | 0 | 729 | 9 | 9 | 243 | 243 | 6804 | 81 × 3 × 28 | too large |
| 30 | 0 | 900 | 10 | 10 | 300 | 300 | 8400 | 100 × 3 × 28 | too large |

[1] Each of the new $T_d$ fullerene cages is composed of four regular hexagonal patches and four equilateral triangular patches. Each hexagonal patch can be subdivided into a large equilateral triangle and three isosceles triangles. The table shows the pairs of indices $(i,j)$ and $(k,l)$ for the equilateral triangles and the number of vertices for each of the triangles. The total number of vertices is 4× the number in the large equilateral triangle, 4× the number in the small equilateral triangle, and 12× the number in the isosceles triangle. The total numbers of vertices in the new cages are all multiples of 28. The symbols "+" and "−" mean definite success or failure by Equi to produce a convex equilateral polyhedron from the cage. Cages with ≥2800 vertices are "too large" for the current version of the Equi program to equiplanarize on a conventional computer.

We make these $T_d$ cages with Carbon Generator (CaGe) software [11]. Figure 5 shows two-dimensional (Schlegel) diagrams of the first four of the first series and the first two of the second (leapfrog) series. A geometric "cage" may have nonplanar faces, as can be seen in the equilateral cages in Figure 1c [3].

## 3.3. Production of Equilateral Polyhedra from the New $T_d$ Cages

Geometrically "convex polyhedra" are also cages, but they must have planar faces and point outward at every vertex [4,5]. Thus, if one were to place a flat plane on any face of a convex polyhedron, the plane would not intersect any of the other faces [4,5]. With two exceptions—the dodecahedron and the truncated icosahedron—equilateral icosahedral Goldberg cages with internal angles in hexagons near 120° have nonplanar faces and are therefore not polyhedra [3] (e.g., Figure 1c). However, it is possible to make the faces of these cages planar and produce convex equilateral icosahedral polyhedra by adjusting internal angles in the faces [3].

We attempted to transform all of the new $T_d$ cages into equilateral tetrahedral polyhedra with Equi, a program that is able to numerically solve equations for equal edge lengths and for planarity of faces. Equi uses a numerical method to obtain $x$, $y$ and $z$ coordinates of the V vertices by solving a system of multivariate nonlinear equations in 3V-6 variables (3 coordinates for each vertex minus 6 degrees of freedom corresponding to translation and rotation of the solution in 3-dimensional space). The system consists of two types of equations: There are 3V/2 quadratic "edge length equations" (1 for each edge) and 3V "face planarity equations" ($n$ for each $n$-gonal face). The system is over-determined—$n$-3 equations for each $n$-gonal face would suffice—but using more equations makes the algorithm more stable. The numerical method is a variant of the well-known Gauss-Newton algorithm for finding a minimum of a function [16] with an added symmetrization step after each iteration to ensure the tetrahedral symmetry ($T_d$) of the result. The coordinates we obtain are accurate to 8 digits, but we could improve the accuracy further if there were reason to do so. The current version of the algorithm can handle cages of up to 2800 vertices within a reasonable time frame on a conventional computer. In the future it might be possible to increase this limit by taking full advantage of the tetrahedral symmetry, requiring a major redesign of the algorithm.

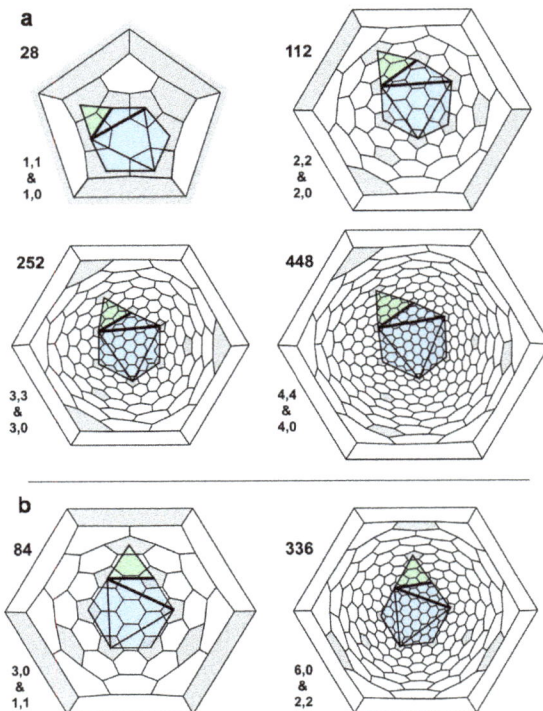

**Figure 5.** Schlegel diagrams of the new cages, each with a (blue) hexagon (containing a large triangle with a bolded edge and three small isosceles triangles around the large triangle), and a (green) triangle (also with a bolded edge). Indices for the equilateral triangles can be obtained for the bolded edges. (**a**) The first four $T_d$ fullerene cages in the first series, with 28, 112, 252 and 448 vertices; (**b**) the first two $T_d$ fullerene cages in the leapfrog series with 84 and 336 vertices.

Although the program is not yet able to solve for the coordinates and angles in very large cages, we can transform all of the first nine cages in the first series into geometrically convex equilateral polyhedra with $T_d$ symmetry ("+" symbols in Table 1). Figure 6a shows the first five. (Details are provided in Figure S1 and Tables S1–S3.) We suggest that all of the new $T_d$ cages in the first series, including larger ones, can be so transformed.

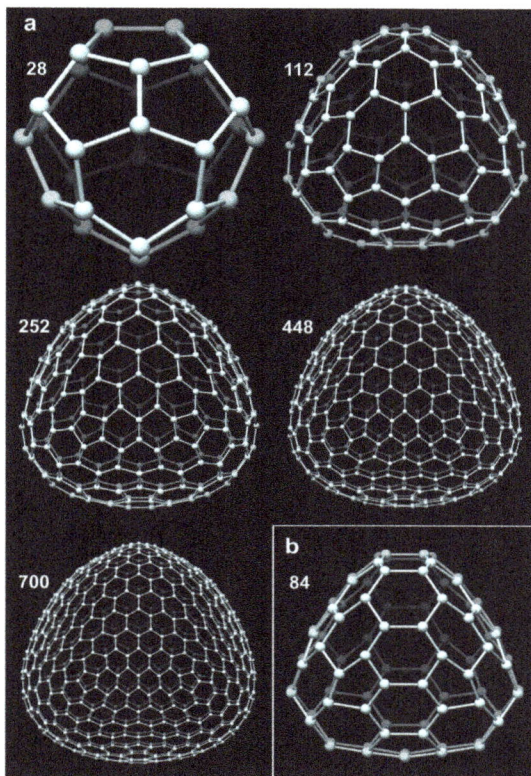

**Figure 6.** Convex equilateral $T_d$ polyhedra from new $T_d$ cages. (**a**) The convex equilateral $T_d$ polyhedra from the first series of $T_d$ cages with 28, 112, 252, 448 and 700 vertices; (**b**) the convex equilateral $T_d$ polyhedron in the leapfrog series of $T_d$ cages with 84 vertices. Figure S1 shows vertex numbers for the polyhedra in (**a**) and (**b**), and Tables S1–S3 show coordinates, internal angles in faces and dihedral angles, respectively, for these polyhedra.

However, for the cages in the second (leapfrog) series, only the smallest with 84 vertices could be transformed into a convex equilateral polyhedron (Figure 6b; Table 1; Figure S1; Tables S1–S3). None of the larger ones, with 336, 756, 1344, and 2100 vertices, could be so transformed. We suggest that none of new leapfrog $T_d$ cages with more than 84 vertices can be transformed into geometrically convex equilateral polyhedra.

We reported that all achiral icosahedral fullerenes with $T$ ($= h^2 + hk + k^2$) $\leq 49$ (980 vertices) and all chiral fullerenes with $T \leq 37$ (740 vertices) can be transformed into geometrically convex equilateral icosahedral polyhedra [3]. Here, we add that with Equi we have been able to so transform all 40 of the icosahedral fullerenes with $T \leq 108$ (2160 vertices)—all of the ($h$,0) ones from (1,0) to (10,0), all of the ($h$,$h$) ones from (1,1) to (6,6) and all of the ($h$,$k$) ones from (2,1) to (9,2). Therefore, we suggest that the failure of the larger leapfrog $T_d$ cages to transform into convex equilateral polyhedra is real and not a failure of our methods.

## 4. Discussion

Above, we suggest that all of the first class of new $T_d$ cages can be transformed into geometrically convex equilateral polyhedra with $T_d$ symmetry. By contrast, only the smallest of their leapfrogs can be so transformed. There are precedents for this situation among the Goldberg cages. On the one hand, we can produce convex equilateral icosahedral polyhedra from all 40 of the icosahedral Goldberg cages that we have tried. On the other hand, among the octahedral Goldberg cages, beyond $h,k = 1,0$ (the Platonic octahedron) and $h,k = 1,1$ (the Archimedean truncated octahedron), we are able to make just one more convex equilateral polyhedron, $h,k = 2,0$. Larger ones have coplanar 4-gonal faces and are thus not convex; correspondingly, for a few of these we are able to show that the only equilateral planar solutions require some vertices with internal angles that sum to 360° [3]. Likewise, among the tetrahedral Goldberg cages, beyond $h,k = 1,0$ (the Platonic tetrahedron) and $h,k = 1,1$ (the Archimedean truncated tetrahedron), we are able to transform just one into a convex equilateral polyhedron, also $h,k = 2,0$ [3], but none of the larger ones.

Could we use any other of the Archimedean polyhedra as hosts? Is it possible to fit cutouts from a two-dimensional tiling of regular hexagons (6.6.6) or from a tiling of regular squares (4.4.4.4) onto the full faces of some other Archimedean polyhedron and have the tiling pattern knit across the edges of the host polyhedron's faces [17]. To see what cutouts are needed, we list the faces (e.g., regular 5-gons, 8-gons, etc.) at each vertex in the remaining 12 Archimedean polyhedra, six with octahedral symmetry (3.4.3.4; 3.4.4.4; 3.8.8; 4.6.6; 4.6.8; 3.3.3.3.4) and six with icosahedral (3.4.5.4; 3.5.3.5; 3.10.10; 4.6.10; 5.6.6; 3.3.3.3.5) (Figure 2). From a 4.4.4.4 tiling, 3-gonal or 6-gonal cutouts do not exhibit 3-fold or 6-fold symmetry and do not permit knitting of the 4.4.4.4 tiling across the edges of the 3-gonal or 6-gonal faces of a host (Figure 4), thus eliminating all of the octahedral Archimedean polyhedra as potential hosts. Likewise, from a 6.6.6 tiling, 4-gonal, 5-gonal or 10-gonal cutouts do not exhibit 4-fold, 5-fold or 10-fold symmetry, respectively, and do not permit knitting of the 6.6.6 tiling across the edges of the 4-gonal, 5-gonal or 10-gonal faces of a host (Figure 4), thus eliminating all of the icosahedral Archimedean polyhedra as potential hosts. Therefore, the only admissible combination of a regular tiling and an Archimedean polyhedron as host is a tiling of hexagons and the truncated tetrahedron.

## 5. Conclusions

The Platonic and Archimedean polyhedra have tetrahedral, octahedral and icosahedral symmetry, lumped together as "polyhedral symmetry" [1]. The polyhedra in these two classes of "convex equilateral polyhedra with polyhedral symmetry" have regular faces. A third class of convex equilateral polyhedra with polyhedral symmetry is comprised of the rhombic polyhedra discovered by Kepler, the rhombic dodecahedron and the rhombic triacontahedron [2]. Like the rhombic polyhedra, members of a fourth class called "Goldberg polyhedral" [3] have faces, the 6-gons, that are not regular. (Because axes of 5-fold, 4-fold and 3-fold symmetry pass through the centers of their 3-gons, 4-gons and 5-gons, the smaller faces of these tetrahedral, octahedral and icosahedral polyhedra are regular.) Here we report another new class, constructed by decorations of the full faces of an Archimedean polyhedron, the truncated tetrahedron, with cutouts from a 6.6.6 tiling of regular hexagons. Although neither the 5-gons nor the 6-gons are regular, they are equilateral and planar.

**Supplementary Materials:** The following are available online at www.mdpi.com/2073-8994/8/8/82/s1. Figure S1: Vertex numbers for the first series of new $T_d$ polyhedra with 28, 112, 252, 448 and 700 vertices and for the other (leapfrog) series with 84 vertices; Table S1: Coordinates of vertices in the new $T_d$ polyhedra, the first series with 28, 112, 252, 448 and 700 vertices, the second (leapfrog) series with 84 vertices. Vertex numbering is shown in Figure S1. Edge lengths are 5 units; Table S2: Internal angles in faces in the new polyhedra. Vertex numbering is shown in Figure S1. The data come directly from Spartan, which provides three angles per vertex, but approximately one angle in each face is duplicated, leaving one other missing. However, the missing angles can be found by taking advantage of the symmetry of the polyhedron and in particular its Schlegel representation in Figure S1; Table S3: Dihedral angles across edges in the new polyhedra. Vertex numbering is shown in Figure S1.

**Acknowledgments:** We thank Gunnar Brinkmann (Department of Applied Mathematics, Computer Science and Statistics, University of Ghent, Belgium) for help with use of the CaGe software to make cages [11]. Molecular graphics in Figure 1a,c, Figure 3a,d and Figure 6 used the UCSF Chimera package [13]. Chimera is developed by the Resource for Biocomputing, Visualization, and Informatics at the UCSF (supported by NIGMS P41-GM103311). The nets in Figure 2 were produced with PolyPro (Pedagoguery Software, Terrace, BC, Canada). All of the other figures were made with Adobe Illustrator in Adobe CS4 (Adobe Systems, Inc., San Jose, CA, USA).

**Author Contributions:** The authors contributed equally to the work.

**Conflicts of Interest:** The University of California, Los Angles has applied for a patent for this work.

## References

1. Cromwell, P.R. *Polyhedra*; Cambridge University Press: Cambridge, UK, 1997.
2. Kepler, J. *Strena, seu de Nive Sexangula Tthe Six-Cornered Snowflake*; Godefridum Tambach: Francofurti ad Moenum, UK, 1611; English translation: Hardie, C. *The Six-Cornered Snowflake*; Clarendon Press: Oxford, UK, 1966.
3. Schein, S.; Gayed, J. A new class of convex equilateral polyhedra with polyhedral symmetry related to fullerenes and viruses. *Proc. Natl. Acad. Sci. USA* **2014**, *111*, 2920–2925. [CrossRef] [PubMed]
4. Coxeter, H.S.M. *Introduction to Geometry*, 2nd ed.; Wiley: New York, NY, USA, 1969.
5. Grünbaum, B. *Convex Polytopes*, 2nd ed.; Springer: New York, NY, USA, 2003.
6. Caspar, D.L.; Klug, A. Physical Principles in the Construction of Regular Viruses. *Cold Spring Harb. Symp. Quant. Biol.* **1962**, *27*, 1–24. [CrossRef] [PubMed]
7. Goldberg, M. A Class of Multi-Symmetric Polyhedra. *Tohoku Math. J.* **1937**, *43*, 104–108.
8. Coxeter, H.S.M. Virus Macromolecules and Geodesic Domes. In *A Spectrum of Mathematics*; Forder, H.G., Butcher, J.C., Eds.; Auckland University Press/Oxford University Press: Oxford, UK, 1971; pp. 98–107.
9. Fowler, P.W.; Manolopoulos, D.E. *An Atlas of Fullerenes*; Clarendon Press: Oxford, UK, 1995.
10. Dutour, M.; Deza, M. Goldberg–Coxeter construction for 3- and 4-valent plane graphs. *Electron. J. Comb.* **2004**, *11*, 20–68.
11. Brinkmann, G.; Friedrichs, O.D.; Lisken, S.; Peeters, A.; Van Cleemput, N. CaGe—A Virtual Environment for Studying Some Special Classes of Plane Graphs—An Update. *Match-Commun. Math. Comput.* **2010**, *63*, 533–552.
12. Hehre, W.; Ohlinger, S. *Spartan'10 for Windows, Macintosh and Linux: Tutorial and User's Guide*; Wavefunction, Inc.: Irvine, CA, USA, 2011.
13. Pettersen, E.F.; Goddard, T.D.; Huang, C.C.; Couch, G.S.; Greenblatt, D.M.; Meng, E.C.; Ferrin, T.E. UCSF chimera—A visualization system for exploratory research and analysis. *J. Comput. Chem.* **2004**, *25*, 1605–1612. [CrossRef] [PubMed]
14. Fowler, P.W.; Cremona, J.E.; Steer, J.I. Systematics of bonding in non-icosahedral carbon clusters. *Theor. Chim. Acta* **1988**, *73*, 1–26. [CrossRef]
15. Kroto, H.W.; Heath, J.R.; O'Brien, S.C.; Curl, R.F.; Smalley, R.E. C$_{60}$: Buckminsterfullerene. *Nature* **1985**, *318*, 162–163. [CrossRef]
16. Björck, A. *Numerical Methods for Least Squares Problems*; SIAM: Philadelphia, PA, USA, 1996.
17. Pawley, G.S. Plane Groups on Polyhedra. *Acta Crystallogr.* **1962**, *15*, 49–53. [CrossRef]

*symmetry*

MDPI

*Article*

# On the Form and Growth of Complex Crystals: The Case of Tsai-Type Clusters [†]

Jean E. Taylor [1], Erin G. Teich [2], Pablo F. Damasceno [3], Yoav Kallus [4] and Marjorie Senechal [5,*]

[1]  Courant Institute of Mathematical Sciences, New York University, New York, NY 10012, USA; jtaylor@cims.nyu.edu
[2]  Applied Physics Program, University of Michigan, Ann Arbor, MI 48109, USA; erteich@umich.edu
[3]  Department of Cellular and Molecular Pharmacology, University of California, San Francisco, CA 94143, USA; pablo.damasceno@ucsf.edu
[4]  Santa Fe Institute, Santa Fe, NM 87501, USA; yoav@santafe.edu
[5]  Mathematics and History of Science and Technology, Smith College, Northampton, MA 01063, USA
*  Correspondence: senechal@smith.edu; Tel.: +1-413-585-3862
†  Dedicated to the memory of Christopher Henley (1955–2015).

Received: 20 July 2017; Accepted: 1 September 2017; Published: 11 September 2017

**Abstract:** Where are the atoms in complex crystals such as quasicrystals or periodic crystals with one hundred or more atoms per unit cell? How did they get there? The first of these questions has been gradually answered for many materials over the quarter-century since quasicrystals were discovered; in this paper we address the second. We briefly review a history of proposed models for describing atomic positions in crystal structures. We then present a revised description and possible growth model for one particular system of alloys, those containing Tsai-type clusters, that includes an important class of quasicrystals.

**Keywords:** complex crystals; quasicrystals; nested clusters; crystal growth

## 1. Introduction

Crystallography inhabits a borderland fruitfully shared by mathematics, physics, chemistry, and geology, among other disciplines. This borderland is an especially fertile field for geometers, as many crystal forms are polyhedral, their atomic patterns poly-complex, and even basic questions are far from settled [1]. One such question concerns the most accurate way to model crystalline form and growth. The discovery of X-ray diffraction early in the last century consolidated the then century-old lattice paradigm, which modeled crystal structures as periodic tilings of $R^3$. This model became the canonical lens through which crystal structure and growth was viewed. It seemed to serve crystallography well: the tiles were metaphorical swatches for atomic patterns, while the regularity of their locations suggested crystal growth intuitively (row upon row, layer upon layer). However, the discovery of aperiodic crystals in 1982 [2] showed that this model of growth and form cannot be the whole story. The new models proposed to fill this gap are still debated [3].

The earliest crystal structure models (a perhaps generous description) explained growth in a rudimentary way. In the 17th century, Johannes Kepler in Austria [4] and Robert Hooke [5] in England imagined crystals as packings of identical spheres and showed how their polyhedral shapes might arise therefrom. Two centuries later, the French abbé R. J. Häuy [6] stacked congruent bricks in various ways to show why and how crystals of the same species could have different external forms. Pyrite crystals, for example, grow as cubes, as octahedra, and sometimes as irregular pentagonal dodecahedra or "pyritohedra".

The centers of the tiles in Häuy's tilings form point lattices, a fact that led to the focus of geometrical crystallography on lattices. With the discovery of X-ray diffraction in 1912 [7], it became possible to locate the atomic positions in periodic geometric patterns precisely. The lattice paradigm was, it seemed, thereby confirmed, and with it a simple corollary [8]: the possible order of rotations for a lattice in dimensions 2 and 3 are 2, 3, 4, and 6. This was known to generations of students as "The Crystallographic Restriction".

Note the absence of the number 5 (and all integers greater than 6). Why? Let L be a plane lattice and $\phi_X$ a 72° rotation about a lattice point X. Choose points A and B such that $|AB| = r$, where r is the minimum distance between points of L. Thus $\phi_A(B)$ is a lattice point $B'$, and $\phi^{-1}_B(A) = A'$. But then $|A'B'| < r$, which is impossible (Figure 1).

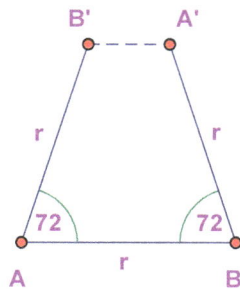

**Figure 1.** A point lattice in the plane cannot have five-fold rotational symmetry (see the discussion above).

A similar argument holds for three-dimensional lattices, since a rotation fixes a plane and an axis orthogonal to it. Thus, icosahedral symmetry is impossible for a three-dimensional point lattice (however, five-fold symmetry is possible for lattices in all dimensions greater than three).

It follows, apparently, that regular dodecahedra and icosahedra are impossible crystal forms. Of course, crystallographers have always understood that the lattice is an abstraction, its points merely placeholders for the motifs—the finite configurations of atoms—that repeat to form the pattern. The Crystallographic Restriction does not forbid *motifs* with five-fold symmetry. This was even known to the ancient Egyptians, at least implicitly (Figure 2).

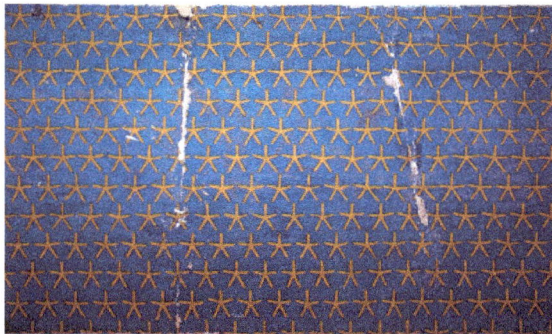

**Figure 2.** Ancient Egyptian tomb ceiling pattern (https://www.pinterest.com/pin/157837161917423903/). Notice that the centers of the stars form a planar point lattice with hexagonal symmetry, although the stars themselves exhibit pentagonal symmetry.

Indeed, as we shall see, the world of intermetallic crystals is replete with analogous patterns in $R^3$. The complexity of these periodic patterns and their close relation to aperiodic crystals is the central theme of this paper.

Dan Shechtman's paradigm-shifting discovery of crystals with icosahedral symmetry [2] showed that the Crystallographic Restriction was merely a theorem about lattices, not a law of nature. A complete understanding of periodic lattices is not sufficient for a complete understanding of crystalline structure.

The skepticism that greeted Shechtman's discovery moulders in history's dustbin, nailed shut by his 2011 Nobel prize. Today we define crystals to be materials with atomic long-range order evidenced not by lattice descriptions, but by sharp spots in their diffraction patterns. Crystals whose diffraction patterns exhibit icosahedral symmetry are called icosahedral quasicrystals (iQCs).

The new diffraction-based definition of "crystal", a major break from the historical lattice paradigm, gives rise to a variety of difficult questions, such as *which* non-lattice atomic arrangements have sharp spots in their diffraction patterns, and how aperiodic crystals self-assemble, if they do not grow by rows and layers. The first of these two questions has given rise to a new field of mathematics, "long range aperiodic order" (see [9,10]). In this paper, we focus on self-assembly.

At the outset of this paradigm shift, in 1984, many (though, significantly, not all) researchers assumed that aperiodic crystals should and could be described, and their growth modeled, by brick-laying, albeit with bricks of a non-Häuy sort, perhaps three-dimensional Penrose-like tiles assembled in accordance with matching rules. But although three-dimensional tiles can be superimposed on many iQC structures, the subdivision can be artificial, and their very complex matching rules do not model physical crystal growth.

Both the subdivisions and the matching rules can be skirted by recasting the structure in higher dimensions. When—as is the case for all aperiodic crystals known so far—the atomic positions can be indexed by $n > 3$ integers, we can consider the n-tuples to be "addresses" in $R^n$ and "lift" the structure to a suitable lattice there [11]. The popularity of this formalism owes much to the fact that higher-dimensional analogues of tools developed earlier for periodic crystals can be used and the three-dimensional positions of the crystal pattern recovered by projection. But the process is complicated, and tells us nothing about actual crystal growth. Very recently, the atomic positions in an iQC have been determined directly from X-ray diffraction data, without the high-dimensional formalism [12].

A third model, nested atomic clusters, has also been studied from the outset, although it received less attention until recently. The cluster approach, which builds on the chemists' concept of coordination polyhedra, holds promise for understanding growth. After a discussion of the features of this model (Sections 2 and 3), we consider one particular example, the Ytterbium-Cadmium iQC whose structure (the first iQC to be "solved"!) was determined in 2007 [13].

## 2. Nested Clusters

Thirty years before the discovery of quasicrystals, the British physicist F.C. Frank suggested that icosahedral atomic clusters, with 12 atoms around a central atom, are common in supercooled liquids, where they may be locally favored structures [14]. That same year, Bergman et al. described icosahedral clusters in certain intermetallic solid crystals [15]; these clusters, with many more atoms in successive shells, are now eponymously known as Bergman clusters. Ten years later, Mackay proposed the nested clusters that bear his name [16]. The use of nested clusters as a descriptive tool for complex crystal structures has steadily grown. A searchable database of multi-shelled nanoclusters in complex intermetallics is currently being developed in Russia [17].

Both the Bergman and Mackay clusters (Figure 3) can be described in several ways; we follow convention in considering them to be three-shelled. In the Bergman cluster, an icosahedron sits inside a dodecahedron which sits inside a larger icosahedron, for a total of $12 + 20 + 12 = 44$ atoms, not counting a (possible) atom at the center. In the Mackay cluster, the second shell is an icosidodecahedron, which has 30 vertices; thus the total number of atoms is 54 (plus, possibly, 1).

To describe the crystal that is the focus of this paper, the Ytterbium-Cadmium (Yb-Cd) iQC, Takakura et al. [13] use a third type of cluster, named for co-author Ang-Pan Tsai (Figure 4). The Tsai cluster has five shells: its innermost shell is a regular tetrahedron, which is encased in a dodecahedron. The dodecahedron sits inside an icosahedron, which sits inside an icosidodecahedron and finally, fifth and outermost, a rhombic triacontahedron. The tetrahedron, the only shell without icosahedral symmetry, has several equivalent lower-symmetry positions within the dodecahedron and flips about among them. Yb atoms are centered at vertices of the icosahedron and Cd atoms at all other vertices of these shells, and also at the edge midpoints of the outer one. Thus each Tsai cluster "has" 146 Cd atoms (4 in the tetrahedron, 20 in the dodecahedron, 30 in the icosidodecahedron, 32 in the rhombic triacontahedron, and 60 in its edges) and 12 Yb, a ratio of 12.1666...: 1. Note that these atoms are not identical in size; rather, the ratio of their radii is approximately 1.2:1, the approximate ratio of the "crystal" atomic radii Yb:Cd (*Tables of Physical & Chemical Constants* (16th edition 1995). 3.1.2 Properties of the Elements. Kaye & Laby Online. Version 1.0 (2005) www.kayelaby.npl.co.uk).

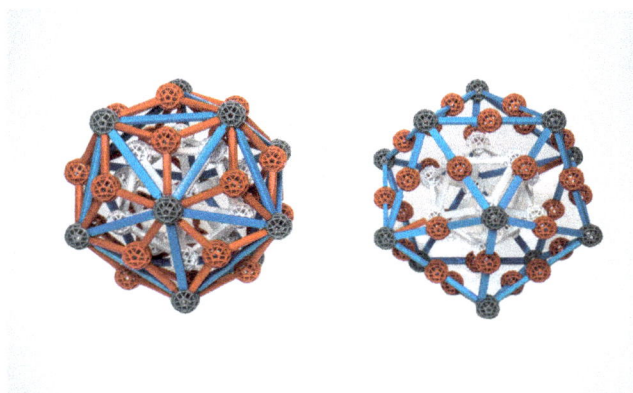

**Figure 3.** Zometools models of Bergman (**left**) and Mackay (**right**) clusters. In both models, the vertices of the inner shell are white nodes, the middle shell red, and the outer shell gray. (The inner and outer icosahedra of the two models should, respectively, be congruent but the limited size of Zometool struts precludes that).

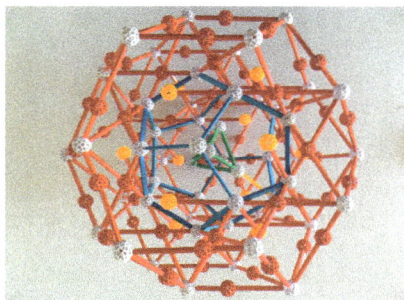

**Figure 4.** A Zometools model of the Tsai cluster. In the center is a tetrahedron (green edges) with vertices (white balls) representing Cd atoms. Surrounding it is a shell composed of a dodecahedron (blue edges) with white balls representing Cd atoms at its vertices plus its dual icosahedron with vertices (yellow balls) representing Yb atoms. Outside this is an icosidodecahedron of Cd atoms (red balls each with two short edges to dodecahedron vertices and two longer ones to three-fold rhombic triaconahedron vertices, these four edges forming an X), and outside that is a rhombic triacontahedron with vertices (white balls) and mid-edges (red balls) of Cd atoms. Parts of these two outer shells have been removed so that the inner structure can be more clearly seen.

## 3. A Conceptual Breakthrough

In the mid-1980s Christopher Henley, reviewing the literature of periodic but complicated intermetallic crystal structures, noticed that they closely approximated the iQC structures of these same or similar materials, much as rational numbers approximate irrationals. He coined the term "approximant" for these periodic structures to "mean (that) the unit cell is indistinguishable from a fragment commonly occurring in the icosahedral phase". In the two cases he studied in detail, "each (of these approximant structures) is a bcc packing of clusters which have two concentric atomic shells with full icosahedral symmetry around their center" [18].

The shorthand "bcc" stands for "body-centered cubic" and refers to the cubic lattice in $R^3$ whose points are not only the cube vertices but also their centers. To visualize the bcc packing by equal spheres, recall that the Voronoi cell V of the $R^3$ bcc lattice is a truncated octahedron with six square facets and eight that are regular hexagons. In this lattice, the distance r between the center o of V and the center of a hexagonal facet is one half the minimum interpoint distance, and the distance s from o to the center of a square facet is larger by a factor of approximately 1.15. Thus, spheres of radius r at the bcc lattice points form a packing of $R^3$ in which the spheres are tangent in the directions of three-fold symmetry but with gaps in the four-fold symmetry directions (Figure 5a). Alternatively, if we place spheres of radius s at the lattice points, we get an arrangement in which spheres are tangent in the four-fold direction but overlap along the three-fold axes (Figure 5b).

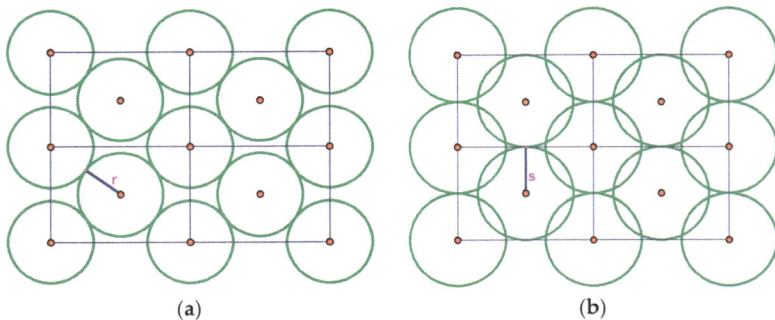

(a)                                                    (b)

**Figure 5.** Two-dimensional cross-sections, perpendicular to a face diagonal, of bcc packings by equal spheres. (**a**) A bcc packing of spheres of radius r; (**b**) A bcc packing of spheres of radius s.

To envision the structures Henley was describing, replace the spheres in Figure 5b by Mackay or Bergman clusters, oriented so that four of their ten three-fold axes are the three-fold rotation axes of the cubic lattice and six edges of the icosahedral shell (which in the Bergman cluster are long diagonals of the rhombic facets) are orthogonal to the cubic lattice's four-fold axes. Depending on the diameter of the clusters and whether they are Bergman or Mackay, this collection could be interpenetrating or not, similarly to the case of Figure 5a,b.

But this "identification of the basic clusters and linkage rules is only the starting point of a complete structure model". Henley observed, because "In reality, a metal structure cannot have large voids; additional, so-called "glue" atoms (about 30% of the total) must occupy sites between clusters" [18].

## 4. Linkages in the Yb-Cd iQC and Its Approximants

In the structures found by Takakura et al. [13] for the Yb-Cd iQC and for its approximants, the Tsai clusters overlap, so if two more small clusters (the AR and OR, described below) are included, no additional glue is needed. But as we explain below, in our view of these structures, more glue is there in disguise.

First let us look closely at the geometry of the rhombic triacontahedron (the outer shell of the Tsai cluster). Its rhombic faces are orthogonal to the two-fold axes of its symmetry group, and its three-fold axes pass through opposite 3-valent vertices. Note (Figure 6) that each 3-valent vertex V is ringed by six vertices, alternately 5- and 3-valent. The vertices of this ring, together with V, are seven of the eight vertices of an obtuse golden rhombohedron, called OR.

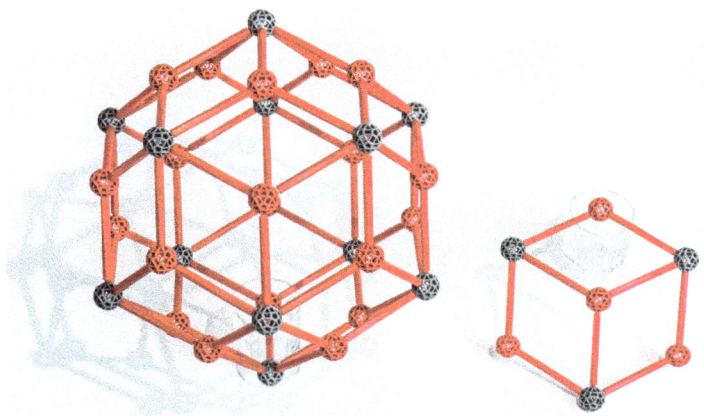

**Figure 6.** (**Left**): A rhombic triacontahedron viewed along a three-fold rotation axis; (**Right**): The six vertices ringing a vertex on the three-fold axis, together with that vertex, are seven of the eight vertices of an obtuse golden rhombohedron (OR).

Following [6], we will say that two congruent rhombic triacontahedra sharing a rhombic face (Figure 7a) are joined by a b-linkage, and are joined by a c-linkage if they share a ring (described above) and thus overlap in an OR (Figure 7b).

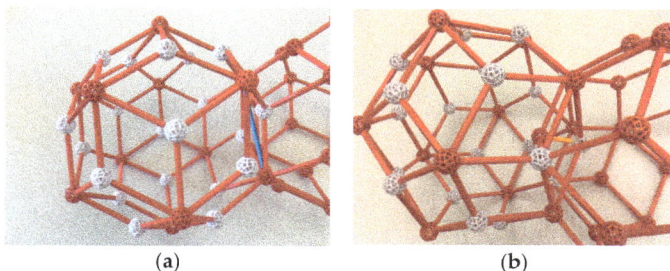

(a)                                        (b)

**Figure 7.** (a) Left: Two rhombic triacontahedra sharing a rhombus (b-linkage). The main diagonal of the shared rhombus is identified by a blue strut; (b) Right: Two rhombic triacontahedra overlapping in an obtuse rhombohedron (c-linkage). A yellow strut marks its short diagonal. Observe that the common vertices that are three-fold for one rhombic triacontahedron are five-fold for the other.

Both OR and acute golden rhombohedron (AR) complexes are found in the iQC structure. In Figure 8, we show the OR (with Cd at its vertices and mid-edges) and the AR (acute golden rhombohedron, with Cd at its vertices and mid-edges and two Yb on its main diagonal).

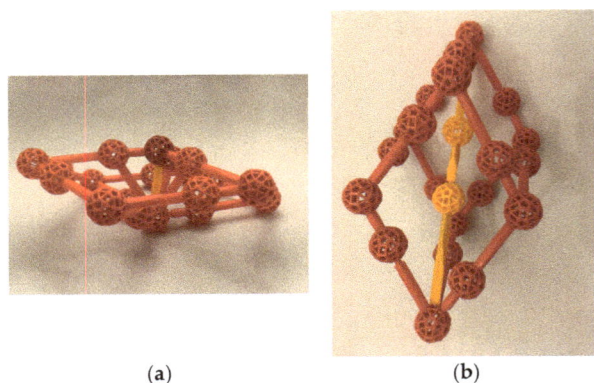

(a)                                    (b)

**Figure 8.** (**a**) The OR complex. Observe from its shadow that the OR looks like a regular hexagon when looking down the yellow strut; (**b**) The AR (acute golden rhombohedron) complex.

The Cd:Yb ratio in the iQC is approximately 5.7:1. Two of its well-studied periodic approximants have ratios of approximately 6:1 and 5.8:1, respectively (the fact that the atomic Cd:Yb ratio in the Yb-Cd iCQ and its approximants (approximately 5.7:1, 5.8:1, and 6:1) is much lower than that of the Tsai cluster is explained by the fact that, in the crystal, the clusters abut and overlap and thus many Cd atoms are shared).

In the 6:1 approximant, the centers of the Tsai clusters are arranged in a bcc lattice (as in Figure 5). The OR occurs only in c-linkages, and the AR does not occur. The AR occurs in the 5.8:1 crystal, but again the OR only occurs in c-linkages.

A key motif in all three structures consists of three linked Tsai clusters, two joined in a b-linkage and the third c-linked to both. It is described and illustrated below.

## 5. A Newly Identified Cluster Suitable for Growth

Interlocking clusters, however, do not suggest a growth mechanism. For pre-formed clusters to interlock, they would have to (partially) disassemble and reassemble, and this seems physically unlikely.

We propose instead an alternative description for Tsai-type crystals and their approximants in terms of a smaller unit that could account for growth and perhaps even nucleation, and could also clarify the role of the inner tetrahedron in those processes. Our unit consists of the inner 36 atoms of a Tsai cluster. We call it a TDI cluster (tetrahedron-dodecahedron-icosahedron) and propose that TDIs are the critical element for a new model of the growth of the Tsai-type quasicrystal and its periodic approximants (Figure 9). In this view, the two outer shells of the Tsai cluster are "glue" that forms spontaneously, depending on the ratio of Yb to Cd atoms in the melt.

Like the Bergman and Mackay clusters, the outer shell of the TDI cluster is a regular icosahedron. One can arrange TDI clusters with their centers in a bcc lattice, again like the Bergman and Mackay clusters, but with space between the clusters. Our suggestion is that, to form the 6:1 alloy, the TDIs join by constructing c-linkages, and the Tsai clusters arise as remaining Cd atoms slip into the voids being created. We suggest a similar underlying mechanism for growth in the case of the 5.8:1 periodic crystal and the iQC.

To see how formation of TDI clusters may lead to growth, we model the TDI with spheres of two sizes located at the atomic positions. We use "Cd" and "Yb" as names for the smaller and larger atoms, respectively.

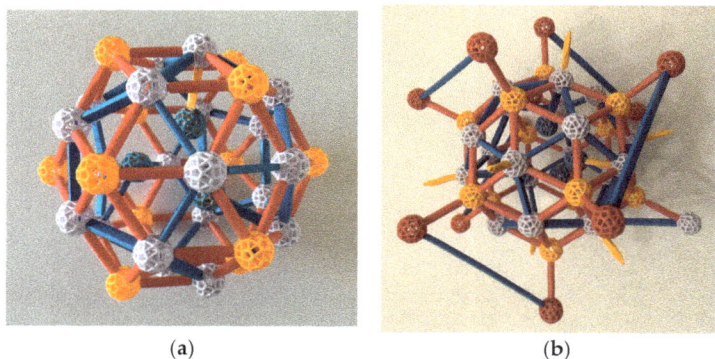

(a)          (b)

**Figure 9.** Zometools models of the tetrahedron-dodecahedron-icosahedron (TDI) cluster and its surroundings. (**a**) A single TDI. The white balls represent Cd atoms at the vertices of a dodecahedron, while blue struts represent the dodecahedron edges. The yellow balls represent Yb atoms at the vertices of an icosahedron. This picture shows all five of the red bonds from a yellow ball to vertices of the dodecahedron. In the center are four blue balls, also representing Cd atoms, at the corners of a tetrahedron; (**b**) The essential structure of $Cd_6Yb$ around each of its TDIs: yellow struts indicate c-linkages and long blue struts indicate the longer diagonal of the shared rhombus for each b-linkage. The blue struts are on the faces of a cube and the yellow struts point into the cube's corners. This combination of linkages occurs for each TDI in $Cd_6Yb$, and always in the same orientation.

TDIs formed in the liquid (why they might do so is discussed in the next section) have a Cd:Yb ratio of 24:12, or 2, so the liquid around TDI clusters would be a soup of mainly Cd atoms. Of course, the structure would be dynamic, forming and disassociating.

What glues the TDI's together? Suppose that two TDIs come close to each other in the melt, so that a Cd (call it V1) in the outer shell of one comes within a natural bond distance to a Cd (call it V2) in the outer shell of the other, forming a temporary bond. We can speculate that the two TDIs might bond such that they have the same orientation, since this is the distance and configuration of TDIs that is in fact seen in all three crystal structures. Note that, in this configuration, the three Yb atoms nearest V1 in its TDI are maximally distant from the three Yb nearest V2 in its TDI. See Figure 10a; the bond is the yellow strut.

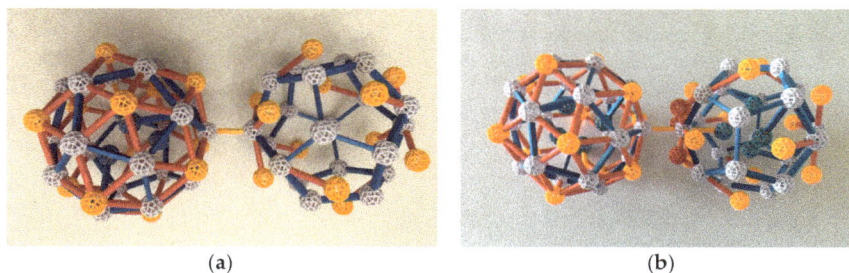

(a)          (b)

**Figure 10.** *Cont.*

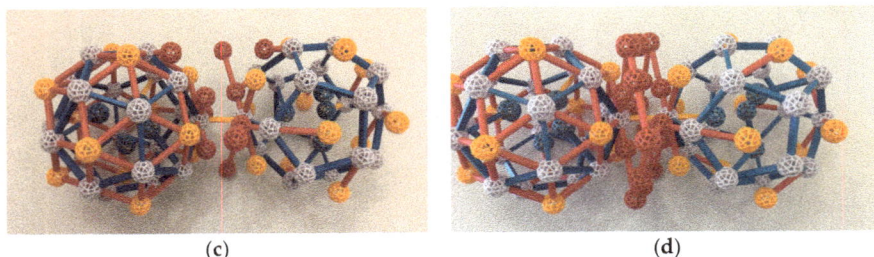

(c)    (d)

**Figure 10.** Development of a two-TDI cluster. (**a**) The initial bond between two TDIs. Only one of the five red bonds around each yellow Yb is shown for the right TDI; they are otherwise identical and have the same orientation; (**b**) Three red atoms from the icosidodecahedral shell have gelled into position on the right TDI at vertex V2; (**c**) Three red atoms have likewise gelled into position on V2, and a red bond and red ball have also been added to each of the three added in (**b**) to V2; (**d**) The entire OR complex has assembled around the yellow strut.

Notice that, in this configuration, there is room for glue, that is, for six Cd atoms in the liquid to slot into the positions in the icosidodecahedral shells that will eventually surround both TDIs, just outside the six TDI dodecahedral edges attached to V1 and V2. Three of them are in place in Figure 10b as red balls, with the two short bonds to dodecahedral atoms that they would have in Tsai clusters. These are presumed to be low potential energy positions, as they are observed experimentally.

Not only do these added atoms have these two short bonds, but also each of the three additions to the TDI of V1 are in the right spot to complete a third bond with V2. Likewise, each addition to V2 is in the right spot for a bond to V1. These six criss-crossing bonds (not shown in the figures) should help stabilize the two TDI clusters.

The fourth edges of the three red balls added to V2 are shown in Figure 10c, with red balls at the ends; three edges with balls are also added to V1. These six extra vertices are vertices of the outer shells of both Tsai clusters. Adding edges between them, with balls mid-edge, completes an OR, forming a c-linkage between the two developing Tsai clusters (Figure 10d). It also creates edges with mid-edge balls that will be in both Tsai clusters.

If now a third TDI forms a similar bond to the first TDI, it might have one of its dodecahedral Cd atoms (call it V2′) bond to a Cd (call it V1′) which is a next-nearest atom in the dodecahedral shell to V1. Then, a full OR can likewise develop between these TDI, and in fact one of the dodecahedral Cd atoms is already in the correct position (Figure 11a). Additionally, a b-type linkage (shared face) of the growing Tsai clusters of the second and third TDIs is automatically generated (Figure 11b). The three-unit cluster is shown schematically in Figure 12.

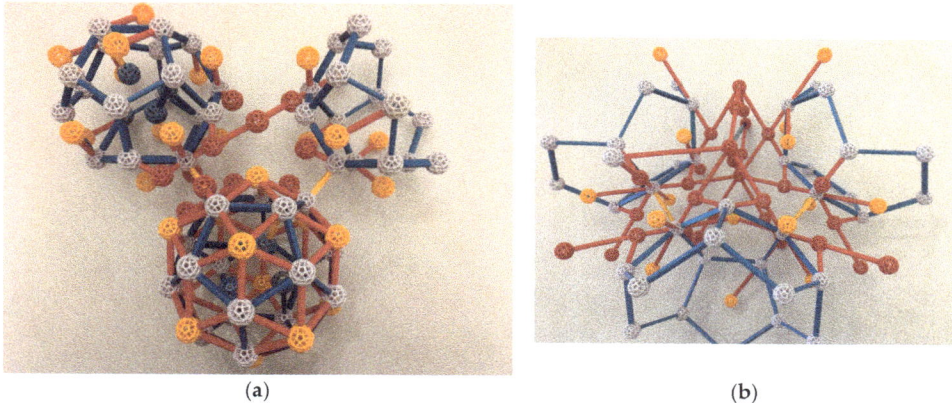

**(a)**          **(b)**

**Figure 11.** Development of a three-TDI structure. (**a**) Initial bond to the third TDI. V1′ and V2′ are the white balls at the ends of the yellow strut on the right. One additional red ball has gelled into place between the second and third TDIs. (**b**) A close-up of the enforced b-linkage of a shared rhombus between the developing Tsai clusters of the second and third TDIs (marked by a long blue strut, seen almost end-on in the reflection-symmetry plane between those two TDIs).

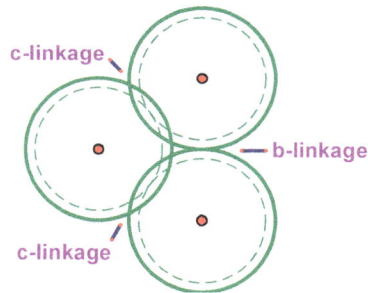

**Figure 12.** A schematic illustration of a two-dimensional section of the three-unit cluster, with dotted circles denoting the TDIs and solid circles the Tsai clusters. The single point of contact marked as a b-linkage in the figure becomes a shared face.

When the three Tsai clusters are completed, they make the key structure identified in Section 4, which occurs in the iQC and the two approximants we are discussing. Its core is the two c-linkages and the resulting b-linkage. We conjecture that the assembly of this core via the above mechanism might even be a mechanism of nucleation, if that cluster of three is stable enough. More TDIs can be added in an identical way. When the Cd:Yb ratio is 6:1, it is sufficient to build the entire bcc periodic crystal, with no leftover Yb or Cd atoms.

This cluster of three TDIs and their additional Cd "glue" is fundamental and appears to be a universal way of constructing local ordering. When the Cd:Yb ratio is 5.8:1, the extra Yb (compared to the 6:1 alloy) makes room for itself on the long diagonal of an isolated AR cluster sharing its faces with Tsai clusters (no longer with their centers in a BCC lattice but with a different periodicity). When the ratio is about 5.7:1, the extra Yb contributes to more elaborate collections of ARs with ORs, but still embedded in a network of TDIs joined by c-linkages and b-linkages with "glue" consistent with the outer shells of Tsai clusters; the result is an iQC. For detailed illustrations and a description of how this iQC structure was determined, see Takakura et al. and [12]. Thus, the focus on c-linkages would yield a reasonable model of growth for both the complex periodic crystals and the icosahedral one. Note that the nature of the b-linkage and especially the c-linkage enforces the same orientation on each TDI and thereby on each Tsai cluster in the completed structure. The long-range order is automatic. Further details of the growth model will be included in our longer, more detailed paper [19].

## 6. Why Might TDIs Exist in the Melt?

The TDI is more than a geometric motif around which soup atoms coalesce; it seems to have physical significance in its own right. Such a structure with all positions occupied by one kind of atom closely resembles the densest known packing of 36 identical balls within a spherical container [20,21]. This configuration consists of an outer shell of 32 balls lying tangent to the confining sphere, 20 of those located at the vertices of a distorted regular dodecahedron and the other 12 at the vertices of a dual distorted regular icosahedron. The remaining four balls are trapped inside the sphere, their centers at the vertices of an approximately regular tetrahedron.

Of course, unlike a real TDI, all atomic radii (ball sizes) in this packing are identical, and atoms (balls) in the outer shell are equidistant from the center of the cluster. Teich et al. found this densest packing configuration via Monte Carlo simulation: they sampled configurations of hard spheres inside a spherical container in the isobaric ensemble while increasing imposed pressure to a putatively high value, thereby compressing the container [20]. They found that in the densest packing formed this way, the inner tetrahedron was not completely sterically immobilized by the outer shell, but instead had some "wiggle room" and vibrated slightly even as the cluster reached its densest possible configuration.

This connection may give credence to the idea that the TDI cluster would have a special stability in the melt. Confinement to a sphere may be argued to be a minimal physically reasonable model of the effects of the surrounding liquid on a given set of atoms, since the surrounding liquid will manifest in an isotropic pressure, on average, about that set of atoms.

Alas, as we noted above, the densest known packing of 36 identical balls within a spherical container is not quite a TDI, because 12 atoms of the TDI are bigger than the others. Moreover, attempts to mimic reality more closely by producing dense clusters of spheres of two appropriate radii via the simulation technique described in [20] did not result in a TDI-like cluster structure in 50 attempts. This indicates that the explanation of the TDI cluster as a purely volume-minimizing one under very high isotropic pressure in the melt is incomplete. The effect of the surrounding melt needs to be considered more carefully, and the atoms modeled more accurately, perhaps including energetic interactions between them rather than sole volume exclusion.

A key additional factor in making the TDI cluster likely to have some fleeting existence in the liquid is that its free energy is reduced by the additional rotational entropy of the inner tetrahedron. It has now been experimentally and computationally verified [22] that the potential energy barriers to rotation of a tetrahedron inside the dodecahedron-icosahedron combined shells are relatively low, and that the entropy of the extra degrees of freedom when rotation occurs win out at higher temperatures, even when the material remains a solid crystal (Specifically, Euchner et al. found evidence that significant entropy is produced by the rotation of the tetrahedron in the 6:1 periodic approximant to a Tsai-type quasicrystal in the Zinc-Scandium system. They examined closely the structure of single crystals of periodic $Zn_6Sc$ (a structure identical to that of $Cd_6Yb$), and how it changed with temperature. This periodic approximant has one Tsai cluster per site in a bcc lattice, as described above for the Yb-Cd system. It was known (their refs [9,10,19]) that there is a critical temperature above which the central tetrahedron appears as disordered. Below that temperature, the crystal transforms to a low-temperature phase where the tetrahedra are ordered). At the even higher temperatures of a melt, the entropy contribution will be even greater, and inner tetrahedron rotation may well be sufficient to stabilize a TDI in the melt.

## 7. Conclusions and Questions

To summarize, we have proposed a growth mechanism for (at least) one type of complex crystal whose structure has been described by overlapping nested atomic clusters. The fundamental units in this structure might not be the larger (Tsai) clusters by which they are usually described. Instead, growth is an arrangement of identical and identically oriented smaller clusters (TDIs), with additional atoms filling in the interstices. The ratio of the two types of atoms in the melt determines the interstitial geometry and hence whether the crystal structure as a whole is one of several periodic possibilities, or is instead an icosahedral quasicrystal.

Further studies are of course needed to elucidate the thermodynamics of this conjectured growth mechanism, as our model thus far primarily rests on geometrical arguments. Beyond that, can our hypothesis shed any light on how other iQCs and their approximants grow? Many questions remain open, and our work is just beginning.

**Acknowledgments:** We would like to thank the American Institute of Mathematics for its support, initially through a week-long workshop and subsequently through its SQuaRes program.

**Author Contributions:** Jean E. Taylor, Erin G. Teich, Pablo F. Damasceno, Yoav Kallus and Marjorie Senechal wrote this paper together.

**Conflicts of Interest:** The authors declare no conflict of interest.

## References

1. Radin, C. The Open Mathematics of Crystallization. *Not. AMS* **2017**, *64*, 551–556. [CrossRef]
2. Shechtman, D.; Blech, I.; Gratias, D.; Cahn, J.W. Metallic Phase with Long-Range Orientational Order and No Translational Symmetry. *Phys. Rev. Lett.* **1984**, *53*. [CrossRef]
3. Senechal, M.; Taylor, J. Quasicrystals: The view from Stockholm. *Math. Intell.* **2013**, *35*, 1–9. [CrossRef]
4. Kepler, J. *Strena, seu de Nive Sexangula*; Godefridum Tampach: Frankfurt am Main, Germany, 1611.
5. Hooke, R. *Micrographia: Or Some Physiological Descriptions of Minute Bodies Made by Magnifying Glasses, with Observations and Inquiries Thereupon*; The Royal Society: London, UK, 1665.
6. Häuy, R.J. *Essai D'une Théorie sur la Structure des Crystaux, Appliqée a Plusiers Genres de Substances Crystallisées, Gogué & Née de la Rochelle*; Libraries, MDCCLXXIV: Paris, France, 1784.
7. Nobel Foundation Website. Available online: https://www.nobelprize.org/nobel_prizes/physics/laureates/1914/laue-facts.html (accessed on 3 September 2017).
8. Bravais, A. Mémoire sur les systèmes formés par les points distribués sur un plan ou dans l'espace. *J. l'École Polytech. Cah.* **1850**, *33*, 1–128.
9. Senechal, M. *Quasicrystals and Geometry*; Cambridge University Press: Cambridge, UK, 1996.
10. Baake, M.; Grimm, U. *Aperiodic Order*; Cambridge University Press: Cambridge, UK, 2013.
11. Lagarias, J. Geometric Models for Quasicrystals I. Delone Sets of Finite Type. *Discret. Comput. Geom.* **1999**, *21*, 161–191. [CrossRef]
12. Hong, S.-T. Three-Dimensional Modeling of Quasicrystal Structures from X-ray Diffraction: An Icosahedral Al-Cu-Fe Alloy. *Inorg. Chem.* **2017**, *56*, 7354–7359. [CrossRef] [PubMed]
13. Takakura, H.; Pay Gómez, C.; Yamamoto, A.; De Boissieu, M.; Tsai, A.P. Atomic structure of the binary icosahedral Yb–Cd quasicrystal. *Nat. Mater.* **2007**, *6*, 58–63. [CrossRef] [PubMed]
14. Frank, F.C. Supercooling of Liquids. *Proc. R. Soc. Lond. Ser. A* **1952**, *215*, 43–46. [CrossRef]
15. Bergman, G.; Waugh, J.L.T.; Pauling, L. Crystal structure of the intermetallic compound Mg32 (Al, Zn)49 and related phases. *Nature* **1952**, *169*, 1057–1058. [CrossRef]
16. Mackay, A. A dense non-crystallographic packing of equal spheres. *Acta Crystallogr.* **1962**, *15*, 916–918. [CrossRef]
17. Blatov, V.A.; Shevchenko, A.P.; Proserpio, D.M. Applied topological analysis of crystal structures with the program packageTopos. *Proc. Cryst. Growth Des.* **2014**, *14*, 3576–3586. [CrossRef]

18. Henley, C. Progress on the atomic structure of quasicrystals in quasicrystals. In Proceedings of the 12th Taniguchi Symposium, Shima Mie Prefecture, Japan, 14–19 November 1989; Fujiwara, T., Ogawa, T., Eds.; Springer: Berlin/Heidelberg, Germany, 1990; pp. 38–47. Available online: http://www.lassp.cornell.edu/clh/PUBS/tani90.pdf (accessed on 5 September 2017).

19. Taylor, J.; Teich, E.; Damasceno, P.; Kallus, Y.; Senechal, M. The role of geometric frustration in the self-assembly of complex crystals. **2017**, Unpublished Work.

20. Teich, E.G.; van Anders, G.; Klotsa, D.; Dshemuchadse, J.; Glotzer, S.C. Clusters of polyhedra in spherical confinement. *Proc. Nat. Acad. Sci. USA* **2016**, *113*, E669. [CrossRef] [PubMed]

21. Pfoertner, H. Densest Packing of Spheres in a Sphere. 2013. Available online: www.randomwalk.de/sphere/insphr/spheresinsphr.html (accessed on 27 March 2015).

22. Euchner, H.; Yamada, T.; Schober, H.; Rols, S.; Mihalkovič, M.; Tamura, R.; Ishimasa, T.; de Boissieu, M. Ordering and dynamics of the central tetrahedron in the 1/1 Zn6Sc periodic approximant to quasicrystal. *J. Phys. Condens. Matter* **2012**, *24*, 415403. [CrossRef] [PubMed]

# Section B:
# Abstract Polyhedra, Maps on Surfaces, and Graphs

*symmetry*

MDPI

*Article*

# Self-Dual, Self-Petrie Covers of Regular Polyhedra

Gabe Cunningham

Northeastern University, 360 Huntington Ave, Boston, MA 02115, USA; E-Mail: gabriel.cunningham@gmail.com; Tel.: +1 617-406-9067

Academic Editor: name

Received: 17 January 2012; in revised form: 21 February 2012 / Accepted: 23 February 2012 / Published: 27 February 2012

**Abstract:** The well-known duality and Petrie duality operations on maps have natural analogs for abstract polyhedra. Regular polyhedra that are invariant under both operations have a high degree of both "external" and "internal" symmetry. The mixing operation provides a natural way to build the minimal common cover of two polyhedra, and by mixing a regular polyhedron with its five other images under the duality operations, we are able to construct the minimal self-dual, self-Petrie cover of a regular polyhedron. Determining the full structure of these covers is challenging and generally requires that we use some of the standard algorithms in combinatorial group theory. However, we are able to develop criteria that sometimes yield the full structure without explicit calculations. Using these criteria and other interesting methods, we then calculate the size of the self-dual, self-Petrie covers of several polyhedra, including the regular convex polyhedra.

**Keywords:** abstract polyhedron; convex polyhedron; duality; map operations; mixing; Petrie polygon; Petrie dual

---

## 1. Introduction

Abstract polyhedra are partially-ordered sets that generalize the face-lattices of convex polyhedra. They are closely related to maps on surfaces (*i.e.*, 2-cell decompositions of surfaces), and indeed, every abstract polyhedron has a natural realization as such a map. The regular polyhedra are the most extensively studied. These are the polyhedra such that the automorphism group acts transitively on the flags (which consist of a vertex, an edge, and a face that are all mutually incident). There is a natural way to build a regular polyhedron from a finitely generated group satisfying certain properties, yielding a bijection between (isomorphism classes of) regular polyhedra and such groups. The ability to view a regular polyhedron as either a map on a surface or as a group has fostered the development of a rich theory.

The well-known duality and Petrie duality operations on maps can be easily applied to polyhedra, though in some rare cases, the Petrie dual of a polyhedron is not a polyhedron. A regular polyhedron that is invariant under both operations has a high degree of "external" symmetry in addition to the "internal" symmetry that regularity measures. In [1], the authors describe a way to build a self-dual, self-Petrie cover of a regular map. Using a similar construction, we show how to build a self-dual, self-Petrie cover of a regular polyhedron. Determining the local structure of the cover is simple. However, finding a presentation for the automorphism group of the cover is generally infeasible. Therefore, our goal is to develop criteria that give us information about the cover without requiring direct calculation.

We start by giving some background on abstract polyhedra, regularity, and the duality operations in Section 2. In Section 3, we present the mixing operation for regular polyhedra and show how to use it to construct self-dual, self-Petrie covers. We then show how to calculate the size of these covers in certain cases. Finally, in Section 4 we calculate the size of the self-dual, self-Petrie covers of several polyhedra, including the regular convex polyhedra.

*Symmetry* **2012**, *4*, 208–218

## 2. Abstract Polyhedra

Let $\mathcal{P}$ be a ranked partially ordered set of *vertices*, *edges*, and *faces*, which have rank 0, 1, and 2, respectively. If $F \leq G$ or $G \leq F$, we say that $F$ and $G$ are *incident*. A *flag* of $\mathcal{P}$ is a maximal chain, and two flags are *adjacent* if they differ in exactly one element. We say that $\mathcal{P}$ is an *(abstract) polyhedron* if it satisfies the following properties:

1. Each flag of $\mathcal{P}$ consists of a vertex, an edge, and a face.
2. Each edge is incident on exactly two vertices and two faces.
3. If $F$ is a vertex and $G$ is a face such that $F \leq G$, then there are exactly two edges that are incident to both $F$ and $G$.
4. $\mathcal{P}$ is *strongly flag-connected*, meaning that if $\Phi$ and $\Psi$ are two flags of $\mathcal{P}$, then there is a sequence of flags $\Phi = \Phi_0, \Phi_1, \ldots, \Phi_k = \Psi$ such that for $i = 0, \ldots, k-1$, the flags $\Phi_i$ and $\Phi_{i+1}$ are adjacent, and each $\Phi_i$ contains $\Phi \cap \Psi$.

As a consequence of the second and third properties above, every flag $\Phi$ has a unique flag $\Phi^i$ that differs from $\Phi$ only in its element of rank $i$. We say that $\Phi^i$ is *$i$-adjacent* to $\Phi$.

For polyhedra $\mathcal{P}$ and $\mathcal{Q}$, an *isomorphism* from $\mathcal{P}$ to $\mathcal{Q}$ is an incidence- and rank-preserving bijection, and an isomorphism from $\mathcal{P}$ to itself is an *automorphism*. We denote the group of all automorphisms of $\mathcal{P}$ by $\Gamma(\mathcal{P})$. There is a natural action of $\Gamma(\mathcal{P})$ on the flags of $\mathcal{P}$, and we say that $\mathcal{P}$ is *regular* if this action is transitive. The faces of a regular polyhedron all have the same number of sides, and the vertices all have the same valency. In general, we say that a polyhedron is of *type* $\{p, q\}$ if every face is a $p$-gon and every vertex is $q$-valent.

Given a regular polyhedron $\mathcal{P}$, fix a *base flag* $\Phi$. Then the automorphism group $\Gamma(\mathcal{P})$ is generated by involutions $\rho_0$, $\rho_1$, and $\rho_2$, where $\rho_i$ maps the base flag $\Phi$ to $\Phi^i$. This completely determines the action of each $\rho_i$ on all flags, since for any automorphism $\varphi$ and flag $\Psi$,

$$(\Psi^i)\varphi = (\Psi\varphi)^i$$

If $\mathcal{P}$ is of type $\{p, q\}$, then these generators satisfy (at least) the relations

$$(\rho_0\rho_1)^p = (\rho_0\rho_2)^2 = (\rho_1\rho_2)^q = \epsilon$$

If these are the only defining relations, then we denote $\mathcal{P}$ by $\{p, q\}$, and $\Gamma(\mathcal{P})$ is a Coxeter group, denoted $[p, q]$. We note that whenever $\mathcal{P}$ is the face-lattice of a regular convex polyhedron, then its denotation is the same as the usual Schläfli symbol for that polyhedron (see [2]). For $I \subseteq \{0, 1, 2\}$ and a group $\Gamma = \langle \rho_0, \rho_1, \rho_2 \rangle$, we define $\Gamma_I := \langle \rho_i \mid i \in I \rangle$. The strong flag-connectivity of polyhedra induces the following *intersection property* in the group:

$$\Gamma_I \cap \Gamma_J = \Gamma_{I \cap J} \text{ for } I, J \subseteq \{0, 1, 2\} \tag{1}$$

In general, if $\Gamma = \langle \rho_0, \rho_1, \rho_2 \rangle$ is a group such that each $\rho_i$ has order 2 and such that $(\rho_0\rho_2)^2 = \epsilon$, then we say that $\Gamma$ is a *string group generated by involutions on 3 generators* (or *sggi*). If $\Gamma$ also satisfies the intersection property given above, then we call $\Gamma$ a *string C-group on 3 generators*. There is a natural way of building a regular polyhedron $\mathcal{P}(\Gamma)$ from a string C-group $\Gamma$ such that $\Gamma(\mathcal{P}(\Gamma)) = \Gamma$ [3] (Theorem 2E11). Therefore, we get a one-to-one correspondence between regular polyhedra and string C-groups on 3 specified generators.

Let $\mathcal{P}$ and $\mathcal{Q}$ be two polyhedra (not necessarily regular). A function $\gamma : \mathcal{P} \to \mathcal{Q}$ is called a *covering* if it preserves adjacency of flags, incidence, and rank; then $\gamma$ is necessarily surjective, by the flag-connectedness of $\mathcal{Q}$. We say that $\mathcal{P}$ *covers* $\mathcal{Q}$ if there exists a covering $\gamma : \mathcal{P} \to \mathcal{Q}$. If $\mathcal{P}$ and $\mathcal{Q}$ are both regular polyhedra, then their automorphism groups are both quotients of

$$W := [\infty, \infty] = \langle \rho_0, \rho_1, \rho_2 \mid \rho_0^2 = \rho_1^2 = \rho_2^2 = (\rho_0\rho_2)^2 = \epsilon \rangle$$

Therefore there are normal subgroups $M$ and $K$ of $W$ such that $\Gamma(\mathcal{P}) = W/M$ and $\Gamma(\mathcal{Q}) = W/K$. Then $\mathcal{P}$ covers $\mathcal{Q}$ if and only if $M \leq K$.

## 2.1. Duality Operations

There are two well-known duality operations on maps on surfaces, described in [1] and earlier articles. These naturally give rise to corresponding operations on polyhedra. The first is known simply as *duality*, and the *dual of* $\mathcal{P}$ (denoted $\mathcal{P}^\delta$) is obtained from $\mathcal{P}$ by reversing the partial order. If a polyhedron is isomorphic to its dual, then it is called *self-dual*.

In order to formulate the second duality operation, we need to define the *Petrie polygons* of a polyhedron. Consider a walk along edges of the polyhedron such that at each successive step, we alternate between taking the first exit on the left and the first exit on the right. When we start with a finite polyhedron, such a walk will eventually take us to a vertex we have already visited, leaving in the same direction as we have before. This closed walk is one of the Petrie polygons of the polyhedron.

We can now describe the second duality operation. Given a polyhedron $\mathcal{P}$, its *Petrie dual* $\mathcal{P}^\pi$ consists of the same vertices and edges as $\mathcal{P}$, but its faces are the Petrie polygons of $\mathcal{P}$. Taking the Petrie dual of a polyhedron also forces the old faces to be the new Petrie polygons, so that $\mathcal{P}^{\pi\pi} \simeq \mathcal{P}$. If $\mathcal{P}$ is isomorphic to $\mathcal{P}^\pi$, then we say that $\mathcal{P}$ is *self-Petrie*.

Since Petrie polygons play a central role in this paper, we expand some of our earlier terminology. If $\mathcal{P}$ is a regular polyhedron, then its Petrie polygons all have the same length, and that length is the order of $\rho_0\rho_1\rho_2$ in $\Gamma(\mathcal{P})$. A regular polyhedron of type $\{p,q\}$ and with Petrie polygons of length $r$ is also said to be of type $\{p,q\}_r$. If $\mathcal{P}$ is of type $\{p,q\}_r$ and it covers every other polyhedron of type $\{p,q\}_r$, then we call it *the universal polyhedron of type* $\{p,q\}_r$ and we denote it by $\{p,q\}_r$. The automorphism group of $\{p,q\}_r$ is denoted by $[p,q]_r$, and this group is the quotient of $[p,q]$ by the single extra relation $(\rho_0\rho_1\rho_2)^r = \epsilon$. We will also extend our notation and use $[p,q]_r$ for the group with presentation

$$\langle \rho_0, \rho_1, \rho_2 \mid \rho_0^2 = \rho_1^2 = \rho_2^2 = (\rho_0\rho_1)^p = (\rho_0\rho_2)^2 = (\rho_1\rho_2)^q = (\rho_0\rho_1\rho_2)^r = \epsilon \rangle$$

even when there is no universal polyhedron of type $\{p,q\}_r$.

The operations $\delta$ and $\pi$ form a group of order 6, isomorphic to $S_3$. A regular polyhedron that is self-dual and self-Petrie is invariant under all 6 operations; such a polyhedron must be of type $\{n,n\}_n$ for some $n \geq 2$. In general, if $\mathcal{P}$ is of type $\{p,q\}_r$, then $\mathcal{P}^\delta$ is of type $\{q,p\}_r$, and $\mathcal{P}^\pi$ is of type $\{r,q\}_p$. Furthermore, the dual and the Petrie dual of a universal polyhedron is again universal.

## 3. Mixing Polyhedra

The mixing construction on polyhedra is analogous to the join of two maps or hypermaps [4]. Using it, we can find the minimal common cover of two or more polyhedra, which will enable us to describe the self-dual, self-Petrie covers of regular polyhedra.

We begin by describing the mixing operation on groups (also called the parallel product in [5]). Let $\mathcal{P}$ and $\mathcal{Q}$ be regular polyhedra with $\Gamma(\mathcal{P}) = \langle \rho_0, \rho_1, \rho_2 \rangle$ and $\Gamma(\mathcal{Q}) = \langle \rho_0', \rho_1', \rho_2' \rangle$. Let $\alpha_i = (\rho_i, \rho_i') \in \Gamma(\mathcal{P}) \times \Gamma(\mathcal{Q})$ for $i \in \{0,1,2\}$. Then we define *the mix of* $\Gamma(\mathcal{P})$ *and* $\Gamma(\mathcal{Q})$ to be the group

$$\Gamma(\mathcal{P}) \diamond \Gamma(\mathcal{Q}) := \langle \alpha_0, \alpha_1, \alpha_2 \rangle$$

Note that the order of any word $\alpha_{i_1} \ldots \alpha_{i_t}$ is the least common multiple of the orders of $\rho_{i_1} \ldots \rho_{i_t}$ and $\rho_{i_1}' \ldots \rho_{i_t}'$. In particular, each $\alpha_i$ is an involution, and $(\alpha_0\alpha_2)^2 = \epsilon$. Therefore, $\Gamma(\mathcal{P}) \diamond \Gamma(\mathcal{Q})$ is a string group generated by involutions. As we shall see, it also satisfies the intersection property (Equation (1)). Recall that $\Gamma_I := \langle \rho_i \mid i \in I \rangle$.

**Proposition 1.** *Let* $\mathcal{P}$ *and* $\mathcal{Q}$ *be regular polyhedra, and let* $I, J \subseteq \{0,1,2\}$. *Let* $\Lambda = \Gamma(\mathcal{P})$, $\Delta = \Gamma(\mathcal{Q})$, *and* $\Gamma = \Lambda \diamond \Delta$. *Then* $\Gamma_I \cap \Gamma_J \subseteq \Lambda_{I \cap J} \times \Delta_{I \cap J}$.

**Proof.** Let $g \in \Gamma_I \cap \Gamma_J$, and write $g = (g_1, g_2)$. Then $g_1 \in \Lambda_I \cap \Lambda_J$ and $g_2 \in \Delta_I \cap \Delta_J$. Now, since $\mathcal{P}$ and $\mathcal{Q}$ are polyhedra, we have that $\Lambda_I \cap \Lambda_J = \Lambda_{I \cap J}$ and $\Delta_I \cap \Delta_J = \Delta_{I \cap J}$. Therefore, $g \in \Lambda_{I \cap J} \times \Delta_{I \cap J}$. $\square$

**Corollary 1.** *Let $\mathcal{P}$ and $\mathcal{Q}$ be regular polyhedra. Let $\Lambda = \Gamma(\mathcal{P})$, $\Delta = \Gamma(\mathcal{Q})$, and $\Gamma = \Lambda \diamond \Delta$. Then $\Gamma$ satisfies the intersection property (Equation (1)).*

**Proof.** Let $\Lambda = \langle \rho_0, \rho_1, \rho_2 \rangle$ and let $\Delta = \langle \rho'_0, \rho'_1, \rho'_2 \rangle$. Let $\Gamma = \langle \alpha_0, \alpha_1, \alpha_2 \rangle$. We need to show that for subsets $I$ and $J$ of $N = \{0, 1, 2\}$, $\Gamma_I \cap \Gamma_J \leq \Gamma_{I \cap J}$. If $I \cap J = \emptyset$, the claim follows immediately from Proposition 1. If $I \subseteq J$, the claim also clearly holds. The only remaining case is when $I = N \setminus \{i\}$ and $J = N \setminus \{j\}$ for $i \neq j$. We will prove the case $I = \{0, 1\}$ and $J = \{1, 2\}$; the other cases are similar. Now, from Proposition 1, we know that $\Gamma_I \cap \Gamma_J \subseteq \langle \rho_1 \rangle \times \langle \rho'_1 \rangle$. We want to show that $(\rho_1, \epsilon)$ and $(\epsilon, \rho'_1)$ are not in $\Gamma_I \cap \Gamma_J$. We have that $\Lambda_I = \langle \rho_0, \rho_1 \rangle$, which is a dihedral group. In particular, all relations of $\Lambda_I$ have even length. The same is true of $\Delta_I$. Therefore, when we reduce a word $(\rho_{i_1} \cdots \rho_{i_k}, \rho'_{i_1} \cdots \rho'_{i_k})$ in $\Gamma_I$, the length of each component must have the same parity. In particular, we cannot have $(\rho_1, \epsilon) \in \Gamma_I$ or $(\epsilon, \rho'_1) \in \Gamma_I$. Therefore, $\Gamma_I \cap \Gamma_J \leq \langle \alpha_1 \rangle$, which is what we wanted to show. $\square$

Since the group $\Gamma(\mathcal{P}) \diamond \Gamma(\mathcal{Q})$ satisfies the intersection property, we can build a regular polyhedron from the group. We call this polyhedron *the mix of $\mathcal{P}$ and $\mathcal{Q}$*, and we denote it $\mathcal{P} \diamond \mathcal{Q}$. By construction, $\Gamma(\mathcal{P} \diamond \mathcal{Q}) = \Gamma(\mathcal{P}) \diamond \Gamma(\mathcal{Q})$.

Note that whether we mix $\mathcal{P}$ and $\mathcal{Q}$ as regular polyhedra, or take their join as maps, we get the same structure. Therefore, Corollary 1 tells us that when we take the join of two maps that correspond to polyhedra, we get another map that corresponds to a polyhedron.

There is another way to describe the mix of $\Gamma(\mathcal{P})$ and $\Gamma(\mathcal{Q})$ using quotients of the group $W$, which was described in Section 2. Let $\mathcal{P}$ and $\mathcal{Q}$ be regular polyhedra with $\Gamma(\mathcal{P}) = W/M$ and $\Gamma(\mathcal{Q}) = W/K$. Then the homomorphism from $W$ to $W/M \times W/K$, sending a word $w$ to the pair of cosets $(wM, wK)$, has kernel $M \cap K$ and image $W/M \diamond W/K$. Thus we see that $\Gamma(\mathcal{P}) \diamond \Gamma(\mathcal{Q}) \simeq W/(M \cap K)$. Therefore, $\mathcal{P} \diamond \mathcal{Q}$ is the minimal regular polyhedron that covers both $\mathcal{P}$ and $\mathcal{Q}$.

Now we can describe how to construct the self-dual, self-Petrie cover of a polyhedron. Let $G = \langle \delta, \pi \rangle$, the group of polyhedron operations generated by duality and Petrie duality, and let $\mathcal{P}$ be a regular polyhedron with $\Gamma(\mathcal{P}) = W/M$. For any $\varphi \in G$, define $\mathcal{P}^\varphi$ to be the regular poset (usually a polyhedron) built from the group $W/\varphi(M)$. Now consider

$$\mathcal{P}^* := \mathcal{P} \diamond \mathcal{P}^\delta \diamond \mathcal{P}^\pi \diamond \mathcal{P}^{\delta\pi} \diamond \mathcal{P}^{\pi\delta} \diamond \mathcal{P}^{\delta\pi\delta}$$

Then $\Gamma(\mathcal{P}^*)$ is the quotient of $W$ by

$$M \cap \delta(M) \cap \pi(M) \cap \delta\pi(M) \cap \pi\delta(M) \cap \delta\pi\delta(M)$$

and since this subgroup is fixed by both $\delta$ and $\pi$, it follows that $\mathcal{P}^*$ is self-dual and self-Petrie. Now, if $\mathcal{P}^\pi$ and $\mathcal{P}^{\delta\pi}$ are both polyhedra, then Corollary 1 tells us that $\mathcal{P}^*$ is also a polyhedron. In the rare cases where $\mathcal{P}^\pi$ is not a polyhedron, we can use the "quotient criterion" [3, Theorem 2E17] to show that $\mathcal{P} \diamond \mathcal{P}^\pi$ is nevertheless a polyhedron, since the natural epimorphism from $\Gamma(\mathcal{P}) \diamond \Gamma(\mathcal{P}^\pi) = \langle \alpha_0, \alpha_1, \alpha_2 \rangle$ to $\Gamma(\mathcal{P})$ is one-to-one on the subgroup $\langle \alpha_1, \alpha_2 \rangle$. In any case, $\mathcal{P}^*$ is the minimal regular, self-dual, self-Petrie polyhedron that covers $\mathcal{P}$.

If $\mathcal{P}$ is a regular polyhedron of type $\{p, q\}_r$, then $\mathcal{P}^*$ is of type $\{n, n\}_n$, where $n$ is the least common multiple of $p$, $q$, and $r$. This gives us a full picture of the local structure of $\mathcal{P}^*$. Determining the global structure, such as the size, isomorphism type, and presentation of $\Gamma(\mathcal{P}^*)$, is much more challenging. With the tools presented so far, the only way we can determine $\Gamma(\mathcal{P}^*)$ is by looking at the intersection of six subgroups of $W$, or finding the diagonal subgroup of the direct product of six groups. Neither alternative is feasible in general.

There is another construction, dual to mixing, that helps us calculate the size of $\Gamma(\mathcal{P}) \diamond \Gamma(\mathcal{Q})$. If $\Gamma(\mathcal{P})$ has presentation $\langle \rho_0, \rho_1, \rho_2 \mid R \rangle$ and $\Gamma(\mathcal{Q})$ has presentation $\langle \rho_0', \rho_1', \rho_2' \mid S \rangle$, then we define *the comix of* $\Gamma(\mathcal{P})$ *and* $\Gamma(\mathcal{Q})$, denoted $\Gamma(\mathcal{P}) \,\square\, \Gamma(\mathcal{Q})$, to be the group with presentation

$$\langle \rho_0, \rho_0', \rho_1, \rho_1', \rho_2, \rho_2' \mid R, S, \rho_0^{-1}\rho_0', \rho_1^{-1}\rho_1', \rho_2^{-1}\rho_2' \rangle$$

Informally speaking, we can just add the relations from $\Gamma(\mathcal{Q})$ to those of $\Gamma(\mathcal{P})$, rewriting them to use $\rho_i$ in place of $\rho_i'$. As a result, the order of any word $\rho_{i_1} \ldots \rho_{i_t}$ in $\Gamma(\mathcal{P}) \,\square\, \Gamma(\mathcal{Q})$ divides the order of the corresponding word in $\Gamma(\mathcal{P})$ and in $\Gamma(\mathcal{Q})$.

Like the mix of two groups, the comix has a natural interpretation in terms of quotients of $W$. In particular, if $\Gamma(\mathcal{P}) = W/M$ and $\Gamma(\mathcal{Q}) = W/K$, then $\Gamma(\mathcal{P}) \,\square\, \Gamma(\mathcal{Q}) = W/MK$. That is, $\Gamma(\mathcal{P}) \,\square\, \Gamma(\mathcal{Q})$ is the maximal common quotient of $\Gamma(\mathcal{P})$ and $\Gamma(\mathcal{Q})$ (with respect to the natural covering maps).

**Proposition 2.** *Let $\mathcal{P}$ and $\mathcal{Q}$ be finite regular polyhedra. Then*

$$|\Gamma(\mathcal{P}) \diamond \Gamma(\mathcal{Q})| \cdot |\Gamma(\mathcal{P}) \,\square\, \Gamma(\mathcal{Q})| = |\Gamma(\mathcal{P})| \cdot |\Gamma(\mathcal{Q})|$$

*Furthermore, if $\Gamma(\mathcal{P}) \,\square\, \Gamma(\mathcal{Q})$ is trivial, then $\Gamma(\mathcal{P}) \diamond \Gamma(\mathcal{Q}) = \Gamma(\mathcal{P}) \times \Gamma(\mathcal{Q})$.*

**Proof.** Let $\Gamma(\mathcal{P}) = W/M$ and $\Gamma(\mathcal{Q}) = W/K$. Then $\Gamma(\mathcal{P}) \diamond \Gamma(\mathcal{Q}) = W/(M \cap K)$ and $\Gamma(\mathcal{P}) \,\square\, \Gamma(\mathcal{Q}) = W/MK$. Let $f : \Gamma(\mathcal{P}) \diamond \Gamma(\mathcal{Q}) \to \Gamma(\mathcal{P})$ and $g : \Gamma(\mathcal{Q}) \to \Gamma(\mathcal{P}) \,\square\, \Gamma(\mathcal{Q})$ be the natural epimorphisms. Then $\ker f \simeq M/(M \cap K)$ and $\ker g \simeq MK/K \simeq M/(M \cap K)$. Therefore, we have that

$$
\begin{aligned}
|\Gamma(\mathcal{P}) \diamond \Gamma(\mathcal{Q})| &= |\Gamma(\mathcal{P})||\ker f| \\
&= |\Gamma(\mathcal{P})||\ker g| \\
&= |\Gamma(\mathcal{P})||\Gamma(\mathcal{Q})|/|\Gamma(\mathcal{P}) \,\square\, \Gamma(\mathcal{Q})|
\end{aligned}
$$

and the result follows.  □

Using Proposition 2, it is often possible to determine the size of $\Gamma(\mathcal{P}) \diamond \Gamma(\mathcal{Q})$ by hand or with the help of a computer algebra system. However, since finding the comix of two groups usually requires that we know their presentations, the result is somewhat less useful for determining the size of $\Gamma(\mathcal{P}^*)$, which is the mix of six groups. In a few nice cases, though, we can determine the size or structure of $\Gamma(\mathcal{P}^*)$ without any difficult calculations. We present a few such results here.

**Theorem 1.** *Let $\mathcal{P}$ be a self-Petrie polyhedron of type $\{p, q\}_p$. Suppose $p$ is odd and that $p$ and $q$ are coprime. Then*

$$\Gamma(\mathcal{P}^*) = \Gamma(\mathcal{P}) \times \Gamma(\mathcal{P}^\delta) \times \Gamma(\mathcal{P}^{\delta\pi})$$

**Proof.** First, we note that $\mathcal{P}^\delta$ is of type $\{q, p\}_p$. In $\Gamma(\mathcal{P}) \,\square\, \Gamma(\mathcal{P}^\delta)$, the order of $\rho_0\rho_1$ divides $p$ and $q$, and since $p$ and $q$ are coprime, we get $\rho_0\rho_1 = \epsilon$; that is, $\rho_0 = \rho_1$. Similarly, $\rho_1\rho_2 = \epsilon$, and so $\rho_1 = \rho_2$. Now, the order of $\rho_0\rho_1\rho_2$ divides $p$. On the other hand, $\rho_0\rho_1\rho_2 = \rho_0^3 = \rho_0$, and so the order divides 2 as well. Therefore, we must have that $\rho_0\rho_1\rho_2 = \epsilon$. This forces all of the generators to be trivial, and thus $\Gamma(\mathcal{P}) \,\square\, \Gamma(\mathcal{P}^\delta)$ is trivial and $\Gamma(\mathcal{P}) \diamond \Gamma(\mathcal{P}^\delta) = \Gamma(\mathcal{P}) \times \Gamma(\mathcal{P}^\delta)$.

Now, $\mathcal{P} \diamond \mathcal{P}^\delta$ is of type $\{pq, pq\}_p$, and $\mathcal{P}^{\delta\pi}$ is of type $\{p, p\}_q$. Then in their comix, we get that $\rho_0\rho_1\rho_2$ has order dividing $p$ and $q$, and thus it is trivial. Thus $\rho_0 = \rho_1\rho_2$, and so $(\rho_1\rho_2)^2 = \epsilon$. On the other hand, we also have that $(\rho_1\rho_2)^p = \epsilon$ in the comix, and since $p$ is odd, this means that $\rho_1\rho_2$ is trivial. So $\rho_0$ is trivial, and $\rho_1 = \rho_2$. Similarly, $\rho_2 = \rho_0\rho_1$, and thus $(\rho_0\rho_1)^2 = \epsilon$. But again, we also have that $(\rho_0\rho_1)^p = \epsilon$, and thus $\rho_0\rho_1 = \epsilon$. So $\rho_0 = \rho_1 = \rho_2 = \epsilon$ and we see that the comix is trivial. Therefore, the mix is the direct product $\Gamma(\mathcal{P}) \times \Gamma(\mathcal{P}^\delta) \times \Gamma(\mathcal{P}^{\delta\pi})$.

Finally, we note that since $\mathcal{P}$ is self-Petrie, we have $\mathcal{P}^\pi = \mathcal{P}$, $\mathcal{P}^{\pi\delta} = \mathcal{P}^\delta$, and $\mathcal{P}^{\pi\delta\pi} = \mathcal{P}^{\delta\pi}$. Therefore, the self-dual, self-Petrie cover of $\mathcal{P}$ consists of just the three distinct polyhedra we have mixed. $\square$

**Corollary 2.** *Let $\mathcal{P}$ be a regular polyhedron of type $\{p, q\}_r$. Suppose $p$ and $r$ are odd and both coprime to $q$. Let $\mathcal{Q} = \mathcal{P} \diamond \mathcal{P}^\pi$. Then*

$$\Gamma(\mathcal{P}^*) = \Gamma(\mathcal{Q}) \times \Gamma(\mathcal{Q}^\delta) \times \Gamma(\mathcal{Q}^{\delta\pi})$$

**Proof.** We start by noting that any self-dual, self-Petrie polyhedron that covers $\mathcal{P}$ must also cover $\mathcal{Q}$; therefore, $\mathcal{P}^* = \mathcal{Q}^*$. Now, the polyhedron $\mathcal{Q}$ is a self-Petrie polyhedron of type $\{\ell, q\}_\ell$, where $\ell$ is the least common multiple of $p$ and $r$. Since $p$ and $r$ are both odd and coprime to $q$, $\ell$ is also odd and coprime to $q$. Then we can apply Theorem 1 and the result follows. $\square$

The condition in Theorem 1 that $p$ is odd is essential. When $p$ is even, we cannot tell from the type alone whether $\Gamma(\mathcal{P}) \square \Gamma(\mathcal{P}^\delta)$ has order 1 or 2. However, if $\mathcal{P}$ is the universal polyhedron of type $\{p, q\}_p$, then we can still determine $|\Gamma(\mathcal{P}^*)|$.

**Theorem 2.** *Let $\mathcal{P} = \{p, q\}_p$, the universal polyhedron of type $\{p, q\}_p$. Suppose $p$ is even and that $p$ and $q$ are coprime. Then*

$$|\Gamma(\mathcal{P}^*)| = |\Gamma(\mathcal{P})|^3 / 8$$

**Proof.** Since $p$ and $q$ are coprime, a presentation for $\Gamma(\mathcal{P}) \square \Gamma(\mathcal{P}^\delta)$ is given by

$$\langle \rho_0, \rho_1, \rho_2 \mid \rho_0^2 = \rho_1^2 = \rho_2^2 = \rho_0\rho_1 = (\rho_0\rho_2)^2 = \rho_1\rho_2 = (\rho_0\rho_1\rho_2)^p = \epsilon \rangle$$

and direct calculation shows that this is a group of order 2. Then by Proposition 2, $|\Gamma(\mathcal{P}) \diamond \Gamma(\mathcal{P}^\delta)| = |\Gamma(\mathcal{P})|^2 / 2$.

Now we consider $(\Gamma(\mathcal{P}) \diamond \Gamma(\mathcal{P}^\delta)) \square \Gamma(\mathcal{P}^{\delta\pi})$. In this group, the order of $\rho_0\rho_1\rho_2$ divides both $p$ and $q$, and thus $\rho_0\rho_1\rho_2 = \epsilon$. Therefore, $\rho_0\rho_1 = \rho_2$, which forces $\rho_0\rho_1$ to have order dividing 2, and similarly $\rho_0 = \rho_1\rho_2$, which forces $\rho_1\rho_2$ to have order dividing 2. Therefore, the comix is a (not necessarily proper) quotient of the group $[2, 2]_1$, a group of order 4. Now, since $\Gamma(\mathcal{P}^{\delta\pi}) = [p, p]_q$ and $p$ is even, we see that $\Gamma(\mathcal{P}^{\delta\pi})$ covers $[2, 2]_1$. We similarly see that $\Gamma(\mathcal{P})$ covers $[2, 1]_1$ and that $\Gamma(\mathcal{P}^\delta)$ covers $[1, 2]_1$, and thus $\Gamma(\mathcal{P}) \diamond \Gamma(\mathcal{P}^\delta)$ covers $[2, 1]_1 \diamond [1, 2]_1$, which is equal to $[2, 2]_1$. Thus we see that the group $[2, 2]_1$ is the maximal group covered by both $\Gamma(\mathcal{P}) \diamond \Gamma(\mathcal{P}^\delta)$ and $\Gamma(\mathcal{P}^{\delta\pi})$, and therefore it is their comix. Therefore,

$$|\Gamma(\mathcal{P}) \diamond \Gamma(\mathcal{P}^\delta) \diamond \Gamma(\mathcal{P}^{\delta\pi})| = \frac{|\Gamma(\mathcal{P}) \diamond \Gamma(\mathcal{P}^\delta)| \cdot |\Gamma(\mathcal{P}^{\delta\pi})|}{4} = \frac{|\Gamma(\mathcal{P})|^3}{8} \quad \square$$

We note here that the arguments used can be generalized to give bounds on $|\Gamma(\mathcal{P}^*)|$ even when we cannot calculate the exact value. In the next section, however, we will only consider cases where we can calculate $|\Gamma(\mathcal{P}^*)|$ exactly.

## 4. The Covers of Universal Polyhedra

In this section, we will consider several of the finite polyhedra $\mathcal{P} = \{p, q\}_r$ and calculate $|\Gamma(\mathcal{P}^*)|$. The results are summarized in Table 1. In most cases, the sizes are easily calculated by applying Corollary 2 to $\mathcal{P}$ or one of its images under the duality operations. A few others can be found by applying Theorem 2 or by calculating the size directly using GAP [6]. We cover the remaining cases here.

**Table 1.** Self-dual, self-Petrie covers of finite polyhedra $\{p,q\}_r$.

| $\mathcal{P}$ | $|\Gamma(\mathcal{P})|$ | $|\Gamma(\mathcal{P}^*)|$ | Method |
|---|---|---|---|
| $\{2, 2k+1\}_{4k+2}$ | $4(2k+1)$ | $8(2k+1)^3$ | Hand |
| $\{2, 4k\}_{4k}$ | $16k$ | $32k^3$ | Hand |
| $\{2, 4k+2\}_{4k+2}$ | $8(2k+1)$ | $8(2k+1)^3$ | Hand |
| $\{3, 3\}_4$ | $24$ | $(24)^3$ | Thm. 3.4 |
| $\{3, 4\}_6$ | $48$ | $(48)^3/8$ | GAP |
| $\{3, 5\}_5$ | $60$ | $(60)^3$ | Thm. 3.4 |
| $\{3, 5\}_{10}$ | $120$ | $(120)^3$ | GAP |
| $\{3, 6\}_6$ | $108$ | $(108)^3/216$ | GAP |
| $\{3, 7\}_8$ | $336$ | $(336)^6/8$ | Cor. 3.5 |
| $\{3, 7\}_9$ | $504$ | $(504)^6$ | Cor. 3.5 |
| $\{3, 7\}_{13}$ | $1,092$ | $(1,092)^6$ | Cor. 3.5 |
| $\{3, 7\}_{15}$ | $12,180$ | $(12,180)^6$ | Cor. 3.5 |
| $\{3, 7\}_{16}$ | $21,504$ | $(21,504)^6/8$ | Cor. 3.5 |
| $\{3, 8\}_8$ | $672$ | $(672)^3/8$ | Thm. 3.6 |
| $\{3, 8\}_{11}$ | $12,144$ | $(12,144)^6/8$ | Cor. 3.5 |
| $\{3, 9\}_9$ | $3,420$ | $(3,420)^3$ | GAP |
| $\{3, 9\}_{10}$ | $20,520$ | $(20,520)^6/216$ | Cor. 3.5 |
| $\{4, 4\}_{4k}$ | $64k^2$ | $64k^6$ | Hand |
| $\{4, 4\}_{4k+2}$ | $16(2k+1)^2$ | $32(2k+1)^6$ | Hand |
| $\{4, 5\}_5$ | $160$ | $(160)^3$ | Thm. 3.4 |
| $\{4, 5\}_9$ | $6,840$ | $(6,840)^6/8$ | Cor. 3.5 |

We start with $\mathcal{P} = \{2, 2s\}$. The order of $\rho_0\rho_1\rho_2$ in $\Gamma(\mathcal{P})$ is $2s$, and therefore $\mathcal{P} = \{2, 2s\}_{2s}$, which has an automorphism group of order $8s$. If $s = 1$, then $\mathcal{P}$ is already self-dual and self-Petrie. In any case, there are only 3 distinct images of $\mathcal{P}$ under the duality operations, and thus

$$\mathcal{P}^* = \mathcal{P} \diamond \mathcal{P}^\delta \diamond \mathcal{P}^\pi$$

Now, $\mathcal{P} \diamond \mathcal{P}^\delta$ is a self-dual regular polyhedron of type $\{2s, 2s\}_{2s}$. We note that for any $s$, the group $\Gamma(\mathcal{P}) \,\square\, \Gamma(\mathcal{P}^\delta)$ is equal to $[2, 2]_{2s}$. In fact, this group is equal to $[2, 2]_2$, the group of order 8 generated by 3 commuting involutions (*i.e.*, the direct product of three cyclic groups of order 2). Then by Proposition 2, we see that $|\Gamma(\mathcal{P}) \diamond \Gamma(\mathcal{P}^\delta)| = 8s^2$. In order to determine whether $\mathcal{P} \diamond \mathcal{P}^\delta$ is self-Petrie, we would like to determine the group $\Gamma(\mathcal{P}) \diamond \Gamma(\mathcal{P}^\delta)$. A computation with GAP [6] suggests that the group is always the quotient of $[2s, 2s]_{2s}$ by the extra relation $(\rho_1\rho_0\rho_1\rho_2)^2 = \epsilon$. Since this extra relation also holds in $\Gamma(\mathcal{P})$ and $\Gamma(\mathcal{P}^\delta)$, this quotient must cover $\Gamma(\mathcal{P}) \diamond \Gamma(\mathcal{P}^\delta)$. Therefore, to prove that this is in fact the mix, it suffices to show that this group has order $8s^2$.

Start by considering the Cayley graph $G$ of

$$\langle \rho_0, \rho_1, \rho_2 \mid \rho_0^2 = \rho_1^2 = \rho_2^2 = (\rho_0\rho_2)^2 = (\rho_1\rho_0\rho_1\rho_2)^2 = \epsilon \rangle$$

. Starting at a vertex and building out from it, we see that the Cayley graph of this group is the uniform tiling 4.8.8 of the plane by squares and octagons. Figure 1 gives a local picture of $G$.

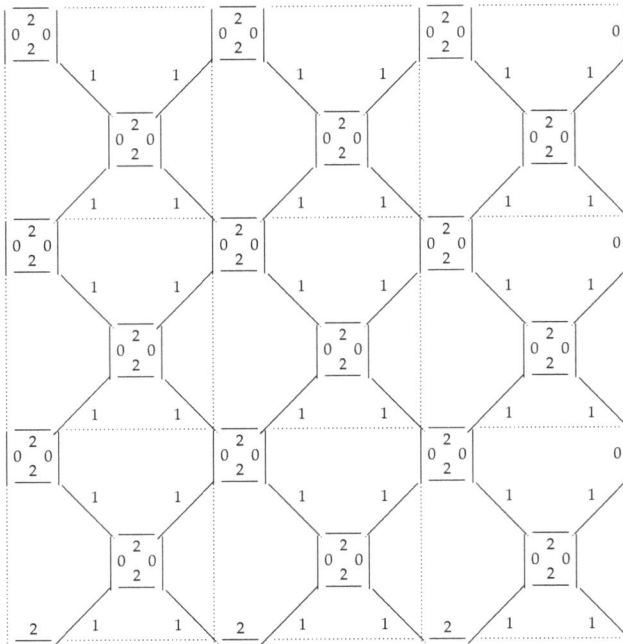

**Figure 1.** Local picture of the Cayley graph $G$.

Now, let us see what happens when we introduce the remaining relations $(\rho_0\rho_1)^{2s} = \epsilon$ and $(\rho_1\rho_2)^{2s} = \epsilon$. We note that the components of $G$ induced by edges labeled 0 and 1 are vertical zigzags, while the components of $G$ induced by edges labeled 1 and 2 are horizontal zigzags. Therefore, adding the relation $(\rho_0\rho_1)^{2s} = \epsilon$ forces an identification between points that are $s$ tiles away vertically, and adding the relation $(\rho_1\rho_2)^{2s} = \epsilon$ forces an identification between points that are $s$ tiles away horizontally. Therefore, the Cayley graph of $\Gamma(\mathcal{P} \diamond \mathcal{P}^\delta)$ consists of an $s \times s$ grid of tiles, with opposite sides identified. Each tile has 8 vertices, so there are a total of $8s^2$ vertices, which shows that the given group has $8s^2$ elements. Therefore the given group is indeed the mix.

Now we consider $(\Gamma(\mathcal{P}) \diamond \Gamma(\mathcal{P}^\delta)) \square \Gamma(\mathcal{P}^\pi)$, which is the quotient of $[2s, 2s]_2$ by the extra relation $(\rho_1\rho_0\rho_1\rho_2)^2 = \epsilon$. Put another way, we get the quotient of $\Gamma(\mathcal{P}) \diamond \Gamma(\mathcal{P}^\delta)$ by the extra relation $(\rho_0\rho_1\rho_2)^2 = \epsilon$. Using the Cayley graph from before, we see that the relation $(\rho_0\rho_1\rho_2)^2 = \epsilon$ forces us to identify each tile with the tiles that touch it at a corner. When $s$ is odd, this forces every tile to be identified, leaving us with a Cayley graph with 8 vertices. When $s$ is even, we instead get 2 distinct tiles, and the Cayley graph has 16 vertices. Therefore, $|\Gamma(\mathcal{P}) \diamond \Gamma(\mathcal{P}^\delta) \diamond \Gamma(\mathcal{P}^{\delta\pi})| = 8s^3$ if $s$ is odd, and $4s^3$ if $s$ is even. In other words, if $s = 2k + 1$, then $|\Gamma(\mathcal{P}^*)| = 8(2k+1)^3$, and if $s = 2k$, then $|\Gamma(\mathcal{P}^*)| = 32k^3$.

Now suppose that $\mathcal{P} = \{2, 2k + 1\}$. This polyhedron is covered by $\mathcal{Q} = \{2, 4k + 2\}$ (which is equal to $\{2, 4k + 2\}_{4k+2}$, as we observed earlier); therefore, $\mathcal{Q} \diamond \mathcal{Q}^\delta$ covers $\mathcal{P} \diamond \mathcal{P}^\delta$. Calculating the size of the mix, we find that $|\Gamma(\mathcal{Q}) \diamond \Gamma(\mathcal{Q}^\delta)| = |\Gamma(\mathcal{P}) \diamond \Gamma(\mathcal{P}^\delta)|$, and thus these two groups must be equal. From this it easily follows that $\mathcal{Q}^* = \mathcal{P}^*$, and thus $|\Gamma(\mathcal{P}^*)| = 8(2k+1)^3$.

Next we consider the polyhedra $\{4, 4\}_{2s}$, which are the torus maps $\{4, 4\}_{(s,s)}$ with $16s^2$ flags [7]. First, suppose that $s = 2k$, so that $\mathcal{P} = \{4, 4\}_{4k}$. Then $\Gamma(\mathcal{P}) \square \Gamma(\mathcal{P}^\pi) = [4, 4]_4$, which has order 64. Therefore, by Proposition 2,

$$|\Gamma(\mathcal{P}) \diamond \Gamma(\mathcal{P}^\pi)| = |\Gamma(\mathcal{P})||\Gamma(\mathcal{P}^\pi)|/64 = 64k^4$$

Now, $\mathcal{P} \diamond \mathcal{P}^{\pi}$ is of type $\{4k, 4\}_{4k}$, and $\mathcal{P}^{\pi\delta}$ is the universal polyhedron of type $\{4, 4k\}_4$. Thus $(\Gamma(\mathcal{P}) \diamond \Gamma(\mathcal{P}^{\pi})) \square \Gamma(\mathcal{P}^{\pi\delta})$ is $[4, 4]_4$ or a proper quotient. Since $\Gamma(\mathcal{P})$ covers $[4, 4]_4$, so does $\Gamma(\mathcal{P}) \diamond \Gamma(\mathcal{P}^{\pi})$. Then since $\Gamma(\mathcal{P}^{\pi\delta})$ also covers $[4, 4]_4$, so does $(\Gamma(\mathcal{P}) \diamond \Gamma(\mathcal{P}^{\pi})) \square \Gamma(\mathcal{P}^{\pi\delta})$, and thus the comix is the whole group $[4, 4]_4$. Thus we see that

$$|\Gamma(\mathcal{P}) \diamond \Gamma(\mathcal{P}^{\pi}) \diamond \Gamma(\mathcal{P}^{\pi\delta})| = |\Gamma(\mathcal{P}) \diamond \Gamma(\mathcal{P}^{\pi})| \cdot |\Gamma(\mathcal{P}^{\pi\delta})|/64 = 64k^6$$

Now suppose that $s = 2k + 1$, so that $\mathcal{P} = \{4, 4\}_{4k+2}$. Then $\Gamma(\mathcal{P}) \square \Gamma(\mathcal{P}^{\pi}) = [2, 4]_2$, which has size 8. Therefore,

$$|\Gamma(\mathcal{P}) \diamond \Gamma(\mathcal{P}^{\pi})| = |\Gamma(\mathcal{P})| \cdot |\Gamma(\mathcal{P}^{\pi})|/8 = 32(2k + 1)^4$$

Now, $\mathcal{P} \diamond \mathcal{P}^{\pi}$ is of type $\{8k + 4, 4\}_{8k+4}$ and $\mathcal{P}^{\pi\delta}$ is the universal polyhedron $\{4, 4k + 2\}_4$. Thus $(\Gamma(\mathcal{P}) \diamond \Gamma(\mathcal{P}^{\pi})) \square \Gamma(\mathcal{P}^{\pi\delta})$ is $[4, 2]_4$ or a proper quotient. Clearly, $\Gamma(\mathcal{P}^{\pi\delta})$ covers $[4, 2]_4$. Furthermore, $\Gamma(\mathcal{P})$ covers $[4, 4]_2$ and $\Gamma(\mathcal{P}^{\pi})$ covers $[2, 4]_4$; therefore, their mix covers $[4, 4]_2 \diamond [2, 4]_4$, which covers $[4, 2]_4$. Therefore, the comix is the whole group $[4, 2]_4$ of order 16, and we see that

$$|\Gamma(\mathcal{P}) \diamond \Gamma(\mathcal{P}^{\pi}) \diamond \Gamma(\mathcal{P}^{\pi\delta})| = |\Gamma(\mathcal{P}) \diamond \Gamma(\mathcal{P}^{\pi})| \cdot |\Gamma(\mathcal{P}^{\pi\delta})|/16 = 32(2k + 1)^6$$

It would be natural here to consider the torus maps $\{3, 6\}_{2s} = \{3, 6\}_{(s,0)}$. However, in this case there are 6 distinct polyhedra under the duality operations, and the problem seems to be intractable.

Finally, we note that the self-dual, self-Petrie covers of $\{3, 3\}_4$ and $\{3, 4\}_6$ have groups of the same order. In fact, since $\{3, 4\}_6$ covers $\{3, 4\}_3$, these two polyhedra have the same self-dual, self-Petrie cover.

## References

1. Jones, G.A.; Thornton, J.S. Operations on maps, and outer automorphisms. *J. Comb. Theory Ser. B* **1983**, *35*, 93–103.
2. Coxeter, H.S.M. *Regular Polytopes*, 3rd ed.; Dover Publications Inc.: New York, NY, USA, 1973; pp. xiv+321.
3. McMullen, P.; Schulte, E. Abstract Regular Polytopes. In *Encyclopedia of Mathematics and Its Applications*; Cambridge University Press: Cambridge, MA, USA, 2002; Volume 92, pp. xiv+551.
4. Breda D'Azevedo, A.; Nedela, R. Join and intersection of hypermaps. *Acta Univ. M. Belii Ser. Math.* **2001**, *9*, 13–28.
5. Wilson, S.E. Parallel products in groups and maps. *J. Algebra* **1994**, *167*, 539–546.
6. The GAP Group. *GAP—Groups, Algorithms, and Programming, Version 4.4.12*; The GAP Group: Glasgow, UK, 2008.
7. Coxeter, H.S.M.; Moser, W.O.J. Generators and Relations for Discrete Groups. In *Ergebnisse der Mathematik und ihrer Grenzgebiete [Results in Mathematics and Related Areas]*, 4th ed.; Springer-Verlag: Berlin, Heidelberg, Germany, 1980; Volume 14, pp. ix+169.

*symmetry*

MDPI

*Article*

# On Center, Periphery and Average Eccentricity for the Convex Polytopes

**Waqas Nazeer [1], Shin Min Kang [2,3], Saima Nazeer [4], Mobeen Munir [1], Imrana Kousar [4], Ammara Sehar [4] and Young Chel Kwun [5,*]**

[1]   Division of Science and Technology, University of Education, Lahore 54000, Pakistan; nazeer.waqas@ue.edu.pk (W.N.); mmunir@ue.edu.pk (M.M.)
[2]   Department of Mathematics and Research Institute of Natural Science, Gyeongsang National University, Jinju 52828, Korea; smkang@gnu.ac.kr or sm.kang@mail.cmuh.org.tw
[3]   Center for General Education, China Medical University, Taichung 40402, Taiwan
[4]   Department of Mathematics, Lahore College for Women University, Lahore 54000, Pakistan; saimanazeer123@yahoo.com (S.N.); imrana.kousar@hotmial.com (I.K.); seharammara93@gmail.com (A.S.)
[5]   Department of Mathematics, Dong-A University, Busan 49315, Korea
*   Correspondence: yckwun@dau.ac.kr; Tel.: +82-51-200-7216

Academic Editor: Egon Schulte
Received: 3 November 2016; Accepted: 24 November 2016; Published: 2 December 2016

**Abstract:** A vertex $v$ is a peripheral vertex in $G$ if its eccentricity is equal to its diameter, and periphery $P(G)$ is a subgraph of $G$ induced by its peripheral vertices. Further, a vertex $v$ in $G$ is a central vertex if $e(v) = rad(G)$, and the subgraph of $G$ induced by its central vertices is called center $C(G)$ of $G$. Average eccentricity is the sum of eccentricities of all of the vertices in a graph divided by the total number of vertices, i.e., $avec(G) = \{\frac{1}{n}\sum e_G(u); \ u \in V(G)\}$. If every vertex in $G$ is central vertex, then $C(G) = G$, and hence, $G$ is self-centered. In this report, we find the center, periphery and average eccentricity for the convex polytopes.

**Keywords:** eccentricity; center; periphery; average eccentricity

## 1. Introduction

In the facility location problem, we select a site according to some standard judgment. For example, if we want to find out the exact location for an emergency facility, such as a fire station or a hospital, we reduce the distance between that facility and the area where the emergency happens, and if we are to decide the position for a service facility, like a post office, power station or employment office, we try to reduce the traveling time of all people who have been living in that district. In the construction of a railway line, a pipeline and a superhighway, we will reduce the distance of the constructing unit for the people living in that area. All of these situations illustrate the concept of centrality but each of these three examples deals with different types of centers. Nowadays, centrality questions are being studied with the help of distance and graphs. We shall observe that many kinds of centers are helpful in facility location problems.

The most important and fundamental concept that extends to the whole of graph theory is distance. The distance is applicable in many fields, such as graph operation, extremal problems on connectivity, diameter and isomorphism testing. The theme of distance is used to check the symmetry of graphs. It also provides a base for many useful graph parameters, like radius, diameter, metric dimension, eccentricity, center and periphery, etc.

The eccentricity of the vertices in $G$ has a fundamental importance. Recently, many indices related to eccentricity have been derived, i.e., eccentric connectivity index, adjacent eccentric sum index, Wiener index and eccentric distance sum [1]. The center and periphery is also based on minimum and maximum eccentricity, respectively. W.Goddard and O. R. Oellermann in [2] have shown that if $G$ is an undirected graph, then,

$$rad(G) \leq diam(G) \leq 2rad(G)$$

They also examined the radius and diameter of certain families of graphs in the same paper, as follows:

1. $rad(K_n) = diam(K_n) = 1$ for $n \geq 2$,
2. $rad(C_n) = diam(C_n) = \frac{n}{2}$,
3. $rad(K_{m,n}) = diam(K_{m,n}) = 2$ if $m$ and $n$ is at least two,
4. $rad(P_n) = \frac{n-1}{2}$, $diam(P_n) = n - 1$.

This implies that complete graphs $K_n$ for $n \geq 2$, complete bipartite graphs $K_{m,n}$ where $m, n \geq 2$ and all cycles are self-centered. Jordan [3] determined the diameter of a tree. Bela Bollobas [4] discussed the diameter of random graphs. The radius and diameter of a bridge graph are determined by Martin Farber in [5]. More general results were presented by V. Klee and D. Larman [6] and Bela Bollobas [4]. B. Hedman determined the sharp bounds for the diameter of the clique graph $K(G)$ in terms of the diameter of $G$. The idea of self-centered graphs is presented and elaborated by Ando, Akiyama and Avis individually [7]. These self-centered graphs are extensively studied in [7–11]. The extremal size of a connected self-centered graph with $p$ vertices and $r$ radius is explained by F. Buckely [12]. The center in maximal outer planar graphs is demonstrated by A. Proskurowski in [13]. Hedetniemi [14] has shown that every graph is the center of some graph. The center of graph $G$ is the full graph if and only if $rad(G) = diam(G)$ [15]. F. Buckely and F. Harary [16] gave the concept of average eccentricity. Average eccentricity is the sum of eccentricities of all of the vertices in a graph divided by the total number of vertices, i.e.,

$$avec(G) = \frac{1}{n} \sum_{u \in V(G)} e_G(u).$$

The upper bounds of average eccentricity are determined by P. Dankelman, W. Goddard and C.S. Swart [17]. Average eccentricity is most important in communication networks. The average eccentricity of Sierpinski graphs $S_n^p$ is determined by Andreas, M. Hinz and Daniele Parisse [18]. Since 1980, the average eccentricity has had a great roll as a molecular descriptor in mathematical chemistry. This is attributed to V.A. Skorobogatov and A.A. Dobrynin [19]. For more details, please see [20–22] and the references therein.

**Definition 1.** *For a connected graph $G$, the eccentricity $e(v)$ of a vertex $v$ is its distance to a vertex farthest from $v$. Thus,*

$$e(v) = Max\{d(u, v) : u \in V(G)\}.$$

**Definition 2.** *The radius $rad(G)$ of $G$ is the minimum eccentricity among all vertices of $G$.*

**Definition 3.** *The diameter $diam(G)$ of $G$ is the maximum eccentricity among all vertices of $G$.*

**Definition 4.** *Average eccentricity is the sum of eccentricities of all of the vertices in a graph divided by the total number of vertices, i.e.,*

$$avec(G) = \frac{1}{n} \sum_{u \in V(G)} e_G(u).$$

**Definition 5.** *A vertex $u$ is eccentric to a vertex $v$ if $d(u, v) = e(v)$.*

**Definition 6.** *A vertex $v$ is a peripheral vertex in $G$ if its eccentricity is equal to its diameter, and periphery $P(G)$ is a subgraph of $G$ induced by its peripheral vertices. Further, a vertex $v$ in $G$ is a central vertex if $e(v) = rad(G)$, and the subgraph of $G$ induced by its central vertices is called center $C(G)$ of $G$. If every vertex in $G$ is a central vertex, then $C(G) = G$, and hence, $G$ is self-centered.*

In the present report, we discuss the center, periphery and average eccentricity for families of convex polytope graphs, $A_n$, $S_n$ and $T_n$.

## 2. The Center and Periphery for Convex Polytope $A_n$

In this section, we determine the center and periphery for convex polytope $A_n$.

**Definition 7.** *The graph of convex polytope (double antiprism) $A_n$ can be obtained from the graph of convex polytope $R_n$ by adding new edges $b_{i+1}c_i$, i.e.,*

$$V(A_n) = V(R_n) \text{ and } E(A_n) = E(R_n) \cup \{b_{i+1}c_i : 1 \le i \le n\}.$$

**Theorem 1.** *For the family of convex polytope $A_n$, $n = 2k$, $Cen(A_n)$ and $Per(A_n)$ are subgraphs induced by the vertices $(b_1, b_2, ..., b_{2k})$ and $\{a_i \cup c_i : 1 \le i \le 2k\}$, respectively.*

**Proof.** For all even values of $n$, select a vertex $a_1$ on the cycle $(a_1 a_2 a_3 ... a_i ... a_{2k})$. Then:

$$d(a_1, a_i) = i - 1, \quad 1 \le i \le k+1 \tag{1}$$

when $i = k + 2$, $d(a_1, a_i) = k - 1$ and for $i = 2k$, $d(a_1, a_i) = 1$.

In addition, for every value of $i$ within $k + 2$ to $2k$, $d(a_1, a_i)$ must lie between $k - 1$ and one, i.e.,

$$d(a_1, a_i) = 2k + 1 - i; k + 2 \le i \le 2k.$$

Thus, to find the vertices farthest from $a_1$ in $A_n$, consider only $1 \le i \le k + 1$.

As each $a_i$ is adjacent to $b_i$, $b_{i-1}$ and each $b_i$ adjacent to $c_i, c_{i-1}$, therefore, (1) implies,

$$d(a_1, b_i) = i, \quad 1 \le i \le k$$

$$d(a_1, c_i) = i + 1, \quad 1 \le i \le k$$

For $k + 1 \le i \le 2k$, consider the cycle $(b_1 b_2 ... b_{k+1} ... b_i ... b_{2k})$. In this cycle, the distance between $b_i$ and $b_{2k}$ is $2k - i$, and $b_{2k}$ is adjacent to $a_1$, therefore, the distance between $a_1$ and $b_i$ is $2k - i + 1$.

$$d(a_1, b_i) = 2k - i + 1, \quad k + 1 \le i \le 2k.$$

Now, consider the cycle $(c_1 c_2 ... c_{k+1} ... c_i ... c_{2k})$. The distance between $c_i$ and $c_{2k-1}$ is $2k - 1 - i$ and the vertex $c_{2k-1}$ is adjacent to $b_{2k}$ and $b_{2k}$ adjacent to $a_1$. It shows,

$$d(a_1, c_i) = 2k - i + 1, \quad k + 1 \le i \le 2k - 1.$$

For $i = 2k$, $d(a_1, c_i) = 2$.

Hence, $c_k$ is a vertex farthest from $a_1$.

$$e(a_1) = k + 1 \tag{2}$$

Thus, the eccentricity of each vertex on inner cycle $(a_1 a_2 a_3 ... a_i ... a_{2k})$ is $k + 1$.

In the same way, take cycle $(b_1 b_2 ... b_i .... b_{2k})$; the distance between $b_1$ and $b_i$ in this cycle is,

$$d(b_1, b_i) = i - 1; 1 \le i \le k + 1 \tag{3}$$

Each $b_i$ is adjacent to $a_i$ and $a_{i+1}$. Therefore,

$$d(b_1, a_1) = 1, \quad (b_1, a_2) = 1$$

For $3 \le i \le k + 1$, consider the path $b_1 \to a_2 \to a_3 \to ... \to a_i$. Then, the distance between $a_i$ and $a_2$ is $i - 2$. $a_2$ is also adjacent to $b_1$. Therefore, the distance between $b_1$ and $a_i$ is as follows,

$$d(b_1, a_i) = i - 1, \quad 3 \leq i \leq k + 1$$

For $k + 2 \leq i \leq 2k$, consider the cycle $(a_1 a_2 ... a_{k+2} ... a_i ... a_{2k})$. The distance between $a_i$ and $a_{2k}$ is $2k - i$. Further, $a_{2k}$ is adjacent to $a_1$ and $a_1$ adjacent to $b_1$. Therefore, the distance between $b_1$ and $a_i$ is $2k - i + 2$.

$$d(b_1, a_i) = 2k - i + 2, \quad k + 2 \leq i \leq 2k$$

Further, $b_i$ is also adjacent to $c_i$ and $c_{i-1}$; it follows from (3):

$$d(b_1, c_i) = i, \quad 1 \leq i \leq k$$

For $k + 1 \leq i \leq 2k$, consider the cycle $(c_1 c_2 ... c_{k+1} ... c_i ... c_{2k})$. The distance between $c_i$ and $c_{2k}$ is $2k - i$, where $c_{2k}$ is also adjacent to $b_1$. Therefore,

$$d(b_1, c_i) = 2k - i + 1, \quad k + 1 \leq i \leq 2k$$

Hence, $b_{k+1}$, $a_{k+1}$ and $c_k$ are the vertices farthest from $b_1$. Therefore:

$$e(b_1) = k, \tag{4}$$

Hence, each vertex on the middle cycle $(b_1 b_2 ... b_i .... b_{2k})$ has eccentricity $k$.

Further, to find out the eccentricity of the vertices on the outer cycle $(c_1 c_2 ... c_i ... c_{2k})$, choose a vertex $c_1$ on this cycle. The distance between $c_1$ and $c_i$ is $i - 1$.

$$d(c_1, c_i) = i - 1, \tag{5}$$

Each $c_i$ is adjacent to $b_i$ and $b_{i+1}$, i.e.,

$$d(c_1, b_1) = 1, \tag{6}$$

For $3 \leq i \leq k + 1$, consider the path $c_1 \rightarrow b_2 \rightarrow b_3 \rightarrow ... \rightarrow b_i$. The distance between $b_2$ and $b_i$ is $i - 2$. As $b_2$ is adjacent to $c_1$, therefore, $c_1$ and $b_i$ has the following distance,

$$d(c_1, b_i) = i - 1, \quad 3 \leq i \leq k + 1$$

For $k + 2 \leq i \leq 2k$, consider the cycle $(b_1 b_2 ... b_{k+2} ... b_i ... b_{2k})$. The distance between $b_i$ and $b_{2k}$ is $2k - i$. As $b_{2k}$ is adjacent to $b_1$ and $b_1$ adjacent to $c_1$, therefore, the distance between $c_1$ and $b_i$ is $2k - i + 2$.

$$d(c_1, b_i) = 2k - i + 2, \quad k + 2 \leq i \leq 2k.$$

Each $b_i$ is also adjacent to $a_i$ and $a_{i+1}$, using the result of (6),

$$d(c_1, a_1) = 2, \quad d(c_1, a_2) = 2$$

For $3 \leq i \leq k + 2$, consider the path $c_1 \rightarrow b_2 \rightarrow b_3 \rightarrow ... \rightarrow b_{i-1} \rightarrow a_i$. In this path, the distance between $b_2$ and $b_{i-1}$ is $i - 3$. $b_{i-1}$ is adjacent to $a_i$ and $b_2$ adjacent to $c_1$. Therefore, the distance between $c_1$ and $a_i$ is $i - 1$.

$$d(c_1, a_i) = i - 1, \quad 3 \leq i \leq k + 2$$

For $k + 3 \leq i \leq 2k$, consider the cycle $(a_1 a_2 ... a_{k+2} ... a_i ... a_{2k})$. The distance between $a_i$ and $a_{2k}$ is $2k - i$. $a_{2k}$ is adjacent to $a_1$, $a_1$ adjacent to $b_1$ and $b_1$ adjacent to $c_1$. Therefore, the distance between $c_1$ and $a_i$ is $2k - i + 3$.

$$d(c_1, a_i) = 2k - i + 3, \quad k + 3 \leq i \leq 2k$$

This shows that $a_{k+1}$ is a vertex farthest from $c_1$. Therefore:

$$e(c_1) = k + 1. \tag{7}$$

Therefore, (2), (4) and (7) imply,

$$diam(A_n) = k + 1 = \frac{n}{2} + 1.$$

and:

$$rad(A_n) = k = \frac{n}{2}.$$

Consequently, $Cen(A_n)$ is a subgraph induced by vertices $(b_1, b_2, ..., b_{2k})$, while the set of vertices $\{a_1, a_2, ..., a_{2k}, c_1, c_2, ..., c_{2k}\}$ is the peripheral vertices. Therefore, the periphery of $A_n$ is the subgraph induced by all of these vertices. $\square$

**Theorem 2.** *For the family of convex polytope $A_n$, $n$ is odd.*

$$Cen(A_n) = Per(A_n) = A_n.$$

**Proof.** Consider, $n = 2k + 1$ $k \geq 2$. Select vertex $a_1$ on the cycle $(a_1 a_2 a_3 ... a_i ... a_{2k+1})$. By using this,

$$d(a_1, a_i) = i - 1, \ 1 \leq i \leq k + 1 \tag{8}$$

while $i$ increases from $k + 2$ to $2k + 1$, $d(a_1, a_i)$ reduces from $k$ to one.

$$d(a_1, a_i) = 2k + 2 - i, \quad k + 2 \leq i \leq 2k + 1$$

Thus, to find the vertices farthest from $a_1$ in $A_n$, we have to take only those values of $i$ that lie between one and $k + 1$.

As each $a_i$ is adjacent to $b_i$, $b_{i-1}$ and each $b_i$ adjacent to $c_i$, $c_{i-1}$, therefore, (8) implies,

$$d(a_1, b_i) = i, \quad 1 \leq i \leq k + 1$$

$$d(a_1, c_i) = 1 + i, \quad 1 \leq i \leq k$$

For $k + 2 \leq i \leq 2k + 1$, consider the cycle $(b_1 b_2 ... b_{k+1} ... b_i ... b_{2k+1})$. In this cycle, the distance between $b_i$ and $b_{2k+1}$ is $2k + 1 - i$, and $b_{2k+1}$ is adjacent to $a_1$. Therefore, the distance between $a_1$ and $b_i$ is $2k - i + 2$.

$$d(a_1, b_i) = 2k - i + 2, \quad k + 2 \leq i \leq 2k + 1.$$

Now, consider the cycle $(c_1 c_2 ... c_{k+1} ... c_i ... c_{2k+1})$. The distance between $c_i$ and $c_{2k}$ is $2k - i$. The vertex $c_{2k}$ is adjacent to $b_{2k+1}$ and $b_{2k+1}$ adjacent to $a_1$. It shows that the distance between $a_1$ and $c_i$ is $2k - i + 2$.

$$d(a_1, c_i) = 2k - i + 2, \quad k + 1 \leq i \leq 2k.$$

For $i = 2k + 1$, $d(a_1, c_i) = 2$.

Hence, $c_k$ and $b_{k+1}$ are the vertices farthest from $a_1$. Therefore:

$$e(a_1) = k + 1 \tag{9}$$

Thus, the eccentricity of each vertex on inner cycle $(a_1 a_2 a_3 ... a_i ... a_{2k+1})$ is $k + 1$.

Similarly as above, the vertices $b_1$ and $b_i$ on cycle $(b_1 b_2 ... b_i .... b_{2k+1})$ have the distance as,

$$d(b_1, b_i) = i - 1, \ 1 \leq i \leq k + 1 \tag{10}$$

Each $b_i$ is adjacent to $a_i$ and $a_{i+1}$. Therefore,

$$d(b_1, a_1) = 1, \quad (b_1, a_2) = 1$$

For $3 \leq i \leq k+2$, consider the path $b_1 \to a_2 \to a_3 \to ... \to a_i$. Then, the distance between $a_i$ and $a_2$ is $i-2$. $a_2$ is also adjacent to $b_1$. Therefore, the distance between $b_1$ and $a_i$ is $i-1$, i.e.,

$$d(b_1, a_i) = i - 1, \quad 3 \leq i \leq k+2$$

For $k+3 \leq i \leq 2k+1$, consider the cycle $(a_1 a_2 ... a_{k+3} ... a_i ... a_{2k+1})$. The distance between $a_i$ and $a_{2k+1}$ is $2k - i + 1$. Further, $a_{2k+1}$ is adjacent to $a_1$, and $a_1$ is adjacent to $b_1$. Therefore, the distance between $b_1$ and $a_i$ is $2k - i + 3$.

$$d(b_1, a_i) = 2k - i + 3, \quad k+3 \leq i \leq 2k+1$$

Further, $b_i$ is also adjacent to $c_i$ and $c_{i-1}$; it follows from (10):

$$d(b_1, c_i) = i, \quad 1 \leq i \leq k+1$$

For $k+2 \leq i \leq 2k+1$, consider the cycle $(c_1 c_2 ... c_{k+2} ... c_i ... c_{2k+1})$. The distance between $c_i$ and $c_{2k+1}$ is $2k + 1 - i$. $c_{2k+1}$ is also adjacent to $b_1$. Therefore,

$$d(b_1, c_i) = 2k + 2 - i, \quad k+2 \leq i \leq 2k+1$$

Hence, $a_{k+2}$ and $c_{k+1}$ are the vertices farthest from $b_1$. Therefore:

$$e(b_1) = k + 1. \tag{11}$$

Hence, each vertex on the middle cycle $(b_1 b_2 ... b_i .... b_{2k+1})$ has eccentricity $k + 1$.

Further, to find out the eccentricity of the vertices on the outer cycle $(c_1 c_2 ... c_i ... c_{2k+1})$, choose a vertex $c_1$ on this cycle. The distance between $c_1$ and $c_i$ is $i - 1$.

$$d(c_1, c_i) = i - 1, \quad 1 \leq i \leq k+1 \tag{12}$$

Each $c_i$ is adjacent to $b_i$ and $b_{i+1}$. Therefore,

$$d(c_1, b_1) = 1, \quad d(c_1, b_2) = 1 \tag{13}$$

For $3 \leq i \leq k+2$, consider the path $c_1 \to b_2 \to b_3 \to ... \to b_i$. The distance between $b_2$ and $b_i$ is $i - 2$. As $b_2$ is adjacent to $c_1$, therefore, the distance between $c_1$ and $b_i$ is $i - 1$.

$$d(c_1, b_i) = i - 1, \quad 3 \leq i \leq k+2$$

For $k+3 \leq i \leq 2k+1$, consider the cycle $(b_1 b_2 ... b_{k+3} ... b_i ... b_{2k+1})$. The distance between $b_i$ and $b_{2k+1}$ is $2k + 1 - i$. As $b_{2k+1}$ is adjacent to $b_1$ and $b_1$ adjacent to $c_1$, therefore, the distance between $c_1$ and $b_i$ is $2k - i + 3$.

$$d(c_1, b_i) = 2k - i + 3, \quad k+3 \leq i \leq 2k+1.$$

Each $b_i$ is also adjacent to $a_i$ and $a_{i+1}$, using the result of (13):

$$d(c_1, a_1) = 2, \quad d(c_1, a_2) = 2$$

For $3 \leq i \leq k+2$, consider the path $c_1 \to b_2 \to b_3 \to \dots \to b_{i-1} \to a_i$. In this path, the distance between $b_2$ and $b_{i-1}$ is $i-3$. $b_{i-1}$ is adjacent to $a_i$, and $b_2$ is adjacent to $c_1$. Therefore, the distance between $c_1$ and $a_i$ is $i-1$.

$$d(c_1, a_i) = i-1, \quad 3 \leq i \leq k+2$$

For $k+3 \leq i \leq 2k$, consider the cycle $(a_1 a_2 \dots a_{k+3} \dots a_i \dots a_{2k+1})$. The distance between $a_i$ and $a_{2k+1}$ is $2k+1-i$. $a_{2k+1}$ is adjacent to $a_1$ and $a_1$ adjacent to $b_1$. In addition, $b_1$ is adjacent to $c_1$. Therefore, the distance between $c_1$ and $a_i$ is $2k-i+4$.

$$d(c_1, a_i) = 2k - i + 4, \quad k+3 \leq i \leq 2k+1$$

This shows that $a_{k+2}$ and $b_{k+2}$ are the vertices farthest from $c_1$. Therefore:

$$e(c_1) = k+1. \tag{14}$$

Consequently, (9), (11) and (14) show the smallest, In addition, the greatest eccentricity of these vertices is $k+1$. Therefore:

$$diam(A_n) = rad(A_n) = k+1 = \frac{n-1}{2} + 1 = \frac{n+1}{2}.$$

Implies:

$$Cen(A_n) = Per(A_n) = A_n.$$

Hence, the family of $A_n$ is self-centered for odd values of $n$. $\square$

### 2.1. Average Eccentricity for Convex Polytope $A_n$

Here, we also are concerned with calculating the average eccentricity for the graph of convex polytope $A_n$. The average eccentricity of any graph can be calculated by dividing the sum of the eccentricities of all of the vertices to the total number of vertices ($\acute{n}$). There are three circles in the graph of convex polytope $A_n$, and each circle consists of $n$ vertices. Therefore, $A_n$ has a total of $3n$ vertices; it follows,

$$avec(A_n) = \frac{1}{n} \sum_{u \in V(G)} e_G u \tag{15}$$

By Theorem 1:

$$
\begin{aligned}
avec(A_n) &= \frac{1}{3 \times (\acute{n})} [n \times \{e(a_1) + e(b_1) + e(c_1)\}] \\
&= \frac{1}{3 \times n} [n \times \{(k+1) + (k) + (k+1)\}] = k + \frac{2}{3} = \frac{n}{2} + \frac{2}{3}.
\end{aligned}
$$

and by Theorem 2,

$$
\begin{aligned}
avec(A_n) &= \frac{1}{3 \times n} [n \times \{3(k+1)\}] \\
&= k+1 = \frac{n-1}{2} + 1 = \frac{n+1}{2}.
\end{aligned}
$$

Therefore, we have the following result:

$$
avec(A_n) = \begin{cases} \dfrac{n+1}{2}, & \text{if } n = 2k+1; \\[2mm] \dfrac{n}{2} + \dfrac{2}{3}, & \text{if } n = 2k. \end{cases}
$$

*2.2. Illustration*

Consider the graph of $A_8$. We have labeled each of its vertices by its eccentricities. The center and periphery are shown in Figures 1 and 2.

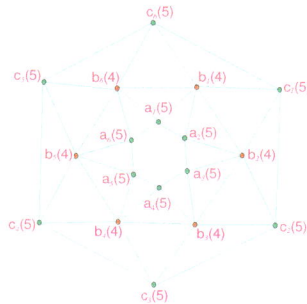

**Figure 1.** The graph of convex polytope $A_8$.

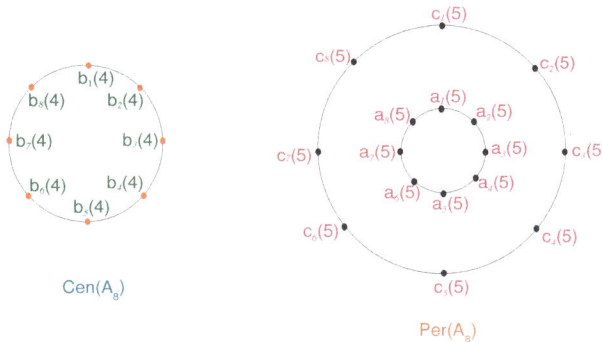

**Figure 2.** Centrality in the graph of convex polytope $A_8$.

## 3. The Center and Periphery for Convex Polytope $S_n$

Here, we examine the center and periphery for convex polytope $S_n$.

**Definition 8.** *The graph of convex polytope (double antiprism) $S_n$ can be obtained from the graph of convex polytope $Q_n$ by adding new edges $c_i c_{i+1}$, i.e.,*

$$V(S_n) = V(Q_n) \quad and \quad E(S_n) = E(Q_n) \cup \{c_i c_{i+1} : 1 \le i \le n\}.$$

For our convenience, we identify the cycle induced by the vertices $(a_1, a_2, ..., a_n)$, $(b_1, b_2, ..., b_n)$, $(c_1, c_2, ..., c_n)$ and $(d_1, d_2, ..., d_n)$ as the inner cycle, interior cycle, exterior cycle and outer cycle, respectively.

**Theorem 3.** *For the family of convex polytope $S_n$, when $n$ is even, we have:*

$$diam(S_n) = \frac{n}{2} + 1.$$

$$rad(S_n) = \frac{n}{2} + 2.$$

**Proof.** Suppose, $n = 2k$, $k \geq 2$. Consider the cycle $(a_1 a_2 \dots a_i \dots a_{2k})$. Here, the eccentricity of only one vertex, i.e., $a_1$, is determined, and due to the symmetry of the graph, all other vertices have the same eccentricity as $a_1$ on this cycle. Using this cycle,

$$d(a_1, a_i) = i - 1, \; 1 \leq i \leq k + 1. \tag{16}$$

For $k + 2 \leq i \leq 2k$, $d(a_1, a_i)$ varies from $k - 1$ to one, i.e.,

$$d(a_1, a_i) = 2k - i + 1, \quad k + 2 \leq i \leq 2k.$$

Thus, to identify a vertex at maximum distance from $a_1$ in $S_n$, take only $1 \leq i \leq k + 1$. As each $a_i$ is adjacent to $b_i$, therefore,

$$d(a_1, b_i) = i, \; 1 \leq i \leq k + 1. \tag{17}$$

For $k + 2 \leq i \leq 2k$, take the interior cycle $(b_1 b_2 \dots b_{k+2} \dots b_i \dots b_{2k})$. In this cycle, the vertices $b_i$ and $b_{2k}$ are at a distance $2k - i$. Further, $b_{2k}$ is adjacent to $b_1$ and $b_1$ adjacent to $a_1$. Therefore, The distance between $a_1$ and $b_i$ is $2k - i + 2$.

$$d(a_1, b_i) = 2k - i + 2, \quad k + 2 \leq i \leq 2k.$$

Each $b_i$ is adjacent to $c_i$ and $c_{i-1}$, by using (17),

$$d(a_1, c_i) = i + 1, \; 1 \leq i \leq k. \tag{18}$$

For $k + 1 \leq i \leq 2k$, consider the exterior cycle $(c_1 c_2 \dots c_{k+1} \dots c_i \dots c_{2k})$. The vertices $c_i$ and $c_{2k}$ are at a distance $2k - i$. As $c_{2k}$ is adjacent to $b_1$ and $b_1$ adjacent to $a_1$, therefore, $a_1$ and $c_i$ are at a distance $2k - i + 2$.

$$d(a_1, c_i) = 2k - i + 2, \quad k + 1 \leq i \leq 2k.$$

$c_i$ is also adjacent to $d_i$, so (18) implies,

$$d(a_1, d_i) = i + 2, \quad 1 \leq i \leq k.$$

For $k + 1 \leq i \leq 2k$, the vertices $d_i$ and $d_{2k}$ are at a distance $2k - i$ in the outer cycle $(d_1 d_2 \dots d_{k+1} \dots d_i \dots d_{2k})$. As $d_{2k}$ is adjacent to $c_{2k}$, $c_{2k}$ adjacent to $b_1$ and $b_1$ adjacent to $a_1$, therefore, $a_1$ and $d_i$ are at a distance $2k - i + 3$.

$$d(a_1, d_i) = 2k - i + 3 \quad k + 1 \leq i \leq 2k.$$

Therefore, $e(a_1) = k + 2$.

In the same manner as above, we calculate the eccentricity of $b_1$ in the cycle $(b_1 b_2 \dots b_i \dots b_{2k})$. The distance between $b_1$ and $b_i$ in this cycle is,

$$d(b_1, b_i) = i - 1, \; 1 \leq i \leq k + 1. \tag{19}$$

and,

$$d(b_1, b_i) = 2k + 1 - i, \quad k + 2 \leq i \leq 2k.$$

Therefore, we only take values of $i$ between one and $k + 1$. As each $b_i$ is adjacent to $c_i$ and $c_{i-1}$, using (19),

$$d(b_1, c_i) = i, \; 1 \leq i \leq k. \tag{20}$$

For $k+1 \leq i \leq 2k$, consider the cycle $(c_1 c_2 ... c_{k+2} ... c_i ... c_{2k})$. The distance between $c_i$ and $c_{2k}$ is $2k - i$. Since, $c_{2k}$ is adjacent to $b_1$. Thus,

$$d(b_1, c_i) = 2k - i + 1, \quad k+1 \leq i \leq 2k.$$

Each $c_i$ is also adjacent to $d_i$. Therefore, (20) shows,

$$d(b_1, d_i) = i + 1, \quad 1 \leq i \leq k.$$

For $k+1 \leq i \leq 2k$, consider the cycle $(d_1 d_2 ... d_i ... d_{2k})$. The distance between $d_i$ and $d_{2k}$ is $2k - i$. As $d_{2k}$ is adjacent to $c_{2k}$ and $c_{2k}$ adjacent to $b_1$, therefore,

$$d(b_1, d_i) = 2k - i + 2, \quad k+1 \leq i \leq 2k.$$

$b_i$ is also adjacent to $a_i$, i.e.,

$$d(b_1, a_i) = i, \quad 1 \leq i \leq k+1.$$

For $k+2 \leq i \leq 2k$, consider the cycle $(a_1, a_2 ... a_{k+2} ... a_i ... a_{2k})$. The distance between the vertices $a_i$ and $a_{2k}$ is $2k - i$. As $a_{2k}$ is adjacent to $a_1$, $a_1$ adjacent to $b_1$, therefore,

$$d(b_1, a_i) = 2k - i + 2, \quad k+2 \leq i \leq 2k.$$

As, $d_k$ and $a_{k+1}$ are farthest from $b_1$, therefore, $e(b_1) = k + 1$. Next, the distance between $c_1$ and $c_i$ in the cycle $(c_1 c_2 ... c_i ... c_{2k})$ is $i - 1$.

$$d(c_1, c_i) = i - 1, \ 1 \leq i \leq k+1. \tag{21}$$

Additionally, for $k+2 \leq i \leq 2k$,

$$d(c_1, c_i) = 2k - i + 1, \quad k+2 \leq i \leq 2k$$

Each $c_i$ is adjacent to $d_i$, from (21):

$$d(c_1, d_i) = i, \quad 1 \leq i \leq k+1.$$

For $k+2 \leq i \leq 2k$, the vertices $d_i$ and $d_{2k}$ are at a distance $2k - i$ in the cycle $(d_1 d_2 ... d_{k+2} ... d_i ... d_{2k})$. The vertex $d_{2k}$ is adjacent to $d_1$ and $d_1$ adjacent to $c_1$. Therefore,

$$d(c_1, d_i) = 2k - i + 2, \quad k+2 \leq i \leq 2k.$$

Each $c_i$ is adjacent to $b_i$ and $b_{i+1}$; Equation (17) implies,

$$d(c_1, b_1) = 1, \ d(c_1, b_2) = 1. \tag{22}$$

For $3 \leq i \leq k+1$, $b_2$ and $b_i$ are at a distance $i - 2$ in the path $c_1 \rightarrow b_2 \rightarrow b_3 \rightarrow ... \rightarrow b_i$. Again, $b_2$ is adjacent to $c_1$; thus, we have:

$$d(c_1, b_i) = i - 1, \ 3 \leq i \leq k+1. \tag{23}$$

For $k+2 \leq i \leq 2k$, consider the cycle $(b_1 b_2 ... b_{k+2} ... b_i ... b_{2k})$. The distance between $b_i$ and $b_{2k}$ is $2k - i$ in this cycle. As, $b_{2k}$ is adjacent to $b_1$, $b_1$ adjacent to $c_1$. Therefore,

$$d(c_1, b_i) = 2k - i + 2, \quad k+2 \leq i \leq 2k.$$

Since $b_i$ is adjacent to $a_i$, it follows from (22) that:

$$d(c_1, a_1) = 2 , \quad d(c_1, a_2) = 2$$

For $3 \leq i \leq k+1$, $b_i$ and $b_2$ are at a distance $i - 2$ in the path $c_1 \to b_2 \to b_3 \to \dots \to b_i \to a_i$. The vertex $b_i$ is adjacent to $a_i$ and $b_2$ adjacent to $c_1$. Therefore,

$$d(c_1, a_i) = i, \quad 3 \leq i \leq k+1. \quad \dots(23)$$

For $k+2 \leq i \leq 2k$, the distance between $a_i$ and $a_{2k}$ is $2k - i$ in the cycle $(a_1 a_2 \dots a_{k+2} \dots a_i \dots a_{2k})$. $a_{2k}$ is again adjacent to $a_1$, $a_1$ adjacent to $b_1$ and $b_1$ adjacent to $c_1$. For that reason,

$$d(c_1, a_i) = 2k - i + 3, \quad k+2 \leq i \leq 2k.$$

Consequently, $d_{k+1}$ and $a_{k+1}$ are farthest from $c_1$. Therefore, $e(c_1) = k+1$.
Next, take a vertex $d_1$ on the outer cycle. In this cycle $(d_1 d_2 \dots d_i \dots d_{2k})$,

$$d(d_1, d_i) = i - 1, \ 1 \leq i \leq k+1. \tag{24}$$

Additionally,

$$d(d_1, d_i) = 2k - i + 1, \quad k+2 \leq i \leq 2k.$$

In addition, each $d_i$ is adjacent to $c_i$,

$$d(d_1, c_i) = i, \quad 1 \leq i \leq k+1.$$

For $k + 2 \leq i \leq 2k$, take a cycle $(c_1, c_2 \dots c_{k+2} \dots c_i \dots c_{2k})$. The vertices $c_i$ and $c_{2k}$ are at a distance $2k - i$ in this cycle. In addition, $c_{2k}$ is adjacent to $c_1$ and $c_1$ adjacent to $d_1$. Then,

$$d(d_1, c_i) = 2k - i + 2, \quad k+2 \leq i \leq 2k.$$

Each $c_i$ is adjacent to $b_i$ and $b_{i+1}$. i.e., $d(d_1, b_1) = 2$ and:

$$d(d_1, b_2) = 2. \tag{25}$$

For $3 \leq i \leq k+1$, the vertices $b_2$ and $b_i$ are at a distance $i - 2$ in the path $d_1 \to c_1 \to b_2 \to b_3 \to \dots \to b_i$. $b_2$ is adjacent to $c_1$ and $c_1$ adjacent to $d_1$ in $S_n$. Therefore,

$$d(d_1, b_i) = i, \ 3 \leq i \leq k+1, \tag{26}$$

For $k + 2 \leq i \leq 2k$, the vertices $b_i$ and $b_{2k}$ are at distance $2k - i$ in the cycle $(b_1, b_2 \dots b_{k+2} \dots b_i \dots b_{2k})$. $b_{2k}$ is adjacent to $b_1$, $b_1$ adjacent to $c_1$ and $c_1$ adjacent to $d_1$; for this,

$$d(d_1, b_i) = 2k - i + 3, \quad k+2 \leq i \leq 2k.$$

In addition, $b_i$ is adjacent to $a_i$. This implies from (25),

$$d(d_1, a_1) = 3 , \quad d(d_1, a_2) = 3.$$

For $3 \leq i \leq k+1$, the vertices $b_2$ and $b_i$ are at a distance $i - 2$ in the path $d_1 \to c_1 \to b_2 \to b_3 \to \dots \to b_i \to a_i$. $b_i$ is adjacent to $a_i$, $b_2$ adjacent to $c_1$ and $c_1$ adjacent to $d_1$ in $S_n$. Therefore,

$$d(d_1, a_i) = i + 1, \quad 3 \leq i \leq k+1.$$

For $k + 2 \leq i \leq 2k$, consider the cycle $(a_1, a_2...a_{k+2}...a_i...a_{2k})$. The vertices $a_i$ and $a_{2k}$ are at a distance $2k - i$ in this cycle. $a_{2k}$ is adjacent to $a_1$, $a_1$ adjacent to $b_1$, $b_1$ adjacent to $c_1$ and $c_1$ adjacent to $d_1$. As a result,

$$d(d_1, a_i) = 2k - i + 4, \quad k + 2 \leq i \leq 2k.$$

Consequently,

$$e(d_1) = k + 2.$$

Thus, it is concluded that maximum eccentricity among all of the vertices of $S_n$ is $k + 2$, and the minimum eccentricity is $k + 1$.

Therefore

$$diam(S_n) = k + 2 = \frac{n}{2} + 2.$$

$$rad(S_n) = k + 1 = \frac{n}{2} + 1.$$

□

The following corollary is straightforward.

**Corollary 1.** *The center and periphery for the family of convex polytope $(S_n)$, when $n$ is even, are subgraphs induced by all of the central vertices $\{b_1, b_2, ..., b_i, ..., b_{2k}, c_1, c_2, ...., c_i, ..., c_{2k}\}$ and peripheral vertices $\{a_1, a_2, ..., a_i, ..., a_{2k}, d_1, d_2, ..., d_i, ..., d_{2k}\}$ of $S_n$, respectively.*

Now, we find out the radius and diameter of $S_n$, when $n$ is odd.

**Theorem 4.** *When $n$ is odd, the family of convex polytope $S_n$ has the radius and diameter as,*

$$diam(S_n) = \frac{n-1}{2} + 3,$$

$$rad(S_n) = \frac{n-1}{2} + 2.$$

**Proof.** Let $n = 2k + 1$, $k \geq 2$. Consider the cycle $(a_1 a_2 .... a_i ... a_{2k+1})$, and select a vertex $a_1$ in it. It is clear that,

$$d(a_1, a_i) = i - 1, \quad 1 \leq i \leq k + 1$$

$$d(a_1, a_i) = 2k + 2 - i, \quad k + 2 \leq i \leq 2k + 1, \tag{27}$$

Thus, the equations above lead to the proof including only $1 \leq i \leq k + 1$ in order to find a vertex having the greatest distance from $a_1$ in $S_n$. Since each $a_i$ is adjacent to $b_i$, therefore, (27) implies that:

$$d(a_1, b_i) = i, \ 1 \leq i \leq k + 1. \tag{28}$$

For $k + 2 \leq i \leq 2k + 1$, the vertices $b_i$ and $b_{2k+1}$ are at a distance $2k - i + 1$ in the cycle $(b_1 b_2 ... b_{k+2} ... b_i ... b_{2k+1})$. $b_{2k+1}$ is adjacent to $b_1$ and $b_1$ adjacent to $a_1$. Therefore, The distance between $a_1$ and $b_i$ is $2k - i + 3$.

$$d(a_1, b_i) = 2k + 3 - i, \quad k + 2 \leq i \leq 2k + 1.$$

Again, each $b_i$ is adjacent to $c_i$ and $c_{i-1}$, by using (28).

$$d(a_1, c_i) = i + 1, \ 1 \leq i \leq k + 1. \tag{29}$$

For $k + 2 \leq i \leq 2k + 1$, the distance between the vertices $c_i$ and $c_{2k+1}$ is $2k + 1 - i$ in the cycle $(c_1 c_2 ... c_{k+1} ... c_i ... c_{2k+1})$. Since, $c_{2k+1}$ is adjacent to $b_1$ and $b_1$ adjacent to $a_1$, therefore, $a_1$ and $c_i$ are at a distance $2k - i + 3$.

$$d(a_1, c_i) = 2k - i + 3, \quad k + 2 \leq i \leq 2k + 1.$$

In addition, $c_i$ is adjacent to $d_i$, therefore, (29) shows,

$$d(a_1, d_i) = i + 2, \quad 1 \leq i \leq k + 1.$$

For $k + 2 \leq i \leq 2k + 1$, the vertices $d_i$ and $d_{2k+1}$ are at a distance $2k + 1 - i$ in the cycle $(d_1 d_2 ... d_{k+1} ... d_i ... d_{2k+1})$. In addition, $d_{2k+1}$ is adjacent to $c_{2k+1}$, $c_{2k+1}$ adjacent to $b_1$ and $b_1$ adjacent to $a_1$. Therefore, $a_1$ and $d_i$ are at a distance $2k - i + 4$.

$$d(a_1, d_i) = 2k - i + 4, \quad k + 2 \leq i \leq 2k + 1.$$

As a result, $d_{k+1}$ is farthest from $a_1$; therefore, $e(a_1) = k + 3$.

In order to find out the eccentricity of the vertices on the cycle $(b_1 b_2 ... b_i ... b_{2k+1})$, the distance between $b_1$ and $b_i$ in this cycle is $i - 1$.

$$d(b_1, b_i) = i - 1, \; 1 \leq i \leq k + 1. \tag{30}$$

In addition,

$$d(b_1, b_i) = 2k - i + 2, \quad k + 2 \leq i \leq 2k + 1.$$

Further, each $b_i$ is adjacent to $c_i$ and $c_{i-1}$, therefore, (30) shows,

$$d(b_1, c_i) = i, \; 1 \leq i \leq k + 1. \tag{31}$$

For $k + 2 \leq i \leq 2k + 1$, consider the cycle $(c_1 c_2 ... c_{k+2} ... c_i ... c_{2k+1})$. The distance between $c_i$ and $c_{2k+1}$ is $2k - i + 1$. Since, $c_{2k+1}$ is adjacent to $b_1$, thus,

$$d(b_1, c_i) = 2k - i + 2, \quad k + 2 \leq i \leq 2k + 1.$$

Each $c_i$ is also adjacent to $d_i$. It is shown from (31),

$$d(b_1, d_i) = i + 1, \quad 1 \leq i \leq k + 1.$$

For $k + 2 \leq i \leq 2k + 1$, consider the cycle $(d_1 d_2 ... d_i ... d_{2k+1})$. The distance between $d_i$ and $d_{2k+1}$ is $2k - i + 1$. As $d_{2k+1}$ is adjacent to $c_{2k+1}$ and $c_{2k+1}$ adjacent to $b_1$, therefore,

$$d(b_1, d_i) = 2k - i + 3, \quad k + 2 \leq i \leq 2k + 1.$$

$b_i$ is also adjacent to $a_i$, i.e.,

$$d(b_1, a_i) = i, \quad 1 \leq i \leq k + 1.$$

For $k + 2 \leq i \leq 2k + 1$, consider the cycle $(a_1 a_2 ... a_{k+2} ... a_i ... a_{2k+1})$. The vertices $a_i$ and $a_{2k+1}$ is $2k + 1 - i$. As $a_{2k+1}$ is adjacent to $a_1$, $a_1$ adjacent to $b_1$, therefore,

$$d(b_1, a_i) = 2k - i + 3, \quad k + 2 \leq i \leq 2k + 1.$$

Since, $d_{k+1}$ is a vertex farthest from $b_1$, therefore, $e(b_1) = k + 2$.

Next, the distance between $c_1$ and $c_i$ in the cycle $(c_1 c_2 ... c_i ... c_{2k+1})$ is,

$$d(c_1, c_i) = i - 1, \; 1 \leq i \leq k + 1. \tag{32}$$

Additionally,

$$d(c_1, c_i) = 2k - i + 2, \quad k + 2 \leq i \leq 2k + 1.$$

Each $c_i$ is adjacent to $d_i$, from (32):

$$d(c_1, d_i) = i, \quad 1 \leq i \leq k + 1.$$

For $k + 2 \leq i \leq 2k + 1$, the vertices $d_i$ and $d_{2k+1}$ are at a distance $2k + 1 - i$ in the cycle $(d_1 d_2 ... d_{k+2} ... d_i ... d_{2k+1})$. The vertex $d_{2k+1}$ is adjacent to $d_1$ and $d_1$ adjacent to $c_1$. Therefore,

$$d(c_1, d_i) = 2k - i + 3, \quad k + 2 \leq i \leq 2k + 1.$$

Each $c_i$ is adjacent to $b_i$ and $b_{i+1}$; Equation (28) implies,

$$d(c_1, b_1) = 1, \; d(c_1, b_2) = 1. \tag{33}$$

For $3 \leq i \leq k + 2$, $b_2$ and $b_i$ are at a distance $i - 2$ in the path $c_1 \rightarrow b_2 \rightarrow b_3 \rightarrow ... \rightarrow b_i$. Again, $b_2$ is adjacent to $c_1$; thus, we have:

$$d(c_1, b_i) = i - 1, \; 3 \leq i \leq k + 2. \tag{34}$$

For $k + 3 \leq i \leq 2k + 1$, consider the cycle $(b_1 b_2 ... b_{k+1} ... b_i ... b_{2k+1})$. The distance between $b_i$ and $b_{2k+1}$ is $2k + 1 - i$ in this cycle. As $b_{2k+1}$ is adjacent to $b_1$ and $b_1$ adjacent to $c_1$, therefore,

$$d(c_1, b_i) = 2k - i + 3, \quad k + 3 \leq i \leq 2k + 1.$$

Since $b_i$ is adjacent to $a_i$, it follows from (33) that:

$$d(c_1, a_1) = 2 \; , \; d(c_1, a_2) = 2$$

For $3 \leq i \leq k + 2$, $b_i$ and $b_2$ are at a distance $i - 2$ in the path $c_1 \rightarrow b_2 \rightarrow b_3 \rightarrow ... \rightarrow b_i \rightarrow a_i$. The vertex $b_i$ is adjacent to $a_i$, $b_2$ adjacent to $c_1$. Therefore,

$$d(c_1, a_i) = i, \quad 3 \leq i \leq k + 2.$$

For $k + 3 \leq i \leq 2k + 1$, the distance between $a_i$ and $a_{2k+1}$ in the cycle $(a_1 a_2 ... a_{k+2} ... a_i ... a_{2k+1})$. $a_{2k+1}$ is again adjacent to $a_1$, $a_1$ adjacent to $b_1$ and $b_1$ adjacent to $c_1$. For that reason,

$$d(c_1, a_i) = 2k - i + 4, \quad k + 3 \leq i \leq 2k + 1.$$

Consequently, $a_{k+2}$ is a vertex farthest from $c_1$. Therefore, $e(c_1) = k + 2$.
Next, take a vertex $d_1$ on the outer cycle. In this cycle, $(d_1 d_2 ... d_i ... d_{2k+1})$,

$$d(d_1, d_i) = i - 1, \; 1 \leq i \leq k + 1. \tag{35}$$

$d(d_1, d_i)$ starts to decrease for $k + 2 \leq i \leq 2k + 1$ as,

$$d(d_1, d_i) = 2k - i + 2, \quad k + 2 \leq i \leq 2k + 1.$$

Each $d_i$ is adjacent to $c_i$,

$$d(d_1, c_i) = i, \quad 1 \leq i \leq k + 1.$$

For $k + 2 \leq i \leq 2k + 1$, take a cycle $(c_1 c_2 ... c_{k+2} ... c_i ... c_{2k+1})$. The vertices $c_i$ and $c_{2k+1}$ are at a distance $2k + 1 - i$ in this cycle. In addition, $c_{2k+1}$ is adjacent to $c_1$ and $c_1$ adjacent to $d_1$. Then,

$$d(d_1, c_i) = 2k - i + 3, \quad k + 2 \leq i \leq 2k + 1.$$

Each $c_i$ is adjacent to $b_i$ and $b_{i+1}$, i.e., $d(d_1, b_1) = 2$ and

$$d(d_1, b_2) = 2. \tag{36}$$

For $3 \leq i \leq k + 2$, the vertices $b_2$ and $b_i$ are at a distance $i - 2$ in the path $d_1 \rightarrow c_1 \rightarrow b_2 \rightarrow b_3 \rightarrow ... \rightarrow b_i$. $b_2$ is adjacent to $c_1$ and $c_1$ adjacent to $d_1$ in $S_n$. Therefore,

$$d(d_1, b_i) = i, \ 3 \leq i \leq k + 2. \tag{37}$$

For $k + 3 \leq i \leq 2k + 1$, consider the cycle $(b_1 b_2 ... b_{k+2} ... b_i ... b_{2k+1})$. The vertices $b_i$ and $b_{2k+1}$ are $2k + 1 - i$. $b_{2k+1}$ is adjacent to $b_1$, $b_1$ adjacent to $c_1$ and $c_1$ adjacent to $d_1$; for this,

$$d(d_1, b_i) = 2k - i + 4, \quad k + 3 \leq i \leq 2k + 1.$$

In addition, $b_i$ is adjacent to $a_i$. Therefore, (36) implies,

$$d(d_1, a_1) = \ , \ d(d_1, a_2) = 3.$$

For $3 \leq i \leq k + 2$, the vertices $b_2$ and $b_i$ are at a distance $i - 2$ in the path $d_1 \rightarrow c_1 \rightarrow b_2 \rightarrow b_3 \rightarrow ... \rightarrow b_i \rightarrow a_i$. $b_i$ is adjacent to $a_i$, $b_2$ adjacent to $c_1$ and $c_1$ adjacent to $d_1$ in $S_n$. Therefore,

$$d(d_1, a_i) = i + 1, \quad 3 \leq i \leq k + 2.$$

For $k + 3 \leq i \leq 2k + 1$, consider the cycle $(a_1 a_2 ... a_{k+2} ... a_i ... a_{2k+1})$. The vertices $a_i$ and $a_{2k+1}$ are $2k + 1 - i$. $a_{2k+1}$ is adjacent to $a_1$, $a_1$ adjacent to $b_1$, $b_1$ adjacent to $c_1$ and $c_1$ adjacent to $d_1$. As a result,

$$d(d_1, a_i) = 2k - i + 5, \quad k + 3 \leq i \leq 2k + 1.$$

This means,

$$e(d_1) = k + 3.$$

It shows that the maximum and minimum eccentricity among all of the vertices of $S_n$ are $k + 3$ and $k + 2$, respectively. Therefore:

$$diam(S_n) = k + 3 = \frac{n - 1}{2} + 3.$$

$$rad(S_n) = k + 2 = \frac{n - 1}{2} + 2.$$

□

Thus, we can summarize the above results as,

**Corollary 2.** *The center for the family of convex polytope $S(n)$ is a subgraph induced by all of the vertices of the interior and exterior cycles, and the periphery is the subgraphs induced by all of the peripheral vertices $\{a_1, a_2, ..., a_i, ..., a_{2k}, d_1, d_2, ..., d_i, ..., d_{2k}\}$ of $S_n$, respectively.*

### 3.1. Average Eccentricity for Convex Polytopes $S_n$

Here, the average eccentricity for the family of $S_n$ is being determined. The graph of $S_n$ consist of four major circles, and there are $n$ vertices in each circle. Therefore, the total number of vertices in $S_n$ (i.e., $\acute{n}$) is $4n$; it follows,

$$avec(S_n) = \frac{1}{\acute{n}} \sum_{u \in V(G)} e_G u$$

By Theorem 3:

$$
\begin{aligned}
avec(S_n) &= \frac{1}{4 \times (\acute{n})}[n \times \{e(a_1) + e(b_1) + e(c_1) + e(d_1)\}] \\
&= \frac{1}{4 \times n}[n \times \{(k+2) + (k+1) + (k+1) + (k+2)\}] \\
&= \frac{1}{4 \times n}[2n \times \{(k+2) + (k+1)\}] \\
&= \frac{1}{2}[2k+3] \\
&= k + \frac{3}{2} \\
&= \frac{n+3}{2}.
\end{aligned}
$$

and by Theorem 4,

$$
\begin{aligned}
avec(S_n) &= \frac{1}{4 \times n}[n \times \{(k+3) + (k+2) + (k+2) + (k+3)\}] \\
&= \frac{1}{4 \times n}[2n \times \{(k+3) + (k+2)\}] \\
&= \frac{1}{2}[2k+5] \\
&= k + \frac{5}{2} \\
&= \frac{n-1}{2} + \frac{5}{2} \\
&= \frac{n+4}{2}.
\end{aligned}
$$

Therefore, we have the following result:

$$
avec(S_n) = \begin{cases} \dfrac{n+3}{2}, & \text{for all even values of } n \\ \dfrac{n+4}{2}, & \text{for all odd values of } n. \end{cases}
$$

### 3.2. Illustration

Consider the graph of $S_6$. Its center and periphery are shown in Figures 3 and 4.

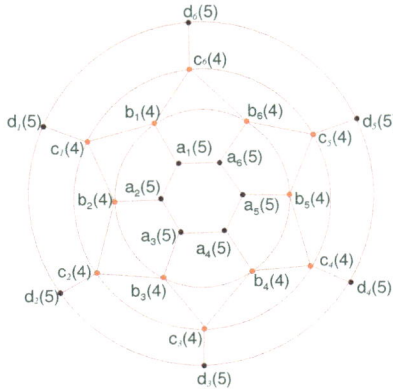

**Figure 3.** The graph of convex polytope $S_6$.

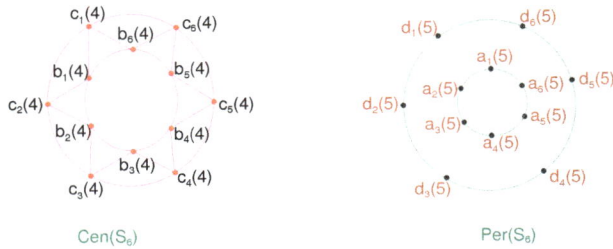

Cen($S_6$)                     Per($S_6$)

**Figure 4.** Centrality for $S_6$.

## 4. The Center and Periphery for Convex Polytopes $T_n$

Here, we established the center and periphery for $T_n$ and show that $T_n$ is not self-centered.

**Definition 9.** *The graph of convex polytope $T_n$ can be obtained from the graph of convex polytope $Q_n$ by adding new edges . It consist of three-sided faces, five-sided faces and n-sided faces. $a_{i+1}b_i$, i.e., $V(T_n) = V(Q_n)$ and $E(T_n) = E(Q_n) \bigcup \{a_{i+1}b_i : 1 \le i \le n\}$.*

This section begins with the following theorem on $T_n$.

**Theorem 5.** *The diameter for the family of convex polytope $T_n$ is,*

$$diam(T_n) = \begin{cases} \dfrac{n}{2}+2, & \text{for } n = 2k; \\ \dfrac{n-1}{2}+2, & \text{for } n = 2k+1. \end{cases}$$

*In addition, its radius,*

$$rad(T_n) = \begin{cases} \dfrac{n}{2}+1, & \text{for } n \text{ to be even}; \\ \dfrac{n+1}{2}, & \text{for } n \text{ to be odd}. \end{cases}$$

**Proof.** Consider, $n = 2k, k \ge 2$. Choose take cycle $(a_1 a_2 ... a_i ... a_{2k})$. In this cycle:

$$d(a_1, a_i) = i - 1, \ 1 \le i \le k+1. \tag{38}$$

For $k + 2 \leq i \leq 2k$, the distance between $a_1$ and $a_i$ decreases from $k - 1$ to one, i.e.,

$$d(a_1, a_i) = 2k - i + 1, \quad k + 2 \leq i \leq 2k.$$

Therefore, we must considered $1 \leq i \leq k + 1$ in order to find the distance of a vertex $a_1$ from a vertex farthest from it in $T_n$.

In the graph of $T_n$, each $a_i$ is adjacent to $b_i$ and $b_{i-1}$; thus, (38) implies,

$$d(a_1, b_i) = i, \ 1 \leq i \leq k. \tag{39}$$

For $k + 1 \leq i \leq 2k$, the vertices $b_i$ and $b_{2k}$ are at a distance $2k - i$ in the cycle $(b_1 b_2 ... b_{k+1} ... b_i ... b_{2k})$. In addition, $b_{2k}$ is adjacent to $a_1$. Therefore, The distance between $a_1$ and $b_i$ is $2k - i + 1$.

$$d(a_1, b_i) = 2k - i + 1 \quad k + 1 \leq i \leq 2k.$$

Further, each $b_i$ is adjacent to $c_i$ and $c_{i-1}$, using (39).

$$d(a_1, c_1) = 2, d(a_1, c_{2k}) = 2$$

for $2 \leq i \leq k$, consider path $a_1 \to b_1 \to b_2 \to ... \to b_i \to c_i$. The distance between $b_1$ and $b_i$ is $i - 1$. Each $b_i$ is adjacent to $c_i$ and $b_1$ adjacent to $a_1$. Therefore,

$$d(a_1, c_i) = i + 1, \ 2 \leq i \leq k. \tag{40}$$

Next, for $k + 1 \leq i \leq 2k - 1$, consider the cycle $(b_1 b_2 ... b_{i+1} ... b_{2k})$. The vertices $b_{2k}$ and $b_{i+1}$ are at a distance $2k - i - 1$. Further, $b_{2k}$ is adjacent to $a_1$ and $b_{i+1}$ adjacent to $c_i$. Therefore,

$$d(a_1, c_i) = 2k - i + 1, \quad k + 1 \leq i \leq 2k - 1.$$

$c_i$ is also adjacent to $d_i$. Therefore, (40) implies

$$d(a_1, d_i) = i + 2, \quad 1 \leq i \leq k.$$

For $k + 1 \leq i \leq 2k$, the vertices $d_i$ and $d_{2k}$ are at a distance $2k - i$ in the cycle $(d_1 d_2 ... d_i ... d_{2k})$. In addition, each $d_{2k}$ is adjacent to $c_{2k}$, $c_{2k}$ adjacent to $b_1$ and $b_1$ adjacent to $a_1$; therefore,

$$d(a_1, d_i) = 2k + 3 - i, \quad k + 1 \leq i \leq 2k.$$

Hence, $d_k$ is a vertex at the largest distance from $a_1$. Therefore, $e(a_1) = k + 2$.

Next, continue this for cycle $(b_1 b_2 ... b_i ... b_{2k})$; we choose a vertex $b_1$, such that,

$$d(b_1, b_i) = i - 1, \ 1 \leq i \leq k + 1. \tag{41}$$

The distance between $b_1$ and $b_i$ decreases from $k - 1$ to one, when $i$ increases from $k + 2$ to $2k$.

$$d(b_1, b_i) = 2k - i + 1, \quad k + 2 \leq i \leq 2k.$$

In addition, each $b_i$ is adjacent to $a_i$ and $a_{i+1}$.

$$d(b_1, a_1) = 1, \quad d(b_1, a_2) = 1$$

and when $3 \leq i \leq k+1$, consider the path $b_1 \rightarrow a_2 \rightarrow a_3 \rightarrow \ldots \rightarrow a_i$. $a_2$ and $a_i$ are at a distance $i-2$ in this path, and $a_2$ is adjacent to $b_1$; therefore,

$$d(b_1, a_i) = i-1, \quad 3 \leq i \leq k+1.$$

For $k+2 \leq i \leq 2k$, consider the cycle $(a_1 a_2 \ldots a_{k+2} \ldots a_i \ldots a_{2k})$. The distance between $a_i$ and $a_{2k}$ is $2k-i$. As $a_{2k}$ is adjacent to $a_1$ and $a_1$ adjacent to $b_1$, therefore,

$$d(b_1, a_i) = 2k-i+2, \quad k+2 \leq i \leq 2k.$$

In addition, $b_i$ is adjacent to $c_i$ and $c_{i-1}$; using (41), we have:

$$d(b_1, c_i) = i, \ 1 \leq i \leq k. \tag{42}$$

For $k+1 \leq i \leq 2k$, consider the path $b_1 \rightarrow b_{2k} \rightarrow b_{2k-1} \rightarrow \ldots \rightarrow b_{i+1} \rightarrow c_i$. The distance between $b_{2k}$ and $b_{i+1}$ is $2k-i-1$. Further, $b_{2k}$ is adjacent to $b_1$. In addition, $b_{i+1}$ is adjacent to $c_i$. Therefore, the distance between $b_1$ and $c_i$ is $2k-i+1$.

$$d(b_1, c_i) = 2k-i+1, \quad k+1 \leq i \leq 2k$$

Further, $c_i$ is adjacent to $d_i$; hence, (42) shows,

$$d(b_1, d_i) = i+1, \quad 1 \leq i \leq k.$$

For $k+1 \leq i \leq 2k$, the vertices $d_i$ and $d_{2k}$ are at a distance $2k-i$ in the cycle $(d_1 d_2 \ldots d_{k+2} \ldots d_i \ldots d_{2k})$. The vertex $d_{2k}$ is adjacent to $c_{2k}$ and $c_{2k}$ adjacent to $b_1$. Therefore,

$$d(b_1, d_i) = 2k-i+2, \quad k+1 \leq i \leq 2k.$$

Hence, $d_k$ is a vertex farthest from $b_1$. Therefore, $e(b_1) = k+1$

Next, to find out the eccentricity of the vertices $\{c_i, 1 \leq i \leq 2k\}$, take a vertex $c_1$ among all $c_i$'s, and each $c_i$ is adjacent to $b_i, b_{i+1}$, i.e.,

$$d(c_1, b_1) = 1, \quad , \quad d(c_1, b_2) = 1$$

and when $3 \leq i \leq k+1$, consider the path $c_1 \rightarrow b_2 \rightarrow b_3 \rightarrow \ldots \rightarrow b_i$. $b_2$ and $b_i$ are at distance $i-2$, and again, $b_2$ is adjacent to $c_1$; thus, we have:

$$d(c_1, b_i) = i-1, \ 3 \leq i \leq k+1. \tag{43}$$

For $k+2 \leq i \leq 2k$, consider the cycle $(b_1 b_2 \ldots b_{k+2} \ldots b_i \ldots b_{2k})$. The distance between $b_i$ and $b_{2k}$ is $2k-i$ in this cycle. As $b_{2k}$ is adjacent to $b_1$ and $b_1$ adjacent to $c_1$, therefore,

$$d(c_1, b_i) = 2k-i+2, \quad k+2 \leq i \leq 2k.$$

Moreover, $b_i$ is adjacent to $a_i$ and $a_{i+1}$; it follows from (43) that:

$$d(c_1, a_1) = 2 \ , \ d(c_1, a_2) = 2, \ d(c_1, a_3) = 2$$

For $4 \leq i \leq k+2$, $a_i$ and $a_2$ are at a distance $i-3$ in the path $c_1 \rightarrow b_2 \rightarrow a_3 \rightarrow \ldots \rightarrow a_i$. Furthermore, $a_2$ is adjacent to $b_2$ and $b_2$ adjacent to $c_1$. Thus,

$$d(c_1, a_i) = i-1, \ 4 \leq i \leq k+2. \tag{44}$$

For $k + 3 \leq i \leq 2k$, the distance between $a_i$ and $a_{2k}$ in the cycle $(a_1 a_2 ... a_{k+3} ... a_i ... a_{2k})$ is $2k - i$, and $a_{2k}$ is adjacent to $a_1$, $a_1$ adjacent to $b_1$ and $b_1$ adjacent to $c_1$. For that reason,

$$d(c_1, a_i) = 2k - i + 3, \quad k + 3 \leq i \leq 2k.$$

Again, $c_i$ is adjacent to $d_i$. Hence,

$$d(c_1, d_i) = i, \quad 1 \leq i \leq k + 1$$

For $k + 2 \leq i \leq 2k$, the vertices $d_{2k}$ and $d_i$ are at a distance $2k - i$ in the cycle $(d_1 d_2 ... d_{k+2} ... d_i ... d_{2k})$, and $d_{2k}$ is adjacent to $d_1$ and $d_1$ adjacent to $c_1$ in $T_n$. Therefore,

$$d(c_1, d_i) = 2k - i + 2, \quad k + 2 \leq i \leq 2k$$

In order to find the distance between $c_1$ and $c_i$, $1 \leq i \leq k + 1$, consider the path $c_1 \to b_2 \to b_3 ... \to b_i \to c_i$. The distance between $b_2$ and $b_i$ is i-2, and $b_i$ is adjacent to $c_i$ and $b_2$ adjacent to $c_1$. Therefore,

$$d(c_1, c_i) = i, \quad 1 \leq i \leq k + 1.$$

For more values of $i$, $d(c_1, c_i)$ begins to reduce as,

$$d(c_1, c_i) = 2k - i + 2, \quad k + 2 \leq i \leq 2k$$

This means that $a_{k+2}$, $c_{k+1}$ and $d_{k+1}$ are the vertices farthest from $c_1$. Therefore, $e(c_1) = k + 1$. Now, we find the eccentricities of the vertices on the cycle $(d_1 d_2 ... d_i ... d_{2k})$. In this cycle,

$$d(d_1, d_i) = i - 1, \quad 1 \leq i \leq k + 1.$$

For $k + 2 \leq i \leq 2k$, the distance between $d_1$ and $d_i$ decreases from $k - 1$ to one.

$$d(d_1, d_i) = 2k + 1 - i, \quad k + 2 \leq i \leq 2k.$$

As $d_i$ adjacent to $c_i$:

$$d(d_1, c_i) = i, \quad 1 \leq i \leq k + 1. \tag{45}$$

When $i$ increases from $k + 2$ to $2k$, the distance between $d_i$ and $d_{2k}$ is $2k - i$ in the cycle $(d_1 d_2 ... d_{k+2} ... d_i ... d_{2k})$. In addition, $d_{2k}$ is adjacent to $d_1$ and $d_i$ adjacent to $c_i$. Thus,

$$d(d_1, c_i) = 2k - i + 2, \quad k + 2 \leq i \leq 2k.$$

As each $c_i$ is adjacent to $b_i$, $b_{i+1}$.

$$d(d_1, b_1) = 2, \quad d(d_1, b_2) = 2, \tag{46}$$

For $3 \leq i \leq k + 1$, consider a path $d_1 \to c_1 \to b_2 \to b_3 \to ... \to b_i$. $b_2$ and $b_i$ are at a distance $i - 2$, and $b_2$ is adjacent to $c_1$ and $c_1$ adjacent to $d_1$ in $T_n$. Therefore,

$$d(d_1, b_i) = i, \quad 3 \leq i \leq k + 1$$

For $k + 2 \leq i \leq 2k$, consider the cycle $(b_1 b_2 ... b_{k+2} ... b_i ... b_{2k})$. The distance between the vertices $b_i$ and $b_{2k}$ is $2k - i$. $b_{2k}$ is adjacent to $b_1$, $b_1$ adjacent to $c_1$ and $c_1$ adjacent to $d_1$; for that reason,

$$d(d_1, b_i) = 2k - i + 3, \quad k + 2 \leq i \leq 2k.$$

In addition, $b_i$ is adjacent to $a_i$ and $a_{i+1}$. Therefore, (46) implies,

$$d(d_1, a_1) = 3, \quad d(d_1, a_2) = 3, \quad d(d_1, a_3) = 3$$

For $4 \leq i \leq k+2$, the vertices $a_3$ and $a_i$ are at a distance $i - 3$ in the path $d_1 \rightarrow c_1 \rightarrow b_2 \rightarrow a_3 \rightarrow a_4 \rightarrow ... \rightarrow a_i$. Further, $a_3$ is adjacent to $b_2$, $b_2$ adjacent to $c_1$ and $c_1$ adjacent to $d_1$ in $T_n$. Therefore,

$$d(d_1, a_i) = i, \quad 4 \leq i \leq k+2.$$

For $k + 3 \leq i \leq 2k$, consider the cycle $(a_1 a_2 ... a_{k+3} ... a_i ... a_{2k})$. The distance between the vertices $a_i$ and $a_{2k}$ is $2k - i$. $a_{2k}$ is adjacent to $a_1$ and $a_1$ adjacent to $b_1$. Further, $b_1$ adjacent to $c_1$ and $c_1$ adjacent to $d_1$. As a result,

$$d(d_1, a_i) = 2k - i + 4, \quad k + 3 \leq i \leq 2k.$$

This shows that $a_{k+2}$ is at the highest distance from $d_1$. Therefore, $e(d_1) = k + 2$.

Thus, it is concluded that the maximum eccentricity among all of the vertices of $T_n$ is $k + 2$, and $k + 1$ is the minimum eccentricity. Therefore, $\text{diam}(T_n) = k + 2 = \frac{n}{2} + 2$.

$$rad(T_n) = k + 1 = \frac{n}{2} + 1.$$

For odd $n$, the proof is analogous to the case discussed above and omitted. $\quad\square$

**Corollary 3.** *The center of $T_n$, when n is even, is the subgraph induced by the central vertices $\{b_i \cup c_i : 1 \leq i \leq n\}$, while the periphery is the subgraph induced by the vertices of inner and outer cycles.*

### 4.1. Average Eccentricity for Convex Polytopes $T_n$

There are four circles in the graph of $T_n$, and each circle has $n$ vertices. The average eccentricity for the graph of convex polytope $T_n$ can be found out by dividing sum of eccentricities of all vertices on each circle to its total number of vertices. Therefore,

$$avec(T_n) = \frac{1}{\acute{n}} \sum_{u \in V(G)} e_G u$$

By Theorem 5:

$$
\begin{aligned}
avec(T_n) &= \frac{1}{4 \times n} [n \times \{(k+2) + (k+1) + (k+1) + (k+2)\}] \\
&= \frac{1}{4 \times n} [2n \times \{(k+2) + (k+1)\}] = \frac{1}{2} [2k+3] = k + \frac{3}{2} = \frac{n+3}{2}.
\end{aligned}
$$

and by Theorem 5,

$$
\begin{aligned}
avec(T_n) &= \frac{1}{4 \times n} [n \times \{(k+2) + (k+2) + (k+1) + (k+2)\}] \\
&= \frac{1}{4 \times n} [3n \times (k+2) + n \times (k+1)] = \frac{1}{4} [4k+7] = \frac{1}{4} [4(\frac{n-1}{2}) + 7] = \frac{n}{2} + \frac{5}{4}.
\end{aligned}
$$

Therefore, we get the following immediate result:

$$
avec(T_n) = \begin{cases} \dfrac{n}{2} + \dfrac{5}{4}, & \text{if } n = 2k+1 \, ; \\ \dfrac{n+3}{2}, & \text{if } n = 2k \, . \end{cases}
$$

### 4.2. Illustration

Consider the graph $T_6$. The center and periphery for $T_6$ are shown in Figures 5 and 6.

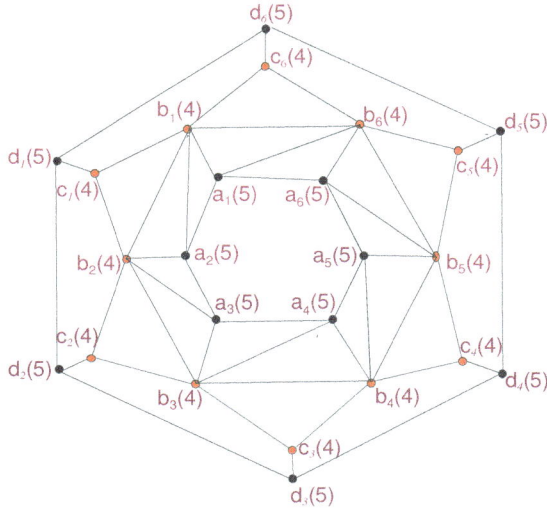

**Figure 5.** The graph of convex polytope $T_6$.

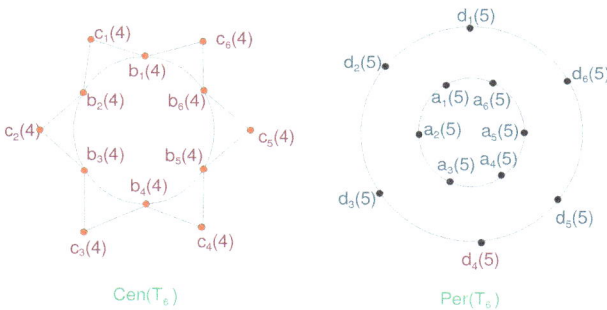

**Figure 6.** Centrality for $T_6$.

## 5. Concluding Remarks

In summary, we have studied the center and periphery of three types of families of convex polytopes via finding a subgraph induced by central and peripheral vertices. The predetermined facts about the eccentricity, radius and diameter of graphs play an important role in order to find the center and periphery for specific families of graphs; the average eccentricity of the above families of graphs has also been demonstrated.

## 6. Open Problems

This paper consist of the center and periphery for families of convex polytope graphs. This is an open problem for new researchers to find the center and periphery for others families of graphs, such as the corona product, composition product and lexicographic product of families of graphs.

**Acknowledgments:** This work was supported by the Dong-A University research fund, Korea.

**Author Contributions:** Waqas Nazeer, Shin Min Kang, Saima Nazeer, Mobeen Munir, Imrana Kausor, Ammara Sehar and Young Chel Kwun contributed equally to the writing of this paper. All authors read and approved the final manuscript.

**Conflicts of Interest:** The authors declare no conflict of interest.

## References

1. Yu, G.; Feng, L. On Connective Eccentricity Index of Graphs. *MATCH Commun. Math. Comput. Chem.* **2013**, *69*, 611–628.
2. Goddar, W.; Oellermann, O.R. Distance in Graphs. Available online: https://people.cs.clemson.edu/goddard/papers/distanceChapter.pdf (accessed on 28 November 2016).
3. Jorden, C.; Lignes, S.L.A.D.; Reine, J. Sur les assemblages de lignes. *J. Ang. Math.* **1869**, *70*, 185–190.
4. Bollobas, B. The Diameter of Random Graph. *Trans. Am. Math. Soc.* **1981**, *268*, 41–52.
5. Farber, M. On Diameters and Radii of Bridge Graphs. *Discret. Math.* **1989**, *73*, 249–260.
6. Klee, V.; Larman, D. Diameter of Random Graph. *Can. J. Math.* **1981**, *33*, 618–640.
7. Akiyama, J.; Ando, K.; Aavis, D. Miscellaneous Properties of Equi-Eccentric Graphs. In *Convexity and Graph Theory (Jerusalem, 1981)*; Elsevier: North Holland, Amsterdam, 1984.
8. Buckely, F. Self-centered Graphs. *Ann. N. Y. Acad. Sci.* **1989**, *576*, 71–78.
9. Buckely, F.; Miller, Z.; Slater, P.J. On Graphs Containing a Given Graphs as Center. *J. Graph Theory* **1981**, *5*, 427–432.
10. Janakiraman, T.N.; Ramanujan, J. On self-centered Graphs. *Math. Soc.* **1992**, *7*, 83–92.
11. Negami, S.; Xu, G.H. Locally Geodesic Cycles in 2-self-centered Graphs. *Discret. Math.* **1986**, *58*, 263–268.
12. Buckely, F. Self-centered Graph With Given Radius. *Congr. Number* **1979**, *23*, 211–215.
13. Proskurowski, A. Centers of Maximal Outer Planar Graph. *J. Graph Theory* **1980**, *4*, 75–79.
14. Hedetniemi, S.T. *Center of Recursive Graphs*; Technical Report CS-TR-79-14; Department of Computer Science, University of Oregon: Engene, OR, USA, 1979; p. 13.
15. Chartrand, G. *Introduction to Graph Theory*; Tata McGraw-Hill Education: New York, NY, USA, 2006.
16. Buckley, F.; Harary, F. *Distance in Graphs*; Addison-Wesley: Red Wood City, CA, USA, 1990.
17. Dankelman, P.; Goddard, W.; Swart, C.S. The Average Eccentricity of a Graph and Its Subgraphs. *Utilitas. Math.* **1988**, *65*, 41–51.
18. Hinz, A.M.; Parisse, D. The Average Eccentricity of Sierpinski Graphs. *Graph Comb.* **2012**, *28*, 671–686.
19. Skorobogatov, V.A.; Dobrynin, A.A. Metric Analysis of Graphs. *MATCH. Commun. Math. Comput. Chem.* **1988**, *23*, 105–151.
20. Takes, F.W.; Kosters, W.A. Computing the Eccentricity Distribution of Large Graphs. *Algorithms* **2013**, *6*, 100–118.
21. Leskovec, J.; Kleinberg, J.; Faloutsos, C. Graph Evolution: Densification and Shrinking Diameters. *ACM Trans. Knowl. Discov. Data* **2007**, *1*, 2.
22. Kang, U.; Tsourakakis, C.E.; Appel, A.P.; Faloutsos, C.; Leskovec, J. HADI: Mining Radii of Large Graphs. *ACM Trans. Knowl. Discov. Data (TKDD)* **2011**, *5*, 8.

*Article*
# Operations on Oriented Maps

**Tomaž Pisanski [1,*], Gordon Williams [2] and Leah Wrenn Berman [2]**

[1] Department of Information Sciences and Technologies (FAMNIT), University of Primorska, 6000 Koper, Slovenia
[2] Department of Mathematics & Statistics, University of Alaska Fairbanks, Fairbanks, AK 99775, USA; giwilliams@alaska.edu (G.W.); lwberman@alaska.edu (L.W.B.)
* Correspondence: Tomaz.Pisanski@upr.si

Received: 31 July 2017; Accepted: 11 November 2017; Published: 14 November 2017

**Abstract:** A map on a closed surface is a two-cell embedding of a finite connected graph. Maps on surfaces are conveniently described by certain trivalent graphs, known as flag graphs. Flag graphs themselves may be considered as maps embedded in the same surface as the original graph. The flag graph is the underlying graph of the dual of the barycentric subdivision of the original map. Certain operations on maps can be defined by appropriate operations on flag graphs. Orientable surfaces may be given consistent orientations, and oriented maps can be described by a generating pair consisting of a permutation and an involution on the set of arcs (or darts) defining a partially directed arc graph. In this paper we describe how certain operations on maps can be described directly on oriented maps via arc graphs.

**Keywords:** map; oriented map; truncation; dual; medial; snub; flag graph; arc graph

**MSC:** 52C20, 05C10, 51M20

## 1. Maps and Oriented Maps

A *map* $\mathcal{M}$ on a closed surface is a two-cell embedding of a finite connected graph $G = (V, E)$ (see, e.g., [1] or [2]). Equivalently, every map can be viewed as a set of three fixed-point-free involutions $r_0, r_1$ and $r_2$ acting on a set of *flags* $\Omega$ with the property that $r_0 r_2$ is also a fixed-point-free involution, in which case we denote the map as $\mathcal{M} = (\Omega, \langle r_0, r_1, r_2 \rangle)$; see, for instance, [3] (p. 415). With this second point of view, each map can be described completely using a three-edge colored cubic graph, called the *flag graph*, whose vertices are elements of $\Omega$, where $\omega_1$ and $\omega_2$ are connected by an edge colored $i$ in the flag graph if and only if $r_i(\omega_1) = \omega_2$. Generally, only connected flag graphs are considered. The graph $G$ is called the *skeleton* or the *underlying graph* of a *map* $\mathcal{M}$.

In some cases, one may relax the conditions and allow fixed points in some of the involutions $r_0, r_1$ and $r_2$. Such a structure describes a map in a surface with a boundary and with its flag graph containing semi-edges. We call such a map and respective flag graph *degenerate*.

If the surface in which the map resides is orientable, then $\mathcal{M}$ is said to be an *orientable map*. There is a well-known test involving flag graphs to determine whether a given map is orientable:

**Proposition 1.** *A map is orientable if and only if its flag graph is bipartite.*

In a bipartite flag graph, the flags (i.e., vertices) of the map come in two color classes; the vertices in a single color class of flags are called *arcs*. Restricting attention to one set of arcs corresponds to choosing an orientation of the orientable map, and we call the restricted graph an *oriented map*; each orientable map gives rise to two oppositely oriented oriented maps. We note that in a map there are four flags per edge of the skeleton, while in an oriented map there are two arcs per edge. That is, if $\Omega$ denotes the set of flags of the flag graph, and $\Sigma$ denotes one of the sets of arcs of the oriented graph, then $|\Omega| = 4|E|$ and $|\Sigma| = 2|E|$, where $E$ is the set of edges of the skeleton of the original map.

Considering only oriented maps, as we do in this paper, allows us to treat each of the two orientations of an orientable map separately. It also allows us to introduce the notion of an *oriented symmetry type graph*, which captures the symmetry properties that are preserved under orientation-preserving automorphisms of an oriented map. Most researchers consider maps and oriented maps separately. A prominent exception is found in the lecture notes of Roman Nedela (see [4]), where both maps and oriented maps are studied using the same tools. The main difference between our approach and Nedela's is that Nedela puts the emphasis on groups, while we focus mainly on graphs (flag graphs and arc graphs).

More technically, we define an oriented map to be a structure $\mathcal{M}_o := \{\Sigma, \langle r, R \rangle\}$, where $r$ is a fixed-point-free involution and $R$ is *any* permutation acting on a set of objects $\Sigma$, called *arcs*. We also view the oriented map $\mathcal{M}_o$ as a partially directed graph, called an *arc graph*, whose vertices are elements of $\Sigma$, having some edges directed and some undirected. The directed edges form a directed 2-factor, corresponding to the action of the permutation $R$ on elements of $\Sigma$. The undirected edges form an undirected 1-factor, corresponding to the action of the involution $r$ on the elements of $\Sigma$. Again, the only other constraint usually imposed on such a structure is the fact that the arc graph must be connected. If the involution $r$ is not fixed-point-free, then we say that the corresponding arc graph is *degenerate*.

We note that each oriented map $\mathcal{M}_o$ gives rise to its *skeleton graph* $G$ in the following way: the vertices of $G$ are the orbits of $R$ on $\Sigma$ and the edges of $G$ are the orbits of $r$ on $\Sigma$. A vertex is incident with an edge if and only if the corresponding orbits have a non-empty intersection. Using this description, we can generate an oriented map $\mathcal{M}_o$ from any orientable map $\mathcal{M} = \{\Omega, \langle r_0, r_1, r_2 \rangle\}$ by first observing that, because $\mathcal{M}$ is orientable, the flag graph is bipartite, with bipartite sets $\Omega^+$ and $\Omega^-$. We define $\Sigma = \Omega^+$, $r = r_0 r_2$ and $R = r_1 r_2$; we note that by the nature of the bipartition of the vertices of $\mathcal{M}$, the elements of $\Sigma$ are in one-to-one correspondence with the endpoints of the edges (i.e., the arcs) of the flag graph of $\mathcal{M}$. (If we instead define $\hat{r} = r_2 r_0$ and $\hat{R} = r_2 r_1$, then we get the "other" orientation of $\mathcal{M}$.) A fragment of an oriented map $\mathcal{M}_o$ is shown in Figure 1, with a portion of the underlying map $\mathcal{M}$ also shown.

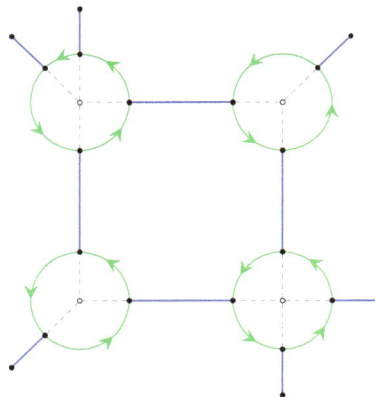

**Figure 1.** A fragment of an oriented map. The arcs are shown as black vertices; the green arrows correspond to the action of $R$ on the arcs, and blue edges, to the action of $r$. The dashed lines and white vertices are from the underlying unoriented map. Orientation is counterclockwise, indicated with arrows.

Conversely, any oriented map $\mathcal{M}_o := \{\Sigma, \langle r, R \rangle\}$ gives rise to the associated orientable map $\mathcal{M} = \{\Omega, \langle r_0, r_1, r_2 \rangle\}$ in the following way: let $\Sigma_0$ and $\Sigma_1$ be two copies (colored using colors 0 and 1) of $\Sigma$, with elements $\sigma_0$ and $\sigma_1$ respectively, and define

$$\Omega := \Sigma_0 \cup \Sigma_1$$

$$r_0(\sigma_0) := r(\sigma)_0 \qquad\qquad r_0(\sigma_1) := r(\sigma)_1$$
$$r_2(\sigma_0) := r(\sigma)_1 \qquad\qquad r_2(\sigma_1) := r(\sigma)_0$$
$$r_1(\sigma_0) := R(\sigma)_1 \qquad\qquad r_1(\sigma_1) := R^{-1}(\sigma)_0$$

It is straightforward to show that the skeleton of an oriented map is isomorphic as a graph to the skeleton of the underlying orientable map.

**Example 1.** *The cube may be considered either as*

- *an oriented map:* $\mathcal{M}_o := \{\Sigma, \langle r, R \rangle\}$, *or*
- *as an (orientable) map:* $\mathcal{M} = \{\Omega, \langle r_0, r_1, r_2 \rangle\}$.

*These two representations are depicted in Figure 2.*

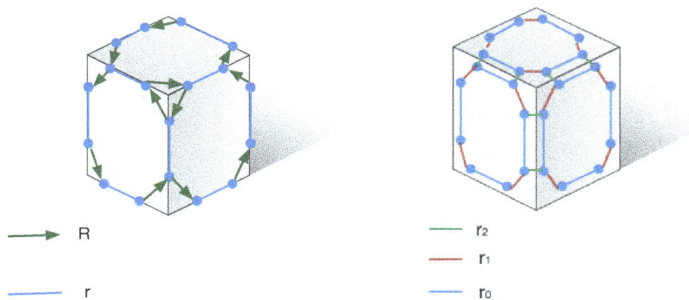

**Figure 2.** The cube may be considered as an oriented map determined by $R$ and $r$, or as a map determined by $r_0, r_1, r_2$.

We recall that two maps $\mathcal{M}^{(1)} = \{\Omega^{(1)}, \langle r_0^{(1)}, r_1^{(1)}, r_2^{(1)} \rangle\}$ and $\mathcal{M}^{(2)} = \{\Omega^{(2)}, \langle r_0^{(2)}, r_1^{(2)}, r_2^{(2)} \rangle\}$ are isomorphic if there exists a bijection $h$ between $\Omega^{(1)}$ and $\Omega^{(2)}$ such that $h r_i^{(1)} = r_i^{(2)} h$ for $i = 0, 1, 2$, and we call such a bijection a *map isomorphism*.

Similarly, we say that two oriented maps $\mathcal{M}_o^{(1)} = \{\Sigma^{(1)}, \langle r^{(1)}, R^{(1)} \rangle\}$ and $\mathcal{M}_o^{(2)} = \{\Sigma^{(2)}, \langle r^{(2)}, R^{(2)} \rangle\}$ are isomorphic if there exists a bijection $h$ between $\Sigma^{(1)}$ and $\Sigma^{(2)}$ such that $h r^{(1)} = r^{(2)} h$ and $h R^{(1)} = R^{(2)} h$. Such a bijection is called an *oriented map isomorphism*.

As usual, the isomorphisms of a structure to itself are called *automorphisms* and form a group. By Aut $\mathcal{M}$ we denote the group of *map automorphisms*, and similarly by Aut $\mathcal{M}_o$ we denote the group of *oriented map isomorphisms*.

**Lemma 1** (Fundamental Lemma of Symmetries of Maps)**.** *The action of the automorphism group* Aut $\mathcal{M}$ *of the map* $\mathcal{M} = \{\Omega, \langle r_0, r_1, r_2 \rangle\}$ *is semi-regular on the set of flags* $\Omega$.

**Lemma 2** (Fundamental Lemma of Symmetries of Oriented Maps)**.** *The action of the automorphism group* Aut $\mathcal{M}_o$ *of the oriented map* $\mathcal{M} = \{\Sigma, \langle r, R \rangle\}$ *is semi-regular on the set of arcs* $\Sigma$.

The validity of both lemmas follows from the fact that for connected structures, any element (flag or arc) can be mapped to any other element in at most one way. Namely, if the image $w$ of a given element $v$ is chosen, then the images of all its neighbours are uniquely determined. By repeating the argument, either the full automorphism is constructed or a contradiction proves that no automorphism mapping $v$ to $w$ exists.

**Definition 1.** *If the automorphism group of a map has k orbits, the map is called a k-orbit map.*

**Definition 2.** *If the automorphism group of an oriented map has k orbits, the oriented map is called a k-orbit oriented map.*

The Fundamental Lemmas of Symmetries of Maps and Oriented Maps have several interesting consequences that follow by the application of basic permutation group theory:

**Corollary 1.** *Let $\mathcal{M} = \{\Omega, \langle r_0, r_1, r_2 \rangle\}$ be a map.*

- *The cardinality of each orbit of $\operatorname{Aut}\mathcal{M}$ on $\Omega$ is equal to the order of $\operatorname{Aut}\mathcal{M}$.*
- *$|\operatorname{Aut}\mathcal{M}|$ is a divisor of $|\Omega|$.*
- *The projection $\mathcal{M} \to \mathcal{M}/\operatorname{Aut}\mathcal{M}$ is a regular covering projection.*
- *The quotient $T(\mathcal{M}) = \mathcal{M}/\operatorname{Aut}\mathcal{M} = \{\Omega', \langle r_0', r_1', r_2' \rangle\}$ is a degenerate flag graph, called the symmetry type graph.*
- *The order $k = |\Omega'|$ of $T(\mathcal{M})$ is equal to $k = |\Omega|/|\operatorname{Aut}\mathcal{M}|$, and $\mathcal{M}$ is a k-orbit map.*

For the definition of a regular covering projection, the reader is referred, for example, to [1].

Symmetry type graphs have been used previously in the analysis of maps; see, for instance, [5–9] . We extend these results of maps to oriented maps in the obvious way by introducing a useful tool that we call an oriented symmetry type graph (see Figure 3):

**Corollary 2.** *Let $\mathcal{M}_o = \{\Sigma, \langle R, r \rangle\}$ be an oriented map.*

- *The cardinality of each orbit of $\operatorname{Aut}\mathcal{M}_o$ on $\Sigma$ is equal to the order of $\operatorname{Aut}\mathcal{M}_o$.*
- *$|\operatorname{Aut}\mathcal{M}_o|$ is a divisor of $|\Sigma|$.*
- *The projection $\mathcal{M}_o \to \mathcal{M}_o/\operatorname{Aut}\mathcal{M}_o$ is a regular covering projection.*
- *The quotient $T_o(\mathcal{M}_o) = \mathcal{M}_o/\operatorname{Aut}\mathcal{M}_o = \{\Sigma', \langle R', r' \rangle\}$ is a degenerate arc graph, called the oriented symmetry type graph.*
- *The order $k_o = |\Sigma'|$ of $T_o(\mathcal{M}_o)$ is equal to $k_o = |\Sigma|/|\operatorname{Aut}\mathcal{M}_o|$, and $\mathcal{M}_o$ is a $k_o$-orbit oriented map.*

We now demonstrate an application of symmetry type graphs and oriented symmetry type graphs. A map is regular (in the strong sense, i.e., the order of its automorphism group is $|\operatorname{Aut}\mathcal{M}| = |\Omega| = 4|E|$) if and only if its symmetry type graph is a one-vertex graph ($k = 1$). Similarly, an oriented map is regular (i.e., the order if its orientation-preserving automorphism group is $|\operatorname{Aut}\mathcal{M}_o| = |\Sigma| = 2|E|$) if and only if its symmetry type graph is a one-vertex graph ($k_o = 1$). We note that a regular oriented map may be either a regular or chiral (orientable) map.

In general, the underlying orientable map of a $k_o$-orbit oriented map is either a $k_o$- or $2k_o$-orbit map.

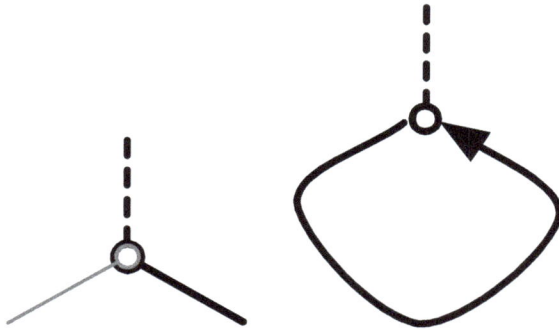

**Figure 3.** An example of a symmetry type graph of a regular map and an oriented symmetry type graph of an oriented regular map; in this case, they may be viewed as the symmetry/oriented symmetry type graph of the cube.

**Example 2.** *The n-sided pyramids for $n > 3$ all have the the same symmetry type graph and the same oriented symmetry type graph. The case $n = 4$ is depicted in Figure 4.*

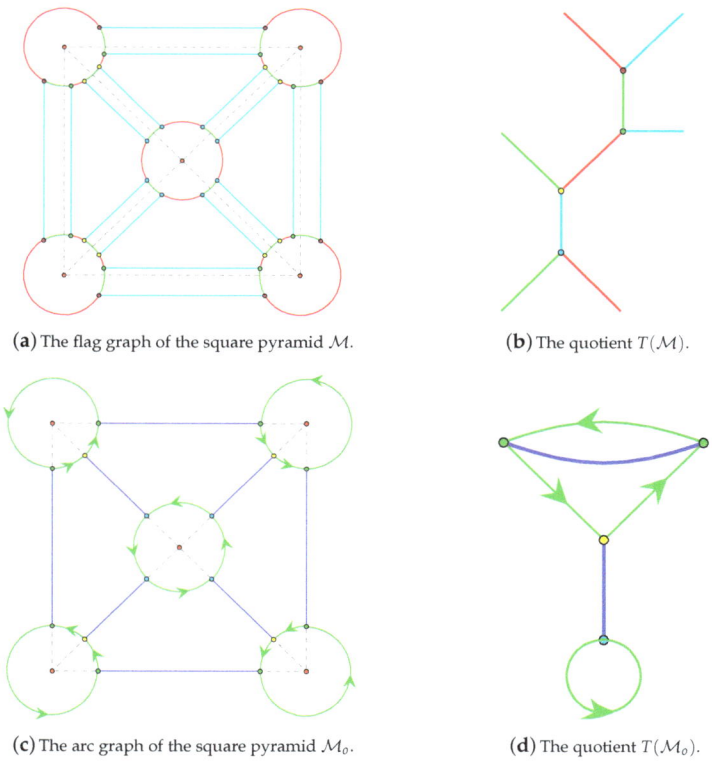

(**a**) The flag graph of the square pyramid $\mathcal{M}$.

(**b**) The quotient $T(\mathcal{M})$.

(**c**) The arc graph of the square pyramid $\mathcal{M}_o$.

(**d**) The quotient $T(\mathcal{M}_o)$.

**Figure 4.** The four-sided pyramid represented as a map $\mathcal{M}$ and as an oriented map $\mathcal{M}_o$, with the corresponding symmetry type graph $T(\mathcal{M})$ and oriented symmetry type graph $T_o(\mathcal{M}_o)$.

*Symmetry* **2017**, *9*, 274

## 2. Operations On Oriented Maps

Operations on maps with the property that the new map resides in the same surface as the original map have been the subject of active investigation (see [6,7,10–13]). If the underlying surface is orientable, we may choose one orientation that induces an orientation of the corresponding map, making it an oriented map. If we keep the surface orientation, the operation may be shifted from maps to oriented maps. In the context of this paper, we work entirely with the generating permutations $r$ and $R$ of the oriented map to describe these operations.

Each operation $\mathtt{Op}$ in this paper can be described in the following way. Each arc $\sigma$ of the original oriented map $\mathcal{M}_o = \{\Sigma, \langle r, R \rangle\}$ gives rise to $k$ arcs $\sigma_0, \sigma_1, \dots, \sigma_{k-1}$ of the resulting oriented map $\mathtt{Op}(\mathcal{M}_o) := \{\Sigma', \langle r', R' \rangle\}$, where $\Sigma' = \Sigma_0 \cup \Sigma_1 \cup \dots \Sigma_{k-1}$ and each $\Sigma_i$ consists of copies of arcs of $\Sigma$ with subscript $i$. If $r$ is the original involution and $R$ is the original permutation of $\mathcal{M}_o$, we denote by $r'$ and $R'$ the respective permutations of the resulting map $\mathtt{Op}(\mathcal{M}_o)$, which reside in the same surface as $\mathcal{M}_o$ with the same orientation as $\mathcal{M}_o$. Because $|\Sigma'| = k|\Sigma|$ we call $k$ the *edge-multiplier* of $\mathtt{Op}$. We derive the action of five operations, Dual ($\mathtt{Du}$), Medial ($\mathtt{Me}$), Truncation ($\mathtt{Tr}$), Snub ($\mathtt{Sn}$), and Chamfer ($\mathtt{Ch}$), by considering directly their local effect on the vertices and edges of an oriented map $\mathcal{M}_o$. Each of these operations is well known from the study of polyhedra. We present the definition of $r'$ and $R'$ for each of these operations, by describing their action on the $k$ arcs $\sigma_1, \dots, \sigma_k$ of $\mathtt{Op}(\mathcal{M}_o)$ associated with each arc $\sigma$ of $\mathcal{M}_o$. Illustrations corresponding to the derivation of each of these new $r'$ and $R'$ are given in the corresponding figures.

### 2.1. Orientation Reversal $\mathtt{Re}$ or *

An operation with edge-multiplier 1 that can be defined on oriented maps but has no counterpart on maps is the *orientation reversal*, $\mathtt{Re}$. Given an oriented map $\mathcal{M}_o = \{\Sigma, \langle r, R \rangle\}$, the reversal is given by $\mathcal{M}_o^* := \{\Sigma, \langle r, R^{-1} \rangle\}$. In other words, we have the following:

$$r'(\sigma_0) = r(\sigma)_0 \qquad\qquad R'(\sigma_0) = R^{-1}(\sigma)_0 \qquad\qquad \mathtt{Re} \qquad (1)$$

### 2.2. Dual $\mathtt{Du}$ and Improper Dual $\mathtt{IDu}$

The definition of dual seems that it should be straightforward. We are tempted to define it as follows:

$$r'(\sigma_0) = r(\sigma)_0 \qquad\qquad R'(\sigma_0) = R(r(\sigma))_0 \qquad\qquad \mathtt{IDu}$$

However, if we draw the corresponding figure, it becomes obvious that the map defined by the above rules is oriented in the opposite direction from the original map. Hence, we call it the *improper dual* $\mathtt{IDu}$. A local portion of a map, along with the derivation of its $\mathtt{Du}$ and $\mathtt{IDu}$, is shown in Figure 5.

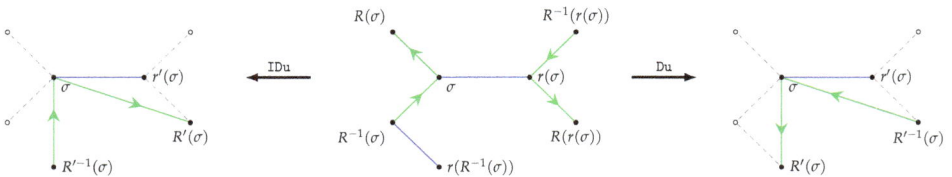

**Figure 5.** Improper dual ($\mathtt{IDu}$) and dual ($\mathtt{Du}$) of an oriented map, local figure. In the center is a local portion of an oriented map, to the left is is the local picture of $\mathtt{IDu}$, and to the right is $\mathtt{Du}$. Because there is only one copy of $\Sigma$ used in the construction of $\mathtt{IDu}$ and $\mathtt{Du}$, we have suppressed the 0 subscript on the arcs for clarity.

The correct definition of the orientation preserving *dual* Du is as follows:

$$r'(\sigma_0) = R(r(R^{-1}(\sigma)))_0 \qquad\qquad R'(\sigma_0) = r(R^{-1}(\sigma))_0 \qquad\qquad \text{Du}$$

### 2.3. Truncation (Tr)

The truncation operation is a well-known operation that has been applied to polyhedra and maps on many occasions; see, for instance, [14]. It is derived from the geometric operation in which the vertices of a convex polyhedron are cut off shallowly to form a new polyhedron. Here, we present it in the oriented version. Each arc $\sigma$ of the oriented map $\mathcal{M}_o := \{\Sigma, \langle r, R \rangle\}$ gives rise to three arcs $\sigma_0, \sigma_1$ and $\sigma_2$ of the truncated oriented map $\text{Tr}(\mathcal{M}_o) := \{\Sigma', \langle r', R' \rangle\}$. Hence the edge-multiplier of Tr is 3. See Figure 6. We define $r'$ and $R'$ as follows:

$$r'(\sigma_0) = r(\sigma)_0 \qquad\qquad R'(\sigma_0) = \sigma_1 \qquad\qquad \text{Tr}$$
$$r'(\sigma_1) = R(\sigma)_2 \qquad\qquad R'(\sigma_1) = \sigma_2 \qquad\qquad \text{Tr}$$
$$r'(\sigma_2) = R^{-1}(\sigma)_1 \qquad\qquad R'(\sigma_2) = \sigma_0 \qquad\qquad \text{Tr}$$

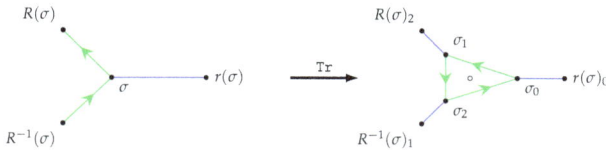

**Figure 6.** Truncated oriented map Tr, local figure.

### 2.4. Medial (Me)

Another operation of the same type is the *medial* Me, with edge-multiplier 2. It is derived from the geometric operation in which the vertices of a convex polyhedron are cut off at the midpoints of the edges (sometimes called full truncation or rectification) to form a new polyhedron. For example, the medial of a cube is the cuboctahedron. The medial of ordinary maps has been studied in [5]. The medial of an oriented map is defined as follows (see Figure 7):

$$r'(\sigma_0) = R(\sigma)_1 \qquad\qquad R'(\sigma_0) = \sigma_1 \qquad\qquad \text{Me}$$
$$r'(\sigma_1) = R^{-1}(\sigma)_0 \qquad\qquad R'(\sigma_1) = r(\sigma)_0 \qquad\qquad \text{Me}$$

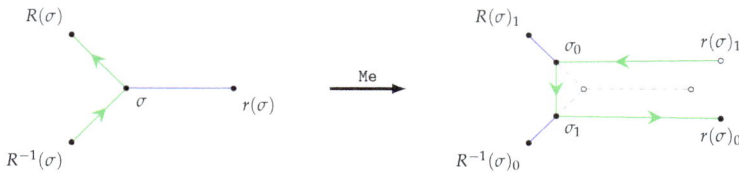

**Figure 7.** Medial Me of an oriented map, local figure.

### 2.5. Snub (Sn)

The snub is an operation with edge-multiplier 5 that can only be defined on oriented (or orientable) maps; see Figure 8. Geometrically, it corresponds to the new polyhedron formed by moving all faces of a convex polyhedron outwards, twisting each face about its center, and adding pairs of triangular faces in place of the original edges. We note that definitions provided by Conway (discussed in [10]) and Coxeter [14] differ slightly.

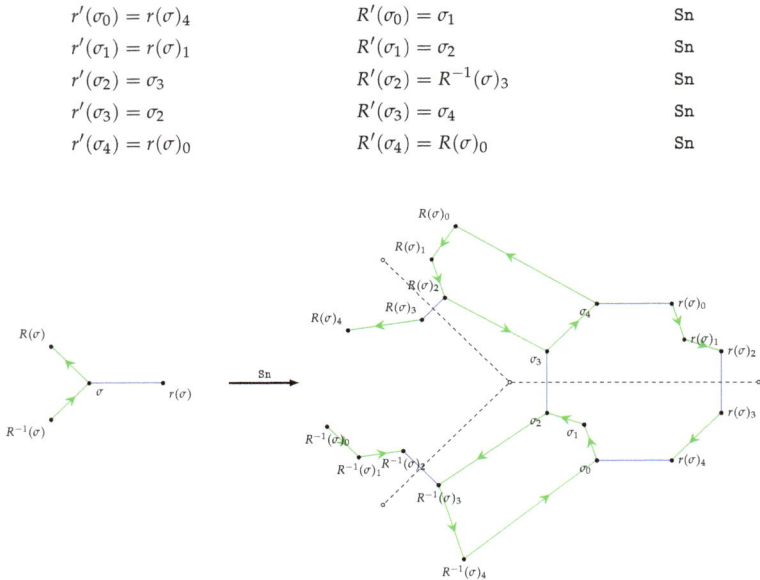

$$r'(\sigma_0) = r(\sigma)_4 \qquad\qquad R'(\sigma_0) = \sigma_1 \qquad\qquad\qquad \text{Sn}$$
$$r'(\sigma_1) = r(\sigma)_1 \qquad\qquad R'(\sigma_1) = \sigma_2 \qquad\qquad\qquad \text{Sn}$$
$$r'(\sigma_2) = \sigma_3 \qquad\qquad\quad R'(\sigma_2) = R^{-1}(\sigma)_3 \qquad\qquad \text{Sn}$$
$$r'(\sigma_3) = \sigma_2 \qquad\qquad\quad R'(\sigma_3) = \sigma_4 \qquad\qquad\qquad \text{Sn}$$
$$r'(\sigma_4) = r(\sigma)_0 \qquad\qquad R'(\sigma_4) = R(\sigma)_0 \qquad\qquad \text{Sn}$$

**Figure 8.** Snub Sn of an oriented map, local figure.

### 2.6. Chamfer Ch

The chamfer operation is an operation with edge-multiplier 4; see Figure 9. Geometrically, it is derived from shallowly slicing off the edges of a convex polyhedron. As a map operation, it was used, for instance, in [8].

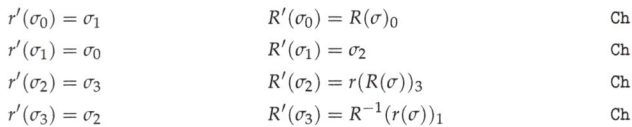

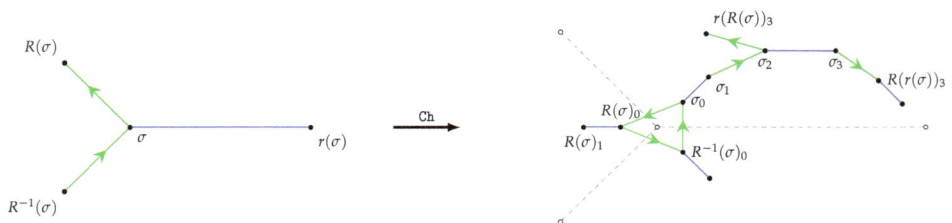

$$r'(\sigma_0) = \sigma_1 \qquad\qquad\quad R'(\sigma_0) = R(\sigma)_0 \qquad\qquad\quad \text{Ch}$$
$$r'(\sigma_1) = \sigma_0 \qquad\qquad\quad R'(\sigma_1) = \sigma_2 \qquad\qquad\qquad\quad \text{Ch}$$
$$r'(\sigma_2) = \sigma_3 \qquad\qquad\quad R'(\sigma_2) = r(R(\sigma))_3 \qquad\qquad \text{Ch}$$
$$r'(\sigma_3) = \sigma_2 \qquad\qquad\quad R'(\sigma_3) = R^{-1}(r(\sigma))_1 \qquad\quad \text{Ch}$$

**Figure 9.** Chamfer Ch of an oriented map, local figure.

### 2.7. One-Dimensional Subdivision Su1

The one-dimensional subdivision is an operation with edge-multiplier 2. It is a simple operation that subdivides each edge of the oriented map. For maps, it has been used in several places in the literature (e.g., [11–13]). It is a building block for composite operations. The following is a formal definition:

$$r'(\sigma_0) = \sigma_1 \qquad\qquad R'(\sigma_0) = R(\sigma)_0 \qquad\qquad \text{Su1}$$
$$r'(\sigma_1) = \sigma_0 \qquad\qquad R'(\sigma_1) = r(\sigma)_1 \qquad\qquad \text{Su1}$$

*2.8. Composite Operations*

Given two operations Op1 and Op2, it is natural to consider the operation Op that is obtained by first applying Op1 and then Op2; that is, Op = Op2 Op1. It is not hard to see that the edge-multiplier of Op is the product of edge-multipliers of Op1 and Op2. The following are some examples:

| | |
|---|---|
| Su2 = Du Tr Du | two-dimensional subdivision |
| BS = Su2 Su1 | barycentric subdivision |
| Le = Tr Du | leapfrog |
| Co = Du BS | combinatorial map |
| An = Du Me | angle map |
| Go = Du Me Tr | gothic operation |

Each of these operations has been used for maps. One can use the same formulae to define them for oriented maps. Two-dimensional subdivision is sometimes called omni-capping. Each face of the original map is subdivided into triangles by placing a new vertex in the center of the face and joining it to each of the original vertices. Leapfrogging is an operation that has been studied extensively in theoretical chemistry; see, for instance, [11]. For instance, the leapfrog of a dodecahedron is a truncated icosahedron. In general, a leapfrog of a fullerene is another fullerene (see, e.g., [11]). In a similar way, An has been used in [15] and Go has been used in [7]. The edge-multiplier for Su2 is 3; for BS, Co and Go, the edge-multiplier is 6; and the edge-multiplier for An is 2. An early use of An can be found in [16], where it is called the *web graph*. In [17], it is called the *radial graph*, and its dual Me is also discussed. In [11–13], several operations on maps are discussed.

## 3. Some Properties of the Operations and *k*-Orbit Oriented Maps

There is an important paper by Orbanić, Pellicer and Weiss [18] in which, among other things, the following theorem is proved.

**Theorem 1.** *If $\mathcal{M}$ is a k-orbit map, then its truncation $Tr(\mathcal{M})$ is either a 3k-, 3k/2- or k-orbit map.*

Motivated by [18], we define the following:

**Definition 3.** *Let Op be a map operation. If there exists a k-orbit map $\mathcal{M}$ such that $Op(\mathcal{M})$ is a $\lambda k$-orbit map, then $\lambda$ is called a* **flag-orbit multiplier.** *By $\Lambda(Op)$ we denote the set of flag-orbit multipliers of Op.*

For instance, $\Lambda(\text{Tr}) = \{1, 3/2, 3\}$.

**Problem 1.** *Investigate sets of flag-orbit multipliers for various operations on maps.*

So far, flag-orbit multipliers have been investigated for truncation, medials and chamfering [5,8,9,18]. One could consider these operations on oriented maps, that is, as follows:

**Definition 4.** *Let Op be an oriented-map operation. If there exists a k-orbit oriented map $\mathcal{M}_o$ such that $Op(\mathcal{M}_o)$ is a $\lambda_o k$-orbit oriented map, then $\lambda_o$ is called an* **arc-orbit multiplier.** *By $\Lambda_o(Op)$ we denote the set of arc-orbit multipliers of Op.*

**Problem 2.** *Investigate sets of arc-orbit multipliers for various operations on oriented maps.*

In the case of maps, the main tools are *symmetry type graphs*. In the case of oriented maps, it seems natural that we should use *oriented symmetry type graphs*.

We note that the operations Op, such as Tr, Me, Du, and Ch, have been studied for general maps, and in the context of oriented maps, they satisfy the following condition: $\text{Op}(\mathcal{M}_o) \cong \text{Op}(\mathcal{M}_o^*)^*$. We call such operations *amphicheiral*—equivalent to their mirror image. On the other hand, this is not the case for the snub operation. We may define a *different* snub operation $\text{Sn}^*(\mathcal{M}_o) := \text{Sn}(\mathcal{M}_o^*)^*$. Operations that are not amphicheiral are called *chiral*. Therefore, the snub is a chiral operation. The snub can only be defined for oriented maps. It certainly cannot be defined on a non-orientable map. If we try to define it for an orientable map, there is no way to distinguish between $Sn^*$ and $Sn$; hence we have to define them both simultaneously and name them arbitrarily.

We may carry this idea to oriented maps and to their oriented symmetry type graphs. We say that an oriented map or oriented symmetry type graph is *amphicheiral* if $\mathcal{M}_o \cong \mathcal{M}_o^*$, or $T(\mathcal{M}_o) \cong T(\mathcal{M}_o^*)$. Otherwise, it is *chiral* (in the sense of Conway). In practice, we may check whether a map is chiral by reversing the directions of arrows and checking whether the resulting map is orientably isomorphic to the original or not. All oriented symmetry type graphs up to four vertices are amphicheiral. The smallest chiral graph is depicted in Figure 10.

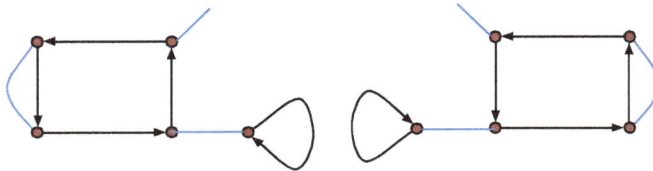

**Figure 10.** Smallest chiral oriented symmetry type graph (both versions).

## 4. Edge-Transitive Oriented Maps

Recently it has been shown that symmetry type graphs are quite a powerful tool when studying certain types of maps. For instance, they provide combinatorial, group-free approaches to the question of classifying edge-transitive maps. In [19], Graver and Watkins classified edge-transitive maps into 14 distinct types. In [20], Širáň, Tucker and Watkins showed that each of the 14 types admits a realization by an oriented map. In [6], Orbanić et al. showed that the 14 types can naturally be described by 14 symmetry type graphs, shown in Figure 11.

The main idea of the proof is to use the fact that in the edge-transitive case, the spanning subgraph of the symmetry type graph determined by $r_0, r_2$ must be connected. There are five possibilities, depicted in Figure 12. By inserting the edges of $r_1$ in all possible ways, one obtains exactly the 14 types.

Karabáš [21] gives a list of low-genus orientable edge-transitive maps. Eight types of these maps are mentioned. Karabáš and Pisanski [22] have shown that these eight classes correspond to eight different oriented symmetry type graphs.

We note that for an oriented map, the number of edge orbits according to the orientation-preserving automorphism group corresponds to the number of edges arising from $r'$ in the respective oriented symmetry type graph. This limits the search to oriented symmetry type graphs on one or two vertices. We call such oriented maps *edge-transitive in the strong sense*.

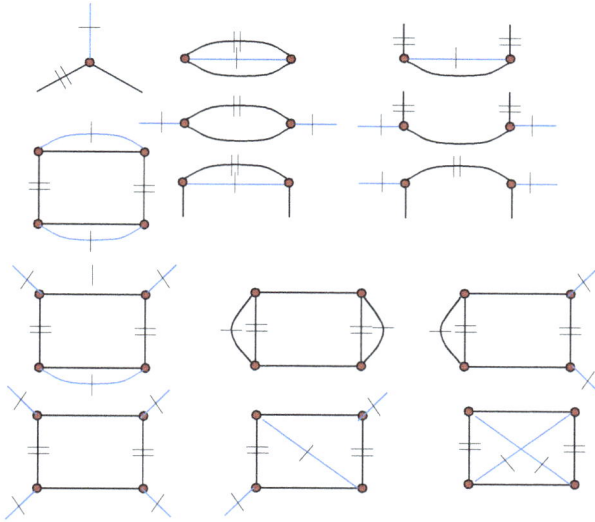

**Figure 11.** The 14 symmetry type graphs of edge-transitive maps.

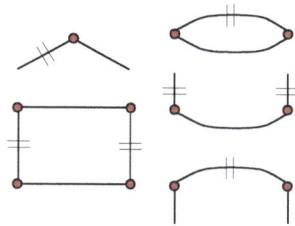

**Figure 12.** Five possible quotients of an edge quadrangle.

On the other hand, an oriented map may not be edge-transitive in the strong sense, but it may have an edge-transitive underlying (orientable) map. Such a map is *edge-transitive in the weak sense*. One can classify such maps.

**Theorem 2.** *([22]) An oriented map is edge-transitive in the weak sense if and only if its oriented symmetry type graph has two edges corresponding to r and there exists an extended automorphism reversing the arrows of R that interchanges the two r-edges.*

This, in turn, implies that a weak edge-transitive oriented map has at most four arc symmetry classes, or equivalently, that its oriented symmetry type graph has at most four vertices. Figure 13 depicts all 17 oriented symmetry type graphs on at most four vertices. In addition to three strong edge-transitive types, we obtain five weak edge-transitive types.

This rather simple combinatorial approach complements the much more sophisticated group-theoretic approach by Jones [23].

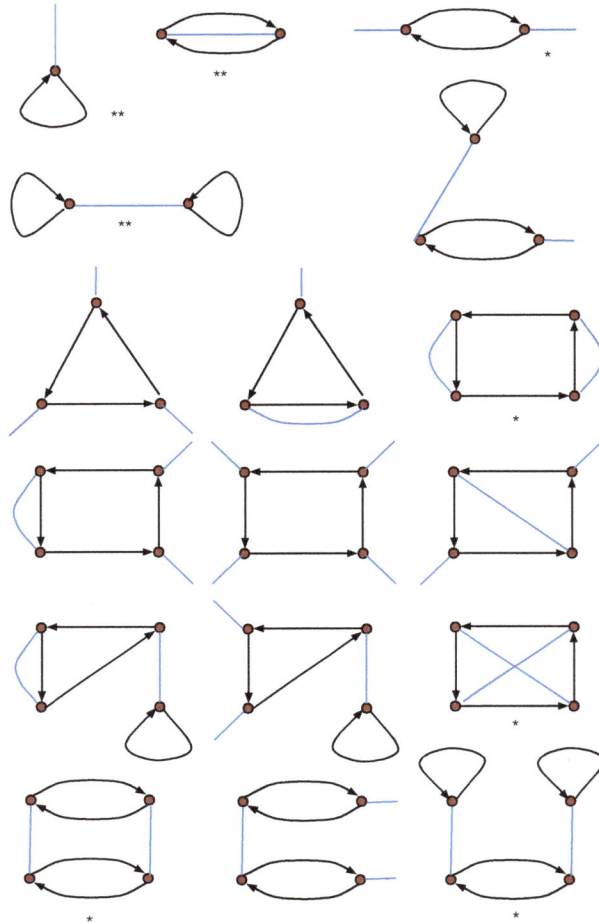

**Figure 13.** The 17 oriented symmetry type graphs on at most four vertices. Eight of these marked by (**) or (*) correspond to edge-transitive maps. The three marked by (**) correspond to the strong edge-transitive maps.

## 5. Conclusions

In this paper, we described the basics of operations on oriented maps and their oriented symmetry type graphs. It would be interesting to carry over the investigations that were performed for similar operations on maps and their symmetry types, such as those by Orbanić et al. [6], del Rio Francos [8,9], and Hubard et al. [5] to oriented maps and their oriented symmetry type graphs. Another context in which it would be interesting to investigate similar questions would be in the higher ranks provided by maniplexes and oriented maniplexes (cf., Cunningham et al. [24]).

Finally, there exists a collection of operations on polyhedra, described using *Conway polyhedron notation* [10,25], originally developed by John H. Conway but later promoted and expanded by George W. Hart, that can be readily carried over to maps and oriented maps. We believe that our approach via flag graphs and arc graphs should be applied to Conway operations to give them rigorous definitions. Namely, to the best of our knowledge, currently only verbal and pictorial descriptions of Conway operations are available. A dictionary that would translate between the two notations would certainly be useful. The reader is referred to [26–30] for related work and for further references. In particular, this theory can be extended to hypermaps [28,31].

**Acknowledgments:** T. Pisanski's research was supported in part by the ARRS of Slovenia under grants: P1-0294 and N1-0032. Leah W. Berman's research was supported by the Simons Foundation (Grant # 209161 to L. Berman). Gordon Williams and Leah Berman would like to thank the Faculty of Mathematics and Physics at the University of Ljubljana for its hospitality and support during our sabbatical stay in Slovenia.

**Author Contributions:** Co-authors are listed alphabetically as usual in mathematics. Each author contributed substantially to the work.

**Conflicts of Interest:** The authors declare no conflict of interest.

## References

1. Gross, J.L.; Tucker, T.W. *Topological Graph theory*; Dover Publications Inc.: Mineola, NY, USA, 2001.
2. Pisanski, T.; Potočnik, P. *Handbook of Graph Theory*; Chapter Graphs on Surfaces; CRC Press: Boca Raton, FL, USA, 2003.
3. Godsil, C.; Royle, G. Algebraic graph theory. In *Graduate Texts in Mathematics*; Springer: New York, NY, USA, 2001; Volume 207.
4. Nedela, R. Maps, Hypermaps and Related Topics, July 2007. Available online: http://www.savbb.sk/~nedela/CMbook.pdf (accessed on 3 July 2017).
5. Hubard, I.; del Río Francos, M.; Orbanić, A.; Pisanski, T. Medial symmetry type graphs. *Electron. J. Comb.* **2013**, *20*, 28.
6. Orbanić, A.; Pellicer, D.; Pisanski, T.; Tucker, T.W. Edge-transitive maps of low genus. *Ars Math. Contemp.* **2011**, *4*, 385–402.
7. Pisanski, T.; Servatius, B. *Configurations from a Graphical Viewpoint*; Springer: New York, NY, USA, 2013.
8. Del Río Francos, M. Chamfering operation on *k*-orbit maps. *Ars Math. Contemp.* **2014**, *7*, 507–524.
9. Del Río Francos, M. Truncation symmetry type graphs. *Ars Comb.* **2017**, *134*, 135–167.
10. Conway, J.H.; Burgiel, H.; Goodman-Strauss, C. *The Symmetries of Things*; A K Peters Ltd.: Wellesley, MA, USA, 2008.
11. Fowler, P.; Pisanski, T. Leapfrog transformations and polyhedra of clar type. *J. Chem. Soc. Faraday Trans.* **1994**, *90*, 2865–2871.
12. Pisanski, T.; Randić, M. Bridges between geometry and graph theory. In *Geometry at Work*; Mathematical Association of America: Washington, DC, USA, 2000; Volume 53, pp. 174–194.
13. Pisanski, T.; Žitnik, A. Representing graphs and maps. In *Topics in Topological Graph Theory*; Cambridge University Press: Cambridge, UK, 2009; Volume 128, pp. 151–180.
14. Coxeter, H.S.M. *Regular Polytopes*; Dover Publications: Mineola, NY, USA, 1973.
15. Mohar, B.; Rosenstiel, P. Tessellation and visibility representations of maps on the torus. *Discret. Comput. Geom.* **1998**, *19*, 249–263.
16. Pisanski, T.; Malnič, A. *The Diagonal Construction and Graph Embeddings*; Graph Theory (Novi Sad, 1983); University of Novi Sad: Novi Sad, Serbia, 1984; pp. 271–290.
17. Negami, S.; Tucker, T.W. Bipartite polyhedral maps on closed surfaces are distinguishing 3-colorable with few exceptions. *Graphs Comb.* **2017**, doi:10.1007/s00373-017-1788-1.
18. Orbanić, A.; Pellicer, D.; Weiss, A.I. Map operations and *k*-orbit maps. *J. Comb. Theory Ser. A* **2010**, *117*, 411–429.
19. Graver, J.E.; Watkins, M.E. *Locally Finite, Planar, Edge-Transitive Graphs*; American Mathematical Soc.: Providence, RI, USA, 1997.
20. Širáň, J.; Tucker, T.W.; Watkins, M.E. Realizing finite edge-transitive orientable maps. *J. Graph Theory* **2001**, *37*, 1–34.
21. Karabáš, J. Edge Transitive Maps on Orientable Surfaces, 2012. Available online: http://www.savbb.sk/~karabas/science.html (accessed on 3 July 2017)
22. Karabáš, J.; Pisanski, T. *Symmetries of Maps and Oriented Maps via Action Graphs*; 2015, in preparation.
23. Jones, G.A. Automorphism groups of edge transitive maps. *arXiv* **2016**, arXiv:1605.09461.
24. Cunningham, G.; del Río-Francos, M.; Hubard, I.; Toledo, M. Symmetry type graphs of polytopes and maniplexes. *Ann. Comb.* **2015**, *19*, 243–268.

25. Wikipedia. Conway Polyhedron Notation. Available online: https://en.wikipedia.org/wiki/Conway_polyhedron_notation (accessed on 3 July 2017).

26. Breda d'Azevedo, A.; Catalano, D.; Karabáš, J.; Nedela, R. Maps of Archimedean class and operations on dessins. *Discrete Math.* **2015**, *338*, 1814–1825.

27. Girondo, E. Multiply quasiplatonic Riemann surfaces. *Exp. Math.* **2003**, *12*, 463–475.

28. James, L.D. Operations on hypermaps, and outer automorphisms. *Eur J. Comb.* **1988**, *9*, 551–560.

29. Karabáš, J.; Nedela, R. Archimedean maps of higher genera. *Math. Comput.* **2012**, *81*, 569–583.

30. Singerman, D.; Syddall, R.I. The Riemann surface of a uniform dessin. *Beiträge Algebra Geom.* **2003**, *44*, 413–430.

31. Singerman, D. Riemann surfaces, Belyi functions and hypermaps. In *Topics on Riemann surfaces and Fuchsian groups (Madrid, 1998)*; Cambridge University Press: Cambridge, UK, 2001; pp. 43–68.

*symmetry*

MDPI

*Article*

# Convex-Faced Combinatorially Regular Polyhedra of Small Genus

**Egon Schulte [1,†,*] and Jörg M. Wills [2]**

[1] Department of Mathematics, Northeastern University, Boston, MA 02115, USA
[2] Department Mathematik, University of Siegen, Emmy-Noether-Campus, D-57068 Siegen, Germany

[†] Supported by NSF-Grant DMS–0856675.
[*] E-Mail: schulte@neu.edu; Tel.: +1-617-373-5511; Fax: +1-617-373-5658.

Academic Editor: name
Received: 28 November 2011; in revised form: 15 December 2011 / Accepted: 19 December 2011 /
Published: 28 December 2011

**Abstract:** Combinatorially regular polyhedra are polyhedral realizations (embeddings) in Euclidean 3-space $\mathbb{E}^3$ of regular maps on (orientable) closed compact surfaces. They are close analogues of the Platonic solids. A surface of genus $g \geqslant 2$ admits only finitely many regular maps, and generally only a small number of them can be realized as polyhedra with convex faces. When the genus $g$ is small, meaning that $g$ is in the historically motivated range $2 \leqslant g \leqslant 6$, only eight regular maps of genus $g$ are known to have polyhedral realizations, two discovered quite recently. These include spectacular convex-faced polyhedra realizing famous maps of Klein, Fricke, Dyck, and Coxeter. We provide supporting evidence that this list is complete; in other words, we strongly conjecture that in addition to those eight there are no other regular maps of genus $g$, with $2 \leqslant g \leqslant 6$, admitting realizations as convex-faced polyhedra in $\mathbb{E}^3$. For all admissible maps in this range, save Gordan's map of genus 4, and its dual, we rule out realizability by a polyhedron in $\mathbb{E}^3$.

**Keywords:** Platonic solids; regular polyhedra; regular maps; Riemann surfaces; polyhedral embeddings; automorphism groups

## 1. Introduction

A *polyhedron P*, for the purpose of this paper, is a closed compact surface in Euclidean 3-space $\mathbb{E}^3$ made up from finitely many *convex* polygons, called the *faces* of $P$, such that any two polygons intersect, if at all, in a common vertex or a common edge (see McMullen, Schulz and Wills [1,2], Brehm and Wills [3], and Brehm and Schulte [4]). The *vertices* and *edges* of $P$ are the vertices and edges of the faces of $P$, respectively. Throughout we insist that any two *adjacent* faces, which share a common edge, do not lie in the same plane. We usually identify $P$ with the underlying map (cell complex) on the surface, or with the abstract polyhedron consisting of the vertices, edges, and faces (as well as $\varnothing$ and $P$ as improper elements), partially ordered by inclusion (see Coxeter and Moser [5] and McMullen and Schulte [6]). In any case, since $P$ is embedded in $\mathbb{E}^3$, the underlying surface is free of self-intersections and is necessarily orientable; and since the faces of $P$ are convex, the underlying abstract polyhedron is necessarily a lattice, meaning here that any two distinct faces meet, if at all, in a common vertex or a common edge.

We are particularly interested in higher-genus analogues of the Platonic solids, the combinatorially regular polyhedra in $\mathbb{E}^3$. A polyhedron $P$ is said to be *combinatorially regular* if its combinatorial automorphism group $\Gamma(P)$ is transitive on the flags (incident triples consisting of a vertex, an edge, and a face) of $P$. Thus a combinatorially regular polyhedron is a polyhedral realization in $\mathbb{E}^3$ of a regular map on an orientable surface of some genus $g$ (see [5]). Each such polyhedron or map has a

(*Schläfli*) *type* $\{p,q\}$ for some $p,q \geqslant 3$, describing the fact that the faces are $p$-gons, $q$ meeting at each vertex. In constructing polyhedral realizations $P$ of a given regular map we are most interested in those that have a large Euclidean symmetry group $G(P)$. Ideally we wish to achieve maximum possible geometric symmetry among polyhedral embeddings. Clearly, $G(P)$ is a subgroup of $\Gamma(P)$. Generally we must expect $G(P)$ to be rather small compared with $\Gamma(P)$.

Regular maps on surfaces have been studied for well over 120 years, and deep connections have been discovered between maps and other branches of mathematics, including hyperbolic geometry, Riemann surfaces, automorphic functions, number fields, and Galois theory. They can be studied from a combinatorial and topological viewpoint as cell-complexes (tessellations) on surfaces, essentially as abstract polyhedra (see [6]), or be viewed as complex algebraic curves over algebraic number fields (see Jones and Singerman [7]).

The Platonic solids provide the only regular maps on the 2-sphere, with $g = 0$. It is well-known that there are infinitely many regular maps on the 2-torus, with $g = 1$, each of type $\{3,6\}$, $\{6,3\}$, or $\{4,4\}$ (see [5]). However, a surface of genus $g \geqslant 2$ can support only finitely many regular maps. For $g \leqslant 6$, a full classification was first obtained by Sherk [8] (for genus 3) and Garbe [9]. Historically, genus 6 provides an important upper bound on the genera for two reasons: first, two of Coxeter's [10] classical regular skew polyhedra discovered in the early days of the study of maps on surfaces have genus 6; and second, and more importantly, genus 6 turned out to be the threshold for a by-hand enumeration of regular maps by genus.

In the past few years there has been great progress in the computer-aided enumeration of regular maps by genus, leading in particular to the creation of a complete census of regular maps on orientable surfaces of genus up to 101 (see Conder [11]). It is quite surprising that there exist infinitely many genera $g$, beginning with $g = 2$, for which the surface of genus $g$ does not admit any regular map with a simple underlying graph at all (see Conder, Siran and Tucker [12] and Breda d'Azevedo, Nedela and Siran [13]).

In this paper we study combinatorially regular polyhedra in $\mathbb{E}^3$ of small genus $g$, meaning that $g$ lies in the historically motivated range $2 \leqslant g \leqslant 6$. There are only finitely many regular maps in this genus range, and we wish to determine those which admit realizations as (convex-faced, combinatorially regular) polyhedra. Only eight regular maps in this range are known to have polyhedral realizations, and two were only discovered quite recently. Some of these polyhedra are quite spectacular and realize famous maps of Klein, Fricke, Dyck, and Coxeter (see [5,10,14–19]). In Section 2 we briefly review these polyhedra and their maps. Then, in Section 3, we provide supporting evidence that this list is complete; in other words, we strongly conjecture that in addition to those eight there are no other regular maps of genus $2 \leqslant g \leqslant 6$ admitting realizations as (convex-faced) polyhedra in $\mathbb{E}^3$. In particular, for all admissible maps in this range, save Gordan's [20] classical map of genus 4, and its dual, we rule out realizability by a polyhedron in $\mathbb{E}^3$. We conclude the paper with a brief discussion of some open problems. For a discussion of regular toroidal polyhedra see also Schwörbel [21].

## 2. The Eight Maps and Their Polyhedra

For a regular map $P$ of type $\{p,q\}$ with $f_0$ vertices, $f_1$ edges, and $f_2$ faces on a closed surface, the order of its automorphism group $\Gamma(P)$ is linked to the Euler characteristic $\chi$ of the surface by the equation

$$\chi = f_0 - f_1 + f_2 = \frac{|\Gamma(P)|}{2}\left(\frac{1}{p} + \frac{1}{q} - \frac{1}{2}\right) \tag{1}$$

If $\{p,q\}$ is of hyperbolic type, this immediately leads to the classical Hurwitz inequality,

$$|\Gamma(P)| \leqslant 84|\chi| \tag{2}$$

with equality occurring if and only if $P$ is of type $\{3,7\}$ or $\{7,3\}$. In particular, this shows that there can only be finitely many regular maps on a given closed surface of non-zero Euler characteristic. For the genus range under consideration, the inequality in Equation (2) also establishes, a priori, an upper bound for the order of the automorphism group of the map, and hence for the geometric symmetry group of any of its polyhedral realizations in $\mathbb{E}^3$.

The combinatorial automorphism group $\Gamma(P)$ is generated by three involutions $\rho_0, \rho_1, \rho_2$ satisfying the Coxeter relations

$$\rho_0^2 = \rho_1^2 = \rho_2^2 = (\rho_0\rho_1)^p = (\rho_1\rho_2)^q = (\rho_0\rho_2)^2 = 1 \tag{3}$$

but in general also some further, independent relations. On the underlying surface, these generators can be viewed as "combinatorial reflections" in the sides of a fundamental triangle of the "barycentric subdivision" of $P$. For a finite regular map, the relations in Equation (3) suffice for a presentation of $\Gamma(P)$ if and only if $P$ is a Platonic solid; that is, if and only if $g = 0$. Thus at least one, but generally more additional relations are needed if $g > 0$. Two kinds of extra relations, usually occurring separately, are of particular importance to us and suffice to describe seven of the eight regular maps under consideration, namely the *Petrie relation*

$$(\rho_0\rho_1\rho_2)^r = 1 \tag{4}$$

and the *hole relation*

$$(\rho_0\rho_1\rho_2\rho_1)^h = 1 \tag{5}$$

These relations are inspired by the notions of a Petrie polygon and of a hole of a regular map, respectively. Recall that a *Petrie polygon* of a regular map (on any surface) is a zigzag along its edges such that every two, but no three, successive edges are the edges of a common face. A *hole* of a regular map (on any surface) is a path along the edges which successively takes the second exit on the right (in a local orientation), rather than the first, at each vertex. The automorphism $\rho_0\rho_1\rho_2$ of $P$ occurring in Equation (4) shifts a certain Petrie polygon of $P$ one step along itself, and hence has period $r$ if the Petrie polygon is of length $r$. Similarly, the automorphism $\rho_0\rho_1\rho_2\rho_1$ of Equation (5) shifts a certain hole of $P$ one step along itself, and hence has period $h$ if the hole is of length $h$. Thus, if the relation in Equation (4) or Equation (5) holds, with $r$ or $h$ giving the correct period of $\rho_0\rho_1\rho_2$ or $\rho_0\rho_1\rho_2\rho_1$, respectively, then $P$ has Petrie polygons of length $r$ or holes of length $h$.

There are a number of standard operations on maps which create new regular maps from old (see [6,22]). The *duality operation* $\delta$ replaces a map $P$ by its dual $P^\delta$ (often denoted $P^*$); algebraically this corresponds to reversing the order of the generators $\rho_0, \rho_1, \rho_2$ of the group. The *Petrie operation* $\pi$ preserves the edge graph of the map $P$ but replaces the faces by the Petrie polygons; the resulting map is the *Petrie dual* $P^\pi$ of $P$. Algebraically, $\pi$ corresponds to replacing $\rho_0, \rho_1, \rho_2$ by the new generators $\rho_0\rho_2, \rho_1, \rho_2$. There is also an operation that substitutes the holes for the faces, but we will not require it here (see [6] Section 7B).

It is well-known that every orientable regular map of type $\{p,q\}$ is a quotient of the corresponding regular tessellation $\{p,q\}$ of the 2-sphere (if $g = 0$), the Euclidean plane (if $g = 1$), or the hyperbolic plane (if $g \geqslant 2$). If a map is the quotient of the regular tessellation $\{p,q\}$ obtained by identifying those pairs of vertices which are separated by $r$ steps along a Petrie polygon, for a specified value of $r$, then it is denoted $\{p,q\}_r$. Similarly, $\{p,q|h\}$ denotes the map derived from the tessellation $\{p,q\}$ by identifying those pairs of vertices which are separated by $h$ steps along a hole, for a specified value of $h$. These identification processes generally do not result in finite maps, but when they do, the corresponding maps and their automorphism groups usually have nice properties. The Platonic solids are particular instances arising as the maps $\{3,3\}_4$, $\{3,4\}_6$, $\{4,3\}_6$, $\{3,5\}_{10}$, and $\{5,3\}_{10}$. The duals of $\{p,q\}_r$ and $\{p,q|h\}$ are $\{q,p\}_r$ and $\{q,p|h\}$, respectively. The Petrie dual of $\{p,q\}_r$ is $\{r,q\}_p$ (there is no simple rule for the Petrie dual of $\{p,q|h\}$).

It is important to point out here that our notation $\{p,q\}_r$ designates only those regular maps for which the relations in Equations (3) and (4) actually form a complete presentation for the automorphism group; in other words, $\{p,q\}_r$ is the "universal" regular map of type $\{p,q\}$ with Petrie polygons of length $r$. A similar remark applies to $\{p,q|h\}$.

We now discuss the eight regular maps that can be realized as polyhedra in $\mathbb{E}^3$. A summary of their basic properties is given in Table 1. In each case we list the genus $g$, the name of the map (if any) or simply its Schläfli type $\{p,q\}$, the order of the combinatorial automorphism group, the face vector $(f_0, f_1, f_2)$, and the original reference. The third last column refers to the complete list of all regular maps of genus at most 101 mentioned earlier. In particular, $Rg.k$ designates the $k^{th}$ map among the orientable (reflexible) regular maps of genus $g$, in the labeling of [11], and $Rg.k^*$ denotes its dual (note that only one map from a pair of duals is listed in [11]). Thus $R3.1$ is the 1st orientable regular map of genus 3 in this list.

**Table 1.** The Eight Regular Maps.

| Genus $g$ | Name or Type | Group Order | Face Vector $(f_0, f_1, f_2)$ | References | Map of [11] | Symmetry Group | Max. |
|---|---|---|---|---|---|---|---|
| 3 | $\{3,7\}_8$ | 336 | (24,84,56) | Klein [17,18] | R3.1 | $[3,3]^+$ | |
| 3 | $\{3,8\}_6$ | 192 | (12,48,32) | Dyck [15,16] | R3.2 | $D_3$ | * |
| 5 | $\{3,8\}$ | 384 | (24,96,64) | Klein and Fricke [19] | R5.1 | $[3,4]^+$ | * |
| 5 | $\{4,5|4\}$ | 320 | (32,80,40) | Coxeter [10] | R5.3 | $C_2^3$ | |
| 5 | $\{5,4|4\}$ | 320 | (40,80,32) | Coxeter [10] | R5.3* | $C_2^3$ | |
| 6 | $\{3,10\}_6$ | 300 | (15,75,50) | Coxeter and Moser [5] | R6.1 | $D_3$ | * |
| 6 | $\{4,6|3\}$ | 240 | (20,60,30) | Coxeter [10], Boole Stott [23] | R6.2 | $[3,3]$ | * |
| 6 | $\{6,4|3\}$ | 240 | (30,60,20) | Coxeter [10], Boole Stott [23] | R6.2* | $[3,3]$ | * |

The next-to-last column of Table 1 records the geometric symmetry group of the most symmetric polyhedral embedding of the map currently known, with an asterisk in the last column indicating if this group has maximum possible order for a polyhedral embedding. Here we let $[3,3]^+$ and $[3,4]^+$, respectively, denote the rotation subgroup of the full symmetry group $[3,3]$ and $[3,4]$ of the tetrahedron $\{3,3\}$ and octahedron $\{3,4\}$.

There are two pairs of duals among the eight maps in Table 1. The other four maps have triangular faces; our analysis in Section 3 will imply that their duals do not admit a polyhedral embedding with convex faces (though at least one, namely $\{7,3\}_8$, does with non-convex faces).

Note that the Klein–Fricke map in the third row of Table 1 does not have a simple designation as a map $\{p,q\}_r$ or $\{p,q|h\}$. The length of its Petrie polygons or 2-holes is not sufficient to define it.

## 2.1. Klein's Map

Klein's map $\{3,7\}_8$ of genus 3 is arguably the most famous regular map of positive genus. It was constructed by Klein [17] to illustrate the importance of the simple transformation group $PSL(2,7)$ for the solution of equations of degree 7, as well as to highlight the analogy with the appearance of the icosahedron and icosahedral group in the study of quintic equations (see [18]). The group $PSL(2,7)$ occurs as the group of orientation preserving automorphisms of $\{3,7\}_8$, and its order 168 maximizes the order among groups of orientation preserving automorphisms on a Riemann surface of genus 3 (see Hurwitz [24] and Jones and Singerman [7]). The surface can also be represented by Klein's quartic

$$x^3y + y^3z + z^3x = 0$$

a plane algebraic curve of order 4 in homogeneous complex variables (see [17]).

A beautiful polyhedron for $\{3,7\}_8$, discovered by the present authors in [25], can be constructed from a pair of homothetic vertex-truncated tetrahedra by removing their hexagonal faces and connecting each of the four pairs of resulting hexagonal circuits by a suitable tunnel made up of triangles. The resulting polyhedron has 56 triangular faces, 7 meeting at each of the 24 vertices, and the geometric symmetry group is the tetrahedral rotation group (isomorphic to the alternating group $A_4$). For illustrations see [26–28]. (Since the symmetry group is that of a Platonic solid, the polyhedron is a Leonardo polyhedron in the sense of [29].)

There are a number of other interesting polyhedral models for Klein's map or its dual, which either relax the strong assumption on the faces to be convex, or have self-intersections (like the Kepler–Poinsot polyhedra) and hence no longer give an embedding of the underlying surface in $\mathbb{E}^3$. Particularly remarkable is the polyhedral model for the dual map $\{7,3\}_8$ discovered in McCooey [30]. This has 24 heptagonal faces, all simply-connected but non-convex, and its symmetry group is the tetrahedral rotation group.

In realizing regular maps, shortcomings such as self-intersections can occasionally be compensated by higher symmetry. For example, in [25] we describe a polyhedral immersion for $\{3,7\}_8$ with self-intersections featuring octahedral rotation symmetry. Other polyhedral immersions of $\{3,7\}_8$ with high symmetry are discussed in [31]. Polyhedral models with self-intersections and non-convex faces are also known for the dual map $\{7,3\}_8$ (see [28]).

### 2.2. Dyck's Map

Dyck's map $\{3,8\}_6$ of genus 3 is closely related to the torus map $\{3,6\}_8$ with 16 vertices, which is just $\{3,6\}_{(4,0)}$ in the notation of [5]. In fact, $\{3,8\}_6 = (\{3,6\}_8)^{\delta\pi\delta}$, where again $\delta$ and $\pi$, respectively, denote the duality operation and Petrie operation. Thus $\{3,8\}_6$ is dual to the Petrie dual of the dual of $\{3,6\}_8$. It also can be represented by Dyck's quartic

$$x^4 + y^4 + z^4 = 0$$

Dyck's map is the smallest regular map of genus $g \geqslant 2$ for which a polyhedral embedding with convex faces is known. As the map is quite small, having just 12 vertices and 32 faces, its polyhedral realizability is far from being obvious. The challenge here is the polyhedral embeddability; the convexity condition on the faces is trivial since they are triangles. The first realization of Dyck's map as a polyhedron (with trivial geometric symmetry group) was discovered by Bokowski [32]. A more symmetric polyhedron, due to Brehm [33], has a symmetry group $D_3$ and exhibits maximum possible geometric symmetry for a polyhedral embedding.

An appealing polyhedral model for $\{3,8\}_6$ with self-intersections but full tetrahedral symmetry can be obtained from a pair of homothetic octahedra by first omitting on each octahedron a set of alternate faces, different sets on the two octahedra, and then joining the resulting triangular circuits by tunnels, one tunnel for each pair of circuits in parallel planes (see [34]). This model has maximum possible symmetry among all polyhedral models of $\{3,8\}_6$ and is significantly more symmetric than the polyhedral embedding with maximal symmetry. For the dual map $\{8,3\}_6$, which is a dodecahedron with octagonal faces, a polyhedral model with self-intersecting faces was described in [35]. It does not seem to be known if a polyhedral embedding of $\{8,3\}_6$ with non-convex faces is possible.

### 2.3. Coxeter's Geometric Skew Polyhedra

The pair of dual regular maps $\{4,6|3\}$ and $\{6,4|3\}$ of genus 6 can be embedded in Euclidean 4-space $\mathbb{E}^4$ such that all combinatorial symmetries are realized as geometric symmetries. The corresponding 4-dimensional polyhedra are two of Coxeter's regular skew polyhedra in $\mathbb{E}^4$ and have (convex) squares or regular hexagons as faces. These polyhedra in $\mathbb{E}^4$ actually trace back to Alicia Boole Stott (see Stott [23] and also p. 45 of Reference [10]), but it was Coxeter who fully explored and popularized them. They are closely related to the regular 4-simplex. Their underlying maps inherit their combinatorial symmetries from two sources, the symmetries and the dualities of the (self-dual) 4-simplex; in fact, the combinatorial automorphism group is isomorphic to $S_5 \times C_2$. It was observed by McMullen, Schulz and Wills [1,2] that, via projection from $\mathbb{E}^4$, they can be polyhedrally embedded with convex faces in $\mathbb{E}^3$. For illustrations of these polyhedra in $\mathbb{E}^3$ see [27,36].

There are similar such polyhedral embeddings for Coxeter's regular skew polyhedra $\{4,8|3\}$ and $\{8,4|3\}$, a dual pair of genus 73, but they fall outside the genus range under consideration (see [1,2,10,23,36]).

### 2.4. Coxeter's Topological Analogues of Skew Polyhedra

The pair of dual maps $\{4,5|4\}$ and $\{5,4|4\}$ are the first instances, with $g \geqslant 2$, in two infinite series of regular maps of type $\{4,q\}$ and $\{q,4\}$, with $q \geqslant 3$, where the member of type $\{4,q\}$ in the first series is dual to the member of type $\{q,4\}$ in the second series. (The members for $q = 3,4$ are the 3-cube $\{3,3\}$ and the torus map $\{4,4\}_{(4,0)}$.) For each $q \geqslant 3$, the combinatorial automorphism group of either map is $C_2^q \rtimes D_q$ ($\cong C_2 \wr D_q$), and the genus is $2^{q-3}(q-4)+1$. The two series were first discovered by Coxeter [10,14] in 1937, but have been rediscovered several times since then (see pp. 260, 261 of Reference [6] for more details). The map of type $\{4,q\}$ can be realized (with square faces) in the 2-skeleton of the $q$-dimensional cube in $\mathbb{E}^q$. In McMullen, Schulz and Wills [1,2], two remarkable infinite series of polyhedra in $\mathbb{E}^3$ of types $\{4,q\}$ and $\{q,4\}$ were described, and in [37] were proved to be polyhedral embeddings of Coxeter's regular maps of these types. Thus all of Coxeter's maps can be realized as polyhedra in $\mathbb{E}^3$. Each of these polyhedra, with $q \geqslant 5$, has a geometric symmetry group $C_2^3$ generated by three reflections in mutually orthogonal planes. (For $q = 3,4$, the symmetry group is larger.) For illustrations of the polyhedra $\{4,5|4\}$ and $\{5,4|4\}$ see [38].

The maps in each series also admit polyhedral embeddings in $\mathbb{E}^4$ as subcomplexes of the boundary complexes of certain convex 4-polytopes: more precisely, of weakly neighborly polytopes, for the maps of type $\{4,q\}$ (see [39]); and of wedge polytopes, for the dual maps of type $\{q,4\}$ (see [40,41]).

### 2.5. The Klein–Fricke Map

The Klein–Fricke map of type $\{3,8\}$ and genus 5 is a double cover of Dyck's map $\{3,8\}_6$ of genus 3, and has 24 vertices and a group of order 384 (see [19]). The Petrie polygons have length 12; however, this is not the universal map $\{3,8\}_{12}$, which is known to be infinite (see p. 399 of Reference [6]).

In their search for vertex-transitive polyhedra in 3-space in [42], Grünbaum and Shephard discovered an equivelar polyhedron of type $\{3,8\}$ with 24 vertices and octahedral rotation symmetry. This shares the same numbers of vertices, edges, and faces, the same genus, and the same group order, but not the full combinatorics, with the Klein–Fricke map, and in particular is not combinatorially regular. Grünbaum [43] describes how this equivelar polyhedron can be altered (by changing 12 edges) to obtain a polyhedral embedding of the Klein–Fricke map. This new polyhedron again has octahedral rotation symmetry and is vertex-transitive. Just recently, this same embedding of the Fricke–Klein map was rediscovered by Brehm and Wills; more details, along with a proof of isomorphism with the Fricke–Klein map, will appear in a forthcoming article [44].

The polyhedron admits a nice description based on the geometry of the snub cube. Its vertices are those of the snub cube, and the boundaries of the six square faces of the snub cube appear as the holes of the polyhedron. The polyhedron consists of an outer shell and an inner shell connected at the holes (but nowhere else). The entire polyhedron can be pieced together from the orbits of four particular triangles, two adjacent triangles in each shell, under the octahedral symmetry group; the triangles in these orbits then are the faces of the polyhedron. The vertices of the polyhedron can be represented by small integer coordinates. The smallest integer coordinates arise when the four triangles are chosen as follows (see the forthcoming article by Brehm, Grünbaum and Wills [44]. The two basic triangles for the outer shell have vertices

$$(1,2,6),\ (2,6,1),\ (6,1,2)$$

and

$$(1,2,6),\ (2,6,1),\ (-2,1,6)$$

respectively; under the standard octahedral rotation group (generated by 4-fold rotations about the coordinate axes and 3-fold rotation about the main space diagonals), these triangles determine 8 regular triangles and 24 non-regular triangles making up the outer shell. The inner shell similarly consists of 8 regular triangles and 24 non-regular triangles obtained under the standard octahedral rotation group from the two basic triangles for the inner shell with vertices

$$(2, -1, 6), \ (-1, 6, 2), \ (6, 2, -1)$$

and

$$(2, -1, 6), \ (-1, 6, 2), \ (-2, 6, -1)$$

respectively.

The Klein–Fricke map also admits a polyhedral realization with self-intersections that highlights the relationship with Dyck's map. The model is constructed from a pair of homothetic icosahedra by removing all their faces and joining suitable hexagonal circuits by tunnels (see [45]); identifying antipodal vertices in this realization gives exactly the previously mentioned polyhedral model with self-intersections for Dyck's map derived from a pair of homothetic octahedra.

*2.6. The Map $\{3, 10\}_6$*

Just like Dyck's map, $\{3, 10\}_6$ is closely related to a torus map, in this case the map $\{3, 6\}_{10} = \{3, 6\}_{(5,0)}$ with 25 vertices (see [5]). Now $\{3, 10\}_6 = (\{3, 6\}_{10})^{\delta \pi \delta}$, with $\delta$ and $\pi$ as before, so $\{3, 10\}_6$ is dual to the Petrie dual of the dual of $\{3, 6\}_{10}$.

A polyhedral embedding for $\{3, 10\}_6$ in $\mathbb{E}^3$ with maximum possible symmetry group $D_3$ was only recently discovered [46]. It is the only polyhedron among the eight with an odd number of vertices.

The map $\{3, 10\}_6$ is the fourth in an infinite series of finite regular maps $\{3, 2s\}_6$, with $s \geqslant 2$, closely related to the unitary reflection groups $[1\,1\,1]^s$ in complex 3-space (see pp. 389, 399 of Reference [6]). The automorphism group of $\{3, 2s\}_6$ is the semidirect product of $[1\,1\,1]^s$ by $C_2$ and has order $12s^2$. The first three maps in this sequence are the octahedron $\{3, 4\}_6 = \{3, 4\}$, the torus map $\{3, 6\}_{(3,0)}$, and Dyck's map $\{3, 8\}_6$. Note that, as above, each map $\{3, 2s\}_6$ is dual to the Petrie dual of the dual of $\{3, 6\}_{2s}$. The sequence of maps $\{3, 2s\}_6$ was first discovered by Coxeter [47].

## 3. Completeness of the List

In this section we give supporting evidence for our conjecture that the eight maps listed in Table 1 are the only (orientable) regular maps of genus $2 \leqslant g \leqslant 6$ that admit realizations as (convex-faced) polyhedra in $\mathbb{E}^3$. In fact, the only maps we have not been able to conclusively exclude as possibly admitting a polyhedral embedding are Gordan's map $\{4, 5\}_6$ and its dual $\{5, 4\}_6$ (see [20]).

We rely on the recent enumeration in [11] of regular maps of genus up to 101. It is immediately clear that we can eliminate maps with multiple edges from consideration, since those certainly cannot be realized as polyhedra. Similarly, by the convexity assumption on the faces, adjacent faces cannot share more than one edge. Using the notation of [11], then this leaves only maps with $m_V = m_F = 1$, where $m_V$ and $m_F$, respectively, denote the edge-multiplicities of the underlying graph of the map and its dual. Now inspection of the list for genus $2 \leqslant g \leqslant 6$ shows that the only possible candidates for polyhedral embeddings are the maps

$$R3.1, \ R3.2, \ R4.2, \ R4.3, \ R4.6, \ R5.1, \ R5.3, \ R5.4, \ R5.9, \ R6.1, \ R6.2$$

and their duals; since R4.6 and R5.9 are self-dual, this gives a total of twenty maps.

Among those twenty, the maps R3.1, R3.2, R5.1 and R6.1 have triangular faces, so their duals have 3-valent vertices. However, a polyhedron (without self-intersection) in $\mathbb{E}^3$ with only 3-valent vertices and convex faces is necessarily a convex polyhedron and hence has genus 0 (see [3]). Thus the duals of R3.1, R3.2, R5.1 and R6.1 can be eliminated as well, leaving at most sixteen possible candidates, including the eight maps of Table 1 admitting polyhedral embeddings as described in the previous section. Finally, then, this leaves the following eight maps

$$R4.2,\ R4.2^*,\ R4.3,\ R4.3^*,\ R4.6,\ R5.4,\ R5.4^*,\ R5.9$$

which need to be ruled out by other means.

First observe that none of the maps $R4.6$, $R5.4$ and $R5.4^*$ has a face-poset which is a lattice; in other words, none of these maps is a cell-complex. Thus polyhedral embeddings cannot exist. In particular, $R4.6$ of type $\{5,5\}$ is the underlying (combinatorially self-dual) map of the (geometrically) dual pair of Kepler–Poinsot polyhedra $\{5,\frac{5}{2}\}$ and $\{\frac{5}{2},5\}$ in $\mathbb{E}^3$, which have self-intersections (see [48]). Here it is easily seen that, for example, two non-adjacent pentagonal faces of $\{5,\frac{5}{2}\}$ with a common vertex, share in fact another vertex which is not adjacent to the first vertex. Hence $R4.6$ is not a lattice. For $R5.4$ of type $\{4,6\}$, discovered by Sherk [49], we can appeal to the planar diagram of the map given on p. 161 of Reference [45] to show that opposite square faces in the vertex-star of a vertex also share the vertex opposite the first vertex in these faces. Hence neither $R5.4$ nor its dual $R5.4^*$ is a lattice.

The three maps $R4.3$ of type $\{4,6\}$, $R4.3^*$ of type $\{6,4\}$, and $R5.9$ of type $\{5,5\}$ can be eliminated by the following lemma, since they have Petrie polygons of length 4. This lemma is also of independent interest.

**Lemma 1.** *If $P$ is a combinatorially regular polyhedron in $\mathbb{E}^3$ (with convex faces and without self-intersection), and if $P$ has Petrie polygons of length 4, then $P$ is the tetrahedron $\{3,3\}$.*

**Proof.** The proof is simple. Let $e_1, e_2, e_3, e_4$ denote the edges of a Petrie polygon, taken in order. Then any pair $e_j, e_{j+1}$ of successive edges determines a face $F_j$ of $P$ (indices taken modulo 4). Clearly, the Petrie polygon cannot lie in a plane, since otherwise (for example) $F_1$ and $F_2$ would have relative interior points in common. Let $T$ denote the convex hull of $e_1, e_2, e_3, e_4$, which is a tetrahedron. Then the affine hulls of the faces $F_j$ of $P$ are just the faces of $T$. Now, if $P$ does not have triangular faces, then (for example) $F_1$ and $F_3$ would have to intersect in relative interior points, namely in one of the two edges of $T$ not among $e_1, e_2, e_3, e_4$. Hence the faces of $P$ must be triangles. However, then the two edges of $T$ not among $e_1, e_2, e_3, e_4$ must also be edges of $P$, so $P$ must reduce to the tetrahedron $T$. $\qquad\square$

Finally, then, we need to exclude the possibility of a polyhedral embedding for the two well-known regular maps $R4.2$ and $R4.2^*$ of genus 4. These two maps, respectively, are Gordan's $\{4,5\}_6$ with 24 vertices, 60 edges, and 30 faces, and its dual $\{5,4\}_6$, with automorphism groups isomorphic to a semidirect product of the full icosahedral group by $C_2$ (see [20]). In this case it is considerably more difficult to prove non-existence of a polyhedral embedding with convex faces. Here we have the following conjecture. If confirmed, this would establish definitively that the eight maps in Table 1 are the only regular maps of genus $2 \leqslant g \leqslant 6$ admitting realizations as convex-faced polyhedra in $\mathbb{E}^3$.

Gordan's maps $\{4,5\}_6$ and $\{5,4\}_6$ do not admit a realization as a (convex-faced) polyhedron in $\mathbb{E}^3$.

In fact, it may be rconjectured more strongly that $\{4,5\}_6$ and possibly also $\{5,4\}_6$ do not even admit a polyhedral embedding with simply-connected planar faces.

On the other hand, Gordan's maps do admit beautiful polyhedral realizations of Kepler–Poinsot type (that is, with self-intersections) with high symmetry. In our paper [45], we described a realization of $\{4,5\}_6$ with full octahedral symmetry based on the geometry of the truncated octahedron. However, even self-intersecting polyhedral models with full icosahedral symmetry were discovered for both $\{4,5\}_6$ and $\{5,4\}_6$, and were described as "skeletal polyhedra of index 2" in Wills [50], Cutler and Schulte [51], and Cutler [52] (see also [53,54]). More details will appear in a forthcoming paper [55]. The high degree of self-intersections in these models, as well as the smallness of the underlying maps, are strong indications that non-self-intersecting polyhedral embeddings with convex faces are impossible. However, we have not been able to confirm this rigorously.

*Symmetry* **2012**, *xx*, 4

## 4. Open Problems

In addition to settling Conjecture 3, the following problems seem particularly interesting in connection with the topics discussed here.

**Genus 7 or 8.** The maps $R7.1$ of genus 7, as well as $R8.1$ and $R8.2$ of genus 8, are the only regular maps of genus 7 or 8 that could possibly admit a realization as a convex-faced polyhedron in $\mathbb{E}^3$ (all other maps violate the condition that $m_V = m_F = 1$). The map $R7.1$ of type $\{3,7\}$ has Petrie polygons of length 18 and a group of order 1008; however, most likely, $R7.1$ itself is not the universal map $\{3,7\}_{18}$. By contrast, $R8.1$ is the universal map $\{3,8\}_8$; its automorphism group is a semidirect product of the unitary reflection group $[1\,1\,1^4]^4$ in complex 3-space by $C_2$, and has order 672 (see pp. 296, 399 of Reference [6]). The map $R8.2$ of type $\{3,8\}$ has Petrie polygons of length 14 and again a group of order 672; however, $R8.2$ itself is not the universal map $\{3,8\}_{14}$, which is known to be infinite (see p. 399 of Reference [6]). Thus all three maps have much larger automorphism groups than the maps in Table 1. It would be desirable to decide whether or not $R7.1$, $R8.1$, and $R8.2$ admit a polyhedral embedding in $\mathbb{E}^3$, and if so, to explicitly construct a polyhedron with maximum possible geometric symmetry.

**Leonardo polyhedra.** Among the eight combinatorially regular maps of Table 1, exactly four admit realizations as *Leonardo polyhedra* in the sense of [29], meaning that the symmetry group of the polyhedron is either the full symmetry group of a Platonic solid or its rotation subgroup. Very few Leonardo polyhedra with large combinatorial automorphism groups seem to be known, and it would be a worthwhile task to find more, particularly Leonardo polyhedra which are combinatorially regular.

**Icosahedral symmetry.** Finally, among all combinatorially regular polyhedra in $\mathbb{E}^3$ discovered so far, none except the icosahedron has full icosahedral symmetry or icosahedral rotation symmetry. Any combinatorially regular polyhedra with icosahedral symmetry would likely be very interesting. As mentioned earlier, when self-intersections are permitted, full icosahedral symmetry can be obtained, for example, for Gordan's maps.

**Acknowledgments:** We are grateful to Ulrich Brehm, Gabor Gévay, and an anonymous referee for a number of valuable comments.

## References

1. McMullen, P.; Schulz, Ch.; Wills, J.M. Equivelar polyhedral manifolds in $E^3$. *Isr. J. Math.* **1982**, *41*, 331–346.
2. McMullen, P.; Schulz, Ch.; Wills, J.M. Polyhedral manifolds in $E^3$ with unusually large genus. *Isr. J. Math.* **1983**, *46*, 127–144.
3. Brehm, U.; Wills, J.M.; Polyhedral Manifolds. In *Handbook of Convex Geometry*; Gruber, P.M., Wills, J.M., Eds.; North-Holland: Amsterdam, The Netherlands, 1993; Volume A, pp. 535–554.
4. Brehm, U.; Schulte, E.; Polyhedral Maps. In *Handbook of Discrete and Computational Geometry*; Goodman, J.E., O'Rourke, J., Eds.; CRC Press: Boca Raton, FL, USA, 1997; pp. 345–358.
5. Coxeter, H.S.M.; Moser, W.O.J. *Generators and Relations for Discrete Groups*, 4th ed.; Springer: Berlin, Heidelberg, Germany, 1980.
6. McMullen, P.; Schulte, E. *Abstract Regular Polytopes*; Cambridge University Press: Cambridge, UK, 2002; Volume 92.
7. Jones, G.A.; Singerman, D. *Complex Functions*; Cambridge University Press: Cambridge, UK, 1987.
8. Sherk, F.A. The regular maps on a surface of genus three. *Can. J. Math.* **1959**, *11*, 452–480.
9. Garbe, D. Über die regulären Zerlegungen geschlossener orientierbarer Flächen. *J. Reine Angew. Math.* **1969**, *237*, 39–55.
10. Coxeter, H.S.M. Regular skew polyhedra in three and four dimensions and their topological analogues. *Proc. Lond. Math. Soc.* **1937**, *s2-43*, 33–62.
11. Conder, M. Regular maps and hypermaps of Euler characteristic −1 to −200. *J. Comb. Theory Ser. B* **2009**, *99*, 455–459. Associated lists available online: http://www.math.auckland. ac.nz/~conder (accessed on 21 December 2011).
12. Conder, M.; Siran, J.; Tucker, T. The genera, reflexibility and simplicity of regular maps. *J. Eur. Math. Soc.* **2010**, *12*, 343–364.

13. Breda d'Azevedo, A.; Nedela, R.; Siran, J. Classification of regular maps of negative prime Euler characteristic. *Trans. Am. Math. Soc.* **2005**, *357*, 4175–4190.

14. Coxeter, H.S.M. The abstract groups $G^{m,n,p}$. *Trans. Am. Math. Soc.* **1939**, *45*, 73–150.

15. Dyck, W. Über Aufstellung und Untersuchung von Gruppe und Irrationalität regulärer Riemannscher Flächen. *Math. Ann.* **1880**, *17*, 473–508.

16. Dyck, W. Notiz über eine reguläre Riemannsche Fläche vom Geschlecht 3 and die zugehörige Normalkurve 4. Ordnung. *Math. Ann.* **1880**, *17*, 510–516.

17. Klein, F. Über die Transformationen siebenter Ordnung der elliptischen Functionen. *Math. Ann.* **1879**, *14*, 428–471; Revised version in *Gesammelte Mathematische Abhandlungen*; Springer: Berlin, Heidelberg, Germany, 1923; Volume 3.

18. Klein, F. *Vorlesungen über das Ikosaeder und die Auflösung der Gleichungen fünften Grades*; B.G. Teubner: Leipzig, Germany, 1884.

19. Klein, F.; Fricke, R. *Vorlesungen über die Theorie der elliptischen Modulfunktionen*; B.G. Teubner: Leipzig, Germany, 1890.

20. Gordan, P. Über die Auflösung der Gleichung vom fünften Grade. *Math. Ann.* **1878**, *13*, 375–405.

21. Schwörbel, J. *Die kombinatorisch regulären Tori*; Diplom-Arbeit, University of Siegen: Siegen, Germany, 1988.

22. Wilson, S.E. Operators over regular maps. *Pac. J. Math.* **1979**, *81*, 559–568.

23. Stott, A.B. Geometrical deduction of semiregular from regular polytopes and space fillings. *Verh. K. Akad. Wet. (Eerste sectie)* **1910**, *11.1*, 3–24.

24. Hurwitz, A. Über algebraische Gebilde mit eindeutigen Transformationen in sich. *Math. Ann.* **1893**, *41*, 403–442.

25. Schulte, E.; Wills, J.M. A polyhedral realization of Felix Klein's map $\{3, 7\}_8$ on a Riemann surface of genus 3. *J. Lond. Math. Soc.* **1985**, *s2-32*, 539–547.

26. Bokowski, J.; Sturmfels, B. *Computational Synthetic Geometry*; Springer-Verlag: Berlin, Heidelberg, Germany, 1989; Volume 1355.

27. Bokowski, J.; Wills, J.M. Regular polyhedra with hidden symmetries. *Math. Intell.* **1988**, *10*, 27–32.

28. Scholl, P.; Schürmann, A.; Wills, J.M. Polyhedral models of Felix Klein's group. *Math. Intell.* **2002**, *24*, 37–42.

29. Gévay, G.; Wills, J.M. On regular and equivelar Leonardo polyhedra. *Ars Math. Contemp.* **2011**, in press.

30. McCooey, D.I. A non-self-intersecting polyhedral realization of the all-heptagon Klein map. *Symmetry Cult. Sci.* **2009**, *20*, 247–268.

31. Richter, D.A. How to make the Mathieu group $M_{24}$. Available online: http://homepages.wmich. edu/~drichter/mathieu.htm (accessed on 26 December 2011).

32. Bokowski, J. A geometric realization without self-intersections does exist for Dyck's regular map. *Discret. Comput. Geom.* **1989**, *4*, 583–589.

33. Brehm, U. Maximally symmetric polyhedral realizations of Dyck's regular map. *Mathematika* **1987**, *34*, 229–236.

34. Schulte, E.; Wills, J.M. Geometric realizations for Dyck's regular map on a surface of genus 3. *Discret. Comput. Geom.* **1986**, *1*, 141–153.

35. Wills, J.M. A combinatorially regular dodecahedron of genus 3. *Discret. Math.* **1987**, *67*, 199–204.

36. Schulte, E.; Wills, J.M. On Coxeter's regular skew polyhedra. *Discret. Math.* **1986**, *60*, 253–262.

37. McMullen, P.; Schulte, E.; Wills, J.M. Infinite series of combinatorially regular maps in three-space. *Geom. Dedicata* **1988**, *26*, 299–307.

38. Schulte, E.; Wills, J.M. Combinatorially Regular Polyhedra in Three-Space. In *Symmetry of Discrete Mathematical Structures and Their Symmetry Groups*; Hofmann, K.H., Wille, R., Eds.; Heldermann: Berlin, Heidelberg, Germany, 1991; pp. 49–88.

39. Joswig, M.; Ziegler, G.M. Neighborly cubical polytopes. *Discret. Comput. Geom.* **2000**, *24*, 325–344.

40. Joswig, M.; Rörig, T. Neighborly cubical polytopes and spheres. *Isr. J. Math.* **2007**, *159*, 221–242.

41. Rörig, T.; Ziegler, G.M. Polyhedral surfaces in wedge products. *Geom. Dedicata* **2011**, *151*, 155–173.

42. Grünbaum, B.; Shephard, G.C. Polyhedra with transitivity properties. *C. R. Math. Rep. Acad. Sci. Can.* **1984**, *6*, 61–66.

43. Grünbaum, B. Acoptic Polyhedra. In *Advances in Discrete and Computational Geometry: Proceedings of the 1996 AMS-IMS-SIAM joint summer research conference, Discrete and Computational Geometry–Ten Years Later, July 14–18, 1996, Mount Holyoke College*; Chazelle, B., Goodman, J.E., Pollack, R., Eds.; American Mathematical Society: Providence, RI, USA, 1999; pp. 163–199.

44. Brehm, U.; Grünbaum, B.; Wills, J.M. A polyhedral embedding of the Klein-Fricke map (tentative title). In preparation, 2012.

45. Schulte, E.; Wills, J.M. Kepler-Poinsot-type realizations of regular maps of Klein, Fricke, Gordan and Sherk. *Can. Math. Bull.* **1987**, *30*, 155–164.

46. Brehm, U. Dresden University, Dresden, Germany. Private communication, 2010.

47. Coxeter, H.S.M. Groups generated by unitary reflections of period two. *Can. J. Math.* **1957**, *9*, 243–272.

48. Coxeter, H.S.M. *Regular Polytopes*, 3rd ed.; Dover: New York, NY, USA, 1973.

49. Sherk, F.A. A family of regular maps of type {6,6}. *Can. Math. Bull.* **1962**, *5*, 13–20.

50. Wills, J.M. Combinatorially regular polyhedra of index 2. *Aequ. Math.* **1987**, *34*, 206–220.

51. Cutler, A.; Schulte, E. Regular polyhedra of index two, I. *Contrib. Algebr. Geom.* **2011**, *52*, 133–161.

52. Cutler, A. Regular polyhedra of index two, II. *Contrib. Algebr. Geom.* **2011**, *52*, 357–387.

53. Grünbaum, B. Regular polyhedra—old and new. *Aequ. Math.* **1977**, *16*, 1–20.

54. Grünbaum, B. Realizations of symmetric maps by symmetric polyhedra. *Discret. Comput. Geom.* **1998**, *20*, 19–33.

55. Cutler, A.; Schulte, E.; Wills, J.M. Gordan's maps as skeletal polyhedra with icosahedral symmetry (tentative title). In preparation, 2012.

symmetry

MDPI

*Article*

# Maniplexes: Part 1: Maps, Polytopes, Symmetry and Operators [†]

## Steve Wilson

Department of Mathematics and Statistics, Northern Arizona University, NAU Box 5717, Flagstaff, AZ 86011, USA; E-Mail: Stephen.Wilson@nau.edu; Tel.: +1-928-523-6890; Fax: +1-928-523-5847

[†] For the Amusement of Tomaž Pisanski on the Occasion of one of his 60th Birthdays.

Received: 29 January 2012; in revised form: 26 March 2012 / Accepted: 5 April 2012 / Published: 16 April 2012

**Abstract:** This paper introduces the idea of a maniplex, a common generalization of *map* and of *polytope*. The paper then discusses operators, orientability, symmetry and the action of the symmetry group.

**Keywords:** symmetry; polytope; maniplex; map; flag; transitivity

## 1. Introduction

The primary purpose of this paper is to introduce the idea of a *maniplex*. This notion generalizes the idea of *map* to higher dimensions, while at the same time slightly generalizing (and perhaps simplifying) the idea of *abstract polytope*.

Part of the motivation for proposing these definitions is the desire to unify the topics of Map and Polytope, to demonstrate their similarity and commonalities.

We will first give definitions for map and (abstract) polytope. We will compare and contrast the two notions. The two topics have in common two simple ideas. One is the idea of a *flag* being the unit brick of construction. The other is that we make items of one dimension by assembling items of one dimension less. We define *maniplex* to generalize both. We then discuss: (1) operators which make one maniplex from another; (2) orientability, symmetry and the action of the symmetry group.

A sequel to this paper will deal with (3) the idea of consistent cycles in graphs (see [1,2]) and its applications to maps, polytopes and maniplexes, and (4) a construction which expands the possibilities for semisymmetric graphs as medial layer graphs (see [3,4]).

First, we examine the foundations of both maps and polytopes.

## 2. Maps

We begin by defining a *map* $\mathcal{M}$ to be an embedding of a connected graph into a compact connected surface. Beneath the apparent simplicity of this definition lie a number of caveats of two types: We are being loose about the word "graph" and restrictive about "embedding".

First, when we use the word "graph", we include what some sources call "multigraphs" or "pseudographs". In particular, we allow such aberrations as loops, multiple edges and semi-edges. Secondly, we want an "embedding" to be a continuous, one-to-one function from a topological form of a graph into a surface which is nice in several ways: (1) the image of the interior of each edge has a neighborhood which is homeomorphic to a disk; (2) each connected component of the complement of the image of the graph is itself homeomorphic to a disk. This is sometimes called a "cellular embedding". We can then think of the cube as an embedding of a graph of eight trivalent vertices onto the sphere. Maps exist on all surfaces.

*Symmetry* **2012**, *4*, 265–275

How can we think of a map combinatorially? Our first step is to choose some point in the interior of each face to be its "center". We choose a point in the (relative) interior of each edge to be its "midpoint". We then subdivide the map into *flags*; these are triangles made by connecting each face-center to each of its surrounding vertices and edge-midpoints, as in Figure 1.

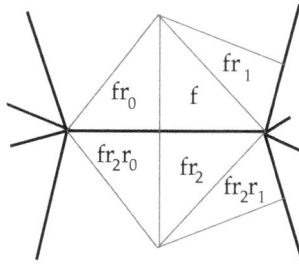

**Figure 1.** Flags in a map.

While these "triangles" are purely topological, by considering the correct Riemann surface, we can regard them as rigid triangles with straight line segments for sides, and then the corners at the edge-midpoints actually are right angles. Each flag $f$ shares a side with three others: one across the face-center-to-midpoint side, one across the hypotenuse and one across the midpoint-to-vertex side. We call these flags $fr_0, fr_1$ and $fr_2$, respectively, as in Figure 1. This defines relations $r_0, r_1$ and $r_2$ on the set $\Omega$ of flags. These permutations on $\Omega$ generate a group $C = C(\mathcal{M})$, called the *connection group* of the map $\mathcal{M}$. (In some contexts, $C(\mathcal{M})$ is called the *monodromy* group of $\mathcal{M}$.) It is clear that, because of the right angles in the flags, $r_0 r_2$ must be the same as $r_2 r_0$; that is, $(r_0 r_2)^2$ is the identity.

## 3. Polytopes

A convex geometric polytope $\mathcal{P}$ is the convex hull of a finite set $S$ of points in some Euclidean space. A *face* in $\mathcal{P}$ is the intersection of $\mathcal{P}$ with some hyperplane which does not separate $S$; it is an *i*-face if its affine hull has dimension $i$. The set $\mathcal{F}$ of faces of $\mathcal{P}$ (of all dimensions) is partially ordered by inclusion. This partial ordering has certain properties, and these form the axiomatics for abstract polytopes. An *abstract polytope* is a partial ordering $\leq$ on a set $\mathcal{F}$ (whose elements are called *faces*) satisfying the following axioms:

(1) $\mathcal{F}$ contains a unique maximal and a unique minimal element.
(2) All maximal chains (these are called *flags*) have the same length. This allows us to assign a "rank" or "dimension" to each face. The unique minimal face (usually called "$\emptyset$") is given rank $-1$.
(3) If faces $a < b < c$ are consecutive in some chain, then there exists exactly one $b' \neq b$ such that $a < b' < c$. (This axiom is usually called the *diamond condition*).
(4) For any $a \leq c$, the *section* $[a, c]$ is the sub-poset consisting of all faces $b$ such that $a \leq b \leq c$. We require it to be true in any section that if $f_1$ and $f_2$ are any two flags of the section, then there is a sequence of flags of the section, beginning at $f_1$ and ending at $f_2$, such that any two consecutive flags differ in exactly one rank. This condition is called *strong flag connectivity*.

See [3,5–13] for illuminating work on polytopes and their symmetry.

In particular, if the rank of the maximal element is $n$, we call $\mathcal{P}$ an *n*-polytope. If $f$ is any flag, then for $-1 \leq i \leq n$, let $f_i$ be its face of rank $i$. For $0 \leq i \leq n - 1$, let $f_i'$ be the unique face other than $f_i$ such that $f_{i-1} \leq f_i' \leq f_{i+1}$, and let $f^i$ be the flag identical to $f$ except that the face of rank $i$ is $f_i'$. From a given *n*-polytope, we can form its *flag graph* in the following way: The vertex set is $\Omega$, the set of all flags in $\mathcal{P}$. It has edges of colors $0, 1, 2, \ldots, n - 1$. The edges of color $i$ are all pairs of the form $\{f, f^i\}$

for $f \in \Omega$. Thus, two vertices are joined (by an edge colored $i$) if they are flags which are identical except at rank $i$. Let $r_i$ be the set of all edges colored $i$. Because all flags have the same entry at rank $-1$ and at rank $n$, $r_i$ will be defined only for $i = 0, 1, \ldots, n-1$.

Now, in an $n$-polytope, if $n > i > j + 1$ and $f$ is any flag, then $(f^i)_{j+1} = f_{j+1}, (f^i)_j = f_j$, and $(f^i)_{j-1} = f_{j-1}$. Therefore $(f^i)'_j = f'_j$. The same equalities hold with $i$ and $j$ reversed, and so $(f^i)^j = (f^j)^i$. Thus, in the flag graph of an $n$-polytope, if $i$ and $j$ differ by more than one, then the subgraph of edges colored $i$ or $j$ consists of 4-cycles spanning the graph.

## 4. Maps and Polytopes

We now generalize both maps and polytopes by asking what they have in common. We have already remarked the similarity of their flag structures. In addition, they are made from lower dimensional structures in a similar way. One assembles a map from a collection of polygons, identifying each edge of one polygon with one edge of one polygon. One assembles a 4-polytope from a collection of facets, which are 3-polytopes, identifying each 2-face of each facet with some 2-face of some facet. Of course, in each identification, the faces must be congruent: we cannot glue a triangle to a square.

## 5. Complexes and Maniplexes

An *n-complex* is a pair $\mathcal{M} = (\Omega, \mathcal{A})$, where $\Omega$ is a set of things called *flags* and $\mathcal{A}$ is a sequence $\mathcal{A} = [r_0, r_1, \ldots, r_n]$ in which the $r_i$'s satisfy:

(1) Each $r_i$ is a partition of $\Omega$ into sets of size 2.
(2) If $i \neq j$, then $r_i$ and $r_j$ are disjoint.
(3) $\mathcal{M}$ is *connected*. We will explain this after some discussion of notation.

We will write "$xr_i = y$" to mean that $\{x, y\} \in r_i$. We will also say that "$x$ and $y$ are *i-adjacent*". Thus we can think of each $r_i$ in several ways:

(1) as a partition,
(2) as a function,
(3) as a permutation,
(4) as a set of edges colored $i$ in a graph $\Gamma$,
(5) in particular, a perfect matching on the set $\Omega$.

The third requirement in the definition, of connectedness, is the assertion that the graph $\Gamma$ is connected. This is equivalent to the statement that the group $C(\mathcal{M})$ generated by the permutations $r_i$ (called the *connection* group) is transitive on $\Omega$.

If $\mathcal{M} = (\Omega, \mathcal{A})$, where $\mathcal{A} = [r_0, r_1, \ldots, r_n]$ and $\mathcal{M}' = (\Omega', \mathcal{B})$, where $\mathcal{B} = [s_0, s_1, \ldots, s_n]$ are n-complexes, an *isomorphism* from one to the other is a function $\phi$ mapping $\Omega$ one-to-one and onto $\Omega'$ such that for all $i$, $r_i \phi = s_i$.

Let $\mathcal{R}_i$ be the set of all of the $r_j$'s for $j \neq i$. Thinking of the $r_i$'s as sets of edges, $\Omega$ has a number of connected components under the union of $\mathcal{R}_i$, and we will call such a component an *i-face*. Alternatively, thinking of the $r_i$'s as permutations on $\Omega$, we can define an *i*-face to be an orbit under $< \mathcal{R}_i >$. A 0-face is a *vertex*, a 1-face is an *edge*, and an n-face is a *facet*. A facet of a facet is a *subfacet*, also called a *bridge*. To be clear: a subfacet is a component or orbit under $r_0, r_1, \ldots, r_{n-2}$. (Notice that if n is less than 2, this list of $r$'s is empty, and so a subfacet in that case is simply a flag.)

What we are calling an "*n-complex*" here is essentially Vince's "combinatorial map". Compare the current paper to [14].

Now we are ready to define *n-maniplex*, (the name is meant to suggest a complex which behaves like a manifold) and we will do that recursively. Every 0-complex is a 0-maniplex. Connectedness implies that $\Omega$ has size exactly 2. Each of its two flags is both a vertex and a facet. Once $n$-maniplexes are defined, an $(n+1)$-maniplex is defined to be any $(n+1)$-complex in which:

(1) Each facet is an n-maniplex.
(2) For any subfacet $a$, $r_{n+1}$ (considered as a function), restricted to $a$, is an isomorphism from $a$ to some subfacet $a'$ ($a'$ need not be different from $a$, though it often is).

Intuitively, any $n+1$-maniplex is made from a collection of $n$-maniplexes by pairing up isomorphic $(n-1)$-faces and defining $r_{n+1}$ to be the union of those isomorphisms.

In the examples that follow, we show each maniplex as a sequence of partitions and then as the graph $\Gamma$.

**Example:** Every 0-complex is a 0-maniplex, isomorphic to this one: $\Omega = \{1,2\}, r_0 = \{\{1,2\}\}$.

**Figure 2.** A 0-maniplex.

**Example:** Every 1-complex is a 1-maniplex, and can be regarded geometrically as a polygon or 2-polytope. $\Omega = \{1,2,3,4,5,6,7,8\}$, $r_0 = \{\{1,2\},\{3,4\},\{5,6\},\{7,8\}\}$, $r_1 = \{\{2,3\},\{4,5\},\{6,7\},\{8,1\}\}$. This maniplex has four vertices ($\{2,3\},\{4,5\}$, *etc.*), and four edges ($\{1,2\},\{3,4\},\dots$); thus it is a 4-gon.

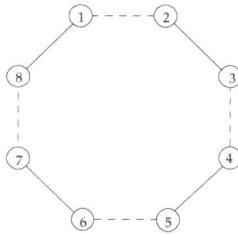

**Figure 3.** A 1-maniplex.

**Example:** A 2-maniplex. The 2-maniplex shown in Figure 4 has four faces (the 0-1 cycles), four vertices (the 1-2 cycles) and 6 edges (the 0-2 cycles). Each vertex shares an edge with each other vertex and so has degree 3. Similarly each face has three sides and so this maniplex is the tetrahedron.

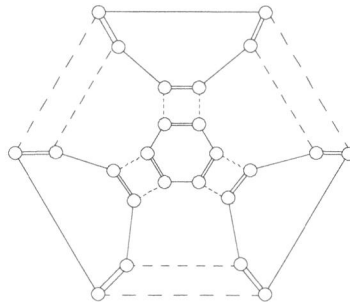

**Figure 4.** The tetrahedron as a 2-maniplex.

One of the consequences of the definition is that if $i > j + 1$, then within the $i$-maniplex $[r_0, r_1, \ldots, r_i]$, two flags (1 and 2 in Figure 5) which are $j$-adjacent are in the same subfacet. Since $r_i$ is an isomorphism of subfacets, these must be $i$-adjacent to two flags which are $j$-adjacent, as in Figure 5.

Thus, whenever $i$ and $j$ differ by more than 1, then $r_i$ and $r_j$ must commute. Any group defined by having generators $[r_0, r_1, \ldots, r_n]$ and relations which require that $|i - j| > 1$ implies that $r_i$ and $r_j$ commute, is called a *Coxeter string group*. Thus, a maniplex is essentially a transitive action of a Coxeter string group. Conversely, any transitive action of a Coxeter string group on a set is a maniplex (as long as no generator has a fixed point). We could, in fact, use that as the definition of a maniplex.

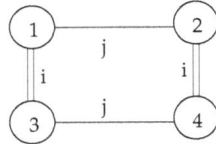

**Figure 5.** Within an i-face.

Let $\Lambda$ be the vertex set of any one component of the distance-2 graph of $\Gamma$. To say that another way, $\Lambda$ is the orbit of some flag under the group $C^+(\mathcal{M})$ generated by all products of the form $r_i r_j$. If $\Lambda = \Omega$, we will call the maniplex *non-orientable*, and we call it *orientable* if otherwise. More briefly, $\mathcal{M} = (\Omega, \mathcal{A})$ is orientable if and only if $\Gamma$ is bipartite.

**Fact:** Every map can be considered as a 2-maniplex.

Moreover, the converse is true as well: Every finite 2-maniplex has an embodiment as a map. We see this by the following construction: Given a finite 2-maniplex $\mathcal{M} = (\Omega, [r_0, r_1, r_2])$, assemble a set of right triangles in which each one has one leg darkened. Label each triangle with an element of $\Omega$. For each $x \in \Omega$, identify the light leg of triangle $x$ with the light leg of triangle $xr_0$, the dark leg of triangle $x$ with the dark leg of triangle $xr_2$, and the hypotenuse of $x$ with the hypotenuse of $xr_1$. (Yes, yes, we must be more careful than that: In the $r_0$ and $r_2$ attachments, the right angles must meet and in the $r_1$ attachment, the two darkened legs must adjoin.) The result is a surface, and the darkened edges form a graph on the surface. The complement of the graph in the surface has components which are connected by $r_0$ and $r_1$, and hence each forms a disc. A similar construction shows that an infinite 2-maniplex in which $r_0 r_1$ and $r_1 r_2$ are of finite orders has an embodiment as a tessellation of some non-compact surface.

Thus maniplexes generalize maps to higher dimensions.

On the other hand, the flag graph of any abstract $(n + 1)$-polytope satisfies the requirements for an $n$-maniplex, and so maniplexes generalize polytopes as well. Notice, incidentally, the disparity in dimension in these two notations. The word "cube", for example, can refer to the three-dimensional solid or to its 2-dimensional surface. It is thus a 3-polytope and a 2-maniplex.

## 6. Non-Polytopal Maps

It is important to note that abstract polytopes do not themselves generalize maps. To show that they do not, we will present two examples of maps, and hence maniplexes, which are not polytopes.

First, consider the map shown in Figure 6.

By examining the identifications making the map, one can see that it has the one face $F$, the five edges 1, 2, 3, 4, 5 and just one vertex $v$. That says that the poset of faces is as in Figure 7.

We see that the poset is not sufficiently detailed to capture the structure of the map. For instance, we cannot, from the poset, determine which edges are consecutive around $F$. This map also fails the diamond condition, as $F$ and $v$ have more than two edges in common.

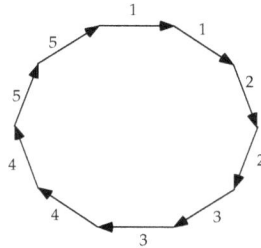

**Figure 6.** A map with one face.

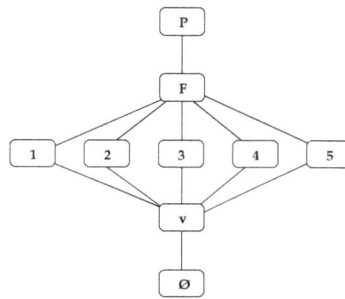

**Figure 7.** The poset of the map.

Figure 8 shows an orientable map made of two octagons, *A* and *B*, with sides identified as indicated by the edge numbers and the colors of the vertices. If we follow the identifications, we see that there are exactly 4 vertices, each of degree 4.

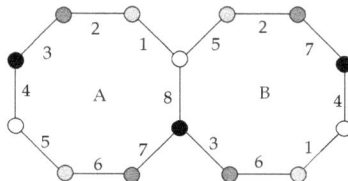

**Figure 8.** Faces and vertices in the map $M'_{8,3}$.

This map is small, but very interesting for its size. It has many symmetries and is in fact a reflexible map (defined in the next section). It is thus something that we would like to include in our study. However, the map does not qualify as an abstract polytope, as the diamond condition fails: Face *A* and the white vertex share not only edges 1 and 8 but also 4 and 5. Thus it is a non-polytopal map. This is not at all uncommon among maps.

## 7. Symmetry in Maniplexes

A *symmetry* or *automorphism* of the maniplex $\mathcal{M} = (\Omega, \mathcal{A})$ is an isomorphism of $\mathcal{M}$ onto itself. It is thus a color-preserving symmetry of the graph $\Gamma$. Therefore, $\sigma \in S_\Omega$ is a symmetry of $\mathcal{M}$ if and only if $\sigma$ commutes with each of the permutations $r_i$ (and thus with all of the connection group). The symmetries form a group, $Aut(\mathcal{M})$. Recall that $\Lambda$ is an orbit under the group $C^+(\mathcal{M})$, which is generated by all words in the $r_i$'s of length 2. Define $Aut^+(\mathcal{M})$ to be the subgroup of $Aut(\mathcal{M})$ which sends $\Lambda$ to itself. We will say that $\mathcal{M}$ is *rotary* provided that $Aut^+(\mathcal{M})$ is transitive on $\Lambda$ and *reflexible* provided that $Aut(\mathcal{M})$ is transitive on $\Omega$. Clearly, any non-orientable rotary maniplex is reflexible. Within the domain of rotary maniplexes, *chiral* means *not reflexible*.

The word "regular" has been used to describe maps and polytopes with large symmetry groups. Unfortunately, the word has meant different things in the two topics. Map theorists, beginning, perhaps, with Brahana, use "regular" to describe maps with rotational symmetries, *i.e.*, rotary maps. Polytope theorists have used "regular" to mean flag-transitive, *i.e.*, reflexible. The compromise language of maniplexes is intended to use words that mean what they say.

If $\mathcal{M}$ is any rotary $n$-maniplex, then for $0 < i \leq n$, the subgraph of $\Gamma$ consisting of edges in $r_{i-1} \cup r_i$ will be a collection of disjoint cycles. If $\mathcal{M}$ is rotary, these cycles will be of the same length, $2p_i$. Then the n-tuple $\{p_1, p_2, p_3, \ldots, p_n\}$ is the *Schläfli symbol* or, more briefly, the *type* of $\mathcal{M}$. For example, the dodecahedron is of type $\{5, 3\}$, the n-dimensional cube has type $\{4, 3, 3, \ldots, 3\}$, and the 600-cell is of type $\{3, 3, 5\}$.

If $\mathcal{M} = (\Omega, [r_0, r_1, \ldots, r_n])$ is a reflexible n-maniplex and $I$ is any one fixed flag, then for each $i$, there is a unique symmetry $\alpha_i$ taking $I$ to $Ir_i$. Because symmetries commute with the connection group, we have that for all $i$ and $j$, $I\alpha_i\alpha_j = Ir_i\alpha_j = I\alpha_j r_i = Ir_j r_i$. This implies that the correspondence $\alpha_i \leftrightarrow r_i$ extends to an antimorphism between $Aut(\mathcal{M})$ and $C(\mathcal{M})$. Thus, for a reflexible maniplex, the connection group and the symmetry group are isomorphic. For more on symmetry in maps see [15,16].

## 8. Operators on maniplexes

In maps, we often find it illuminating to consider relations between two maps. It is useful to describe *operators*, constructions which, given one map, make another, related map. The paper [17] considers many such operators in detail, as does [18]. We wish to generalize these to maniplexes; fortunately, this is straightforward:

If $\mathcal{M} = (\Omega, \mathcal{A})$, and $\mathcal{A} = [r_0, r_1, \ldots, r_n]$, the *dual* of $\mathcal{M}$, denoted $D(\mathcal{M})$, has the same flag set $\Omega$, but the reversed sequence $\mathcal{A}' = [r_n, r_{n-1}, \ldots, r_1, r_0]$. It is not obvious from the recursive definition that this is a maniplex, but the alternative definition as an action of a Coxeter string group with fixed-point free generators makes that clear.

We define the *Petrie* of $\mathcal{M}$, $P(\mathcal{M})$, to have sequence $\mathcal{A}'$ identical to $\mathcal{A}$ except at the entry of index $n - 2$: $[r_0, r_1, \ldots, r_{n-3}, r'_{n-2} = r_n r_{n-2}, r_{n-1}, r_n]$. In order for this to qualify as a maniplex, $r'_{n-2}$ must be an involution and commute with all other generators except, possibly, $r_{n-3}$ and $r_{n-1}$. Clearly, $r'_{n-2}$ is an involution because $r_n$ and $r_{n-2}$ commute with each other. It commutes with $r_n$ because $r_{n-2}$ does. And it commutes with all $r_i$ for $i \leq n - 4$ because they commute with $r_n$ and $r_{n-2}$.

Similarly, we define the *opposite* of $\mathcal{M}$, denoted $opp(\mathcal{M})$, to have sequence $[r_0, r_1, r'_2 = r_0 r_2, r_3, \ldots, r_n]$.

Now, $opp$ has an interesting effect on maps: if $\mathcal{M}$ is a map, then $opp(\mathcal{M})$ has the same faces but at each edge, the meeting of the faces is reversed, as in Figure 9.

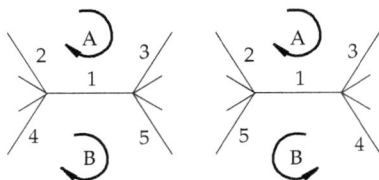

**Figure 9.** An edge in a map and its opposite.

For example, if $\mathcal{M}$ is the tetrahedron, then $opp(\mathcal{M})$ is the hemi-octahedron of type $\{3,4\}$ on the projective plane.

The effect of $opp$ on higher-dimensional maniplexes is interesting and clear: if the facets of $\mathcal{M}$ are copies of the $(n-1)$-maniplex $\mathcal{F}$ for $n > 2$, then the facets of $opp(\mathcal{M})$ are copies of the $(n-1)$-maniplex $opp(\mathcal{F})$, and the $r_n$ connections are the same in both maniplexes. Because the change between $\mathcal{M}$ and $opp(\mathcal{M})$ happens at the $i = 2$ level only, all $i$-faces in $opp(\mathcal{M})$ for $i > 2$ are $opp$ of their counterparts in $\mathcal{M}$.

The general effect of the Petrie operator is less easy to visualize; our best hope here is to realize that $P = D\, opp\, D$. The maniplexes related to $\mathcal{M}$ via $D, P, opp$ and their products are the *direct derivates* of $\mathcal{M}$. We note that the class of maniplexes is closed under these operations.

If $n = 2$, so that the maniplex is a map, $opp$ is also equal to $PDP$, and so $\mathcal{M}$ has at most 6 direct derivates. For any higher value of $n$, we have $Popp = oppP$, and so we get at most 8 direct derivates from one maniplex. Thus the operators act as $D_3$ or $D_4$ on the direct derivates of each map.

Even in seemingly simple cases these operators can produce surprising results. For example, let $\mathcal{M} =$ the 4-simplex. This is a 4-polytope, a 3-maniplex, of type $\{3,3,3\}$. It consists of 5 tetrahedra, assembled so that three facets surround each edge. It has 5 facets, 10 triangular faces, 10 edges and 5 vertices, and is isomorphic to its dual.

For this maniplex, $P(\mathcal{M})$ is a 3-maniplex of type $\{6,4,3\}$; it has the same 5 vertices, and the same 10 edges. It has 10 hexagonal 2-faces, and, shockingly, *just one facet*. The facet is isomorphic to the map $\mathcal{N} = \{6,4|3\}_5$. This map has 10 hexagonal faces; these are the subfacets of the maniplex. The map has 30 edges and 15 vertices. It is formed from Coxeter's regular skew polyhedron $\{6,4|3\}$ by identifying vertices, edges, faces which are antipodal in some Petrie path. The connector $r_3$ joins each flag of $\mathcal{N}$ to the one diametrically opposite to it in its face. This surprising non-polytopal maniplex would not have been found without the use of the operator $P$.

We also define a "bubble" operator $B_j$, analogous to the "hole" operator, $H_j$, defined for maps. The effect of this on $\mathcal{M}$ is to replace $r_{n-1}$ with $r_{n-1}(r_n r_{n-1})^{j-1}$. As in the map case, the result might not be connected and so we define $B_j(\mathcal{M})$ to be any one component of the result.

To visualize the effect of the operator $B_j$, it may help to consider Figure 10, which shows a simplified view of facets and subfacets arranged about a subsubfacet.

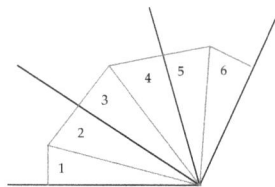

**Figure 10.** Facets and subfacets.

In this figure, the solid lines represent subfacets, the common point of these lines is the subsubfacet, and the wedges of space between the solid lines represent facets. Then triangle 1 represents a typical flag $f$, triangle 2 is $fr_{n-1}$, and 3 is $fr_{n-1}r_n$. Similarly, 4 and 6 are $fr_{n-1}(r_n r_{n-1})^1$ and $fr_{n-1}(r_n r_{n-1})^2$. Thus the connector $r_{n-1}(r_n r_{n-1})^{j-1}$ links each subfacet to those which are separated from it by $j$ facets. These subfacets join by the new $(n-1)$-connector to form "bubbles" which are the facets of the new maniplex.

For instance, in the 600-cell, which is composed of 600 regular tetrahedra arranged so that 5 of them surround every edge, each vertex $v$ is incident with 20 tetrahedral facets. The 20 triangles opposite to $v$ in this configuration form an icosahedron. Thus $B_2$ of the 600-cell is a maniplex whose facets are icosahedra, meeting 5 at each edge. In the geometric 600-cell, $B_2$ produces the star-polytope called the *icosahedral 120-cell*.

The maniplexes related to $\mathcal{M}$ through all products of $D, P, opp, B_j$ are the *derivates* of $\mathcal{M}$.

The derivates of a maniplex give a potentially very large class of maniplexes which are related to each other. Understanding any one of them, then, will increase our understanding of the others.

## 9. Conclusions

There are many interesting open questions regarding maniplexes. Here are just a few of them:

(1) Can we classify all rotary maniplexes having one facet? Two facets? In maps, the classifications of 1-face and 2-face maps are easy and all such maps are reflexible. In maniplexes, several examples of chiral 1-facet and 2-facet 3-maniplexes are known.
(2) What conditions on a polytope will guarantee that all of its derivates are polytopal as well?
(3) Given a rotary map $\mathcal{M}$, what are all the rotary 3-maniplexes having facets isomorphic to $\mathcal{M}$?
(4) Given a graph $\Gamma$, what are all the rotary maniplexes whose 1-skeleton is isomorphic to $\Gamma$?

## 10. Acknowledgement

The author wishes to thank all three referees of this paper for insightful suggestions for its improvement.

## References

1. Miklivič, S.; Potpčnik, P.; Wilson, S. Consistent cycles in graphs and digraphs. *Graphs Combin.* **2007**, *23*, 205–216.
2. Miklivič, S.; Potočnik, P.; Wilson, S. Overlap in consistent cycles. *J. Graph Theory* **2007**, *55*, 55–71.
3. Monson, B.; Schulte, E.; Pisanski, T.; Weiss, A.I. Semisymetric graphs from polytopes. *J. Comb. Theory A* **2007**, *114*, 421–435.
4. Monson, B.; Weiss, A.I. Medial Layer Graphs of Equivelar 4-polytopes. *Eur. J. Combin.* **2007**, *28*, 43–60.
5. Conder, M.; Hubard, I.; Pisanski, T. Constructions for chiral polytopes. *J. London Math. Soc.* **2008**, *77*, 115–129.
6. Coxeter, H.S.M. The Edges and faces of a 4-dimensional polytope. *Congres. Numerantium* **1980**, *28*, 309–334.
7. Hubard, I. Two-orbit polyhedra from groups. *Eur. J. Combin.* **2010**, *31*, 943–960.
8. Hubard, I.; Orbanič, A.; Weiss, A.I. Monodromy groups and self-invariance. *Can. J. Math.* **2009**, *61*, 1300–1324.
9. Hubard, I.; Weiss, A.I. Self-duality of chiral polytopes. *J. Comb. Theory A* **2005**, *111*, 128–136.
10. McMullen, P.; Schulte, E. *Abstract Regular Polytopes*, 1st ed.; Cambridge University Press: Cambridge, UK, 2002.
11. Pellicer, D.A. Construction of higher rank polytopes. *Discret. Math.* **2010**, *310*, 1222–1237.
12. Schulte, E. *Regulàre Inzidenzkomplexe*, Universität DortmE. und Dissertation, 1980.
13. Schulte, E. Symmetry of polytopes and polyhedra. In *Handbook of Discrete and Computational Geometry*, 2nd Ed.; Goodman, J.E., O'Rourke, J., Eds.; CRC Press: Boca Raton, FL, USA, 2004.
14. Vince, A. Combinatorial maps. *J. Comb. Theory B* **1983**, *34*, 1–21.
15. Coxeter, H.S.M.; Moser, W.O.J. *Generators and Relations for Discrete Groups*, 4th ed.; Springer-Verlag: New York, NY, USA, 1980.
16. Orbanič, A.; Pellicer, D.; Tucker, T. Edge-transitive Maps of Low Genus. *Ars Math. Cont.* **2011**, *4*, 385–402.
17. Wilson, S. Operators over regular maps. *Pacific J. Math.* **1979**, *81*, 559–568.
18. Orbanič, A.; Pellicer, D.; Weiss, A.I. Map Operations and K-orbit Maps. *J. Comb. Theory A* **2010**, *117*, 411–429.

# Section C:
# Polyhedral Structures, Arts, and Architecture

symmetry

MDPI

*Article*

# A Peculiarly Cerebroid Convex Zygo-Dodecahedron is an Axiomatically Balanced "House of Blues": The Circle of Fifths to the Circle of Willis to Cadherin Cadenzas

**David A. Becker**

Department of Chemistry and Biochemistry, Florida International University, Miami, FL 33199, USA; beckerd@fiu.edu; Tel.: +1-305-348-3736; Fax: +1-305-348-3772

Received: 24 August 2012; in revised form: 28 October 2012; Accepted: 6 November 2012; Published: 15 November 2012

**Abstract:** A bilaterally symmetrical convex dodecahedron consisting of twelve quadrilateral faces is derived from the icosahedron via a process akin to Fuller's Jitterbug Transformation. The unusual zygomorphic dodecahedron so obtained is shown to harbor a bilaterally symmetrical jazz/blues harmonic code on its twelve faces that is related to such fundamental music theoretical constructs as the Circle of Fifths and Euler's tonnetz. Curiously, the patterning within the aforementioned zygo-dodecahedron is discernibly similar to that observed in a ventral view of the human brain. Moreover, this same pattern is arguably evident during development of the embryonic pharynx. A possible role for the featured zygo-dodecahedron in cephalogenesis is considered. Recent studies concerning type II cadherins, an important class of proteins that promote cell adhesion, have generated data that is demonstrated to conform to this zygo-dodecahedral brain model in a substantially congruous manner.

**Keywords:** bilateral symmetry; geometrical music theory; dodecahedra; morphogenesis; cephalogenesis; cadherins; brain networks

---

## 1. Introduction

The rhombic dodecahedron and trapezo-rhombic dodecahedron are prominent paradigms in the context of the two reported solutions to the famous Kepler Conjecture concerning the most efficient packing of uniform spheres [1]. Notably, the patterning of the respective duals of both of these fundamental dodecahedra can be derived in a consistent manner from nature's most symmetrical polyhedron, the icosahedron. Thus, as pointed out by R. Buckminster Fuller in his Jitterbug Transformation [2], the removal of a specific set of six edges from the thirty edges of an icosahedron as shown in Figure 1 generates his so-called Vector Equilibrium, the cuboctahedron (the dual of the rhombic dodecahedron). Similarly, as depicted in Figure 2, removal of a different set of six edges from an icosahedron produces the dual of the trapezo-rhombic dodecahedron; namely, the anticuboctahedron (also known as the triangular orthobicupola or Johnson Solid #27). The graphs of the cuboctahedron and anticuboctahedron are both planar quartic graphs on twelve vertices. In terms of connectivity, there are five distinct planar quartic graphs on twelve vertices that can be obtained from an icosahedral graph by removing one of five different sets of six edges. Of these five graphs (shown in Figure 3), only the first three have detectable symmetry.

**(a)**                                                           **(b)**

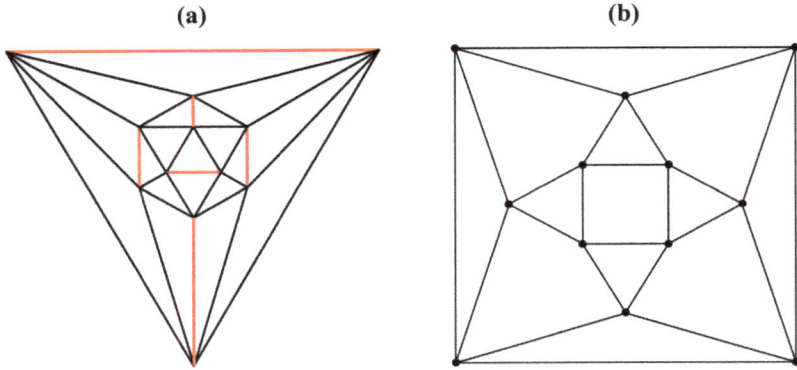

**Figure 1.** From (**a**) icosahedral graph to (**b**) cuboctahedral graph (removal of six red edges).

**(a)**                                                           **(b)**

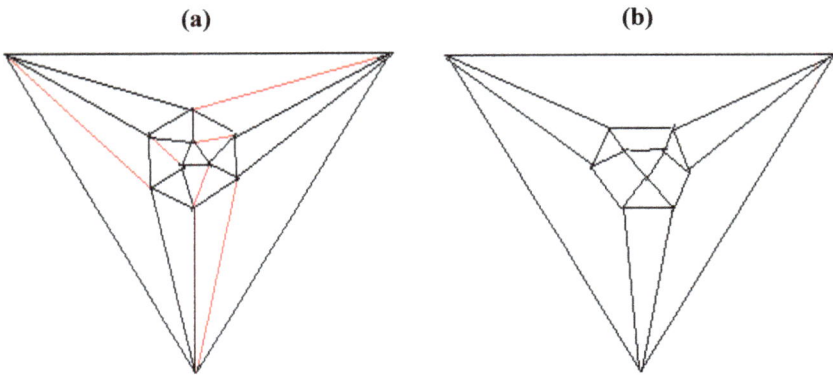

**Figure 2.** Removal of (**a**) the six indicated icosahedral graph edges (shown in red) generates (**b**) the anticuboctahedral graph.

Figure 3a is that of the cuboctahedron, while Figure 3b is that of the anticuboctahedron. Given the importance of the respective dodecahedral duals that correspond to Figure 3a,b, it is logical to surmise that the dodecahedral dual of Figure 3c may also prove to be an interesting geometrical entity. What then is its structure?

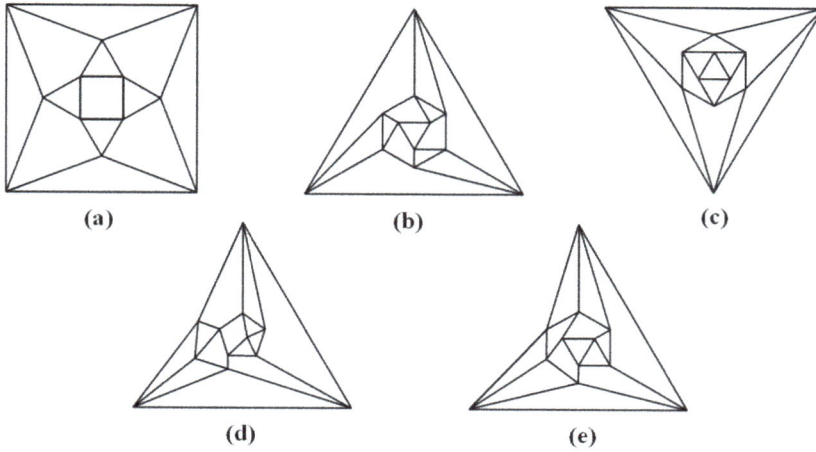

**Figure 3.** The five planar quartic graphs (**a–e**) on twelve vertices that can be obtained from the icosahedral graph by removing a set of six edges.

## 2. Results and Discussion

Good progress toward answering the aforementioned question can be achieved by removing a specific set of twelve edges from a tetrakis hexahedral graph as indicated in Figure 4. Thus, removal of the Figure 4 red edges yields Figure 5a that can be stretched and rotated to yield the isomorphic graph in Figure 5b. Curiously, the graph in Figure 5b possesses only one C2 axis, unlike the graph of the rhombic dodecahedron or that of the trapezo-rhombic dodecahedron—both which can be obtained similarly from the tetrakis hexahedral graph by removing one of two different sets of twelve edges (as shown in Figure 6). A convex dodecahedron that exemplifies the patterning within Figure 5b is shown in Figure 7 and displays striking bilateral symmetry.

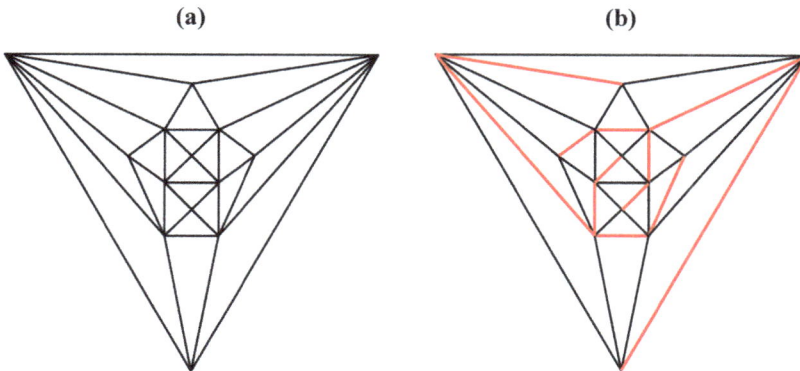

**Figure 4.** The twelve edges to be removed from (**a**) the tetrakis hexahedral graph (in order to arrive at the patterning of the particular dodecahedron that is the dual of Figure 3c) are indicated in (**b**) in red.

**(a)**　　　　　　　　　　　　　　**(b)**

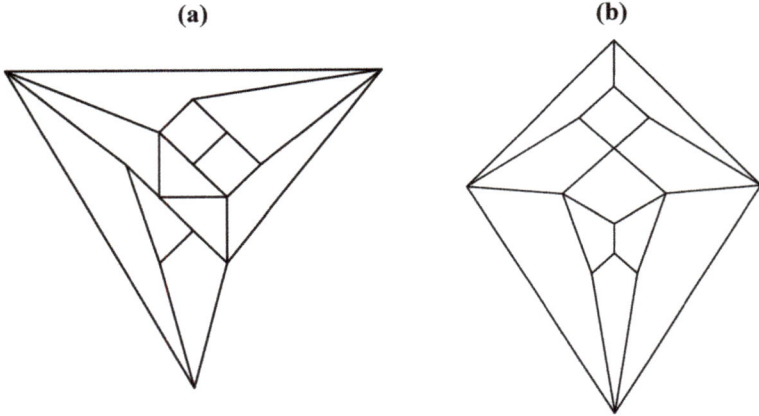

**Figure 5.** (a) The graph obtained from the tetrakis hexahedral graph after removal of the twelve red edges in Figure 4, can be redrawn as (b), the isomorphic graph.

**(a)**　　　　　　　　　　　　　　**(b)**

**(c)**　　　　　　　　　　　　　　**(d)**

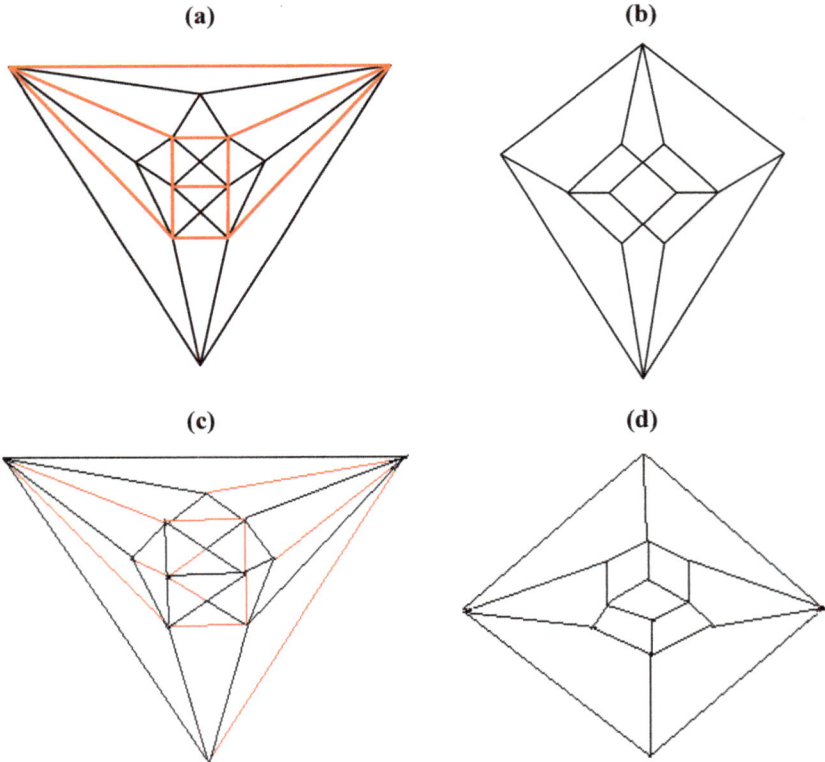

**Figure 6.** Removal of the indicated red edges in (a) the tetrakis hexahedral graph generates (b) the rhombic dodecahedral graph, while removal of the twelve red edges from the tetrakis hexahedral graph as shown in (c) produces (d) the trapezo-rhombic dodecahedral graph.

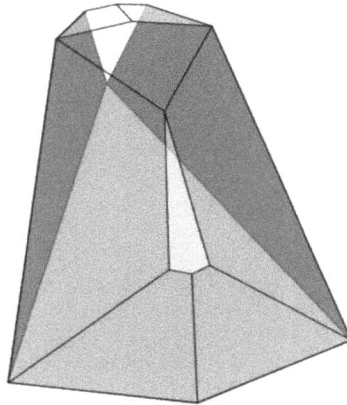

**Figure 7.** A convex, bilaterally symmetrical dodecahedron that exemplifies the patterning in Figure 5.

Remarkably, one of the 3,326,400 harmonically distinct intonations of the Figure 7 zygo-dodecahedron—an intonation that can be produced by assigning each one of the twelve tones to one of each of the twelve quadrilateral faces as illustrated in Figure 8a—exhibits both an important relationship to the Circle of Fifths [3,4] (vide infra) as well as a bilaterally symmetrical network of tones that features all six tritone pair midpoints in this zygo-dodecahedron's single reflection plane.

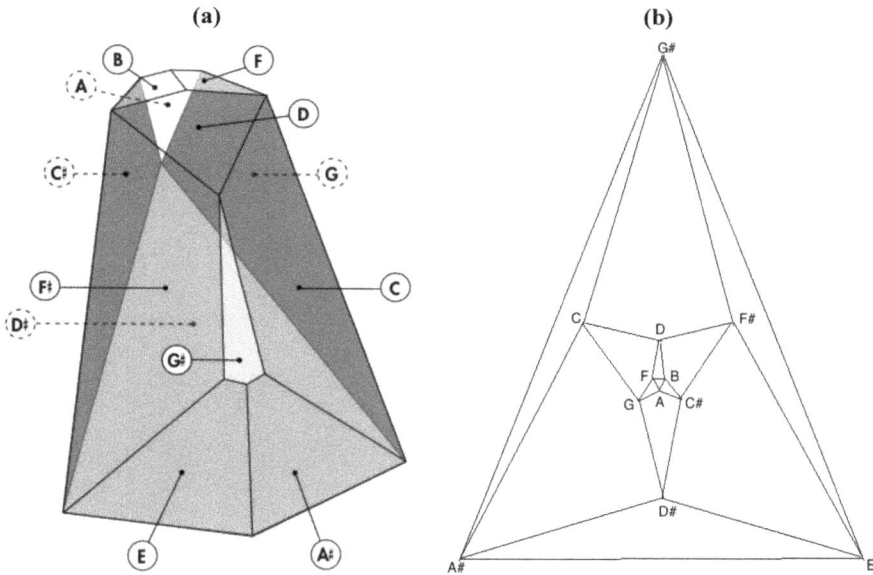

**Figure 8. (a)** A bilaterally symmetrical network of zygo-dodecahedral tones that encodes important jazz/blues sonorities and that is related to the Circle of Fifths; **(b)** The corresponding Schlegel diagram of the dual of the preceding zygo-dodecahedron.

Consider the twelve pentatonic sets that arise from the Figure 8a dodecahedron as seen in Figure 9. Each of these pentatonic sets is formed by combining any given tone with the tones assigned to each of the four faces that share an edge with the quadrilateral face that is assigned to the given tone. Interestingly, all twelve of these pentatonic sets can be construed as chords that are bedrock jazz/blues sonorities in the dominant seventh family. Moreover, as is demonstrated by the original song "Code of Blues", compelling musical compositions in the blues genre can be assembled that employ chord progressions and melodies that stem directly from such pentatonic sets as they are encountered in sequence along a continuous path that moves from face to neighboring face of this intonated zygo-dodecahedron [5].

**(a)**

Figure 9. *Cont.*

**(b)**

**(c)**

| TONE | PENTATONIC SET | | | | | POSSIBLE CLASSIFICATION |
|------|---|---|---|---|---|---|
| C  | C  | D  | G  | G♯ | A♯ | A♯ 13 (no5,no11) |
| C♯ | C♯ | D♯ | F♯ | A  | B  | B9 |
| D  | C  | D  | F  | F♯ | B  | D 13+9 (no5,no11) |
| D♯ | C♯ | D♯ | E  | G  | A♯ | D♯ 7♭9 |
| E  | D♯ | E  | F♯ | G♯ | A♯ | F♯ 13 (no5,no11) |
| F  | D  | F  | G  | A  | B  | G9 |
| F♯ | C♯ | D  | E  | F♯ | G♯ | E 13 (no5,no11) |
| G  | C  | D♯ | F  | G  | A  | F9 |
| G♯ | C  | E  | F♯ | G♯ | A♯ | G♯ 9+5 |
| A  | C♯ | F  | G  | A  | B  | A 9+5 |
| A♯ | C  | D♯ | E  | G♯ | A♯ | C 7+5+9 |
| B  | C♯ | D  | F  | A  | B  | C♯ 7+5♭9 |

**Figure 9.** (a) Sheet Music for the original composition "Code of Blues" (first page); (b) Sheet music for the original composition "Code of Blues" (second page); (c) Pentatonic sets from the intonated zygo-dodecahedron in Figure 8a.

With regard to the relationship between the Figure 8a dodecahedral array of tones and the Circle of Fifths, the following analyses serve to reveal a powerful correspondence. First, it is widely known that Euler's 2-D tonnetz, a music theoretical mainstay that emerged from seminal work reported in 1739 [6], is a fundamental tonal geometry that maps important triads and that clearly features the Circle of Fifths within one of its rows. Second, it can be seen that Euler's 2-D tonnetz is closely related to a three-dimensional arrangement of the twelve tones upon the twelve faces of a rhombic dodecahedron as presented in Figure 10a. Thus, remembering that the dual polyhedron of the rhombic dodecahedron is the cuboctahedron, subtle rearrangement of nine particular tones of Euler's 2-D tonnetz (highlighted by the trapezoidal perimeter in Figure 11a) produces the geometry in Figure 11b that constitutes a 2-D projection of the top half of a cuboctahedron (with correspondingly intonated vertices) that is resting upon one of its triangular faces.

Third, exactly half of the pentatonic sets (six of the twelve sets) that emanate from the rhombic dodecahedral geometry in Figure 10a bear remarkable similarity to exactly half of the Figure 9 pentatonic sets (six of the twelve sets) that originate from the zygo-dodecahedral geometry in Figure 8a. Thus, three of the six pentatonic sets from the rhombic dodecahedral geometry in Figure 10a can be viewed as major seventh chords with the added ninth (fully tertian major ninth chords), while three other such sets can be viewed as major seventh chords with the added thirteenth. The roots of these three sets with added thirteenths are a semitone lower than the roots of the three sets with added ninths. This same pattern is present among six of the Figure 9 pentatonic sets from the Figure 8a zygo-dodecahedral array—with a crucial difference being that the three zygo-dodecahedral seventh chords with added ninths (fully tertian dominant ninth chords) and the three zygo-dodecahedral seventh chords with added thirteenths (also containing ninths but no fifths) possess dominant seventh rather than major seventh cores. Moreover, these six pentatonic sets from Figure 9 and the six corresponding pentatonic sets from Figure 10 recapitulate important harmonic properties of the Circle of Fifths such as the fact that any set composed of five contiguous tones in the Circle of Fifths (the tones of fundamental major/minor pentatonic scales) comprises a major triad along with the respective ninth and thirteenth (typified in the key of F# by the harmonically coherent black keys of the piano).

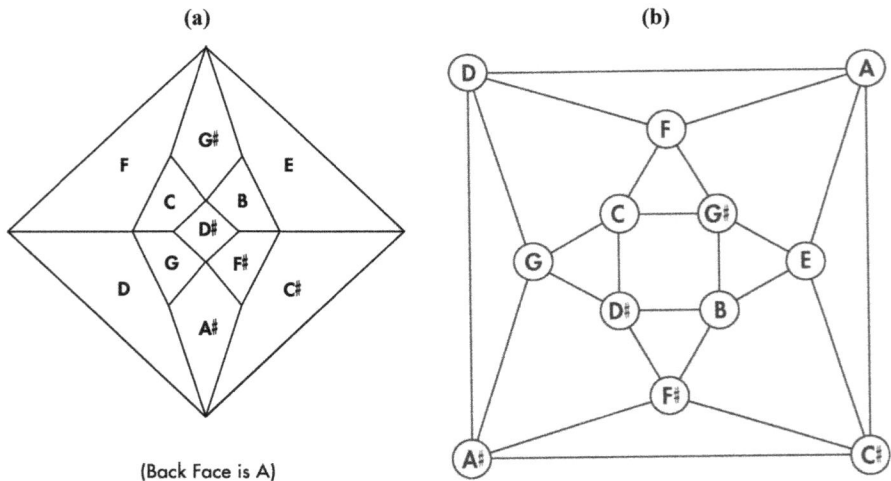

**Figure 10.** (**a**) An important intonation of the rhombic dodecahedron with connections to Euler's tonnetz, the Circle of Fifths, and the zygo-dodecahedral intonation in Figure 8a. (**b**) The corresponding dual cuboctahedral graph of the preceding rhombic dodecahedral graph.

**(a)**          **(b)**

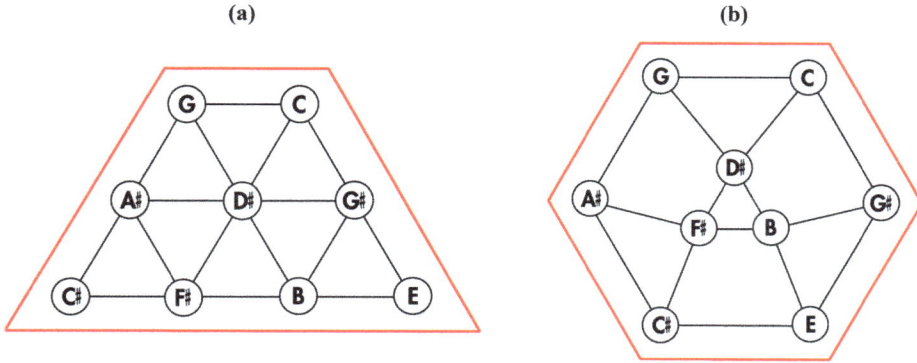

**Figure 11.** (**a**) Nine particular tones from Euler's tonnetz; (**b**) A subtle rearrangement of the nine tones from Euler's tonnetz in Figure 11a reveals a cuboctahedral projection that corresponds to the rhombic dodecahedron in Figure 10a (and the cuboctahedron in Figure 10b).

Additionally, just as any tone in the Circle of Fifths is flanked by its respective dominant and sub-dominant tones to constitute a vital 1,4,5 relationship, so too are several striking 1,4,5 relationships present in the Figure 10a rhombic dodecahedral tonal geometry such as in neighboring major ninths (E major ninth, A major ninth, B major ninth) and minor ninths (C minor ninth, F minor ninth, G minor ninth) in the six aforementioned Figure 10 pentatonic sets. Likewise encoded within Figure 10a are elements of the harmonically important cadence IV to V7 to I as occurs in three neighboring pentatonic sets in Figure 10a (E major ninth, F# 7 + 9, B major ninth). The rhombic dodecahedral array in Figure 10a also sports important 1,4,5 relationships in triads that are encoded by those eight of the fourteen rhombic dodecahedral vertices that define a cube. Thus, four of these eight triads are major (E, F#, A, B with 1,4,5 relationships in the key of E or B)) and four of the eight are minor (C, D, F, G with 1,4,5 relationships in the key of C or G). In three dimensions, the four major triads can be seen to be encoded by vertices comprising the top four cubic vertices while the bottom four such cubic vertices encode the four minor triads (Figure 12a). With regard to the cube, it should be noted that the twelve face centers of the rhombic dodecahedron define the twelve edge centers of an appropriately sized cube such that the geometry in Figure 10 can be redrawn on cube edges as in Figure 12b. It is believed that the representation in Figure 12b is the first tonal geometry that is related to prior music theoretical cornerstones (the Circle of Fifths and Euler's tonnetz) and that captures important major/minor 1,4,5 relationships while utilizing all twelve tones upon a single cube (Douthett and Steinbach's "Cube Dance" [7–9] uses four interconnected cubes with each of the four cubes containing six of the twelve tones). Also noteworthy is the fact that important geometric properties concerning tritone pair midpoints exist in both the Circle of Fifths and in the Schlegel diagram corresponding to the Figure 8a zygo-dodecahedron (shown in Figure 13) such that all six tritone pair midpoints in the Circle of Fifths are copunctal at the center of the circle whereas all six tritone pair midpoints in Figure 13 are centrally collinear. In three dimensions, as has been previously mentioned, all six tritone pair midpoints of the Figure 8a zygo-dodecahedron lie conspicuously in its lone reflection plane.

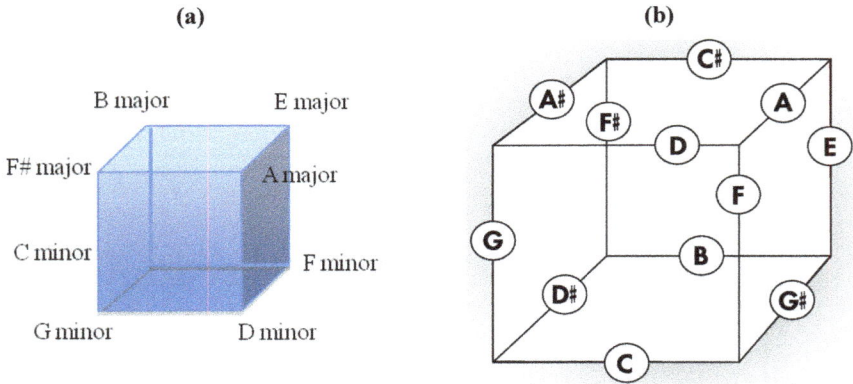

**Figure 12.** (a) The four major triads and four minor triads encoded at those eight of the fourteen vertices of the rhombic dodecahedron in Figure 10a that define a cube; (b) A version of the intonated rhombic dodecahedron in Figure 10a that features intonated cube edges.

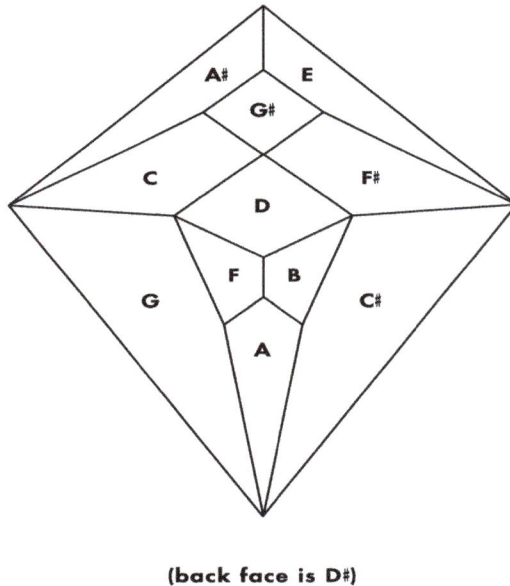

**(back face is D#)**

**Figure 13.** Schlegel diagram corresponding to the Figure 8a zygo-dodecahedron.

Fourth, both the rhombic dodecahedral array of tones in Figure 10a and the zygo-dodecahedral array of tones in Figure 8a exhibit a precise correlation medially and laterally to one of the two possible ways that the Circle of Fifths can be bilaterally sectioned according to specific criteria that will become clear in the following discussion. Thus, for the sake of argument, consider the ways that a circle with twelve evenly spaced vertices can be divided in a bilaterally symmetrical manner with two straight lines to produce a medial group of four vertices, a left lateral group of four vertices, and a right lateral group of four vertices such that any one of these three groups of four vertices possesses at least one vertex on both sides of a third line that bisects the circle. There are exactly two solutions to such criteria as shown in Figures 14 and 15.

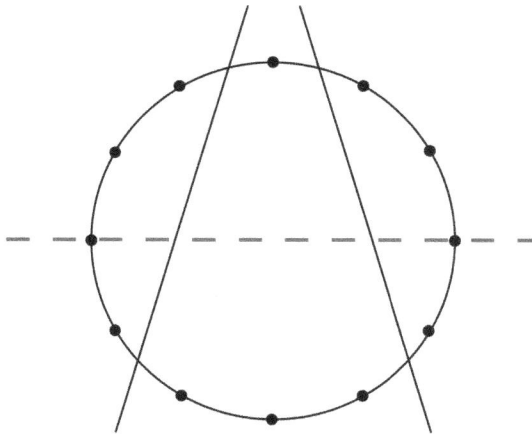

**Figure 14.** One of the two possible modes to section a circle with twelve evenly spaced vertices according to the criteria specified in the text.

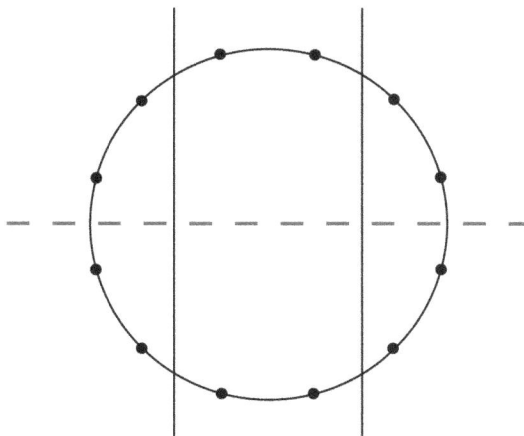

**Figure 15.** The second of the two possible modes to section a circle with twelve evenly spaced vertices according to the criteria specified in the text.

Insertion of the tones of the Circle of Fifths onto Figures 14 and 15 in the manner shown respectively in Figures 16 and 17 allows for the aforementioned correlations to be appreciated. Thus, it is evident that the four medial tones in Figure 16 (D#, G#, A, A#) are indeed the same four tones on the four medial faces of the Figure 10a rhombic dodecahedron when it is seated on the rhombic face assigned to the tone A and oriented such that the rhombic face assigned to the tone A# is closest to the viewer. Moreover, the four left lateral tones in Figure 16 (F, C, G, D) perfectly match the four left lateral tones on this rhombic dodecahedron. The correlation is complete after confirming that the four right lateral tones in Figure 16 (E, B, F#, C#) are the same as this particularly oriented Figure 10a rhombic dodecahedron's four right lateral tones.

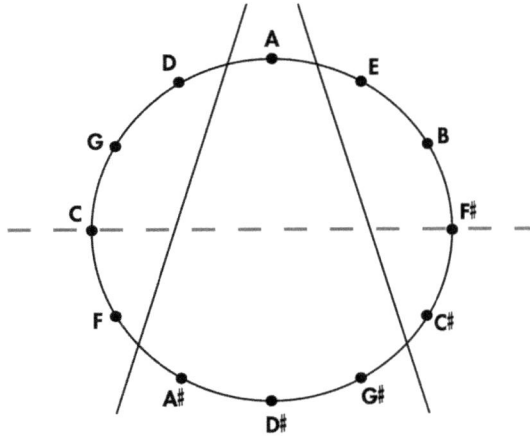

**Figure 16.** An installation of the tones of the Circle of Fifths onto the sectioned circle in Figure 14.

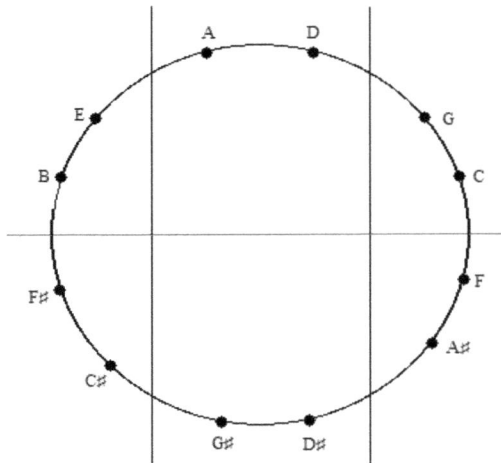

**Figure 17.** An installation of the tones of the Circle of Fifths onto the sectioned circle in Figure 15.

A similar correspondence is gleaned upon comparing the medial and lateral groups of tones in the zygo-dodecahedral tonal geometry in Figure 8a to the medial and lateral groups of tones in Figure 17. As can be seen, the medial tones in both Figures 8a and 17 are identical (D, G#, D#, A) as are the left lateral tones (A#, F, C, G) and the right lateral tones (E, B, F#, C#). In view of the fact that there are only two possible trisections of the Circle of Fifths according to the preceding criteria—both of which showing perfect correspondence to one of the two featured dodecahedral arrays of tones—and in view of the fact that there are only two types of seventh chords that can result via the addition of a seventh on top of a major triad in root position (major seventh or dominant seventh)—both of which are paradigmatically exemplified in one or the other of the two intonated dodecahedral geometries as seen within the respective aforementioned pentatonic ninth and thirteenth chords—it is tempting to conclude that these two dodecahedral tone networks are in fact fundamental twin expansions of the Circle of Fifths into the third dimension.

*Symmetry* **2012**, *4*, 644–666

It has thus far been put forth that an obscure (and perhaps completely overlooked) bilaterally symmetrical dodecahedron consisting of twelve convex quadrilaterals (with eight different shapes and sizes) harbors a bilaterally symmetrical harmonic assembly of the twelve tones in an axiomatic manner. Curiously, close inspection of Figure 18, a rendering of the ventral view of the human brain (an image that appears on an NIH webpage [10]), reveals remarkable similarities to the delineated zygo-dodecahedral paradigm.

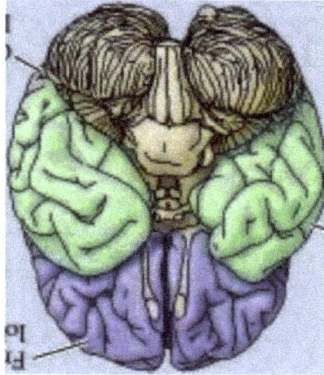

**Figure 18.** Ventral view of the human brain (image reproduced with permission from [10], copyright 2001).

Thus, as indicated in the side-by-side comparison within Figure 19, the four major paired structures in the aforementioned image show substantial congruence with the respectively color-coded regions of the zygo-dodecahedral Schlegel diagram such that the cerebellar lobes are depicted in black, the temporal lobes in green, the frontal lobes in violet, and the olfactory bulbs in white. The three medial structures in the Schlegel diagram in Figure 19 correspond to the medulla/pons (yellow), the diencephalon (black), and the corpus callosum (black). The large quadrilateral region (on the underside of the Schlegel diagram's two-dimensional projection) is attributed to the cerebral cortex. It is interesting to note that certain quadrilateral boundaries define well-recognized anatomical brain structures. For example, the Circle of Willis (highlighted in blue in the image in Figure 20) [11] has been referred to as a diamond-shaped network of vasculature at the diencephalon's perimeter [11] (attributed to the perimeter of the large medial black region in Figure 19). Another pertinent quadrilateral region of the brain is the rhomboid fossa—a structure forming the floor of the fourth ventricle that grossly defines the ventral brain boundaries of the medulla and pons [12] (attributed to the yellow region in Figure 19).

**Figure 19.** A side-by-side comparison between a color-coded ventral view of the human brain and a correspondingly colored Schlegel diagram of the Figure 7 zygo-dodecahedron (ventral view of the human brain reproduced with permission from [10], copyright 2001).

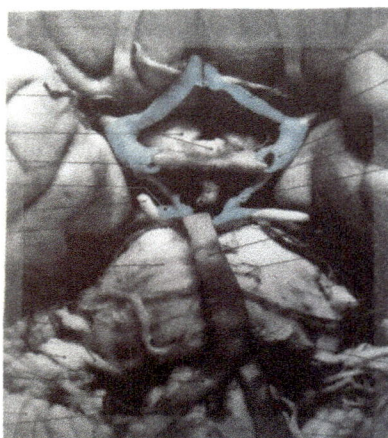

**Figure 20.** The diamond shape of the Circle of Willis (highlighted in blue) is readily apparent (image reproduced with permission from [11], copyright 2005).

The zygo-dodecahedral pattern emphasized herein can also be arguably discerned during the development of the embryonic pharynx as illustrated within vintage diagrams in Gray's Anatomy [13]. Three drawings from Henry Gray's classic tome on human anatomy display the morphogenesis of the pharynx from approximately 26 to 30 days of gestation and examination of Figure 21a–c allows the following putative zygo-dodecahedral correspondences to be made.

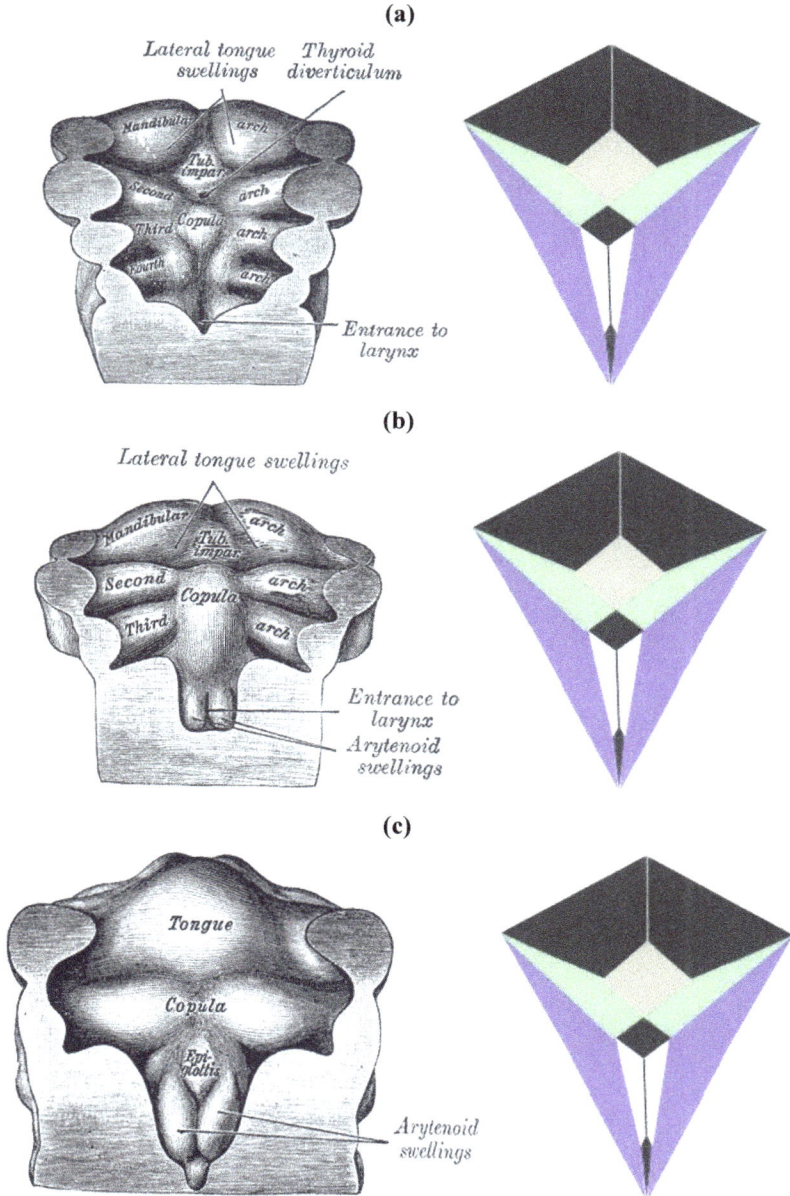

**Figure 21.** (a) A discernible zygo-dodecahedral motif in the developing pharynx; (b) Several days later, the discernible zygo-dodecahedral motif becomes more pronounced with the development of the arytenoid swellings; (c) Several days later still, the entrance to the larynx temporarily closes. Images of the pharynx reproduced from [13].

Thus, the paired black quadrilaterals match the lateral lingual swellings, the medial yellow quadrilateral is assigned to the tuberculum impar, the paired green quadrilaterals are ascribed to the second pharyngeal arch, the paired white quadrilaterals correspond to the arytenoid swellings, the black medial quadrilateral at the bottom is attributed to the larynx, the large black medial quadrilateral is assigned to the copula linguae, and the paired violet quadrilaterals roughly match the remaining pre-fused pharyngeal arches. The large quadrilateral on the underside of the zygo-dodecahedral projection corresponds to the hypopharynx. That quadrilateral regions could well be involved in workings of the tongue beyond embryonic stages is bolstered by a condition known as median rhomboid glossitis [14] that is characterized by the appearance of a red diamond-shaped zone of medial inflammation toward the back of the tongue that is believed to reflect pathology within the remnant tissue of the tuberculum impar (persistent tuberculum impar is another name for the condition).

Similarities between the dolichocephalic skull/head morphology of certain canine breeds and the structure of the Figure 7 zygo-dodecahedron are also worthy of consideration as seen in Figure 22. Assignments that match each one of the dodecahedral quadrilaterals to major bone/cartilaginous structures can be made such that the white medial dodecahedral face is assigned to the maxilla, the large medial light gray face to the mandible, the paired dark gray faces to the temporal bones, the paired white faces to the nasal bones, the paired gray faces at the rear to the auricles, the remaining paired faces to the parietal bones, the thin medial gray quadrilateral toward the rear to the occipital bone, and the remaining medial face to the frontal bone.

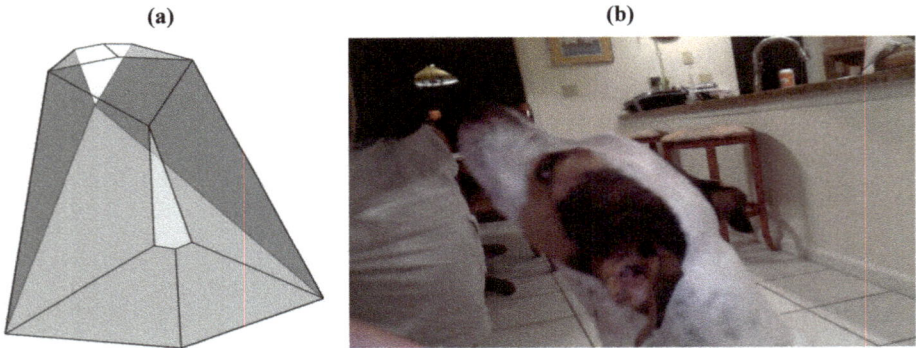

**Figure 22.** Similarity between the overall morphology of (**a**) the Figure 7 zygo-dodecahedron and (**b**) the head of a dolichocephalic dog.

A non-convex version of the Figure 7 zygo-dodecahedron is more akin to the human skull/head as seen in Figure 23. Biologically speaking, it is well known that cephalization is strongly linked to the emergence of a bilaterally symmetrical body plan [15]. It is possible that simple icosahedral modifications that lead to zygomorphic structures—such as those icosahedral modifications leading to the zygo-dodecahedron featured in this work—could underlie primary events that take place during cephalogenesis.

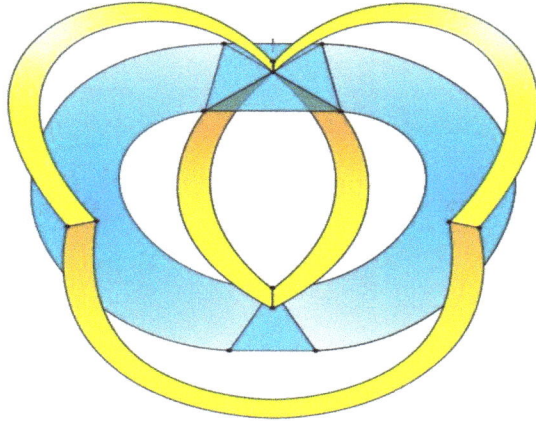

**Figure 23.** A non-convex version of the dodecahedron in Figure 7 is discernibly cephaloid such that the bottom yellow medial face corresponds to the mandible, the upper two yellow faces are assigned to the auricles, the two central yellow faces are attributed to the nasal bones, the lower blue medial triangle is ascribed to the maxilla, the two large central blue faces are assigned to the temporal bones, the top blue medial triangle corresponds to the occipital bone (the hash mark at the top of this face indicates the mid-line), the other blue medial triangle is ascribed to the frontal bone, and the remaining two blue triangles are ascribed to the parietal bones.

Interestingly, a relatively recent paper by Shimoyama *et al.* [16] documents heterophilic binding data (shown in Figure 24a) for certain members of an important class of cell adhesion proteins; namely, the type II cadherins, that, with respect to a group of seven of the eight examined proteins in which distinct binding preferences are displayed, perfectly conforms in four possible modes to the Figure 3c graph, as is detailed, in one of the four possible modes, within Figure 24b.

The particular mode shown in Figure 24b is the only one of the four possible modes in which three prominent cadherins (cadherin-6, cadherin-11, and cadherin-12) occupy medial positions and that places cadherin-6, in view of its documented role in forebrain development [17], at the most anterior medial quadrilateral. The medial quadrilateral corresponding to the medulla/pons has been assigned to cadherin-11 on the basis of data that indicates high cadherin-11 expression in the spinal cord [18]. Cadherin-12, also known as BR-cadherin or brain-cadherin, has been placed at the large medial quadrilateral corresponding to the cerebral cortex and this assignment is supported by the detection of abundant quantities of this particular cadherin in the developing and adult cerebral cortex [19]. At the time of publication of the Shimoyama studies (in the year 2000), it was not possible to investigate the heterophilic binding properties of all type II cadherins because not all of them were then known. It is now known that the number of vertebrate type II cadherin genes is thirteen [20]—a number conspicuously close to the twelve that would be expected in a dodecahedral network. Moreover, the number thirteen is significant in terms of the repeat motif for either the fcc (affiliated with the rhombic dodecahedron) or hcp (affiliated with the trapezo-rhombic dodecahedron) lattices in which uniform spheres aggregate with twelve spheres surrounding a central sphere. The number thirteen may also be significant in the context of the zygo-dodecahedral network wherein a thirteenth type II cadherin could theoretically promote the growth of a cluster of cells at an interior region near or at the center of this bilaterally symmetrical dodecahedron.

The analysis in Figure 24b leaves out one of the eight type II cadherins in the Shimoyama study; namely, cadherin-18 (formerly known as cadherin-14). This cadherin is excluded for several reasons. First, as described above, a presumed dodecahedral network would seem to suggest that one of the thirteen type II cadherins should be removed in order to generate a system with twelve postulated components. Second, the Shimoyama data shows that cadherin-18 has fully redundant binding proclivities when compared to cadherin-7. Third, cadherin-18 is the only one of the type II cadherins with the amino acid sequence nye in a rather highly conserved region of the third extracellular domain (nine of the other twelve possess a dfe sequence, two possess a dye sequence, and one possesses the sequence sfe).

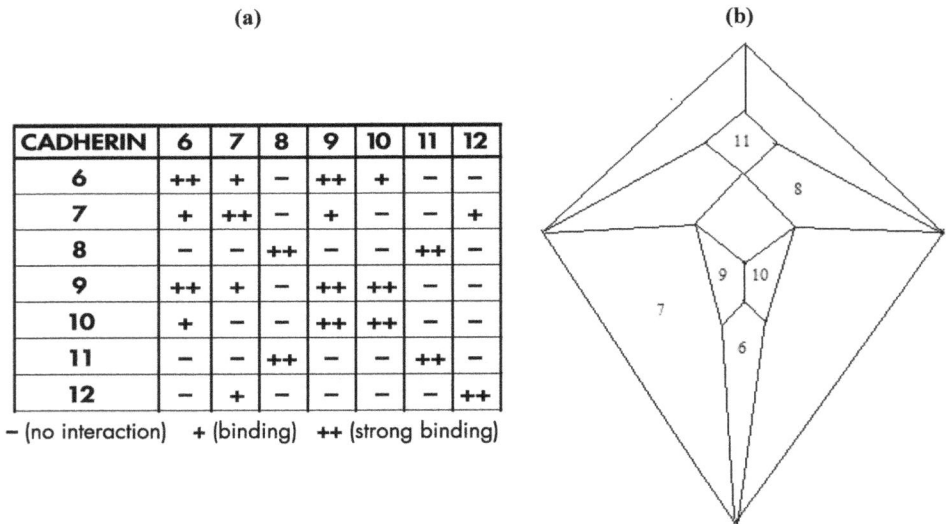

(a)

| CADHERIN | 6 | 7 | 8 | 9 | 10 | 11 | 12 |
|---|---|---|---|---|---|---|---|
| 6 | ++ | + | − | ++ | + | − | − |
| 7 | + | ++ | − | + | − | − | + |
| 8 | − | − | ++ | − | − | ++ | − |
| 9 | ++ | + | − | ++ | ++ | − | − |
| 10 | + | − | − | ++ | ++ | − | − |
| 11 | − | − | ++ | − | − | ++ | − |
| 12 | − | + | − | − | − | − | ++ |

− (no interaction)    + (binding)    ++ (strong binding)

(b)

**Figure 24.** (a) Shimoyama type II cadherin binding data for cadherins six through twelve; (b) An installation of cadherins six through twelve onto the Figure 5 zygo-dodecahedral Schlegel diagram in a manner that is consistent with the Shimoyama binding data in Figure 24a (the back face is assigned to cadherin-12).

An intriguing experimental finding in a 2005 Nature article [21], known cerebellar type II cadherin expression patterns [22], and data within an elegant 2009 article by Hulpiau and van Roy [23] are instrumental in completing the assemblage of the putative zygo-dodecahedral type II cadherin network by assigning the remaining five type II cadherins to the five vacant quadrilaterals in Figure 24b. Thus, affinity capture mass spectrometry has demonstrated a protein-protein interaction between cadherin-19 and cadherin-6 [21]. Strikingly, with respect to experimentally determined protein-protein interactions of cadherin-19, the website interlogfinder.com lists cadherin-6 as the single currently known cadherin-19 binding partner [24]. Within Figure 24b, only one quadrilateral is vacant wherein one of its four edges is shared by the quadrilateral that bears cadherin-6 (the aforementioned vacant quadrilateral has been, as previously discussed, assigned to one of the two lateral frontal lobes), and therefore, cadherin-19 is placed at this particular quadrilateral. In the cerebellum, according to available data such as that in the Body Atlas of the website nextbio.com, expression of cadherin-24 is higher than in any other brain region [22]. Of the two vacant cerebellar lobes, cadherin-24, by virtue of the Hulpiau/van Roy data—a phylogenetic analysis of the first extracellular domains of each of the thirteen type II cadherins such that the hierarchy shown in Figure 25 is obtained to reflect a succession of closest relatives within this group of proteins [23]—is placed into that cerebellar quadrilateral which shares an edge with cadherin-8. Likewise, in view of its pronounced cerebellar expression [22], cadherin-22 is placed into the other cerebellar quadrilateral. Placement of cadherin-5 into the medial quadrilateral representing the diencephalon is made because cadherin-5 and cadherin-8 are neighbors in the Hulpiau/van Roy hierarchy [23]. By process of elimination, cadherin-20 is assigned to the only remaining vacant quadrilateral (that which represents one of the temporal lobes). Cadherin-20 is highly expressed in major temporal lobe substructures such as the tail of the caudate nucleus and hippocampus [22,25].

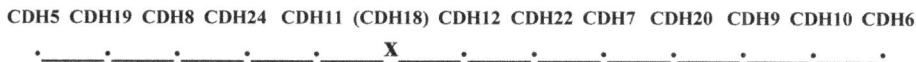

**CDH5 CDH19 CDH8 CDH24  CDH11 (CDH18) CDH12 CDH22 CDH7  CDH20  CDH9 CDH10 CDH6**

$\cdot\underline{\quad}\cdot\underline{\quad}\cdot\underline{\quad}\cdot\underline{\quad}\cdot\underline{\quad}\mathbf{X}\underline{\quad}\cdot\underline{\quad}\cdot\underline{\quad}\cdot\underline{\quad}\cdot\underline{\quad}\cdot\underline{\quad}\cdot\underline{\quad}\cdot$

**Figure 25.** The Hulpiau/van Roy phylogenetic analysis of Type II cadherins yields a linear map of structural relatives (CDH18 has been excluded for reasons explained in the text).

Thus, combination of the Shimoyama data and the Hulpiau/van Roy data allows the zygo-dodecahedral network in Figure 26 to be proposed that utilizes all of the type II cadherins except for cadherin-18. The network in Figure 26 places, on average, each cadherin next to nearly three of four closest neighbors in the Figure 25 Hulpiau/van Roy hierarchy while retaining the binding preferences in the Shimoyama data. It is interesting to note that experimental binding evidence exists to link cadherin-6 to all four of its nearest neighbors in the Figure 26 array [16,21]. It is also noteworthy that Figure 26 places cadherin-8 in a manner such that it is surrounded by all four of its nearest cadherin neighbors in the Hulpiau/van Roy hierarchy. The assignment of cadherin-5 (vascular endothelial cadherin or VE-cadherin), a protein that affects the development of the vasculature, to the diencephalon is reasonable in light of the fact that the Circle of Willis constitutes the major route by which blood is supplied to the brain in a quadrilateral-shaped loop around the diencephalon. Indeed, while some researchers strictly limit the presence of cadherin-5 to endothelial cells, there are recent reports of cadherin-5 expression in astrocytes [26] and within cells of the embryonic diencephalon of the zebrafish [27].

It is important to keep in mind that it is common that several members of the type II cadherin family are expressed simultaneously at most points in the life cycle of many cells [28], but that does not preclude the existence of the proposed zygo-dodecahedral network. While an initial impulse to hastily dismiss the zygo-dodecahedral cadherin map in Figure 26 as far-fetched may be understandable, it can also be said that when this protein array is considered against the backdrop of the highly cited Shimoyama binding studies, the compelling Hulpiau/van Roy hierarchy, the experimentally established binding interaction between cadherin-19 and cadherin-6, the high cerebellar expression of both cadherin-24 and cadherin-22, and the acute resemblance of the aforementioned ventral patterning of the brain to that of the icosahedrally-derived zygo-dodecahedron featured herein, even the most hardened skeptic would likely find it difficult to argue counter to the assertion that the Figure 26 model is well beyond a random fit of the data. Knockout experiments are consistent with some degree of redundancy [29] for some of the type II cadherins but more work with compound knockouts seems necessary to provide further clarity with respect to determining which type II cadherins (or pairs thereof) are indispensable morphogenetic constituents.

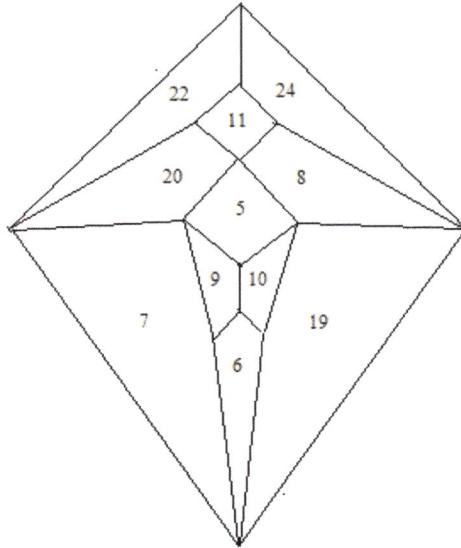

**Figure 26.** A zygo-dodecahedral brain map of type II cadherins that conforms to the Shimoyama binding data and that reflects substantial agreement with the Hulpiau/van Roy cadherin map in Figure 25. CDH22 and CDH24 are assigned to the cerebellar lobes; CDH8 and CDH20 are assigned to the temporal lobes; CDH7 and CDH19 are assigned to the frontal lobes; CDH9 and CDH10 are assigned to the olfactory bulbs; CDH11 is assigned to the medulla/pons; CDH5 is assigned to the diencephalon; CDH6 is assigned to the corpus callosum (the back face has been assigned to cadherin-12 and corresponds to the cerebral cortex).

The proposed network in Figure 26 would require asymmetric brain expression of certain type II cadherins, but such asymmetric expression of cadherins has precedent in that cadherin-2 (a type I cadherin also known as N-cadherin) is expressed to a greater extent in the left hemisphere than in that on the right [30]. In view of recent findings by Monaco *et al.* that link mutations in the type II cadherin gene for cadherin-8 to the incidence of autism [31], the elucidation of cadherin and protocadherin networks in brain morphogenesis is an important matter that continues to be a focus of extensive research efforts [32]. It is interesting to note that recent work suggests that the number twelve may hold special significance in the context of brain connectivity as van den Heuvel and Sporns have posited that the human brain contains a bilaterally symmetrical "Rich Club" of twelve major interconnected neuronal hubs [33]. That the brain's twelve pairs of cranial nerves may also be pertinent to the proposed zygo-dodecahedral model is a topic for further deliberation.

## 3. Conclusions

In summary, it seems appropriate to mention that both Kepler [34] and Euler [6] were so enamored with musical harmony that each of these eminent scientists became devoted to careful contemplation of the subject. Centuries earlier, the Pythagoreans held the Platonic dodecahedron to be a sacred object and the fact that its dual, the icosahedron, can be simply modified to yield a zygomorphic brain-like dodecahedron with cherished harmonic properties that relate to the Circle of Fifths is indeed an unusual chain of associations. Perhaps Sir Thomas Browne said it best as "Nature geometrizeth and observeth order in all things" [35].

**Acknowledgments:** DAB wishes to express infinite thanks to Susan Amy Marcus, for her unwavering support.

# References and Notes

1. Hales, T.C.; Ferguson, S.P. *The Kepler Conjecture: The Hales-Ferguson Proof*; Lagarias, J.C., Ed.; Springer: New York, NY, USA, 2011; p. 39.
2. Verheyen, H.F. The complete set of Jitterbug transformers and the analysis of their motion. *Comput. Math. Appl.* **1989**, *17*, 203–250. [CrossRef]
3. Jensen, C.R. A theoretical work of late seventeenth-century muscovy: Nikolai Diletskii's "Grammatika" and the earliest circle of fifths. *J. Am. Musicol. Soc.* **1992**, *45*, 305–331. [CrossRef]
4. Heinichen, J.D. *Der General-Bass in der Composition*; G. Olms: New York, NY, USA, 1969.
5. Careful inspection of the presented sheet music for the original jazz/blues composition "Code of Blues" reveals that the entire melody and complete chord progression are assembled by executing a continuous path involving six neighboring pentatonic sets from the zygo-dodecahedron in Figure 8. Musical compositions that exhibit Western harmony in a variety of genres can be created with other carefully selected tonal arrays upon the Figure 7 zygo-dodecahedron and upon the rhombic dodecahedron.
6. Euler, L. *Tentamen Novae Theoriae Musicae Ex Certissimis Harmoniae Principiis Dilucide Expositae*; Saint Petersburg Academy: St. Petersburg, Russia, 1739; p. 147.
7. Douthett, J.; Steinbach, P. Parsimonious graphs: A study in parsimony, contextual transformation, and modes of limited transposition. *J. Music Theory* **1998**, *42*, 241–263. [CrossRef]
8. Tymoczko, D. See further discussions of Douthett and Steinbach's "Cube Dance". In *A Geometry of Music*; Oxford University Press: Oxford, UK, 2011; p. 416.
9. Tymoczko, D. The Generalized Tonnetz. *J. Music Theory* **2012**, *56*, 1–52. [CrossRef]
10. Purves, D.; Augustine, G.J.; Fitzpatrick, D.; Katz, L.C.; LaMantia, A.-S.; McNamara, J.O.; Williams, S.M. *Neuroscience*, 2nd ed.; Sinauer Associates: Sunderland, MA, USA, 2001.
11. Afifi, A.K.; Bergman, R.A. *Functional Neuroanatomy*, 2nd ed.; Lange Medical Books/McGraw-Hill: New York, NY, USA, 2005.
12. Fix, J.D. Striae medullares of the rhomboid fossa divide the rhomboid fossa into the superior pontine portion and the inferior medullary portion. In *BRS Neuroanatomy*, 4th ed.; Lippincott, Williams & Wilkins: Philadelphia, PA, USA, 2007; p. 9.
13. Gray, H. *Anatomy of the Human Body*, 20th ed.; Lea & Febiger: Philadelphia, PA, USA, 1918; Available online: http://www.bartleby.com/107/ (accessed on 12 November 2012).
14. Semmet, J.F. Median rhomboid glossitis. *Radiology* **1939**, *32*, 215–220. [CrossRef]
15. Grabowsky, G.L. Symmetry, locomotion, and the evolution of an anterior end: A lesson from sea urchins. *Evolution* **1994**, *48*, 1130–1146. [CrossRef]
16. Shimoyama, Y.; Tsujimoto, G.; Kitajima, M.; Natori, M. Identification of three human type-II classic cadherins and frequent heterophilic interactions between subclasses of type-II classic cadherins. *Biochem. J.* **2000**, *349*, 159–167. [CrossRef] [PubMed]
17. Inoue, T.; Inoue, Y.U.; Asami, J.; Izumi, H.; Nakamura, S.; Krumlauf, R. Analysis of mouse Cdh6 gene regulation by transgenesis of modified bacterial artificial chromosomes. *Dev. Biol.* **2008**, *315*, 506–520. [CrossRef] [PubMed]
18. Simonneau, L.; Thiery, J.P. The mesenchymal cadherin-11 is expressed in restricted sites during the ontogeny of the rat brain in modes suggesting novel functions. *Cell Commun. Adhes.* **1998**, *6*, 431–450. [CrossRef]
19. Mayer, M.; Bercsényi, K.; Géczi, K.; Szabó, G.; Lele, Z. Expression of two type II cadherins, Cdh12 and Cdh22 in the developing and adult mouse brain. *Gene Expr. Patterns* **2010**, *10*, 351–360. [CrossRef] [PubMed]
20. Katsamba, P.; Carroll, K.; Ahlsen, G.; Bahna, F.; Vendome, J.; Posy, S.; Rajebhosale, M.; Price, S.; Jessell, T.M.; Ben-Shaul, A.; Shapiro, L.; Honig, B.H. Linking molecular affinity and cellular specificity in cadherin-mediated adhesion. *Proc. Nat. Acad. Sci. USA* **2009**, *106*, 11594–11599. [CrossRef] [PubMed]
21. Rual, J.F.; Venkatesan, K.; Hao, T.; Hirozane-Kishikawa, T.; Dricot, A.; Li, N.; Berriz, G.F.; Gibbons, F.D.; Dreze, M.; Ayivi-Guedehoussou, N.; *et al.* Towards a proteome-scale map of the human protein-protein interaction network. *Nature* **2005**, *437*, 1173–1178. [CrossRef] [PubMed]
22. Nextbio Home Page. Available online: http://www.nextbio.com (accessed on 12 November 2012).
23. Hulpiau, P.; van Roy, F. Molecular evolution of the cadherin superfamily. *Int. J. Biochem. Cell Biol.* **2009**, *41*, 349–369. [CrossRef] [PubMed]

24. Enter the identifier 28513 for cadherin-19. Avaliable online: http://interologfinder.org (accessed on 12 November 2012).

25. BrainSpan Home Page. Avaliable online: http://www.brainspan.org/static/home (accessed on 12 November 2012).

26. Boda-Heggemann, J.; Régnier-Vigouroux, A.; Franke, W.W. Beyond vessels: Occurrence and regional clustering of vascular endothelial (VE-)cadherin-containing junctions in non-endothelial cells. *Cell Tissue Res.* **2009**, *335*, 49–65. [CrossRef] [PubMed]

27. Sumanas, S.; Jorniak, T.; Lin, S. Identification of novel vascular endothelial-specific genes by the microarray analysis of the zebrafish cloche mutants. *Blood* **2005**, *106*, 534–541. [CrossRef] [PubMed]

28. Oda, H.; Takeichi, M. Structural and functional diversity of cadherin at the adherens junction. *J. Cell Biol.* **2011**, *193*, 1137–1146. [CrossRef] [PubMed]

29. Lefkovics, K.; Mayer, M.; Bercsényi, K.; Szabó, G.; Lele, Z. Comparative analysis of type II classic cadherin mRNA distribution patterns in the developing and adult somatosensory cortex and hippocampus suggests significant functional redundancy. *J. Comp. Neurol.* **2012**, *520*, 1384–1405. [CrossRef] [PubMed]

30. Sun, T.; Patoine, C.; Abu-Khalil, A.; Visvader, J.; Sum, E.; Cherry, T.J.; Orkin, S.H.; Geschwind, D.H.; Walsh, C.A. Early asymmetry of gene transcription in embryonic human left and right cerebral cortex. *Science* **2005**, *308*, 1794–1798. [CrossRef] [PubMed]

31. Pagnamenta, A.T.; Khan, H.; Walker, S.; Gerrelli, D.; Wing, K.; Bonaglia, M.C.; Giorda, R.; Berney, T.; Mani, E.; Molteni, M.; *et al.* Rare familial 16q21 microdeletions under a linkage peak implicate cadherin 8 (CDH8) in susceptibility to autism and learning disability. *J. Med. Genet.* **2011**, *48*, 48–54. [CrossRef] [PubMed]

32. Hertel, N.; Krishna-K; Nuernberger, M.; Redies, C. A cadherin-based code for the divisions of the mouse basal ganglia. *J. Comp. Neurol.* **2008**, *508*, 511–528. [CrossRef] [PubMed]

33. van den Heuvel, M.P.; Sporns, O. Rich-Club organization of the human connectome. *J. Neurosci.* **2011**, *31*, 15775–15786. [CrossRef] [PubMed]

34. Kepler, J. *Harmonices mundi libri V*; J. Planck: Linz, Austria, 1619.

35. Browne, S.T. *The Garden of Cyrus*; Oxford University: London, UK, 1736.

*symmetry*

MDPI

*Article*

# Computer-Aided Panoramic Images Enriched by Shadow Construction on a Prism and Pyramid Polyhedral Surface

Jolanta Dzwierzynska

Department of Architectural Design and Engineering Graphics, Rzeszow University of Technology, Powstancow Warszawy 12, 35-959 Rzeszow, Poland; joladz@prz.edu.pl; Tel.: +48-17-865-1507

Received: 1 August 2017; Accepted: 24 September 2017; Published: 3 October 2017

**Abstract:** The aim of this study is to develop an efficient and practical method of a direct mapping of a panoramic projection on an unfolded prism and pyramid polyhedral projection surface with the aid of a computer. Due to the fact that straight lines very often appear in any architectural form we formulate algorithms which utilize data about lines and draw panoramas as plots of functions in Mathcad software. The ability to draw panoramic images of lines enables drawing a wireframe image of an architectural object. The application of the multicenter projection, as well as the idea of shadow construction in the panoramic representation, aims at achieving a panoramic image close to human perception. The algorithms are universal as the application of changeable base elements of panoramic projection—horizon height, station point location, number of polyhedral walls—enables drawing panoramic images from various viewing positions. However, for more efficient and easier drawing, the algorithms should be implemented in some graphical package. The representation presented in the paper and the method of its direct mapping on a flat unfolded projection surface can find application in the presentation of architectural spaces in advertising and art when drawings are displayed on polyhedral surfaces and can be observed from multiple viewing positions.

**Keywords:** polyhedra; panoramic projection; shadow construction; Computer Aided Design (CAD); descriptive geometry; maps on surfaces

## 1. Introduction

Transforming the reality of the visual world into a flat picture plane has been a challenging task since the early years of architectural design. Perspective as a visual representation of space from a specific view point has become the most popular method throughout a wide variety of fields. However, understanding the concept of perspective, and the approach to it, has changed over the years. The techniques of pictorial perspective were discussed by painters as part of a growing interest in skenographia as long ago as in the fifth century BC [1]. The Greeks and Romans understood perspective and utilized foreshortening techniques in art, but over time, their knowledge was lost. Perspective regarded as a practical method of drawing a scene captured by an artist was rediscovered again and investigated during the Renaissance [2–4]. Then, it developed itself as a kind of science of vision encompassing the nature of functioning of a human eye, as well as the nature and behavior of light. However, only the creation of descriptive geometry as a branch of mathematics in the seventeenth century enabled further development of the theory of perspective as a method of projection, which enabled its comprehensive research [5,6]. A historical evolution of the perspective projection, as well as the perception of architecture in this projection, is presented in [7]. Perspective drawing as a result of the perspective representation is widely discussed as a medium for design and communication in architecture both during preliminary architectural work and at an advanced stage [8–10].

Research on perspective concentrates mostly on a descriptive, as well as a computer-aided, construction of linear perspective. Today, modern graphics software can create and demonstrate various perspective representations, as well as steer and control them. There exists abundant research in this field. Investigation of the processing of linear perspective and binocular information for the perceptual judgment of depth is discussed in [11,12]. Different projection transformations defining various perspective representations are presented in [12]. The kind of perspective, determining different outcomes of the perspective projection, depends both on structure of a perspective apparatus and object location. In turn, the structure of the perspective apparatus determined by a picture plane/a picture surface and a station point/camera orientation, defines the variant of perspective applied (rectilinear or curvilinear). However, the object location establishes the number of vanishing points of three main object directions $x$, $y$, $z$. Due to this fact, the most common categorizations of artificial perspective onto a flat projection plane are one-, two-, and three-point, which refers to the number of vanishing points applied. Establishing vanishing points plays an important role not only in perspective creation but also in the reverse process that is the reconstruction of perspective [13,14]. Therefore, several works deal with automatic detection of the vanishing points in monoscopic image, which is the first step to three dimensional data extraction [15–17]. Much work has been done in the field of perspective analyses and perspective construction of the architectural environment onto a single flat projection plane [2–6]. There is also a great interest in the curvilinear type of perspective, especially in various methods of panorama creation on cylindrical, as well as spherical, surfaces [18]. The idea of the perspective construction onto a non-regular prism surface composed of several flat elements is presented in [19]. There, the descriptive method of drawing perspective on an unfolded prism projection surface is presented, as well as the approach to drawing this perspective with computer aid. In this paper, we develop this idea and present an effective and practical method of constructing a panoramic projection of the polyhedral architectural form onto both prism and pyramid projection surface, which is inscribed properly in a cylindrical or conical surface. In order to achieve panoramic images close to human perception, we develop the idea of perspective creation from the center moving on a circular path. Such an approach was presented in the case of constructing the classical panorama onto a cylindrical surface and in the case of an inverse cylindrical panorama where the centers of projection were dispersed on a circle or on a straight line [20,21]. In the paper we develop effective algorithms, which allow us to draw, with computer aid, a panoramic image on an unfolded regular prism, as well as on a pyramid surface. Moreover, due to the fact that very often the panoramic representation of the architectural form is enriched by shadow construction, we address the problem of natural shadow construction in perspective onto a polyhedral projection surface; in particular, we develop the idea of shadow contour determination in the panoramic image on a regular prism and pyramid polyhedral surface.

Shadows play a very important role in generating an impression of three-dimensionality of the two-dimensional image, as they enhance our perception of space. However, perspective construction of natural (solar) shadows is one of the most complex geometric constructions due to many possible arrangements of a light source, shadow casting edges, as well as a shadow-receiving surface in relation to the viewpoint. The basic rules of constructing shadows have been widely discussed in multiple publications by famous thinkers since the Renaissance. Nowadays, they are mostly considered in the context of soft and hard shadows for computer graphics. In our approach we concentrate on the geometrical aspects of cast and attached shadows. We construct shadow lines in the axonometric view or orthographic views according to the rules of shadow constructing [22]. Shadow lines establish the border between a cast shadow and an illuminated area which is next represented onto the polyhedral surface.

## 2. General Aspects of Shadow Construction

The invention of linear perspective, as well as developments in utilization of light and shadows, revolutionized visual arts and architecture in the Renaissance. Light was, and still is, a tool that artists use to define their subject matters and to add a certain sense of realism to the picture. Therefore, to the artists and art theorists of the Renaissance, the proper depiction of shadows was of great interest. Their works explored the contrasts between light and darkness. This technique of tonal contrasting between light and darkness, called chiaroscuro, demonstrated the skill of an artist in the management of shadows to create a three-dimensional effect in a painting [23,24]. Leonardo da Vinci is regarded as one of the pioneers in research on light and shadow [25]. His research embraced not only a geometric approach to shadow creation, but also optical aspects of it as he provided one of the very first studies of penumbra—the area which receives partial light from the source during lighting. This term was coined and investigated further by Kepler [26].

Shadows vary greatly as a function of the lighting environment. The main factor which determines the shadow's appearance and shape is the type of light source applied: artificial or natural. Shadows can appear as hard-edged or soft-edged and can contain both the umbra and penumbra area. The relative size of the umbra-penumbra is a function of the size and shape of the light source, as well as its distance from the object. The definition of the penumbra rate as well as its significance and application in architecture is discussed in [27]. The work shows future possibilities of incorporating penumbra zones into the architectural design process.

Today, graphics software make it possible to accurately render shadows from a point and directional light sources in various interactive applications. There are several different approaches to rendering shadows with computer aid. The most popular approaches to define shadow regions are: the planar projected shadows approach, an approach which uses shadow volume and shadow maps approach, as well as a combination of them. A planar projected shadow approach is an extremely simple method of generating shadows onto planar surface. The method simply involves drawing the projection of the given object onto a plane. The shadow volume approach deals with objects of polygonal structure. It consists in generating shadows by creating for each object a shadow volume that the object blocks from the light source. It assumes that any object located in the shadow volume is in shadow. This infinite shadow volume is defined by lines emanating from the light source through vertices of the object. The basic idea of the shadow map construction is that the object is in shadow if it is not visible to the light [28].

Our approach to shadow construction is similar to the planar projected shadows approach. In our considerations we concentrate on shadows formed by the natural light. In this case light appears to emanate as straight line rays from the surface area of the light source—the sun. Due to the fact that the light source is at infinity, the light rays are parallel. The angular size of the sun is relatively small and constant, while the majority of shadow casting objects are close to the ground. Therefore, the penumbra, that is, the fuzzy boundaries between shadowed regions and fully-illuminated regions, can be ignored in the perspective image [23]. This also implies a simplification of the construction of shadows in perspective to the geometry of the shadow casting object, the viewer's location, as well as the surface's geometry on which the shadow is casted. In such geometrical terms the edges of the shadows are sharp and clearly defined as boundary lines circumscribed by light rays. Due to this fact they can be determined by geometric methods and rules. However, these rules can be applied when the shadows are viewed from a distance, that is both the object and its shadow are within a $60°$ circle of view.

## 3. Geometrical Aspects of Panoramic Projection onto a Polyhedral Surface

According to its descriptive geometry definition, perspective is a central projection from a real point onto a projection plane/surface, and it is subjected to projective geometry rules. The most common type of perspective which finds application is a vertical linear perspective onto two dimensional plane. It creates an illusion of depth by use of so-called 'vanishing points' to which all parallel lines converge at the level of horizon, which is the eye level. The base elements which determine this kind of perspective are a picture plane, a station point, and a base plane, while invariants of this perspective projection are incidence, collinearity of three non-coinciding points, and division of line segment parallel to the projection plane [22].

In our considerations we take into account panoramic projection, namely wide view perspective onto the projection surface being a regular polyhedral surface that is the surface composed of the several similar flat walls. Due to this fact, the above invariants of perspective projection onto a flat picture plane can find application for each separate polyhedral wall.

According to our assumptions, the projection apparatus in this case is received from an apparatus of a cylindrical or conical panoramic projection by replacing the cylindrical/conical surface with the regular polyhedral surface inscribed in it. Due to this fact, the apparatus of the considered representation is composed of a polyhedral projection surface $\tau$, a viewpoint $S/S_X$ and a base plane $\pi \perp \tau$. The center of projection can be a single stationary point $S$ belonging to the axis $l$ or a point $S_X$ moving on the circular path $s$ included in a horizon plane (Figure 1).

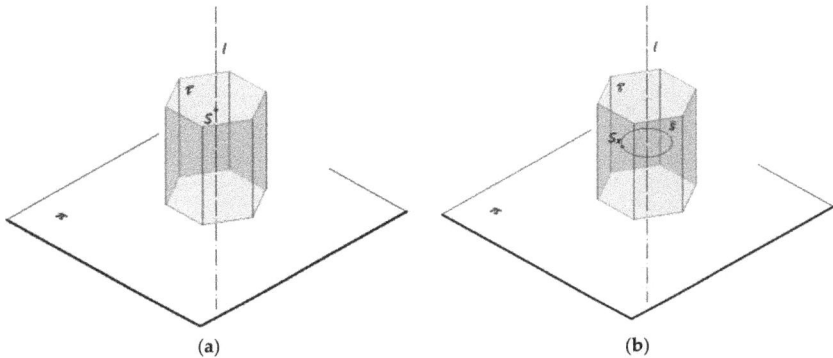

**Figure 1.** Projection apparatus: (**a**) Panorama on a prism surface from a stationary center $S$; and (**b**) panorama on a prism surface from a moving center $S_X$.

In the case of the panoramic projection onto pyramid polyhedral projection surface two variants of the projection apparatus structure can be distinguished depending on the location of the surface's vertex $W$ towards the base plane $\pi$: above, variant A; below, variant B (Figure 2).

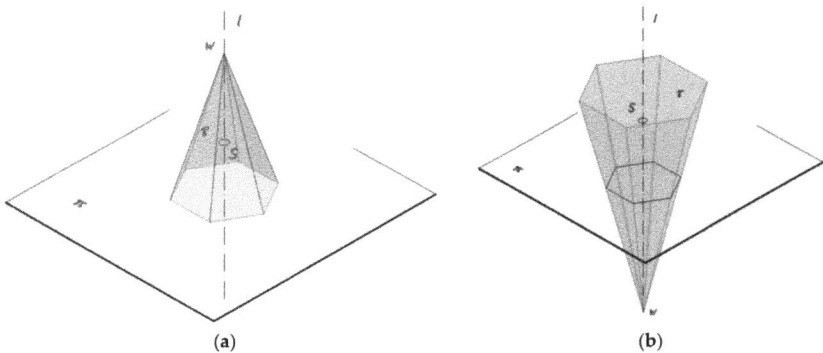

**Figure 2.** Projection apparatus of a panorama on a pyramid surface: (**a**) of version A; and (**b**) of version B.

Defining the apparatus of perspective projection in this way, a perspective image of any proper point $F$ is a pair of two points $(F^S, F^{OS})$, Figure 3. The point $F^S$ is a central projection of $F$ onto $\tau$ from $S/S_F$, whereas $F^{OS}$ is a central projection of $F^O$ (orthogonal projection of $F$ onto a base plane $\pi$) onto $\tau$. Both points $F^S$ and $F^{OS}$ are included in the same generatrix line $t_F$ which goes through a vertex $W/W\infty$ and through a point of a base polygon $p$. The point $F^S$ is the main projection, whereas the point $F^{OS}$ is an auxiliary projection enabling restitution. A changeable center $S_F$ of panoramic projection is attributed to the given real point $F$ by cutting the circle $s$ by half-plane $\lambda$ determined by the edge $l$ and the point $F$ (see Figure 3). The generatrix line $t_F$ is also the main projection of a vertical line $t$ from $S/S_F$ onto $\tau$.

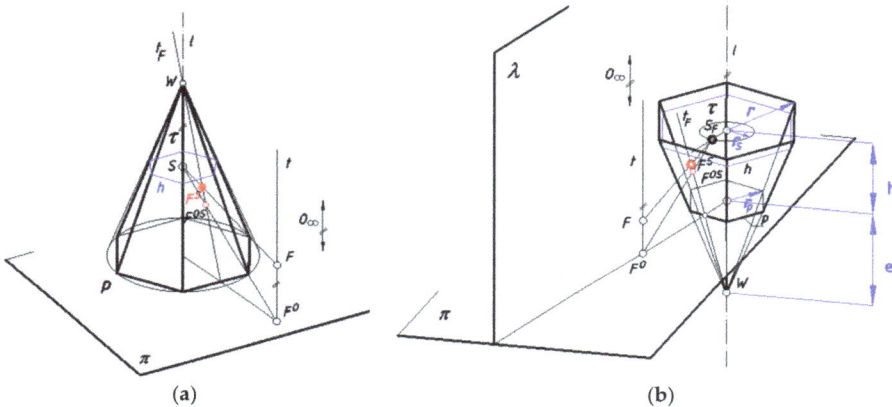

**Figure 3.** Representation of a point $F$ on a pyramid surface: (**a**) of version A from a stationary center $S$; and (**b**) of version B from a moving center $S_X$.

In the considered panoramic projection, we represent all points which are situated behind the projection surface that is all points located on the other side of the projection surface than the center of projection $S$ and the points which are situated on the base plane $\pi$ or above it.

## 4. Mapping Polyhedral Panorama Directly on an Unfolded Projection Surface

Due to the fact that each polyhedral surface can be unfolded on a plane, it is convenient to present the images of our panoramic representation on a flat unfolded surface. In order to do that, we transform the images contained in the projection surface $\tau$ into their counterparts included in the unfolded surface $\tau^R$. Such a transformation is realized by projecting each generatrix line $t_X$ of the polyhedral surface $\tau$ from the center $S/S_X$ onto the base plane $\pi$. Then, it is possible to establish projective relations between the points on the generatrix lines of this degenerate flat surface obtained as a result of projection and their counterparts on the generatrix lines contained on the unfolded surface. A similar approach is presented in [20], where the construction of a cylindrical panorama is presented as well in [29], where construction of a conical panorama is shown.

*Establishing Equations Displaying Geometrical Relations Occurring during Projection*

Let us consider a central projection $^S t_F$ of a generatrix line $t_F$ (included in a prism/pyramid projection surface) from a center $S/S_F$ onto a base plane $\pi$ (Figure 4).

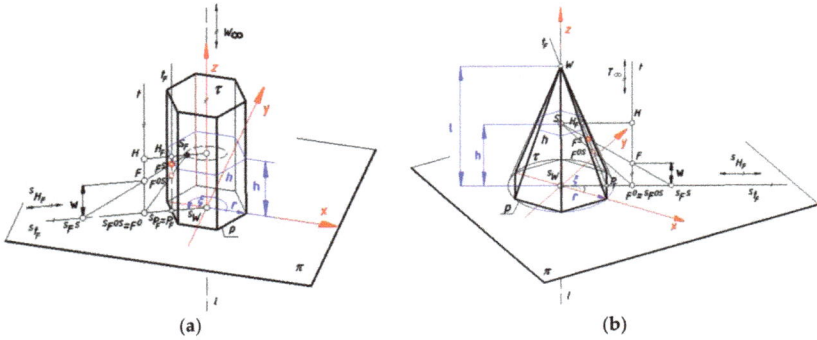

**Figure 4.** Central projection $^{S}t_F$ of a generatrix line $t_F$ onto $\pi$ in order to realize the transformation: (**a**) projection from a moving point $S_F$ in the case of a panorama on a polygonal surface; and (**b**) projection from a stationary point $S$ in the case of a panorama on a pyramid surface.

We distinguish four characteristic points included in a generatrix line $t_F$: $W\infty/W$, $P_F$, $F^{O,S}$, and $H_F$, where a point $P_F$ is included in the base polygon $p$, a point $H_F$ is included in the horizon polygon, and a point $W\infty/W$ is a vertex of the projection surface—a point at infinity or a real point (see Figure 4). After the central projection of $t_F$ from $S/S_F$ onto $\pi$ we respectively receive range of points: $^{S}W_F$, $^{S}P_F$, $^{S}F^{O,S}$, and $^{S}H_F$ included in the line $^{S}t_F$, (see Figure 4). The considered range of points included in $^{S}t_F$ and the range of points contained in a generatrix $t_F$ are homologous. Additionally, the range of points $^{S}P_F$, $^{S}H_F$, $^{S}W_F$, and $^{S}F^{O,S}$ on the line $^{S}t_F$, and the range of points $P_F{}^{R}$, $H_F{}^{R}$, $W^{R}$, and $F^{O,SR}$ on the line $t_F{}^{R}$ contained in the unfolded surface $\tau^{R}$ are related by the projective transformation. This transformation for a prism projection surface is expressed in the graphical way in Figure 5. The transformation, in the case of a pyramid projection surface for both versions A and B, is presented properly in Figures 6 and 7. The mentioned above graphical connection enables drawing a panoramic image $F^{SR}$ of a point $F$ when its projection $^{S}F^{S}$ is given. It is worth nothing that for any point $F \in \pi$, $^{S}F^{S} = {}^{S}F^{O,S} = F$, which simplifies the construction.

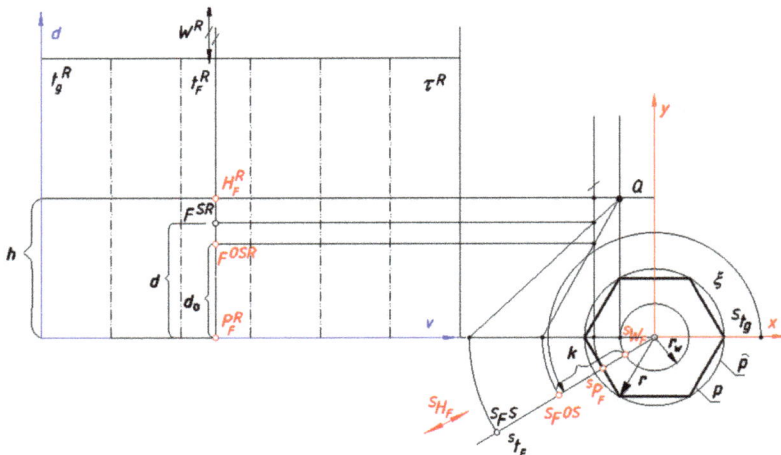

**Figure 5.** Graphical connection between the range of points on the line $^{S}t_F$ and a proper range of points on the line $t_F{}^{R}$ included in the unfolded surface $\tau^{R}$ in the case of panorama on a prism surface from a moving center $S_F$.

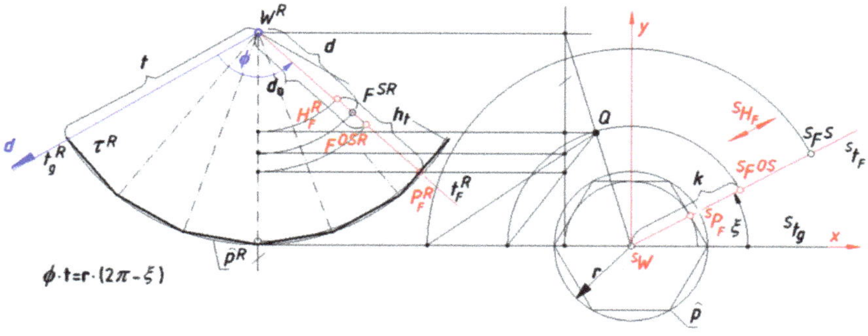

**Figure 6.** Graphical connection between the range of points on the line $^S t_F$ and a proper range of points on the line $t_F{}^R$ included in the unfolded surface $\tau^R$ in the case of single center panorama on a pyramid surface of version A.

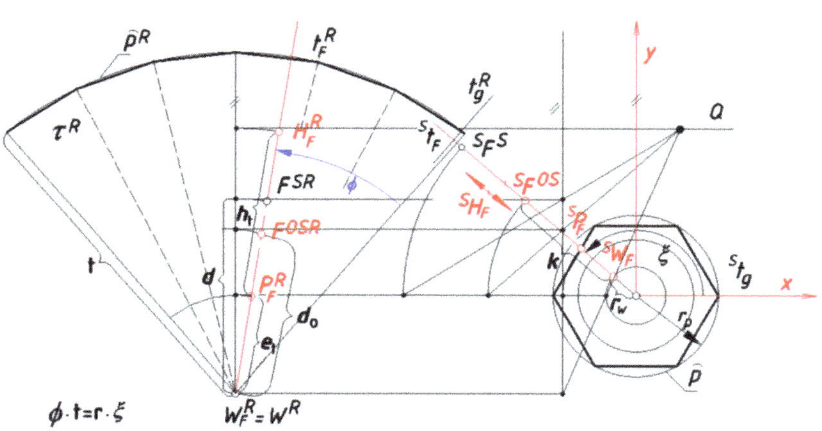

**Figure 7.** Graphical connection between the range of points on the line $^S t_F$ and a proper range of points on the line $t_F{}^R$ included in the unfolded surface $\tau^R$ in the case of a multicenter panorama on a pyramid surface of version B.

Due to above projective relations, the cross ratio of the quadruple of range points on the line $^S t_F$ as well as the cross ratio of the quadruple of proper range points on a line $t_F{}^R$ is preserved during transformation, which can be expressed as follows:

$$\frac{^S F^{O,SS} P_F}{^S W_F{}^S P_F} \div \frac{^S F^{O,SS} H_F}{^S W_F{}^S H_F} = \frac{F^{O,SR} P_F^R}{W^R P_F^R} \div \frac{F^{O,SR} H_F}{W^R H_F^R} \tag{1}$$

Similarly, the range of points on a vertical line $t$ and a proper range of points on a line $t^F$ are homologues, Figure 4. Therefore:

$$\frac{F^O H}{FH} \div \frac{F^O T_\infty}{FT_\infty} = \frac{F^{O,S} H_F}{F^S H_F} \div \frac{F^{O,S} W}{F^S W} \tag{2}$$

where $W$ in the equations can be a real point or a point at infinity.

We determine, (see Figures 4–7):

- the distance of the point $^SF^{O,S} = F^O$ from the point $^SW$ by $k$;
- the distance of the point $F^{O,SR}$ from the point $P_F{}^R$ by $d_o$, if the case of the projection on a prism surface;
- the distance of the point $F^{O,SR}$ from the point $W^R$ by $d_o$, if the case of the projection on a pyramid surface;
- the distance of the point $F^{SR}$ from the point $P_F{}^R$ by $d$, if the case of the projection on a prism surface;
- the distance of the point $F^{SR}$ from the point $W^R$ by $d$, if the case of the projection on a pyramid surface;
- the distance of the point $H_F{}^R$ from the point $P_F{}^R$ by $h_t$;
- the distance of the point $P_F{}^R$ from the point $W^R$ by $t$;
- the distance of the point $P_F$ from the point $W$ by $e_t$;
- the distance of the point $P_F$ from the point $^SW_F$ by $r_p$, if the case of the projection on a pyramid surface of version B;
- the distance of the point $^SW_F$ from the center of the base polygon $p$ by $r_w$.

According to Figures 5–7 and the equations (1) and (2), we derive formulas:

$$d_o = \frac{h \times (k - r)}{k - r_S} \tag{3}$$

$$d = \frac{(w - h) \times (h - d_o)}{h} + h \tag{4}$$

in the case of a prism projection surface,

$$d_o = \frac{t \times (t - h_t) \times (k - r_w)}{h_t \times (r_w - r) + t \times (k - r_w)} \tag{5}$$

$$d = \frac{h \times (d_o - \frac{r_s \times t}{r}) \times (t - h_t) + \frac{r_s \times t}{r} \times (h - w) \times (-h_t + t - d_o)}{h \times (d_o - \frac{r_s \times t}{r}) + (h - w) \times (t - d_o - h_t)} \tag{6}$$

in the case of a pyramid surface of version A,

$$d_o = \frac{-e_t \times (h_t + e_t) \times (k - r_w)}{h_t \times (k - r_p) - (h_t + e_t) \times (k - r_w)} \tag{7}$$

$$d = \frac{h \times (h_t + e_t) \times \left(d_o - \frac{r_s \times t}{r}\right) + \frac{r_s \times t}{r} \times (h - w) \times (h_t + e_t - d_o)}{(h - w) \times (h_t + e_t - d_o) + h \times (d_o - \frac{r_s \times t}{r})} \tag{8}$$

in the case of a pyramid surface of version B.

In the above equations the value of $r$ and $r_p$ change according to base polygon geometry and value of $\zeta$. We apply the same formulas for both panoramic projection from a single center and for multicenter projection. In the case of the application of the stationary view point the radius of the circle of viewpoints $r_s$ equals zero.

## 5. Drawing Perspective with Computer Aid

### 5.1. Methods and Methodology

The projective relations expressed above enable the creation of the panoramic representations on prism and pyramid surfaces with computer aid. Usually computer programs for drawing perspective representations use linear algebra, in particular matrix multiplication to describe transformations of point coordinates of a model to the point coordinates on a screen [30,31].

However, compared to other panorama construction methods that use information about points for computer vision and the CAD system, it is more convenient for us to utilize data about lines, which apply very often in architectural forms. Due to this fact, we place Cartesian coordinate system of axis $x$, $y$, $z$ in such a way that $x$ and $y$ are included in the base plane $\pi$ and $z$ overlaps with an axis $l$. Next, using the equations derived in section four, we create analytical algorithms for drawing a panoramic image of a line $AB$ passing through two different points $A(x_a,y_a,z_a)$ and $B(x_b,y_b,z_b)$ given by their spatial coordinates in the system $x$, $y$, $z$. The image is created directly on the unfolded polyhedral surface $\tau^R$. In the case of the panorama on a prism projection surface, the panoramic image is drawn as a plot of function $d(v)$ in the Cartesian coordinate system of axis $d$, $v$, placed as it is shown in Figure 5. For a given point $F \in AB$ a coordinate $v$ is the distance measured on the unfolded surface between the border generatrix $t_g{}^R$ and the generatrix $t_F{}^R$ containing the panoramic image $F^{SR}$ of this point (see Figure 5). In the case of a panoramic projection onto a pyramid projection surface, the line is drawn as a plot of function $d(\Phi)$ in the polar coordinate system. For a given point $F \in AB$ an angular coordinate $\Phi$ is the angle between the border generatrix $t_g{}^R$ and the generatrix $t_F{}^R$ containing a panoramic image $F^{SR}$ (Figures 6 and 7). The vertex $W^R$ contained in the unfolded surface is chosen as a pole, whereas the border ruling $t_g{}^R$ is taken as a polar axis. In both cases of panoramic projection onto prism and pyramid surfaces, it is convenient to establish the ruling $t_g$ as a border ruling, which projection ${}^S t_g$ onto $\pi$ is included in axis $x$ (see Figures 6 and 7). Both variables $v$ and $\Phi$ occurring respectively in the functions $d(v)$ and $d(\Phi)$ are dependent on $\xi$— the angle between ${}^S t_g$ and ${}^S t_F$ measured on $\pi$ (see Figures 5–7). The basis for creating our algorithms to draw polyhedral panoramas were the algorithms for drawing cylindrical and conical panoramas applied in [20,21]. In our approach we treat panorama onto a prism/pyramid surface as the panorama onto a cylindrical/conical surface with a changeable radius $r_a$ of the base circle $p_r$—the circumcircle of the base polygon $p$ included in $\pi$ (Figure 8).

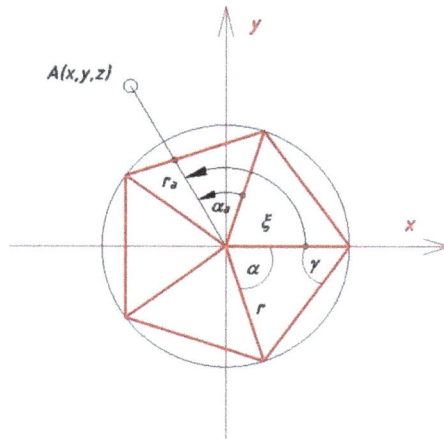

**Figure 8.** The scheme for establishing geometric and analytical relations occurring for each regular polygon.

For each regular polygon is as follows, Figure 8:

$$r_a = \frac{a \cdot \sin \gamma}{\sin(\alpha - \alpha_a) + \sin \alpha_a} \tag{9}$$

Due to the fact that $r_a$ changes periodically dependently on the value of $\xi$, the algorithms for drawing a panoramic image of a straight line onto polyhedral surface are much more complicated than the ones for drawing panoramas on cylindrical and conical surfaces.

In order to draw a panoramic image of the segment *AB* of a straight line on a prism/pyramid surface, the range of function's $d(v)/d(\Phi)$ arguments needs to be specified in advance. The ability to draw panoramic images of line segments make it possible to draw an edge image of an architectural object, provided that the coordinates of its vertices are known.

*5.2. Results—Some Examples of the Application of the Algorithms*

The starting point for any computer aided construction of a panoramic image is establishment of the base elements of panoramic projection, as well as the location of the represented object. The base elements of panoramic projection determined by the structure of the projection apparatus are: radius *r* of the base circle which circumscribes the base polygon *p*, number of vertices of the base polygon, height of horizon *h* and radius of the circle of viewpoints $r_S$, in the case of multicenter projection. For the panoramic projection onto pyramid projection surface also the location of the surface's vertex is necessary. As far as the location of represented object is concerned, it should be located in a cone of good vision during the cone's rotation around the axis *l*. Let us show some examples of the algorithms' application for drawing panoramic images of a simple building form, Figure 9.

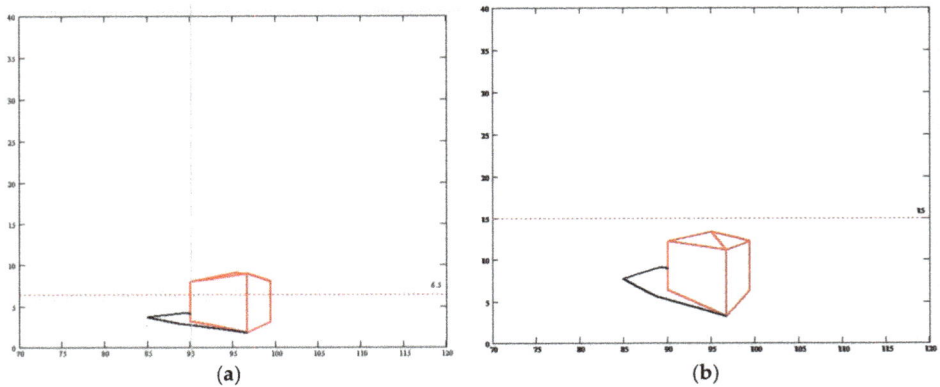

**Figure 9.** Mapping of a single center panorama onto a regular six wall prism surface inscribed in a cylindrical surface with the radius of the base circle of 20 m from various viewer locations: (**a**) the horizon height equals 6 m; and (**b**) the horizon height equals 15 m.

The above figure shows panoramic projections of the same building onto a prism surface from various viewing positions, however, with the same direction of light rays. This is a single center projection onto a regular six-wall prism projection surface and, respectively, with horizon height equal to 6 m and 15 m. The radius of the circumcircle of the base polygon is equal to 20 m, whereas the building height equals 12 m. The multicenter panoramic representation onto a prism projection surface is presented in Figure 10. This is the projection of the same building but with different viewer's location, as well as with a different direction of the sun rays than in the previous case. In the presented image, we can notice changes in the panoramic view of the shadow's border due to projection onto various prism walls.

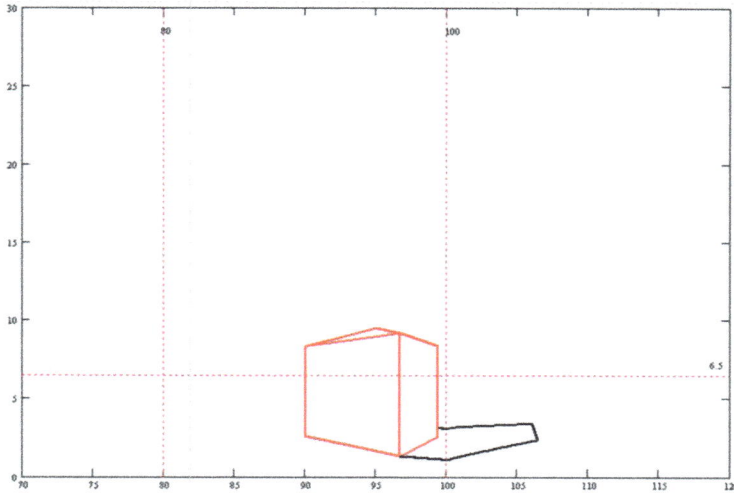

**Figure 10.** Mapping a multicenter panorama onto a regular six-wall prism surface inscribed in a cylindrical surface with the radius of the base circle of 20 m and with a horizon height of 6 m.

A single center panoramic projection onto a regular six wall pyramid polyhedron surface inscribed in a cylindrical surface with the radius of the base circle of 20 m is shown in Figure 11. Considering the most convenient viewing direction (perpendicular to a projection surface) it is recommended to apply the version A of the projection apparatus for frog's eye view images, whereas the version B for bird's eye view images.

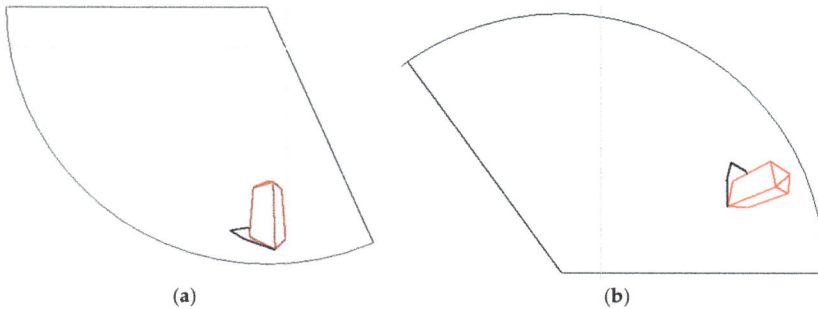

(a)            (b)

**Figure 11.** Mapping of a single center panorama on a pyramid surface inscribed in a cylindrical surface with the radius of the base circle of 20 m from various viewer locations: (**a**) panorama of version A with a horizon height of 6 m—a frog's eye view image; and (**b**) panorama of version B with a horizon height of 15 m—a bird's eye view image.

The representation of a building's cast shadow on the ground in panoramic image requires preliminary determination of the shadow border line on the base plane, which is then treated as any flat object during projection. The issue starts to be more complicated if we need to represent a complex object or several objects located in such a way that intrinsic shadows need to be considered. It is possible to generate the border of the intrinsic shadows in Mathcad software, too, as it is shown in Figure 12.

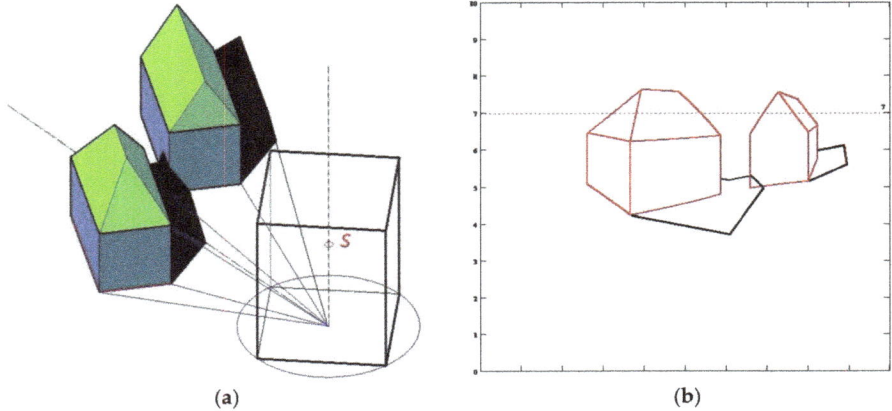

(a)                                                                                  (b)

**Figure 12.** Panoramic projection of several buildings onto a regular four-wall prism surface with horizon height of 7 m: (**a**) the location of buildings towards a projection surface; and (**b**) the result of the generation of the panoramic image in Mathcad.

However, due to the fact that sometimes shadow generation in Mathcad requires prior complicated construction of a shadow border line, it is much more convenient to realize shadow construction directly in the panoramic view on the unfolded projection surface. It can be done with computer aid of AutoCAD. In this case, we can base our analysis on the graphical connections between any point contained in the base plane $\pi$ and its panoramic image on the unfolded projection surface presented in Section 4. These connections are especially useful to establish starting assumptions for drawing perspective. Due to this fact, in order to draw a panoramic image of a buildings' layout presented in the Figure 12, it is first necessary to draw an orthogonal projection onto $\pi$ of both buildings and a projection surface, as well as to place an unfolded projection surface $\tau^R$ on $\pi$ as it is shown in Figure 13.

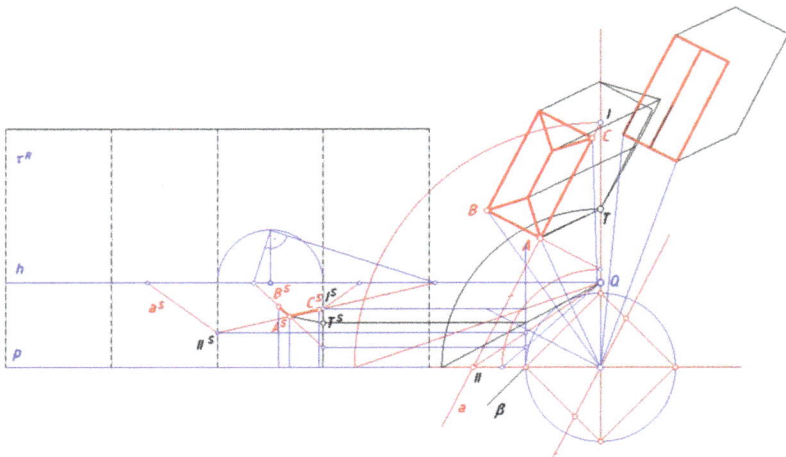

**Figure 13.** Establishing starting assumptions (projections of edges *AB* and *AC* and a shadow line *AT*) for drawing a single center panoramic image on a prism surface with a horizon height of 7 m.

Similarly as in other perspective representations, our starting assumptions for drawing a panoramic image are two edges *AB* and *CD* of one building, which are mutually perpendicular and included in $\pi$, whereas for shadows construction it is a cast shadow line *AT* of one building vertical edge. The central projection of a straight line *a*, which contains the edge *AB*, is an angular line $a^s$. The end points of this angular line are included in a horizon line *h*, whereas its vertices are included in polyhedral edges. However, due to the fact that panoramic images of both edges *AB* and *AC* are contained in the same polyhedral wall, due to the building's location, the drawing can be supported by a classical linear perspective construction onto a single plane. The image of the second building is located in the next polyhedral wall. The most complicated construction is the shadow construction, which enables the establishment of various vanishing points of light rays and shadow lines for each polyhedral wall. Moreover, the shadow's border line changes its shape due to projection on the various walls. The result of the panorama construction is presented in Figure 14.

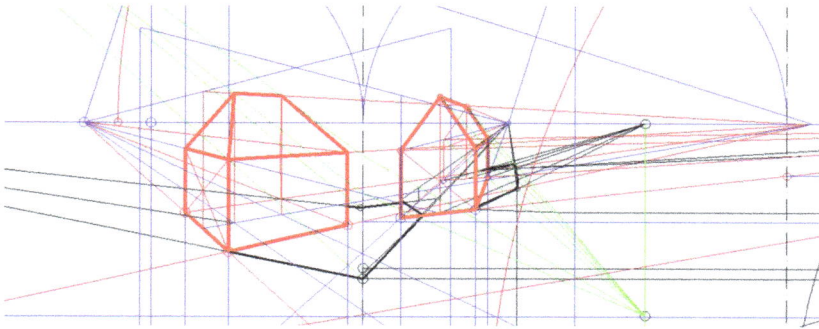

**Figure 14.** Mapping of a panoramic image of several buildings onto a regular four-wall prism surface with horizon height of 7 m in the AutoCAD.

Figure 14 shows the mapping of a single center projection onto regular four-wall prism surface of the buildings presented in Figure 12a. The horizon height equals 7 m, similarly as in the previous representation by Mathcad, therefore, the results presented in Figures 12b and 14 can be compared. Due to limitations of the number of plots which can be created in Mathcad for one drawing, the algorithms' work have been tested only on the simple examples of model buildings. However, the tests showed that the algorithms work well and can be implemented in other graphical package such as for example AutoCAD.

## 6. Discussion

In this paper, we proposed the method of computer aided construction of panoramic image onto regular prism and pyramid surfaces. The main point of this method was effective and practical drawing of a panorama on a flat unfolded projection surface. Due to the fact that straight lines appear very often in any architectural form, we used this information and developed algorithms for drawing them. Curved lines, in order to be represented, should be approximated by certain segments of straight lines. The ability to draw panoramic images of lines enables drawing panoramas of wireframe models of represented figures. The application of the multicenter projection in the panoramic representation aims at achieving the panoramic image close to human perception. However, we are aware that vision and human perception are very complex processes and should be considered from various angles. We present the geometrical approach only. However, as far as the geometric construction and graphical mapping is concerned, we can state that our method works well.

The elaborated algorithms for drawing panoramas were formulated and tested in Mathcad software. However, they can be implemented in the majority of graphical packages to make drawing more efficient. Thanks to the application of the changeable base elements of perspective in the algorithms, they enable the creation of panoramic images from different viewing positions, as well as on a polyhedral projection surface determined by various metric characteristics. Therefore, they can find application in representation of architectural space when drawings are displayed on the polyhedral surfaces and can be observed from stationary or moving viewing positions. The wireframe panoramic image can form the basis for further various advertising and artistic presentations.

**Conflicts of Interest:** The author declares no conflict of interest.

## References

1. Small, J.P. Skenographia in Brief. In *Performance in Greek and Roman Theatre*, 1st ed.; Harrison, G.W.M., Liapēs, V., Liapis, V., Eds.; Brill: Leiden, The Netherlands, 2013; Volume 353, pp. 154–196.
2. Biermann, V.; Borngasser Klein, B.; Evers, B.; Freigang, K.; Gronert, A.; Jobst, K.; Kremeier, J.; Lupfer, G.; Paul, J.; Ruhl, C.; et al. *Architectural Theory from the Renaissance to the Present*; Taschen Gmbh: Koln, Germany, 2003; pp. 8–126.
3. Elkins, J. Piero della Francesca and the Renaissance Proof of Linear Perspective. *Art Bull.* **2014**, *69*, 220–230. [CrossRef]
4. Argan, G.C.; Robb, N.A. The Architecture of Brunelleschi and the Origins of Perspective Theory in the Fifteenth Century. *J. Warbg. Court. Inst.* **1946**, *9*, 96–121. [CrossRef]
5. Rapp, J.B. A Geometrical Analysis of Multiple Viewpoint Perspective in the Work of Giovani Battista Piranesi: An Application of Geometric Restitution of Perspective. *J. Archit.* **2008**, *13*, 701–736. [CrossRef]
6. Cocchiarella, L.E. Perspective between Fiction: Pattern Mutations through Science and Art. *J. Geom. Graph.* **2015**, *19*, 237–256.
7. Dusoiu, E.-C. Architectural Representation between Imagination and Revelation. *J. Civ. Eng. Archit.* **2017**, *11*, 199–211. [CrossRef]
8. Hewitt, M. Representational Forms and Modes of Conception; An Approach to the History of Architectural Drawing. *J. Archit. Educ.* **2014**, *39*, 2–9.
9. Unwin, S. Analyzing Architecture through Drawing. *Build. Res. Inf.* **2007**, *35*, 101–110. [CrossRef]
10. Prokopska, A. *Methodology of Architectural Design, Preliminary Phases of the Architectural Process*; Publishing House of Rzeszow University of Technology: Rzeszow, Poland, 2015; pp. 39–121. (In Polish)
11. Bruggeman, H.; Yonas, A.; Konczak, J. The Processing of Linear Perspective and Binocular Information for Action and Perception. *Neuropsychologia* **2007**, *45*, 1420–1426. [CrossRef] [PubMed]
12. Elias, R. Projections. In *Digital Media*; Springer: Cham, Switzerland, 2014; pp. 319–386.
13. Dzwierzynska, J. Reconstructing Architectural Environment from a Perspective Image. *Procedia Eng.* **2016**, *161*, 1445–1451. [CrossRef]
14. Dzwierzynska, J. Single Image Based Modeling Architecture from a Historical Photograph. *IOP Conf. Ser. Mat. Sci. Eng.* **2017**, *245*, in press.
15. Rojas-Sola, J.I.; Romero-Manchado, A. Use of Discrete Gradient Operators for the Automatic Determination of Vanishing Points: Comparative Analysis. *Expert Syst. Appl.* **2012**, *39*, 11183–11193. [CrossRef]
16. Romero-Manchado, A.; Rojas-Sola, J.I. Application of Gradient-Based Edge Detectors to Determine Vanishing Points in Monoscopic Images: Comparative Study. *Image Vis. Comput.* **2015**, *43*, 1–15. [CrossRef]
17. Andalo, F.A.; Taubin, G.; Goldenstein, S. Efficient Height Measurements in Single Images Based on the Detection of Vanishing Points. *Comput. Vis. Image Underst.* **2015**, *138*, 51–60. [CrossRef]
18. Alashaikh, A.H.; Bilani, H.M.; Alsalman, A.S. Modified Perspective Cylindrical Map Projection. *Arab. J. Geosci.* **2014**, *7*, 1559–1565. [CrossRef]
19. Dzwierzynska, J. Descriptive and Computer Aided Drawing Perspective on an Unfolded Polyhedral Projection Surface. *IOP Conf. Ser. Mat. Sci. Eng.* **2017**, *245*, in press.
20. Dzwierzynska, J. Cylindrical Panorama-Two Approaches. *J. Biul. Pol. Soc. Geom. Eng. Graph.* **2009**, *19*, 9–14.
21. Dzwierzynska, J. Direct Construction of an Inverse Panorama from a Moving View Point. *Procedia Eng.* **2016**, *161*, 1608–1614. [CrossRef]

22. Pottman, H.; Asperl, A.; Hofer, M.; Kilian, A. *Architectural Geometry*, 1st ed.; Bentley Institute Press: Exton, PA, USA, 2007; pp. 35–194.
23. Knill, D.C.; Mammasian, P.; Kersten, D. Geometry of Shadows. *J. Opt. Soc. Am.* **1997**, *14*, 3216–3232. [CrossRef]
24. Mollicone, A. The Theory of Linear Shadows and Chiaro-Scuro. In Proceedings of the Conference: The Ways of Merchants _ X International Forum of Studies, Aversa, Italy, 31 May–1 June 2012.
25. Pepper, S. Leonardo da Vinci and the Perspective of Light. *Fidelino, J. Poetry Sci. Statecraft.* **2001**, *10*, 33–53.
26. Welsh, W.; Orosz, J. (Eds.) The Discovery of Ellipsoidal Variations in the Kepler Light Curve of Hat-P-7. *Astrophys. J. Lett.* **2012**. [CrossRef]
27. Trujillo, S.; Hernan, J. Calculation of the Shadow-Penumbra Relation and Its Application on Efficient Architectural Design. *Sol. Energy* **2014**, *110*, 139–150. [CrossRef]
28. Segal, M.; Korobkin, C.; van Widenfelt, R.; Foran, J.; Haeberli, P. Fast Shadows and Lighting Effects Using Texture Mapping. In SIGGRAPH '92 Proceedings of the 19th Annual Conference on Computer Graphics and Interactive Techniques, Chicago, IL, USA, 26–31 July 1992; Volume 26, pp. 249–252. [CrossRef]
29. Dzwierzynska, J. A Conical Perspective Image of an Architectural Object Close to Human Perception. *IOP Conf. Ser. Mater. Sci. Eng.* **2017**, *245*, in press.
30. Haglund, L.; Fleet, D.J. Stable Estimation of Image Orientation. In Proceedings of the 1st International Conference on Image Processing, Austin, TX, USA, 13–16 November 1994; pp. 68–72.
31. Salomon, D. Perspective Projection. In *Transformations and Projections in Computer Graphics*; Springer-Verlag: London, UK, 2006; pp. 71–144. ISBN 978-1-84628-620-9.

*symmetry*

MDPI

*Article*

# Aesthetic Patterns with Symmetries of the Regular Polyhedron

**Peichang Ouyang [1], Liying Wang [2,*], Tao Yu [1] and Xuan Huang [1]**

[1]   School of Mathematics & Physics, Jinggangshan University, Ji'an 343009, China; g_fcayang@163.com (P.O.);
      yutao@jgsu.edu.cn (T.Y.); huangxuanhx@126.com (X.H.)
[2]   School of Water Conservancy and Electric Power, Hebei University of Engineering, Handan 056021, China
*    Correspondence: 2000wangly@163.com; Tel.: +86-310-857-3129

Academic Editor: Egon Schulte
Received: 14 December 2016; Accepted: 22 January 2016; Published: 3 February 2017

**Abstract:** A fast algorithm is established to transform points of the unit sphere into fundamental region symmetrically. With the resulting algorithm, a flexible form of invariant mappings is achieved to generate aesthetic patterns with symmetries of the regular polyhedra.

**Keywords:** regular polyhedra; reflection group; fundamental region; invariant mapping

## 1. Introduction

Due to the perfect symmetry of regular polyhedra, they have been the subject of wide attention [1–6]. Dutch artist Escher et al. [7] designed several amazing woodcarvings of polyhedral symmetries. His artwork inspired Séquin and Yen to design and manufacture similar spherical artwork semiautomatically [8]. With the development of modern computers, there is considerable research on the automatic generation of aesthetic patterns; see the doctoral dissertation of Kaplan [9] and references therein. Such patterns simultaneously possess complex form and harmonious geometry structure, which exhibit the beauty of math. In this paper, we first review the merit and drawback of strategies used in creating symmetrical patterns. Then, we present a new approach to yield aesthetic patterns with symmetries of the regular polyhedra.

Let $\mathcal{G}$ and $\mathcal{M}$ be, respectively, a symmetry group and a mapping. $\mathcal{M}$ is called invariant with respect to $\mathcal{G}$ if it satisfies:

$$\mathcal{M} \circ \gamma = \mathcal{M}, \ \forall \, \gamma \in \mathcal{G} \tag{1}$$

$\mathcal{M}$ is called equivariant with respect to $\mathcal{G}$ if it satisfies:

$$\mathcal{M} \circ \gamma = \gamma \circ \mathcal{M}, \ \forall \, \gamma \in \mathcal{G} \tag{2}$$

Invariant mapping and equivariant mapping are two important methods adopted to generate symmetrical patterns. Mathematicians have highlighted the importance of such mappings in many situations [10,11]. Field and Golubitsky first proposed equivariant mappings to yield aesthetic patterns with discrete planar symmetries [12]. This idea later inspired Reiter to create chaotic attractors with symmetries of the tetrahedron [13] and octahedron [14] in three-dimensional Euclidean space $\mathfrak{R}^3$. Recently, Lu et al. established several families of invariant mappings to generate similar images [15]. All the mappings used above are polynomials, because invariant or equivariant mappings of the polynomial form are easier to construct. However, polynomials are not appropriate to create visually appealing patterns, since they lack variety. Furthermore, for the symmetry group of complex generators, even polynomials are not easy to construct. This is why polynomials do not appear to yield regular dodecahedron patterns of great complexity. Group summation is a classic technique used in the invariant theory [16]. To generate patterns with symmetries of the regular polyhedra, Chung introduced this technique and constructed a flexible form of equivariant mappings [17]. However, this kind of mapping still has to meet certain requirements. The general summation form of equivariant mappings:

$$\sum_{i=1}^{|G|} \sigma_i f(\sigma_i^{-1}(z)), \ \sigma_i \in G, \ z \in \mathfrak{R}^n \tag{3}$$

was proposed by Dumont, where $f$ is a mapping from $\mathfrak{R}^n$ to $\mathfrak{R}^n$, $G$ is a finite group, and $|G|$ the order of $G$ [18]. They utilized (3) to explore chaotic attractors with symmetries that are close to forbidden symmetries. Since $f$ in (3) can be arbitrary mappings rather than the particular polynomials, one can choose $f$ freely, and the resulting patterns are more beautiful. Following (3), Jones et al. created many appealing attractors [19]; Reiter successfully realized a dodecahedron attractor that possesses complex symmetries [20]. Dumont later improved (3) so that it could be applicable for crystallographic point groups [21].

Although (3) is easy to construct and theoretically feasible for any finite group, this strategy is not appropriate for the symmetry group of large order. Notice that there are $|G|$ terms in the summation; for a group of large order, (3) usually has ill-conditioned sensitivity. This leads to patterns that have unaesthetic noise. For example, regular dodecahedron attractors of 120 symmetries generated by (3) [20] are not as beautiful as images shown in [19]. Computational cost is also a problem of (3) that should not be neglected. Dumont experimented with space group 227 of order 192 [21]. They commented that "finding a visually interesting attractor for this group was most challenging because experiments ran slowly".

Regular polytopes are higher-dimensional generalizations of regular polyhedra. Their structures are similar to that of regular polyhedra, but with more symmetries. For example, symmetries of 24-cell and 600-cell in $\mathfrak{R}^4$ are 1152 and 14,400, respectively, which far exceed the symmetries of any regular polyhedron [22]. In this paper, we present a fast and convenient approach to generating aesthetic patterns with symmetries of the regular polyhedra. The proposed mapping not only has flexible form, but also avoids the order restriction appearing in (3).

## 2. Symmetry Groups of Regular Polyhedra

A regular polyhedron is a convex polyhedron whose faces are regular and equal and whose vertices have similar neighborhoods. A regular polyhedron can be briefly represented as Schläfli symbol of the form $\{p, q\}$, where $p$ is the number of sides of each face and $q$ the number of faces meeting at each vertex [1]. There are five regular polyhedra, better known as Platonic solids: tetrahedron $\{3, 3\}$, octahedron $\{3, 4\}$, cube $\{4, 3\}$, dodecahedron $\{5, 3\}$, and icosahedron $\{3, 5\}$ (Figure 1).

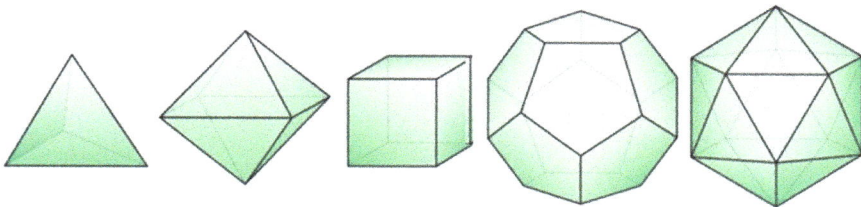

**Figure 1.** The five regular polyhedra.

Groups generated by reflections deserve special consideration for two reasons: (1) there is a general theory covering them all; (2) they contain the remaining point groups as subgroups [1,22]. This section concerns the reflection groups of regular polyhedra. For convenience, we denote the reflection group of $\{p, q\}$ as $[p, q]$. We first introduce some basic concepts.

Given an object $\mathcal{O}$, a *symmetry* of $\mathcal{O}$ is a congruent or isometric transformation. The *symmetry group* $\mathcal{Q}$ of $\mathcal{O}$ comprises all its symmetries. The elements $g_1$, $g_2$, ... $g_n$ of group $\mathcal{Q}$ are called a set of *generators* if every element of $\mathcal{Q}$ is expressible as a finite product of their powers (including negative powers). The *fundamental region* under group $\mathcal{Q}$ is a connected set whose transformed copies under the action of $\mathcal{Q}$ cover the entire space without overlapping, except at boundaries. In Euclidean space, a *reflection group* is a discrete group which is generated by a set of reflections.

Suppose that $\alpha_{[p,q]}$, $\beta_{[p,q]}$ and $\gamma_{[p,q]}$ are generators of $[p,q]$. Then:

$$\alpha_{[p,q]}^2 = \beta_{[p,q]}^2 = \gamma_{[p,q]}^2 = (\alpha_{[p,q]}\beta_{[p,q]})^p = (\beta_{[p,q]}\gamma_{[p,q]})^q = (\gamma_{[p,q]}\alpha_{[p,q]})^2 = I \qquad (4)$$

is an abstract presentation of $[p,q]$, where $I$ is the identity [22]. Since $\{p,q\}$ and $\{q,p\}$ are dual, they share the same symmetry group $[p,q]$.

Let $\{p,q\}$ be a regular polyhedron inscribed in the unit sphere $S^2 = \{(x,y,z)^T \in \Re^3 \mid x^2 + y^2 + z^2 = 1\}$. By joining the center $O$ of $S^2$ and a point $A$ of $\{p,q\}$, the directed line $\overrightarrow{OA}$ intersects $S^2$ at a point $A'$. For a given $\{p,q\}$, this projection establishes an equivalence relation between regular polyhedron $\{p,q\}$ and spherical tiling $\{p,q\}$ projected on $S^2$. Therefore, spherical tiling $\{p,q\}$ can be indiscriminately regarded as $\{p,q\}$. Henceforth, we concentrate on spherical tiling $\{p,q\}$ instead of regular polyhedron $\{p,q\}$ itself.

Geometrically, generators $\alpha_{[p,q]}$, $\beta_{[p,q]}$ and $\gamma_{[p,q]}$ are reflections which can be realized as $3 \times 3$ orthogonal matrixes. Let $\Pi_{[p,q]}^\alpha$, $\Pi_{[p,q]}^\beta$ and $\Pi_{[p,q]}^\gamma$ be, respectively, reflection planes associated with $\alpha_{[p,q]}$, $\beta_{[p,q]}$ and $\gamma_{[p,q]}$. Then, the region surrounded by those planes forms a spherical right triangle $\triangle_{[p,q]}$ on $S^2$. Repeated reflections along sides of $\triangle_{[p,q]}$ will tile $S^2$ exactly once. This suggests that $\triangle_{[p,q]}$ is a fundamental region associated with $[p,q]$. Figure 2a illustrates a fundamental region $\triangle_{[3,4]}$.

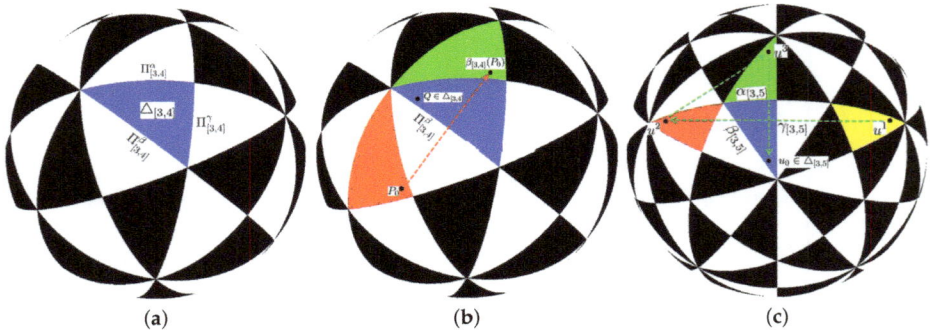

(a)    (b)    (c)

**Figure 2.** (a) The blue spherical right triangle $\triangle_{[3,4]}$ surrounded by planes $\Pi_{[3,4]}^\alpha$, $\Pi_{[3,4]}^\beta$, and $\Pi_{[3,4]}^\gamma$ forms a fundamental region associated with group $[3,4]$; (b) Let $Q \in \triangle_{[3,4]}$ and $P_0 \notin \triangle_{[3,4]}$ be two points on the different sides of $\Pi_{[3,4]}^\beta$. Then, $\beta_{[3,4]}(P_0)$ and $Q$ lie on the same side of $\Pi_{[3,4]}^\beta$, and the distance between them is smaller than $P_0$ and $Q$; and (c) A schematic illustration that shows how Theorem 1 transforms $u^1 \notin \triangle_{[3,5]}$ into $\triangle_{[3,5]}$ symmetrically. In this case, $u^1$ is first transformed by $\gamma_{[3,5]}$ so that $u^2 = \gamma_{[p,q]}(u^1)$ goes into red tile. Then, $u^2$ is transformed by $\beta_{[3,5]}$ so that $u^3 = \beta_{[p,q]}(u^2)$ goes into green tile. At last, $u^3$ is transformed by $\alpha_{[3,5]}$ so that $u_0 = \alpha_{[3,5]}(u^3) \in \triangle_{[3,5]}$.

Generators $\alpha_{[p,q]}$, $\beta_{[p,q]}$, and $\gamma_{[p,q]}$, and fundamental region $\triangle_{[p,q]}$ are important contents of the next section. We summarize them as follows and refer the reader to [17] for more details.

■ Tetrahedral group $[3,3]$:

$$\Pi_{[3,3]}^\alpha : x + y = 0, \ \Pi_{[3,3]}^\beta : y - z = 0, \ \Pi_{[3,3]}^\gamma : x - y = 0 \qquad (5)$$

$$\alpha_{[3,3]} = \begin{bmatrix} 0 & -1 & 0 \\ -1 & 0 & 0 \\ 0 & 0 & 1 \end{bmatrix}, \ \beta_{[3,3]} = \begin{bmatrix} 1 & 0 & 0 \\ 0 & 0 & 1 \\ 0 & 1 & 0 \end{bmatrix}, \ \gamma_{[3,3]} = \begin{bmatrix} 0 & 1 & 0 \\ 1 & 0 & 0 \\ 0 & 0 & 1 \end{bmatrix} \tag{6}$$

$$\triangle_{[3,3]} = \{(x,y,z)^T \in S^2 | x+y \geq 0, \ y-z \leq 0, \ x-y \leq 0\} \tag{7}$$

- Octhedral group [3, 4]:

$$\Pi^{\alpha}_{[3,4]} : \ x-z = 0, \ \Pi^{\beta}_{[3,4]} : \ x-y = 0, \ \Pi^{\gamma}_{[3,4]} : \ y = 0 \tag{8}$$

$$\alpha_{[3,4]} = \begin{bmatrix} 0 & 0 & 1 \\ 0 & 1 & 0 \\ 1 & 0 & 0 \end{bmatrix}, \ \beta_{[3,4]} = \begin{bmatrix} 0 & 1 & 0 \\ 1 & 0 & 0 \\ 0 & 0 & 1 \end{bmatrix}, \ \gamma_{[3,4]} = \begin{bmatrix} 1 & 0 & 0 \\ 0 & -1 & 0 \\ 0 & 0 & 1 \end{bmatrix} \tag{9}$$

$$\triangle_{[3,4]} = \{(x,y,z)^T \in S^2 | x-z \leq 0, \ x-y \geq 0, \ y \geq 0\} \tag{10}$$

- Icosahedral group [3, 5]:

$$\Pi^{\alpha}_{[3,5]} : \ y = 0, \ \Pi^{\gamma}_{[3,5]} : \ x = 0, \ \Pi^{\beta}_{[3,5]} : \ -\zeta x + y - z/\zeta = 0, \ \text{where } \zeta = \frac{1+\sqrt{5}}{2} \tag{11}$$

$$\alpha_{[3,5]} = \begin{bmatrix} 1 & 0 & 0 \\ 0 & -1 & 0 \\ 0 & 0 & 1 \end{bmatrix}, \ \beta_{[3,5]} = \frac{1}{2}\begin{bmatrix} -1/\zeta & \zeta & -1 \\ \zeta & 1 & 1/\zeta \\ -1 & 1/\zeta & \zeta \end{bmatrix}, \ \gamma_{[3,5]} = \begin{bmatrix} -1 & 0 & 0 \\ 0 & 1 & 0 \\ 0 & 0 & 1 \end{bmatrix} \tag{12}$$

$$\triangle_{[3,5]} = \{(x,y,z)^T \in S^2 | y \geq 0, \ x \leq 0, \ \zeta x - y + z/\zeta \geq 0\} \tag{13}$$

### 3. Transform Points of $S^2$ into Fundamental Region Symmetrically

In this section, we present a fast algorithm that transforms points of $S^2$ into fundamental region $\triangle_{[p,q]}$ symmetrically. To this end, we first prove a lemma.

**Lemma 1.** *Let* $\Pi : \ xm_1 + ym_2 + zm_3 = 0$ *be a plane in* $\mathfrak{R}^3$ *with* $m_1^2 + m_2^2 + m_3^2 = 1$,

$$R = \begin{bmatrix} 1 & 0 & 0 \\ 0 & 1 & 0 \\ 0 & 0 & 1 \end{bmatrix} - 2 \begin{bmatrix} m_1 \\ m_2 \\ m_3 \end{bmatrix} [m_1, m_2, m_3] \tag{14}$$

*be the reflection R associated with* $\Pi$. *Assume* $P_0 = (x_0, \ y_0, \ z_0)^T \in S^2$ *and* $P_1 = (x_1, \ y_1, \ z_1)^T \in S^2$ *are points on the different sides of* $\Pi$; *i.e.:*

$$\begin{cases} m_1 x_0 + m_2 y_0 + m_3 z_0 > 0 \\ m_1 x_1 + m_2 y_1 + m_3 z_1 < 0 \end{cases}, \ or \ \begin{cases} m_1 x_0 + m_2 y_0 + m_3 z_0 < 0 \\ m_1 x_1 + m_2 y_1 + m_3 z_1 > 0 \end{cases} \tag{15}$$

*Then:*

$$||P_0 - P_1||_s > ||R(P_0) - P_1||_s \tag{16}$$

*where norm* $|| \bullet ||_s$ *represents spherical distance.*

**Proof.** By the formula of spherical distance, $||P_0 - P_1||_s = \arccos(x_0x_1 + y_0y_1 + z_0z_1)$. Assume

$$R(P_0) = \begin{pmatrix} X_1 \\ Y_2 \\ Z_3 \end{pmatrix}, \text{ then direct computation shows } \begin{pmatrix} X_1 \\ Y_2 \\ Z_3 \end{pmatrix} = \begin{pmatrix} x_1 - 2m_1(m_1x_1 + m_2y_1 + m_3z_1) \\ y_2 - 2m_2(m_1x_1 + m_2y_1 + m_3z_1) \\ z_3 - 2m_3(m_1x_1 + m_2y_1 + m_3z_1) \end{pmatrix}.$$

So the spherical distance between $R(P_0)$ and $P_1$ is:

$$||R(P_0) - P_1||_s = \arccos[(x_0X_1 + y_0Y_1 + z_0Z_1)] = \arccos[(x_0x_1 + y_0y_1 + z_0z_1)$$
$$-2(m_1x_1 + m_2y_1 + m_3z_1)(m_1x_0 + m_2y_0 + m_3z_0)]$$

By (15), we have $-2(m_1x_1 + m_2y_1 + m_3z_1)(m_1x_0 + m_2y_0 + m_3z_0) > 0$. Notice that $|x_0X_1 + y_0Y_1 + z_0Z_1| \le \frac{(x_0^2+y_0^2+z_0^2)+(X_0^2+Y_0^2+Z_0^2)}{2} = \frac{1+1}{2} = 1$, and likewise $|x_0x_1 + y_0y_1 + z_0z_1| \le 1$. Conclusion (16) follows immediately, since arccos is a monotonically decreasing function. □

We use a diagram to explain the geometric meaning of Lemma 1. In Figure 2b, let $Q \in \triangle_{[3,4]}$ and $P_0 \notin \triangle_{[3,4]}$ be points on the different sides of plane $\Pi^\beta_{[3,4]}$. Then, $\beta_{[3,4]}(P_0)$ and $Q$ lie on the same side of $\Pi^\beta_{[3,4]}$. Lemma 1 says that the distance between $\beta_{[3,4]}(P_0)$ and $Q$ is smaller than $P_0$ and $Q$. In other words, for two points on the different sides of a plane, reflection transformation of the plane can shorten their distance.

**Theorem 1.** *Let $\triangle_{[p,q]}$ be the fundamental region with respect to $[p,q]$, $Q$ be an interior point of $\triangle_{[p,q]}$. For a point $u^1 \in S^2$ outside $\triangle_{[p,q]}$, the following algorithm determines a transformation $\Gamma_n \in [p,q]$ and a symmetrically placed point $u_0$ so that $u_0 = \Gamma_n(u^1) \in \triangle_{[p,q]}$.*

*Step 1: let $k = 1$, $\Gamma_0 = \begin{bmatrix} 1 & 0 & 0 \\ 0 & 1 & 0 \\ 0 & 0 & 1 \end{bmatrix}$.*

*Step 2: compute $\mathcal{D}_i = ||Q - R_i(u^{k-1})||_s$, where $R_i \in \{\alpha_{[p,q]}, \beta_{[p,q]}, \gamma_{[p,q]}\}$, $i = 1,2,3$.*
*Step 3: choose $j_k$ so that $j_k$ is the subscript of $\min\{\mathcal{D}_1, \mathcal{D}_2, \mathcal{D}_3\}$, $j_k \in \{1,2,3\}$.*
*Step 4: let $u^k = R_{j_k}(u^{k-1})$, $\Gamma_k = R_{j_k} \times \Gamma_{k-1}$.*
*Step 5: if $u^k \in \triangle_{[p,q]}$, stop; otherwise, set $k = k+1$, repeat Steps 2–5.*
*Step 6: assume $n$ is the number of cycles, then $u_0 = \Gamma_n(u^1) \in \triangle_{[p,q]}$, where:*

$$\Gamma_n = R_{j_n} \times R_{j_{n-1}} \times ...R_{j_2} \times R_{j_1} \tag{17}$$

**Proof.** $R_1$, $R_2$, and $R_3$ are isometrical symmetrical transformations, so $u^k = R_{j_k}(u^{k-1})$ obtained in Step 4 is always a symmetrical point of $u^1$ lying on $S^2$. Recall that fundamental region $\triangle_{[p,q]}$ is a spherical triangle surrounded by planes $\Pi^\alpha_{[p,q]}$, $\Pi^\beta_{[p,q]}$, and $\Pi^\gamma_{[p,q]}$. For $u^{k-1} \notin \triangle_{[p,q]}$, there must exist a plane $\Pi \in \{\Pi^\alpha_{[p,q]}, \Pi^\beta_{[p,q]}, \Pi^\gamma_{[p,q]}\}$ so that $u^{k-1}$ and $Q$ lie on different sides of $\Pi$. By Lemma 1, there exists a reflection $R_{j_k}$ associated with $\Pi$ so that:

$$||Q - R_{j_k}(u^{k-1})||_s = ||Q - u^k||_s < ||Q - u^{k-1}||_s \tag{18}$$

Thus, each time a chosen transformation $R_{j_k}$ is employed, the transformed $u^k = R_{j_k}(u^{k-1})$ will get nearer to $Q$, and eventually fall into $\triangle_{[p,q]}$. Let $\Gamma_n = R_{j_n} \times R_{j_{n-1}} \times ...R_{j_2} \times R_{j_1}$, then $u_0 = \Gamma_n(u^1) \in \triangle_{[p,q]}$. □

Theorem 1 describes an algorithm that transforms points of $S^2$ into $\triangle_{[p,q]}$ symmetrically. Figure 2c illustrates an example of how Theorem 1 works.

By the definition of fundamental region, copies of $\triangle_{[p,q]}$ can tile $S^2$ exactly once; i.e.,

$$S^2 = \bigsqcup_{i=1}^{|[p,q]|} \tau_i(\triangle_{[p,q]}), \ \tau_i \in [p,q]$$

where $|[p,q]|$ is the order of group $[p,q]$. Indeed, for $u^1 = \tau_i(u_0) \in S^2$, Theorem 1 provides a method to find a specific $\Gamma_n$ of the form (17) so that $u_0 = \Gamma_n(u^1) \in \triangle_{[p,q]}$. Thus, we can use Theorem 1 to determine entire elements of $[p,q]$ automatically. In practice, we found that the algorithm of Theorem 1 is a very fast algorithm. On average, each point of $S^2$ will be transformed into $\triangle_{[3,3]}$, $\triangle_{[3,4]}$, $\triangle_{[3,5]}$ within 3.16, 4.70, 7.94 times. Essentially, regular polytopes and polyhedra have the same kind of symmetry groups—finite reflection groups. This means that it should be possible to extend this fast algorithm to treat regular polytopes with thousands of symmetries.

## 4. Colorful Spherical Patterns with $[p,q]$ Symmetry

In this section, we describe how to create colorful spherical patterns with $[p,q]$ symmetry. For a point $u^1 \in S^2$, we define a mapping $\mathcal{M}$ of the form:

$$\mathcal{M}(u^1) = \begin{cases} \mathcal{F}(u^1), & \text{for } u^1 \in \triangle_{[p,q]} \\ \mathcal{F}(\Gamma_n(u^1)), & \text{for } u^1 \notin \triangle_{[p,q]}, \text{ but } \Gamma_n(u^1) \in \triangle_{[p,q]} \end{cases} \tag{19}$$

where $\Gamma_n$ is a transformation determined by Theorem 1, $\mathcal{F}$ is an arbitrary mapping from $\mathfrak{R}^3$ to $\mathfrak{R}^3$. By (1), mapping $\mathcal{M}$ is essentially an invariant mapping associated with $[p,q]$. Let $\mathcal{M}^k(u^1)$ be the the $k$th iteration of $\mathcal{M}$ at $u^1$. For a given positive integer $m$, by the dynamical behavior of iteration sequences $\{\mathcal{M}^k(u^1)\}_{k=1}^{m}$, we assign a certain color to $u^1$. Using this method, point $u_0 \in \triangle_{[p,q]}$ and its symmetrical point $\tau(u_0)$ ($\tau \in [p,q]$) will be assigned the same color. Consequently, the colored $S^2$ obtained by this method will have $[p,q]$ symmetries. Figures 2–5 show six aesthetic patterns obtained in this manner.

The color scheme used above was borrowed from [23]. We have employed this scheme to render fractal [24] and hyperbolic patterns [25], which could enhance the visual appeal of patterns effectively. We refer the reader to [26] for more details.

Equation (19) has the following outstanding features. First, to create symmetrical patterns, one needs to construct mappings that meet certain requirements [12–15,17–21]. However, under certain circumstances, this kind of mapping is not easy to achieve. Mapping (19) has no requirements, and we can construct mappings at will. For example, the mapping used in the left of Figure 3 is:

$$\mathcal{F}(u^1) = \begin{pmatrix} 1.2\cos[y+\sin(yz+x)] \\ 1.65\cos(x-y)[\cos(x+z)+2\sin y] \\ e^{\sin[-1.5+\cos(x+y)]} \end{pmatrix}, \ u^1 = (x,y,z)^T \in \triangle_{[3,4]}$$

Second, as pointed out in Section 1, (3) is not appropriate for the symmetry group of large order. By contrast, (19) avoids such a restriction on symmetry order, so it should be possible to extend the method to treat regular polytopes with thousands of symmetries.

**Figure 3.** Two spherical patterns with [3,3] symmetries.

**Figure 4.** Two spherical patterns with [3,4] symmetries.

**Figure 5.** Two spherical patterns with [3,5] symmetries.

**Acknowledgments:** We produced Figures 2–5 in the VC++ 6.0 programming environment with the aid of OpenGL, a powerful graphics software package. We thank Adobe and Microsoft for their friendly technical support. This work was supported by the Natural Science Foundation of China (No. 11461035) and Doctoral Startup Fund of Jingangshan University (Nos. JZB1303, JZB11002).

**Author Contributions:** Peichang Ouyang conceived the framework and structured the whole paper; Liying Wang performed the experiments and wrote the paper; Tao Yu and Xuan Huang checked the results; Peichang Ouyang, Liying Wang, Tao Yu and Xuan Huang completed the revision of the article.

**Conflicts of Interest:** The authors declare no conflict of interest.

### References

1.  Coxeter, H.S.M. *Regular Polytopes*; Dover: New York, NY, USA, 1973.
2.  Mcmulle, P.; Schulte, E. *Abstract Regular Polytopes*; Cambridge University Press: Cambridge, UK, 2002.

3.  Conway, J.H.; Burgiel, H.; Goodman-Strauss, C. *The Symmetries of Things*; A K Peters Press: Natick, MA, USA, 2008.
4.  Armstrong, V.E. *Groups and Symmetry*; Springer: New York, NY, USA, 1987.
5.  Magnus W. *Noeuclidean Tessellation and Their Groups*; Academic Press: New York, NY. USA, 1974.
6.  Rees, E.G. *Notes on Geometry*; Springer: Berlin/Heidelberg, Germany, 1983.
7.  Escher, M.C.; Ford, K.; Vermeulen, J.W. *Escher on Escher: Exploring the Infinity*; Harry N. Abrams: New York, NY, USA, 1989.
8.  Yen, J.; Séquin, C. Escher Sphere Construction Kit. In Proceedings of the 2001 Symposium on Interactive 3D Graphics, Research Triangle Park, NC, USA, 19–21 March 2001; pp. 95–98.
9.  Kaplan, C.S. Computer Graphics and Geometric Ornamental Design. Ph.D. Thesis, University of Washington, Washington, DC, USA, 2002.
10. Field, M. *Dynamics and Symmetry*; Imperial College Press: London, UK, 2007.
11. Humphreys, J.E.; Bollobas, B.; Fulton, W.; Katok, A.; Kirwan, F.; Sarnak, P.; Simon, B. *Reflection Groups and Coxeter Groups*; Cambridge University Press: Cambridge, UK, 1992.
12. Field, M.; Golubitsky, M. *Symmetry in Chaos*; Oxford University Press: Oxford, UK, 1992.
13. Reiter, C.A. Chaotic Attractors with the Symmetry of the Tetrahedron. *Comput. Graph.* **1997**, *6*, 841–884.
14. Brisson, G.F.; Gartz, K.M.; McCune, B.J.; O'Brien, K.P.; Reiter, C.A. Symmetric Attractors in Three-dimensional Space. *Chaos Soliton Fract.* **1996**, *7*, 1033–1051.
15. Lu, J.; Zou, Y.R.; Liu, Z.Y. Colorful Symmetric Images in Three-dimensional Space from Dynamical Systems. *Fractals* **2012**, *20*, 53–60.
16. Benson, D.J.; Hitchin, N.J. *Polynomial Invariant of Finite Groups*; Cambridge University Press: Cambridge, UK, 1993.
17. Chung, K.W.; Chan, H.S.Y. Spherical Symmetries from Dynamics. *Comput. Math. Appl.* **1995**, *7*, 67–81.
18. Dumont, J.P.; Reiter, C.A. Chaotic Attractors Near Forbidden Symmetry. *Chaos Soliton Fract.* **2000**, *11*, 1287–1296.
19. Jones, K.C.; Reiter, C.A. Chaotic Attractors with Cyclic Symmetry Revisited. *Comput. Graph.* **2000**, *24*, 271–282.
20. Reiter, C.A. Chaotic Attractors with the Symmetry of the Dodecahedron. *Vis. Comput.* **1999**, *4*, 211–215.
21. Dumont, J.P.; Heiss, F.J.; Jones, K.C.; Reiter, C.A.; Vislocky, L.M. *N*-dimensional Chaotic Sttractors with Crystallographic Symmetry. *Chaos Soliton Fract.* **2001**, *4*, 761–784.
22. Coxeter, H.S.M.; Moser, W.O.J. *Generators and Relations for Discrete Groups*; Springer: New York, NY, USA, 1980.
23. Lu, J.; Ye, Z.X.; Zou, Y.R. Colorful Patterns with Discrete Planar Symmetries from Dynamical Systems. *Fractals* **2012**, *18*, 35–43.
24. Ouyang, P.C.; Fathauer, R.W. Beautiful Math, Part 2, Aesthetic Patterns Based on Fractal Tilings. *IEEE Comput. Graph.* **2014**, *1*, 68–75.
25. Ouyang, P.C.; Chung, K.W. Beautiful Math, Part 3, Hyperbolic Aesthetic Patterns Based on Conformal Mappings. *IEEE Comput. Graph.* **2014**, *2*, 72–79.
26. Ouyang, P.C.; Cheng, D.S.; Cao, Y.H.; Zhan, X.G. The Visualization of Hyperbolic Patterns from Invariant Mapping Method. *Comput. Graph.* **2012**, *2*, 92–100.

# Section D:
# Combinatorial Geometry

*symmetry*

MDPI

*Note*

# A Note on Lower Bounds for Colourful Simplicial Depth

**Antoine Deza** [1,*], **Tamon Stephen** [2] and **Feng Xie** [1]

[1]   Advanced Optimization Laboratory, Department of Computing and Software, McMaster University, Hamilton, Ontario L8S 4K1, Canada; xief@mcmaster.ca

[2]   Department of Mathematics, Simon Fraser University, Burnaby, British Columbia V5A 1S6, Canada; tamon@sfu.ca

[*]   Author to whom correspondence should be addressed; deza@mcmaster.ca; Tel.: +1-905-525-9140 (ext. 23750).

Received: 18 October 2012; in revised form: 18 December 2012; Accepted: 31 December 2012; Published: 7 January 2013

**Abstract:** The colourful simplicial depth problem in dimension $d$ is to find a configuration of $(d+1)$ sets of $(d+1)$ points such that the origin is contained in the convex hull of each set, or colour, but contained in a minimal number of *colourful* simplices generated by taking one point from each set. A construction attaining $d^2 + 1$ simplices is known, and is conjectured to be minimal. This has been confirmed up to $d = 3$, however the best known lower bound for $d \geq 4$ is $\lceil \frac{(d+1)^2}{2} \rceil$. In this note, we use a branching strategy to improve the lower bound in dimension 4 from 13 to 14.

**Keywords:** colourful simplicial depth; Colourful Carathéodory Theorem; discrete geometry; polyhedra; combinatorial symmetry

---

A *colourful configuration* is the union of $(d+1)$ sets, or colours, $\mathbf{S}_0, \mathbf{S}_1, \ldots, \mathbf{S}_d$ of $(d+1)$ points in $\mathbb{R}^d$. Let $\mathbf{S} = \cup_{i=0}^d \mathbf{S}_i$. Without loss of generality we assume that the points in $\mathbf{S} \cup \{\mathbf{0}\}$ are in general position. We are interested in the *colourful simplices* formed by taking the convex hull of a set containing one point of each colour. The *colourful simplicial depth* problem is to find a colourful configuration, with each $\mathbf{S}_i$ containing the origin $\mathbf{0}$ in the interior of its convex hull, minimizing the number of colourful simplices containing $\mathbf{0}$. We denote this minimum by $\mu(d)$. We take the simplices to be closed and remark that the minimum should be attained.

Computing $\mu(d)$ can be viewed as refining Bárány's Colourful Carathéodory Theorem [1] whose original version gives $\mu(d) \geq 1$, and $\mu(d) \geq d+1$ when strengthened to show that *every* point of the configuration generates at least one such simplex. The question of computing $\mu(d)$ was studied in Deza *et al.* [2], which showed $\mu(2) = 5$, that $2d \leq \mu(d) \leq d^2 + 1$ for $d \geq 3$ and that $\mu(d)$ is even when $d$ is odd. The lower bound has since been improved by Bárány and Matoušek [3] (who verified the conjecture for $d = 3$), Stephen and Thomas [4] and Deza *et al.* [5], which includes the current strongest bound of $\mu(d) \geq \lceil \frac{(d+1)^2}{2} \rceil$ for $d \geq 4$.

One motivation for colourful simplicial depth is to establish bounds on ordinary simplicial depth. A point $p \in \mathbb{R}^d$ has *simplicial depth* $k$ relative to a set $S$ if it is contained in $k$ closed simplices generated by $(d+1)$ sets of $S$. This was introduced by Liu [6] as a statistical measure of how representative $p$ is of $S$. See [7–10] for recent progress on this problem. We remark also that the colourful simplicial depth of a point is the number of solutions to a colourful linear program in the sense of [11] and [12].

*Octahedral Systems*

Call a $(d+1)$-uniform hypergraph on $\mathbf{S} = \cup_{i=1}^{d+1} \mathbf{S}_i$ a *colourful hypergraph*. A colourful configuration defines a colourful hypergraph by taking hyperedges corresponding to colourful simplices containing

**0** in their interior. We will call a colourful hypergraph that arises from a colourful configuration with $\mathbf{0} \in \cap_{i=1}^{d+1} \operatorname{conv}(\mathbf{S}_i)$ a *configuration hypergraph*. Our strategy, following [13], is to show that a particular configuration hypergraph whose hyperedges correspond to the colourful simplices containing **0** in a configuration cannot exist. The Colourful Carathéodory Theorem gives that any configuration hypergraph must satisfy:

**Property 1.** *Every vertex of a configuration hypergraph belongs to at least one of its hyperedges.*

Fix a colour $i$. We call a set $t$ of $d$ points that contains exactly one point from each $\mathbf{S}_j$ other than $\mathbf{S}_i$ an *i-transversal*. That is to say, an $i$-transversal $t$ has $t \cap \mathbf{S}_i = \varnothing$ and $|t \cap \mathbf{S}_j| = 1$ for $i \neq j$. We call any pair of disjoint $i$-transversals an *i-octahedron*; these may or may not generate a cross-polytope, *i.e.*, a $d$-dimensional octahedron, in the geometric sense that their convex hull is a cross-polytope with same coloured points never adjacent in the skeleton of the polytope.

A key property of colourful configurations is that for a fixed $i$-octahedron $\Omega$, the parity of the number of colourful simplices containing **0** formed using points from $\Omega$ and a point of colour $i$ does not depend on which point of colour $i$ is chosen. This is a topological fact that corresponds to the fact that **0** is either *inside* or *outside* the octahedron, see the *Octahedron Lemma* of [3] for a proof. Figure 1 illustrates this in a two-dimensional case where **0** is at the centre of a circle that contains points of the three colours.

We carry the definitions of *i-transversals* and *i-octahedra* over to the hypergraph setting. Then any configuration hypergraph must satisfy:

**Property 2.** *For any octahedron $\Omega$ of a hypergraph, the parity of the set of hyperedges using points from $\Omega$ and a fixed point $s_i$ for the ith coordinate is the same for all choices of $s_i$.*

Consider a colourful hypergraph whose vertices are $\mathbf{S} = \cup_{i=0}^{d}\mathbf{S}_i$ and whose hyperedges have exactly one element from each set. If the hypergraph satisfies Property 2 we call it an *octahedral system*, if it additionally satisfies Property 1 we call it an *octahedral system without isolated vertex*. A colourful configuration with $\mathbf{0} \in \cap_{i=1}^{d+1} \operatorname{conv}(\mathbf{S}_i)$ and $k$ colourful simplices containing **0** has a configuration hypergraph that is an octahedral system without isolated vertex with $k$ hyperedges. Let $\nu(d)$ be the minimum number of hyperedges in an octahedral system without isolated vertex with $(d+1)$ colours. Then $\nu(d) \leq \mu(d)$. It is an interesting question whether there are any octahedral systems without isolated vertex not arising from any colourful configurations, and if not, whether $\nu(d) < \mu(d)$ for some $d$. This purely combinatorial approach was originally suggested by Bárány [14].

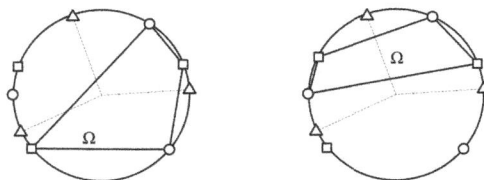

**Figure 1.** Two-dimensional cross-polytopes $\Omega$ containing and not containing 0.

Octahedral systems have the advantage of being combinatorial and finite. In principle, for any particular $d$ and $k$ we can check if there exists an octahedral system without isolated vertex on $\mathbf{S} = \cup_{i=0}^{d}\mathbf{S}_i$ with up to $k$ hyperedges by generating all the—finitely many—hypergraphs with up to $k$ hyperedges, each containing one element from each $\mathbf{S}_i$ and then testing if they satisfy Properties 1 and 2. The difficulty lies in the sheer number of such hypergraphs, and in verifying Property 2 efficiently.

We obtain lower bounds for $\nu(d)$ by trying to build an octahedral system without isolated vertex by adding one hyperedge at a time. We can reduce the search space by exploiting the many combinatorial symmetries in such hypergraphs and considering only configurations that satisfy certain normalizations. However, this alone is not sufficient to improve the known lower bounds even for $d = 4$. We thus turn our attention to how to use Property 2 effectively.

We use two strategies. The first is to look at a particular subset of parity conditions that are relatively independent. The second is to use the following lemma, proved in [15]. Call a hyperedge $e$ of a colourful hypergraph *isolated* if there is no other hyperedge that differs from $e$ only in a single coordinate. Then:

**Lemma 1.** *An octahedral system with $d^2$ or fewer hyperedges must not contain any isolated hyperedges.*

*Enumeration Details*

We begin by fixing an arbitrary colour as colour 0 and an arbitrary 0-transversal. We can label the points in each set from 0 to $d$ and, without loss of generality, take the transversal to contain the 0 point of each set. For convenience we write hyperedges as a string of $(d+1)$ numbers and transversals as string of $d$ numbers with * corresponding to the omitted colour. Thus the 0-transversal considered is $t_0 := *00 \ldots 0$.

Consider the $d$ octahedra generated by transversals $t_k := *kk \ldots k$, for $k = 1, 2, \ldots d$. Note that the initial numberings are arbitrary, and we may fix them as part of our search algorithm. Given a colourful hypergraph, we can form a $d \times (d+1)$ binary table by writing down for each $(r, s)$ the parity of the number of edges using vertices from the octahedron formed by $t_0$ and $t_r$ with initial coordinate $s$. We call this the *parity table*. If a colourful hypergraph satisfies Property 2, its parity table has constant rows.

The advantage of focusing on this table is that the entries are relatively independent. Only hyperedges of the form $x00 \ldots 0$ can change more than one entry of this table. After accounting for such hyperedges, each entry can only be affected by the $2^d - 1$ hyperedges that are on the relevant octahedron with the given initial coordinate.

We now use the results of [5] to break the problem into several cases based on $\ell$, the number of hyperedges containing $t_0$, $b$, the number of the parity table octahedra that have odd parity, and $j$, the minimum number of transversals covering any point of colour 0.

It is clear that for any octahedral system without isolated vertex and with $d^2$ or fewer hyperedges we must have $1 \leq \ell, b, j \leq d$ and that the number of hyperedges is at least $j(d+1)$. Further, [5] shows that we must have $j + b \geq d + 1$, and that the number of hyperedges must be at least $(b + \ell)(d+1) - 2b\ell$, as well as at least $d\ell + 1$ assuming that the colour 0 is chosen to minimize $\ell$ and that $\ell \geq \frac{d+2}{2}$. This last fact allows us to assume that $\ell \leq d - 1$.

To rule out possible octahedral systems without isolated vertex of size 13, it is sufficient to consider cases where $j = 1$ or $j = 2$, which in turn means $b = 3$ or $b = 4$. In the case $b = 3$, we have at least $15 - \ell$ simplices, so $\ell = 2$ or $\ell = 3$, and in the case $b = 4$, we have $20 - 3\ell$ so $\ell = 3$. In summary, we need to rule out systems where the triple $(\ell, b, j)$ is one of $(3, 4, 2), (3, 4, 1), (3, 3, 2), (2, 3, 2)$.

By reordering the points of colour 0, we can take the hyperedges $x0000$ to be in the system for $0 \leq x \leq \ell - 1$, and not in the system for $\ell \leq x \leq 4$. Consider the parity table after including these hyperedges with $\ell = 3$, illustrated in Table 1.

**Table 1.** The parity table with $\ell = 3$.

|        | 0 | 1 | 2 | 3 | 4 |
|--------|---|---|---|---|---|
| *1111  | 1 | 1 | 1 | 0 | 0 |
| *2222  | 1 | 1 | 1 | 0 | 0 |
| *3333  | 1 | 1 | 1 | 0 | 0 |
| *4444  | 1 | 1 | 1 | 0 | 0 |

Now if $b = 4$, then we are requiring that the parity table be comprised entirely of 1's. So in this case the entries in the first three columns are correct, while the entries in the last two columns are incorrect.

For $(\ell, b, j) = (3, 4, 2)$ we proceed to enumerate configurations as follows. Since $\ell = 3$, we include initial hyperedges $00000, 10000, 20000$. We then add hyperedges to correct each of the eight entries of Table 1, which must be fixed to get the correct parity table for $b = 4$. As previously remarked, adding any hyperedge not of the form $x0000$ will change only a single entry in the parity table. For instance, the entry in the first row and fourth column can be changed only by a hyperedge of the form $3abcd$ where $a, b, c, d \in \{0, 1\}$. Given that that $30000$ cannot be added to the configuration without changing $\ell$, there remain only 15 possible hyperedges that change the entry, and one must be in our configuration. In fact, by reordering the colours we can take it to be one of $31000, 31100, 31110$ and $31111$.

We could continue to exploit symmetries in this way—for instance depending on which of the previous 4 hyperedges is chosen, the next hyperedge could be one of 4 to 7 hyperedges fixing the next table entry. However, we did not do this so as to avoid extensive case analysis. Instead, we began branching on all 15 possible hyperedges that switch a given table entry until the table is correct and the partial configuration has 11 hyperedges.

As we branch we check two simple predictors that may indicate that the configuration requires several more hyperedges. First, we look for points that are not currently included in any hyperedge. If some colour still has $k$ uncovered points, then we require $k$ additional hyperedges. Second, since any vertex of colour 0 must be covered by at least $j$ hyperedges, we examine which points of colour 0 are not contained in sufficiently many hyperedges, and get a score $k'$ by summing up the undercounts. At the same time, we may find that all vertices of colour 0 are already covered by more than $j$ hyperedges (especially when $j = 1$), in which case the partial configuration no longer belongs to this subcase and can be excluded. Again, we require $k'$ additional hyperedges. If either $k$ or $k'$ is sufficiently large (in this case 3), then the current partial configuration cannot extend to an octahedral system without isolated vertex with less than 14 hyperedges and is abandoned.

Otherwise, we examine the configuration to see if it has an isolated hyperedge. If it contains an isolated hyperedge $e$, then by Lemma 1, if the configuration is to extend to an octahedral system without isolated vertex with less than 17 hyperedges, it must include a hyperedge adjacent to $e$. That is, it must contain $e'$ differing from $e$ only in a single coordinate. There are only 20 such hyperedges so we can branch on them. We then repeat the process of applying predictors and looking for an isolated hyperedge until we either find an octahedral system without isolated vertex with less than 14 hyperedges, or all partial configurations with fewer hyperedges are exhausted.

If we do arrive at a partial configuration with no isolated hyperedges, then as a last resort we may have to branch on all possible hyperedges. However, this happens infrequently enough that the enumeration ends in a reasonable time.

The remaining cases, where $(\ell, b, j)$ is $(3, 4, 1), (3, 3, 2)$ or $(2, 3, 2)$ are similar. Having exhausted all these cases, we conclude that $\nu(4) \geq 14$, and hence $\mu(d) \geq 14$.

*Final remarks*

This strategy was implemented by Xie [16] in Python version 2.6 on an AMD Opteron Processor 8356 core (2.3G Hz) and is able to prove that $\nu(4) \geq 14$ in about 30 days of CPU time. This improves by 1 the bound of Deza *et al.* [5], from $\mu(4) \geq 13$ to $\mu(4) \geq 14$. Since this article was written, Deza *et al.* [17] have introduced a different approach that shows $\mu(4) = 17$ and improves the bounds in higher dimension as well.

We note that are $\binom{5^5}{13} \approx 4.25 \times 10^{35}$ colourful hypergraphs on 5 points in each of 5 colours with 13 edges that we need to exclude. In our search strategy, after choosing $(l, b, j)$, the first $l$ hyperedges are determined, and the next $5b + 4l - 2bl$ hyperedges are chosen to fix entries in the parity table. Without considering isolated edges, this leaves a search space of size $15^{5b+4l-2bl} 55^{13-5b-5l+2bl}$; in our detailed example with $(l, b, j) = (3, 4, 2)$, this is $15^8 \left(5^5\right)^2 \approx 2.50 \times 10^{16}$. A space of this size is still slightly

*Symmetry* **2013**, *5*, 47–53

beyond the modest computational resources we used. Considering isolated edges further reduces the space substantially, allowing each case to be solved in a few days.

We conclude by mentioning that many aspects of colourful simplices are just beginning to be explored. For instance, the combinatorial complexity of a system of colour simplices is analyzed in [18]. As far as we know, the algorithmic question of computing colourful simplicial depth is untouched, even for $d = 2$ where several interesting algorithms for computing the monochrome simplicial depth have been developed. See for instance the survey [19].

**Acknowledgments:** This work was supported by grants from the Natural Sciences and Engineering Research Council of Canada (NSERC) and MITACS, and by the Canada Research Chairs program. The second author thanks the University of Cantabria for hospitality while working on this paper. The authors would like to thank the referees for comments that improved the paper and Egon Schulte for editing this special issue.

## References

1. Bárány, I. A generalization of Carathéodory's theorem. *Discret. Math.* **1982**, *40*, 141–152. [CrossRef]
2. Deza, A.; Huang, S.; Stephen, T.; Terlaky, T. Colourful simplicial depth. *Discret. Comput. Geom.* **2006**, *35*, 597–604. [CrossRef]
3. Bárány, I.; Matoušek, J. Quadratically many colorful simplices. *SIAM J. Discret. Math.* **2007**, *21*, 191–198.
4. Stephen, T.; Thomas, H. A quadratic lower bound for colourful simplicial depth. *J. Comb. Opt.* **2008**, *16*, 324–327. [CrossRef]
5. Deza, A.; Stephen, T.; Xie, F. More colourful simplices. *Discret. Comput. Geom.* **2011**, *45*, 272–278. [CrossRef]
6. Liu, R.Y. On a notion of data depth based on random simplices. *Ann. Statist.* **1990**, *18*, 405–414. [CrossRef]
7. Gromov, M. Singularities, expanders and topology of maps. Part 2: From combinatorics to topology via algebraic isoperimetry. *Geom. Funct. Anal.* **2010**, *20*, 416–526. [CrossRef]
8. Matoušek, J.; Wagner, U. On Gromov's method of selecting heavily covered points. 2012. Available online: http://arxiv.org/abs/1102.3515 (accessed on 31 December 2012).
9. Karasev, R. A simpler proof of the Boros-Füredi-Bárány-Pach-Gromov theorem. *Discret. Comput. Geom.* **2012**, *47*, 492–495. [CrossRef]
10. Král', D.; Mach, L.; Sereni, J.S. A new lower bound based on Gromov's method of selecting heavily covered points. *Discret. Comput. Geom.* **2012**, *48*, 487–498.
11. Bárány, I.; Onn, S. Colourful linear programming and its relatives. *Math. Oper. Res.* **1997**, *22*, 550–567. [CrossRef]
12. Deza, A.; Huang, S.; Stephen, T.; Terlaky, T. The colourful feasibility problem. *Discret. Appl. Math.* **2008**, *156*, 2166–2177. [CrossRef]
13. Custard, G.; Deza, A.; Stephen, T.; Xie, F. Small octahedral systems. In Proceedings of the 23rd Annual Canadian Conference on Computational Geometry, Toronto, Ontario, Canada, 10–12 August 2011; pp. 267–272.
14. Bárány, I. Hungarian Academy of Sciences, Budapest, Hungary. Personal communication, 2010.
15. Deza, A.; Stephen, T.; Xie, F. Computational lower bounds for colourful simplicial depth. Available online: http://arxiv.org/abs/1210.7621 (accessed on 31 December 2012).
16. Xie, F. Python code for octrahedral system computation. Available online: http://optlab.mcmaster.ca/om/csd/ (accessed on 31 December 2012).
17. Deza, A.; Meunier, F.; Sarrabezolles, P. A combinatorial approach to colourful simplicial depth. Available online: http://arxiv.org/abs//1212.4720 (accessed on 31 December 2012).
18. Schulz, A.; Tóth, C.D. The union of colorful simplices spanned by a colored point set. *Comput. Geom.* **2013**, in press.
19. Aloupis, G. Geometric measures of data depth. In *Data Depth: Robust Multivariate Analysis, Computational Geometry and Applications*; American Mathematical Society: Providence, RI, USA, 2006; Volume 72, pp. 147–158.